Carbamate Insecticides: Chemistry, Biochemistry, and Toxicology

Authors:

Ronald J. Kuhr

Department of Entomology
New York State
Agricultural Experiment Station
Geneva, New York

H. Wyman Dorough

Department of Entomology
University of Kentucky
Lexington, Kentucky

Published by

CRC PRESS

CRC PRESS, Inc.
18901 Cranwood Parkway · Cleveland, Ohio 44128

Library of Congress Cataloging in Publication Data

Kuhr, Ronald J.
Carbamate insecticides.

At head of title: CRC.
Bibliography: p.
Includes index.
1. Carbamates. 2. Insecticides. I. Dorough, H. Wyman, joint author. II. Title.
SB952.C3K84 632'.951 74-25265
ISBN 0-87819-052-X

International Standard Book Number 0-87819-052-X

Library of Congress Card Number 74-25265
Printed in the United States

FOREWORD

The three most important classes of synthetic pesticides are the chlorinated hydrocarbons, organophosphorus compounds, and carbamates. The first two classes of insecticides have been thoroughly covered in the CRC Pesticides Series – the 2-volume work *Chlorinated Insecticides* by G. T. Brooks and *Organophosphorus Pesticides* by M. Eto, published during 1973 and 1974. The present book by two outstanding insect toxicologists, Professors Kuhr and Dorough, represents probably the most thorough treatment of the field of carbamate insecticides written to-date. As the title of the book implies, the authors have striven to cover the chemistry (synthesis and residue methodology), the biochemistry (mode of action, metabolism, and metabolic reactions), and toxicology (both to target and to nontarget organisms) in a single volume. I believe that Dorough and Kuhr, writing as a team, have succeeded very well in producing a modern text and reference work, including considerations for biological and environmental stability of carbamates. The authors have carefully delineated their topic and have not included in their discussion carbamate herbicides and dithiocarbamate fungicides, which will be the subject of other books to be published in this series in the future.

Gunter Zweig, Ph.D.
Editor
CRC Pesticide Series

Washington, D.C.
Spring, 1976

PREFACE

Of the three major groups of synthetic organic insecticides (organochlorine, organophosphorus, and carbamate), the carbamates were the last to be introduced and are generally considered as a "new" chemical family of insect poisons, despite the fact that insecticidally active structures were discovered almost thirty years ago. However, carbamates contributed little to the control of insect pests until the early 1960s, and most of this initial success was due to a single compound. In more recent times, the number of commercial carbamate insecticides has increased substantially, and the prognosis for the future is good.

The growth in carbamate science and technology has been accompanied by a proliferation of numerous research reports in scientific journals throughout the world. Progress in certain areas has been summarized periodically by review articles and occasional chapters in books, but, to the best of our knowledge, no one has presented a thorough discussion of the chemistry, biochemistry, and toxicology of carbamate insecticides within a single volume. This may be due in part to the dynamics of the science itself; that is, new information is appearing almost daily, so that a book on the topic is potentially out of date before it is published. Although there is some truth to this, the basic principles of carbamate insecticide chemistry, synthesis, mode of action, metabolism, environmental degradation, etc., have been established and are not likely to change significantly. Thus, preparation of a book devoted solely to carbamate insecticides seemed justified. In addition, the current governmental banning and restrictions on the uses of certain pesticides, especially organochlorine insecticides, has placed a greater burden of crop and animal protection on other pesticides, including the carbamate chemicals. A state-of-the-art publication at this time could help ensure that these materials are used in a manner which will provide the most effective pest control with a minimum of undesirable effects on man and his environment.

This book represents a careful analysis of thousands of research articles. In most cases original papers were consulted, and the temptation to use review articles as a major source of information was avoided. Our literature search was extensive through 1973, and continued on a somewhat less comprehensive basis during 1974 and 1975 as the chapters were being written. Hopefully, we have not overlooked too many important contributions. The book is meant to be a detailed reference work of carbamate insecticides, and consequently it is augmented with ample literature citations. Nonetheless, an attempt was made to present the material in a style that will allow those outside the area of pesticide chemistry, particularly students, to benefit from a reading of its contents.

There are a number of people who contributed to the publication of this monograph, and we take this opportunity to thank them for their help and patience. Most of the manuscript typing was done by Mary Blackburn, Helen Daniecki, and Helen Melious. Valuable assistance in editing and final proof reading was provided by the editor of the CRC Pesticide Series, Gunter Zweig, by the CRC staff, and by Carabelle Hessney. The following associates of H.W.D. are gratefully acknowledged for their assistance in literature search, preparation of figures, and preliminary proofing of draft copy: Harry Bryant, Ke-Chang Chen, Mary Blackburn, Charles Swanson, and Joseph Thorstenson. A considerable portion of the literature search and some of the data presented are integral parts of research supported by Environmental Protection Agency Grant No. R802005. Much of the writing by R.J.K. was done while he was on sabbatic leave at the University of Guelph, and he thanks Freeman McEwen and other members of the Department of Environmental Biology for arranging his sabbatic and use of the University libraries. Finally, the authors express sincere appreciation to their respective departments for the encouragement and support received during this undertaking.

Ronald J. Kuhr
H. Wyman Dorough
Spring, 1976

This volume is dedicated to J. E. Casida, B. W. Arthur, G. A. Ediger, D. W. Clements, and the many other educators who have so influenced our lives.

THE AUTHORS

Ronald J. Kuhr is Associate Professor of Insect Toxicology at the New York State Agricultural Experiment Station in Geneva, New York. He earned his B.S. degree at the University of Wisconsin, Madison, and his Ph.D. degree at the University of California, Berkeley. Before accepting the position at Geneva, Dr. Kuhr conducted postdoctoral research at the Agricultural Research Council's Pest Infestation Laboratory in Slough, England. He wrote part of this book while on sabbatic leave at the University of Guelph in Ontario, Canada.

H. Wyman Dorough is a toxicologist at the University of Kentucky, where he is Professor in the Department of Entomology, Associate Dean for Research in the Graduate School, and a member of the Interdisciplinary Graduate Toxicology Program faculty. He received his B.S. and M.S. degrees from Auburn University, Alabama, and his Ph.D. degree from the University of Wisconsin, Madison. Prior to his appointment at Kentucky, Dr. Dorough was first Assistant and then Associate Professor in the Department of Entomology at Texas A & M University, College Station.

TABLE OF CONTENTS

Chapter 1
Development and Use . . . 1
Introduction . . . 1
History . . . 2
Applications . . . 5
References . . . 12

Chapter 2
Chemistry and Synthesis . . . 15
Introduction . . . 15
Chemical Properties and Reactions . . . 17
Synthesis of Carbamates . . . 21
Synthesis of Metabolites . . . 27
Radiosynthesis . . . 33
References . . . 37

Chapter 3
Mode of Action . . . 41
Introduction . . . 41
Cholinesterase Inhibition . . . 44
Signs and Symptoms of Poisoning . . . 58
References . . . 66

Chapter 4
Structure – Activity Relationships . . . 71
Introduction . . . 71
Factors Determining Insecticidal Activity of Aromatic Carbamates . . . 74
Factors Determining Insecticidal Activity of Oxime Carbamates . . . 89
Correlations of I_{50}, LD_{50}, and Hydrolytic Stability . . . 93
References . . . 97

Chapter 5
Toxicology . . . 103
Introduction . . . 103
Toxicity to Insects . . . 103
Toxicity to Mammals . . . 106
Toxicity to Nontarget Organisms . . . 117
References . . . 119

Chapter 6
Metabolic Reactions . . . 125
Introduction . . . 125
Hydrolysis . . . 125
Oxidation . . . 127
Conjugation . . . 132
References . . . 136

Chapter 7
Metabolism . . . 143
Introduction . . . 143
Dimethylcarbamates (Dimetilan, Isolan, and Pyramat) . . . 144
Monomethylcarbamates Carbaryl (Sevin, UC 7744, Denapon, Karbosep, Carpolin, NAC) . . . 146
Tsumacide (MTMC, Metacrate, C-3) . . . 160

Meobal (S-1046, MPMC) . . . 162
3,5-Dimethylphenyl N-Methylcarbamate (XMC, Macbal, Cosban) . . . 163
Banol (Carbanolate, Chlorxylam) . . . 164
Landrin (SD 8530) . . . 166
UC 10854 (H 5727, IPFC) . . . 167
HRS-1422 . . . 168
Bux (Ortho 5353, Bufencarb, Metalkamate) . . . 169
RE 11775 and RE 5365 . . . 170
RE 5030 and Butacarb . . . 171
Baygon (Propoxur, Unden, Blattanex, B 39007, Suncide, Arprocarb) . . . 172
Zectran (Mexacarbate, Dowco 139) . . . 173
N-Acetyl Zectran . . . 175
Matacil (Aminocarb, B 44646, A 363) . . . 175
Mesurol (Methiocarb, Mercaptodimethur, Draza, B 37344) . . . 177
Carbofuran (Furadan, NIA 10242, Curaterr) . . . 178
N-(2-Toluenesulfenyl) Carbofuran . . . 182
Mobam (MCA 600) . . . 184
Formetanate (Schering 36056, Fundal Forte 750, Dicarzol, Carzol SP) . . . 186
UC 34096 . . . 186
Aldicarb (Temik, UC 21149) . . . 187
Methomyl (Lannate, Nudrin, Dupont 1179) . . . 190
Thiocarboxime (Talcord, WL 21959, SD 17250-W) . . . 191
DS-15647 . . . 193
References . . . 194

Chapter 8
Biological and Environmental Stability . . . 201
Introduction . . . 201
Animals . . . 201
Plants . . . 205
Soils . . . 214
Water and Air . . . 222
Photodecomposition . . . 228
Human Exposure . . . 232
References . . . 236

Chapter 9
Residue Methodology . . . 243
Introduction . . . 243
Colorimetric Methods . . . 243
Enzymic Methods . . . 244
Fluorescence . . . 246
Gas Chromatography . . . 247
Other Methods . . . 251
References . . . 253

Chapter 10
Resistance and Synergism . . . 259
Introduction . . . 259
Occurrence of Resistance . . . 262
Mechanisms and Genetics . . . 269
Synergism . . . 274
References . . . 286

Index . . . 293

Chapter 1

DEVELOPMENT AND USE

INTRODUCTION

Few people are not impressed by the grace and unequalled beauty of a multicolored butterfly floating through a summer's day. Yet many of the same people shudder with revulsion at the thought of a spiny cockroach darting from corner to corner in their pantry. These two groups of insects, nevertheless, have something in common with each other and with the millions of related species comprising the class Insecta. They all possess an irrepressible ability to adapt, survive, and multiply – an ability which allows them to permeate the earth's land and fresh water from pole to pole in numbers that defy comprehension. Evolution placed them here long before man appeared, but when he did appear, the insects quickly moved in to share his home, his food, his clothing, and even his blood.

For a long time, the balance of nature permitted a peaceful, though not always enjoyable, coexistence of the two classes of animal. However, man's ingenuity and gregarious tendencies upset the status quo when he exchanged his nomadic existence for a stable community life. This, of course, required the husbanding of animals and the cropping of plants to ensure a local supply of food and clothing. It was not long before the farmer realized that, if his monocultures were to survive, certain insect populations would have to be managed. Thus, the war between man and pestiferous insect had begun.

The first weapons were often crude and inefficient, and it was not until man resorted to chemicals that he gained a distinct advantage. The pioneer insecticides, developed and utilized through the late 18th, 19th, and early 20th centuries, were mostly inorganic compounds, such as copper sulfate, lead arsenate, Paris green, sodium fluoride, and sulfur, and the botanical insecticides rotenone, pyrethrum, and nicotine.[1] Despite various problems with phytotoxicity, poor coverage, lack of persistence, and so forth, many pest insect populations were controlled with these materials.

The beginning of the 20th century brought a rapid increase in urbanization, leaving fewer and fewer people in charge of food and fiber production. Seemingly endless acres of grain, corn, cotton, and cattle provided "promised lands" for numerous insect pests; at the same time, the clustering of people in cities and towns made it easy for insects to transfer disease organisms from human to human with alarming efficiency. Worldwide transportation systems promoted the spread of insect species from country to country, often without concomitant transport of their natural enemies. Unmolested, the "foreigners" ravaged our crops at will.

The beginning of the 20th century was also a time of war. Paradoxically, it was during this period of man killing man that modern synthetic insecticides evolved. First (1939) came DDT, which was discovered in a search for more effective moth-proofing agents.[2] Few other chemicals have had as great an impact on man's survival. At about the same time, dwindling supplies of nicotine and pyrethrum forced several countries into a frantic quest for synthetic replacements. The result was the discovery of the organophosphorus insecticides schradan, TEPP, and the still widely used parathion.[3]

Following World War II, the development and commercialization of numerous synthetic organochlorine and organophosphorus insecticides, coupled with the simultaneous availability of more effective herbicides and fungicides, supported both an agricultural boom and a population explosion. However, in the sometimes careless distribution of certain of these chemicals, man contaminated large portions of our environment[4] and also expedited the development of insect resistance to many of our best weapons (see Chapter 10). The call for "alternative" control methods resulted in the discovery of a multitude of insect attractants, repellants, sterilants, pathogens, hormones, predators, and parasites.[5] Some of these proved to be extremely effective; others are, and may always be, only at the experimental stage. But none of these alternatives will solve all of our insect problems, and it is likely that the best approach will be a combination of natural and artificial methods, including the application of synthetic insecticides.

Recent governmental bans and restrictions on the use of organochlorine compounds and the growing number of organophosphorus-resistant strains of insects are placing a greater burden of protection onto the newest major class of insect toxins, the carbamates. Because these materials were developed mostly in the 1960's, after the global distribution of DDT and during a period of environmental awareness, the carbamates tend to be more biodegradable, and some are considerably less toxic to nontarget species. However, certain structures can be fairly persistent. The chemistry, biochemistry, and toxicology of this new family of synthetic insecticides constitute the subject of this book.

HISTORY

The Calabar Bean and Physostigmine

Biologically active carbamate chemicals have a long and colorful history, with reported use reaching back several hundred years. Their origin as a part of African witchcraft trials was retraced recently in a book on medicinal plants.[6] At the beginning of the 17th century, several European nations discovered West Africa as a source of slave labor for use on their plantations in the West Indies and other tropical parts of the New World. Throughout the next two centuries, many European forts and trading posts were established along West Africa's coastline; one of the most prominent of these was Old Calabar, located on the Cross River estuary (now the southeast portion of Calabar Province in Nigeria).

The people inhabiting the area, the Efiks, were fishermen, but they became prosperous traders of palm oil and slaves during the 17th and 18th centuries. Each settlement had a chief who was responsible for enforcing local laws, mediating disputes, leading the army into war, and arranging peace pacts with neighboring tribes. The chiefs, town elders, and a few other privileged citizens together formed the powerful Egbo Society, which functioned as a feared federal government of the Efik people. Laws were enforced by fines, by trade boycotts, and often by capital punishment. The latter was employed particularly in witchcraft trials, where the person accused of practicing witchcraft as well as the accuser was subjected to the deadly *esere*.

Esere is the Efik work for Calabar bean and for the poison ordeal in which the beans were used. If the accused was judged by Egbo to be guilty, he was forced to swallow a milky potion made by macerating several Calabar bean seeds in a small quantity of water. The condemned native was then ordered to walk around until justice was served in one of two ways. Either the prisoner was fortunate enough to regurgitate the "milk," in which case he was declared innocent and allowed to go free, or he would develop shakes, froth at the mouth, and quickly die, thus establishing his guilt.

European curiosity in the poison ordeal brought a handful of Calabar seeds back to Scotland in the mid 1800's, where they were sprouted, and the mother plant was botanically described for the first time.[7] The legume, a runner clinging to bushes and trees along the sides of streams and swampy river banks, was given the name *Physostigma venenosum* (Balfour). Ripe pods apparently burst open throughout the year to free the coffee-colored Calabar beans (slightly larger than a lima bean), which floated downstream and collected on banks, where natives harvested them for trials.

At the same time, in other European laboratories, chemists were seeking clues for characterization of the powerful natural toxin. Tentative identification as a vegetable alkaloid appeared in 1863, together with a suggested name, *eserine,* after the Efik word *esere.*[8] A year later, the main alkaloid was isolated, purified, and renamed *physostigmine.*[9] Many decades of tedious attempts at structure elucidation followed before the molecular configuration of the ordeal poison was finally established in 1925[10] and proved by synthesis in 1935[11] (Figure 1.1). Interestingly, no other naturally occurring carbamate ester has ever been reported.

Studies describing the toxicology of physostigmine were also initiated halfway through the 19th century, when a rabbit was treated subcutaneously with an emulsion of the Calabar bean.[12] What became known as typical carbamate poisoning symptoms were observed and completely documented for the first time. In 1870, the violent neural responses elicited by bean extracts were found to be reversed upon administration of atropine,[13] an effective antidote for today's carbamate insecticides (Chapters 3 and 5). Later, the Calabar alkaloid itself was shown to be an antidote for poisoning by curare.[14] Meanwhile, the pharmacological benefits of physostigmine were discovered to include use as a miotic,[15] as a drug for treatment of glaucoma,[16] and as a drug for relief from the symptoms of the muscle disease myasthenia gravis.[17]

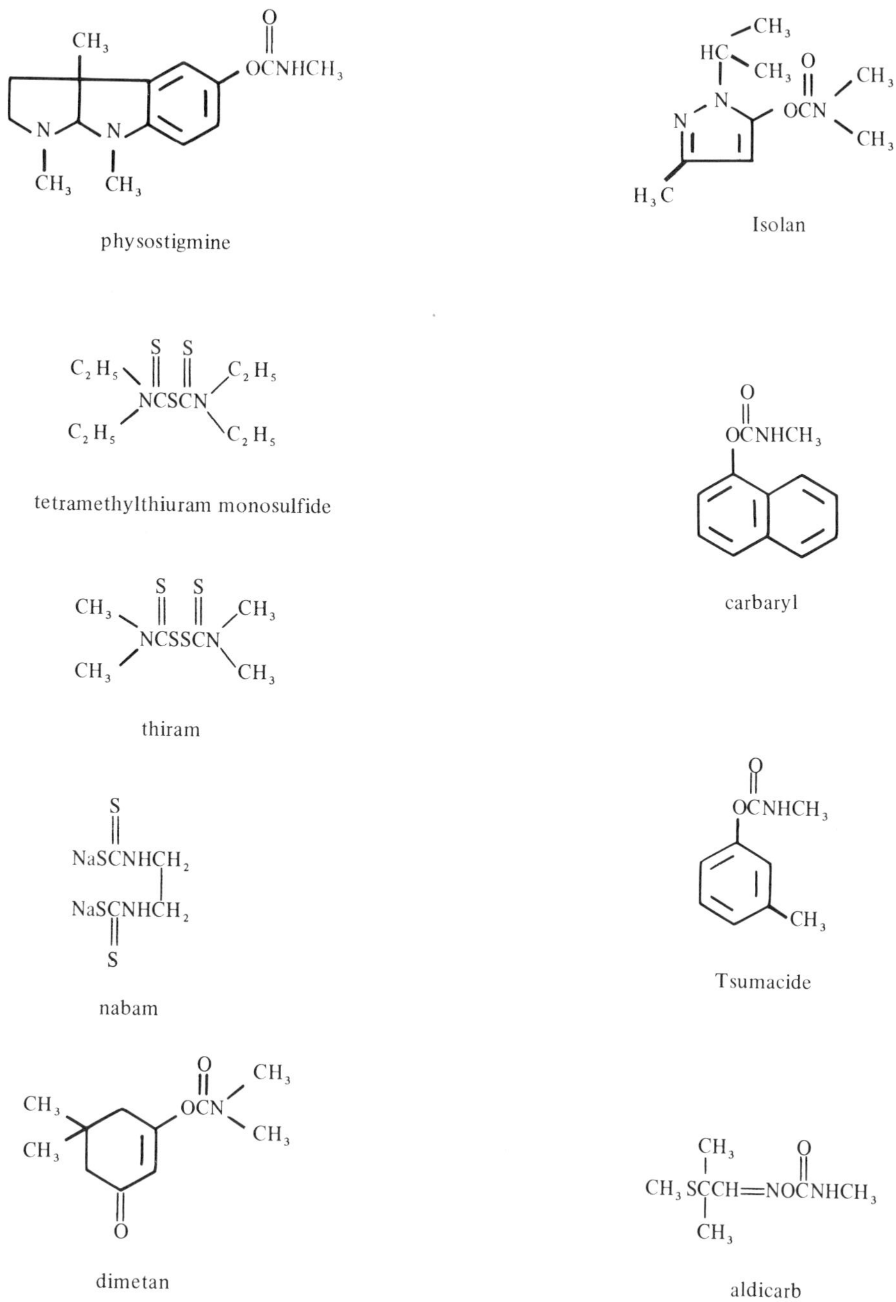

FIGURE 1.1. Structures of biologically active carbamate chemicals.

Biochemically, physostigmine played a significant role in elucidating the mechanism of impulse transmission along the mammalian nervous system. In 1921 it was experimentally proved that movement of a message from a nerve end to a body organ was mediated by chemical means.[18] Shortly thereafter, the transmitting substance "Vagusstoff" was suggested to be acetylcholine, mostly on the evidence that physostigmine blocked the enzymic metabolism of both "Vagusstoff" and acetylcholine, but not of other substances.[19] In addition, the introduction of physostigmine into a perfused sympathetic nerve ganglion caused an accumulation of acetylcholine in the perfusate.[20] Thus, it appeared certain that physostigmine was competing with acetylcholine for the same sites on the enzyme that catalyzed the deactivation of acetylcholine within the nervous system.[21] This description of the mode of action of physostigmine, together with its chemical characterization, prompted the synthesis and evaluation of numerous analogs for use as cholinergic drugs.[22] Nearly all of these compounds, of course, possessed a carbamate moiety, but in addition, they also contained either an ionizable or quaternary amine group, the significance of which will be made apparent in the next section.

Discovery of Insecticidal Activity

The first carbamate esters to exhibit insecticidal potential were derivatives of dithiocarbamic acids investigated by the E. I. du Pont de Nemours Company in 1931.[23] Although the compounds were ineffective as insect stomach poisons, a few analogs, notably tetraethylthiuram monosulfide (Figure 1.1), proved to be toxic by contact to soft-bodied insects, especially aphids. In addition, certain structures such as thiram (Figure 1.1) functioned as antifeeding compounds, preventing foliar damage by tent caterpillars, Japanese beetles, and Mexican bean beetles. Finally, there were also indications that tetraethylthiuram monosulfide and nabam (Figure 1.1) were toxic to mites.[23,24] However, the superior fungicidal action of these carbamates, together with the advent of the more effective organochlorine and organophosphorus insecticides, discouraged development of dithiocarbamates as insecticides, so that their commercial use today is largely as fungicides. Nevertheless, occasional reports of insect and mite control by dithiocarbamates continue to appear in the literature.[25,26]

The first bona fide carbamate insecticide was synthesized by Dr. Hans Gysin, a senior research chemist at the Geigy Chemical Company in Switzerland.[27] In the mid-to-late 1940's the company was interested in the development of insect repellents, especially aromatic amides. Dr. Gysin's search for more active structures included the preparation of a series of cycloaliphatic carbamate esters, one of which was dimetan (Figure 1.1). When this compound was passed through biological screening tests, it performed poorly as a repellent, but was found to be remarkably toxic to houseflies, aphids, and several other small insects.

The discovery of dimetan's insecticidal properties caused a shift in Geigy's objectives away from insect repellents and toward development of carbamate insecticides. The most promising materials proved to be derivatives of heterocyclic enols;[28-31] two of these, Isolan (Figure 1.1) and dimetilan, together with dimetan, reached commercial production in Europe during the 1950's. Isolan was limited mostly to use as a selective systemic aphicide, dimetan as a contact aphicide, and dimetilan as a flypaper contact poison. All of the Geigy insecticides were esters of dimethylcarbamic acid.

Based partly on the findings of the Geigy Company, and also on the knowledge that certain carbamate structures were effective herbicides, Dr. Joseph A. Lambrech of the Union Carbide Corporation in the United States synthesized experimental compound UC 7744 in 1953.[32,33] The carbamate ester differed from Geigy insecticides by substitution of an aryl group for the enol group and replacement of the dimethylcarbamyl moiety with a monomethyl counterpart. From the first bioassay through the next 3 years of intensive laboratory and field testing, UC 7744 appeared to be an unusually promising insecticide.[34] A description of the compound, first given the trade name Sevin and later the common name carbaryl (Figure 1.1), appeared in the literature in 1957.[35] Shortly thereafter, a request for tolerance of carbaryl residues on or in raw agricultural commodities was submitted to the U.S. Food and Drug Administration.[36] In 1959, only 6 years after its genesis, more than one million pounds of the new carbamate insecticide were used around the world.[37] Its low mammalian toxicity, environmental degradability, and broad-spectrum insect toxicity ensured continuous growth in sales, so that by 1971 more than 55 million pounds of the carbamate were produced in the United States, making carbaryl one of the most abundantly manufactured of all pesticides.[38]

Independent from the research at Geigy and Union Carbide, several university and government insect toxicologists began to experiment with the Calabar alkaloid physostigmine and with related carbamate drugs as potential insecticides. One of the most active groups was located at the University of California, Riverside, under the direction of Professor Robert L. Metcalf. Although it was soon established that several carbamate drugs inhibited insect nerve cholinesterase in vitro, none of the compounds were toxic to insects when applied topically to the living organisms (see

Chapter 4). Dr. Metcalf and his associates, notably Dr. T. R. Fukuto, theorized that the reason these compounds were inactive was the permanent charge on the amine salts and quaternary structures which prevented penetration of the drugs through the waxy insect cuticle and fatty insect nervous system. Therefore, they synthesized a series of 49 uncharged lipid-soluble mimics of physostigmine as part of what has become a classic example of a structure-activity study.

The results of the study were published in 1954,[39] the same year in which Gysin's discoveries were reported[27] and about the same time that carbaryl's insecticidal potential was realized.[34] Among the 49 carbamate esters, several substituted-phenyl monomethylcarbamates were shown to be highly toxic by contact to houseflies, greenhouse thrips, and green citrus aphids.[39] Though none of these insecticides have ever been marketed in the United States, several Japanese companies have patented and developed a few of the original structures during the last several years, e.g., Hopcide, Mipcin, Macbal, and Tsumacide (Figure 1.1). But perhaps more important, the studies of Metcalf and Fukuto firmly established the superior insecticidal activity of the aryl N-methylcarbamate structure which has since formed the basis for most of the modern commercial and experimental carbamate insecticides.

One other major structural innovation originated with the chemists of Union Carbide during an attempt to synthesize an N-methylcarbamate possessing some of the electronic characteristics of an aryl compound, but one with a closer spatial resemblance to the nerve cholinesterase substrate acetylcholine.[40] The result was the discovery of the oxime carbamates, in particular Temik or aldicarb (Figure 1.1), which are not only effective contact and systemic insecticides, but tend to be nematicidal and/or miticidal as well.[40-43]

Another name linked with the early development of carbamate insecticides is Dr. John E. Casida, a professor at the University of California, Berkeley, and formerly at the University of Wisconsin. Under Dr. Casida's guidance, several graduate students uncovered the principal chemical and biochemical pathways governing the fate of carbamate molecules in living organisms and in the environment. For example, one of the first contributions, which resulted from the thesis research of H. Wyman Dorough, helped establish the importance of oxidative metabolism in the detoxication of methylcarbamates by insects, mammals, and mammalian liver enzymes.[44,45] Shortly thereafter, the thesis research of Ronald J. Kuhr assisted in defining the metabolism of carbamate insecticides by plants.[46,47] There are certainly many other efforts from research groups around the world which could be cited as landmarks in the evolution of modern carbamate insecticides.

APPLICATIONS

Of the thousands of methylcarbamates synthesized by various chemical companies in the last 20 years, only about a dozen have reached sizable production as commercial insecticides. Several other experimental materials have been the subject of considerable laboratory and field testing, and a few of these may be successfully registered and marketed in the coming years. Some of the characteristics and use patterns for most of these insecticides (listed in alphabetical order) are summarized on the following pages.

Aldicarb

The first oxime carbamate pesticide registered under a trade name was Temik, a product of Union Carbide Corporation.[48-51] A year later, in 1967, the active ingredient was given the common name aldicarb [2-methyl-2-(methylthio)-propionaldehyde O-(methylcarbamoyl)oxime], and in early 1970 Temik 10% granular was registered for use on cotton by the U.S. Environmental Protection Agency. Since then, further applications on sugar beets, sugarcane, sweet potatoes, potatoes, peanuts, and ornamentals have been approved. Outside of the United States, aldicarb is also used on tobacco, onions, and hops. Established U.S. tolerances for aldicarb and its toxic metabolites range from negligible in sugarcane, sweet potatoes, milk, and animal products to a high of 1 ppm in potatoes.

Only granular formulations of aldicarb are marketed, mostly because of the high mammalian toxicity of the parent compound and principal metabolite (rat acute oral LD_{50} for both is 1 mg/kg and rabbit dermal LD_{50} is 5 mg/kg). The granules containing 5 to 15% active ingredient are applied by various means to the soil for systemic control of aphids, thrips, mealybugs, plantbugs, whiteflies, phytophagous mites, and nematodes. The residual properties of the pesticide usually provide from 4 to 12 weeks protection from a single application at recommended rates, which vary from ¼ to 3½ lb active/acre for food crops and 5½ to 11 lb active/acre for ornamentals. Although Temik granules are considerably less toxic dermally than pure aldicarb, extreme caution must be observed when handling this pesticide, and it should not be used around homes.

Aldicarb is not phytotoxic unless direct contact is made with seeds at high dosages. However, recommended procedures avoid any serious problems. With respect to compatibility with other pesticides, no adverse effects have been observed, but one should avoid exposing aldicarb to lime or other alkaline products which quickly decompose the carbamate. In 1971, about two million pounds of Temik were manufactured in the United States.[38]

Aminocarb

See Matacil.

Arprocarb

See Baygon.

Bassa

The insecticide 2-*s*-butylphenyl N-methylcarbamate is marketed by Mitsubishi Chemical Industries under the name Bassa, and by Sumitomo Chemical Company under the name Osbac.[48-50] It is available as a 50% emulsifiable concentrate, 2% dust, and 3% granule for control of leafhoppers and planthoppers on rice outside of the United States. The residual material is applied to plant surfaces or water and kills on contact. When mixed with BHC, it also effectively controls rice stem borers. Toxicology studies indicate that the carbamate is a relatively safe material to handle (rat oral LD_{50} is 410 mg/kg and mouse dermal LD_{50} is 4,200 mg/kg), but some precautions are needed when the insecticide is used around fish.

Baygon

Baygon was introduced by the Bayer Leverkusen Company of West Germany in 1959 and later given the common names propoxur and arprocarb for the active ingredient 2-isopropoxyphenyl N-methylcarbamate.[48-50,52] In addition to the Baygon trademark, other names, including Blattanex, Unden, Sendran, and Suncide, have also been registered. In the United States, Baygon development is under the direction of the Chemagro Division of Baychem Corporation. Although temporary tolerances granted by the U.S. Government allow Baygon residues on a few food crops, its primary use has been against insect pests of man and domestic animals.

Manufactured formulations include a 50% and 80% wettable powder, an 18% emulsifiable concentrate, a 48% ULV sprayable concentrate, 1% and 2% baits, 1% and 5% dusts, and an oil fog concentrate. Recommended application rates vary from 1/8 to 3 lb active/acre, and concentrations in household sprays for control of aphids, flies, mosquitoes, cockroaches, ants, earwigs, crickets, chinch bugs, fleas, ticks, spiders, and millipedes range from 0.5 to 2%. The carbamate gives rapid action for quick knockdown and provides residual effectiveness for 4 to 6 weeks. There have been several reports of phytotoxicity, honeybee toxicity, and bird toxicity, but the compound is relatively safe to use around mammals (rat acute oral LD_{50} is 100 mg/kg and dermal LD_{50} is $>$ 1,000 mg/kg). When applied to soil, some systemic action against aphids, thrips, and other sucking insects has been reported.

Blattanex

See Baygon.

Bufencarb

See Bux.

Butacarb

Butacarb is the common and trade name for 3,5-di-*t*-butylphenyl N-methylcarbamate, a product of Boots Company Limited.[50] Apparently, the only use for the insecticide is as a sheep dip for control of the sheep blowfly. Formulated as a 20% or 25% miscible oil, the chemical offers 10 to 16 weeks persistency with no ill effects to the animal, due to very low mammalian toxicity, i.e., rat oral LD_{50} of more than 4,000 mg/kg.

Bux

In 1966, the Chevron Chemical Company discovered the insecticidal potential of a 3:1 mixture of 3-(1-methylbutyl)phenyl N-methylcarbamate and 3-(1-ethylpropyl)phenyl N-methylcarbamate.[48-50,53] The mixture was introduced for experimentation in 1968 under the trade names Bux and Ortho 5353, and one year later the U.S. Government provided a negligible residue tolerance limit of 0.05 ppm on corn grain, fresh corn, corn fodder and forage, rice, and rice straw. Common names proposed for the mixed active ingredients are metalkamate and bufencarb.

As a residual soil insecticide, Bux holds U.S. registrations for use on field corn, popcorn, and sweet corn when applied at planting or at first cultivation. In other countries, Bux is registered for treatment of rice, pineapple, and other crops. Rates of application range from 1/2 to 2 lb active/acre with 10% granules, 24% or 30% emulsifiable concentrates, and 2% or 4% dusts. The methylcarbamates are effective against larvae of the corn rootworm, the rice water weevil, cotton leaf perforater, rice stem borer, armyworm, rice green leafhopper, and others. Bux exhibits an acute oral LD_{50} to rats of 87 mg/kg and a rabbit dermal LD_{50} of 680 mg/kg. It is not phytotoxic, but it is hazardous to certain fish. Based on U.S. production figures in 1971, Bux sales rank as the third-largest among methylcarbamate insecticides (six million pounds in 1971).[38]

Carbamult

See promecarb.

Carbaryl

Carbaryl is the common name for the active ingredient (1-naphthyl N-methylcarbamate) of the insecticide Sevin, a trademark of Union Carbide Corporation.[48-50,54] The product, also known by numerous other common and trade designations, is a low-hazard, moderately residual contact and stomach insect poison available as a 50% wettable powder, 80% or 85% sprayable powder, 5% or 10% granule, 2 to 10% dust, 5% bait, and various other flowable and emulsifiable formulations. Carbaryl is by far the most widely used carbamate insecticide, as evidenced by the 55 million pounds manufactured by Union Carbide in 1971.[38]

It is almost impossible to list all of the uses for Sevin, but some of the more important include application to cotton, corn, soybeans, various fruits, vegetables, bananas, pineapple, olives, cacao, coffee, rice, sugar cane, and nearly 100 other food, fiber, and forage crops around the world. In addition, the insecticide is applied to forests, range land, livestock, pets, poultry, and buildings. Rates of application usually extend from 1/2 to 4 lb active/acre or from 1/2 to 1½ lb active/100 gal of water to give control of 160 harmful insect species. High residue tolerances of 5 to 10 ppm on many food crops allow use within a few days of harvest, and even higher tolerances on forage make it possible to spray carbaryl the day before harvest.

Carbaryl's effectiveness against a wide range of insects and its low mammalian toxicity (rat acute oral LD_{50} is 500 mg/kg and dermal LD_{50} is >4,000 mg/kg) make it an ideal compound for use around homes, in home gardens, and on pets. With the exception of honybees, carbaryl presents minimal hazard to most nontarget organisms, including farm animals, birds, fish, and other wildlife. Environmentally, deposits of Sevin average half-lives of 3 to 4 days on plant foliage, 7 to 9 days in soil, and 1 to 5 days in water, making this a very biodegradable pesticide.

There are certain plant species that are sensitive to carbaryl sprays, and, in fact, carbaryl is often used as a thinning agent on apple trees. Sevin is compatible with most pesticides, except under alkaline conditions, where the active ingredient, like most carbamates, rapidly degrades.

Carbofuran

According to production figures for 1971, eight million pounds of 2,3-dihydro-2,2-dimethyl-7-benzofuranyl N-methylcarbamate were manufactured in the United States, making carbofuran the second-largest-selling carbamate insecticide.[38] It entered the market in 1967 under the registered trademark Furadan as a result of development by the Niagara Chemical Division of FMC Corporation.[48-50,55] Another trade name, less known, is Curaterr, and at present carbofuran is also manufactured by Bayer Company of West Germany and Baychem Corporation in the United States.

Carbofuran is a broad-spectrum, long-residual insecticide and nematicide effective by contact, stomach, and systemic action. Formulated as a 2%, 3%, 5%, or 10% granule and as a 48% flowable, Furadan is usually applied to foliage for insect control at rates of 1/4 to 1 lb active/acre, placed in or above seed furrows at rates of 2 to 4 lb active/acre, and broadcast at rates of 6 to 10 lb active/acre for nematode control. The half-life for carbofuran in soil averages 30 to 60 days, but within plants the carbamate is much less persistent. When used according to label directions, it is nonphytotoxic and compatible with most other pesticides, but should not be used with certain herbicides, such as propanil.

Registration by the U.S. Environmental Protection Agency has been granted for certain uses of carbofuran: alfalfa, residue tolerance of 5 ppm for green material and 20 ppm for hay; field corn, 0.1 ppm for grain and 5 ppm for fodder and forage; peanuts, 0.1 ppm for nut and 1 ppm for hulls; peppers, 1 ppm; rice, 0.2 ppm for grain and straw; sugar cane, 0.1 ppm; and flue-cured tobacco. Outside of the U.S., Furadan is used in the production of bananas, coffee, and sugar beets. Important insect pests controlled by Furadan include corn rootworm, rice water weevil, wireworm, sugar cane borer, alfalfa weevil, alfalfa snout beetle, armyworms, budworms, European corn borer, flea beetle, pea aphid, thrips, and hornworms. Additionally, several nematode species are also susceptible to carbofuran.

The toxicology of carbofuran is unusual in that the rat oral LD_{50} is very low, namely 11 mg/kg, yet the rabbit dermal LD_{50} is more than 10,000 mg/kg (though the dermal value for a 75% wettable powder formulation is 3,400 mg/kg). Hence, problems from surface contact with Furadan are minimal, but potential hazards from ingestion or inhalation by applicators as well as by game birds and animals require caution during use.

Carpolin
See carbaryl.

Carzol
See formetanate hydrochloride.

Dicarzol
See formetanate hydrochloride.

Dioxacarb
Ciba-Geigy of Switzerland introduced the insecticide 2-(1,3-dioxolane-2-yl)phenyl N-methylcarbamate in 1968 under the trademarks Elocron and Famid and with the common name dioxacarb.[48-50] The experimental compound (code number C-8353) is a rapid-acting contact and stomach poison, particularly effective against cockroaches when applied to surfaces at rates of 0.5 to 2 g/m^2. Its residual activity persists for up to 6 months on walls, but it is rapidly degraded in soils or on plant foliage. Tests on crops at rates of 1/4 to 1½ lb active/acre indicate possible control of a number of sucking and chewing insects, such as aphids, potato beetles, cacao bugs, alfalfa weevils, leafhoppers, sawflies, and bollworms.

Formulations available for testing include a 50% wettable powder, 20% and 40% emulsifiable concentrate, 5% ready-to-use powder, and a 2% granule. The acute rat oral LD_{50} of dioxacarb is about 115 mg/kg and the rabbit dermal LD_{50} is 1,950 mg/kg.

Draza
See Mesurol.

Elocron
See dioxacarb.

Etrofol
See Hopcide.

Etrofolan
See Mipcin.

Famid
See dioxacarb.

Formetanate Hydrochloride
Carzol SP and Dicarzol are trademarks of Schering A.G. of West Germany for the hydrochloride salt of 3-dimethylaminomethyleneiminophenyl N-methylcarbamate.[48-50,56] The compound, known by the common name formetanate hydrochloride, was discovered by Schering in 1962 and has been developed in the United States by NOR-AM Agricultural Products. Formulated as an 82% soluble powder or 50% wettable powder, formetanate hydrochloride controls spider mites, rust mites, certain aphids, thrips, leaf miners, leafhoppers, lygus bugs, slugs, and snails on horticultural, agronomic, and ornamental plants.

Registered uses for Carzol SP in the U.S. at rates of 1/2 to 5 lb active/acre include application to citrus, apples, pears, peaches, plums, nectarines, prunes, grapes, and seed alfalfa. Phytotoxicity may occur at higher-than-recommended rates. The rat oral LD_{50} for formetanate hydrochloride is 20 mg/kg and the rabbit dermal LD_{50} is >10,200 mg/kg. Caution is needed to avoid nontarget kills of honeybees, beneficial mites, and some fish. Livestock should not be grazed in formetanate-treated areas nor fed forage sprayed with this pesticide.

Furadan
See carbofuran.

Hopcide

CPMC is the common name for 2-chlorophenyl N-methylcarbamate, the active ingredient of Kumiai Chemical's Hopcide and Bayer's Etrofol.[50] The products are being developed as 50% wettable powders for use against leafhoppers and planthoppers in rice crops. Results of toxicity tests indicate an acute oral LD_{50} for rats of 648 mg/kg and a dermal LD_{50} for rabbits of more than 500 mg/kg.

HRS-1422

This carbamate, which showed early promise as an insecticide, has never been commercially produced, although it has been included in a number of comparative studies with other carbamate insecticides. The compound is a product of the Hooker Chemical Corporation and has the chemical formula 3,5-diisopropylphenyl N-methylcarbamate.

Isoprocarb

See Mipcin.

Landrin

Landrin is a residual soil insecticide composed of a 3:1 mixture of the 3,4,5- and 2,3,5-isomers of trimethylphenyl N-methylcarbamate.[48,50,57] The compounds originated in the Agricultural Chemicals Division of Shell Chemical Company in 1968 and are formulated as a 50% wettable powder and 15% granule. Applied at the rate of 3/4 to 1 lb active/acre, Landrin gives season-long control of corn rootworm larvae on field corn and popcorn, for which a temporary residue tolerance has been established. The mixed isomers produce an acute oral LD_{50} of 208 mg/kg (rat) and a dermal LD_{50} of more than 2,500 mg/kg (rabbit).

Lannate

See methomyl.

Matacil

One of the oldest experimental methylcarbamate insecticides is aminocarb (4-dimethylamino-3-methylphenyl N-methylcarbamate), which was introduced in 1963 as Matacil by Bayer Leverkusen and researched in the United States by the Chemagro Division of Baychem Corporation.[48-50] Although the compound has not advanced beyond experimental status in the U.S., it is used in Europe on cole crops, lettuce, corn, and ornamentals.

Aminocarb is especially effective as a residual stomach poison against many lepidopterous larvae and biting insects; it also shows promise for use against slugs and snails and for control of certain forest insect pests. Formulations include a 50% and 80% wettable powder, a 5% dust, a 2% bait, and an 18% emulsifiable concentrate. Recommended application rates range from ½ to 2½ lb active/acre. Rat acute oral LD_{50} is 30 mg/kg and dermal LD_{50} is 275 mg/kg.

Meobal

The active ingredient of Meobal insecticide, 3,4-dimethylphenyl N-methylcarbamate, was discovered in 1966 by the Sumitomo Chemical Company.[48-50] Formulations as a 30% emulsifiable concentrate, 50% wettable powder, 2% dust, and 3% granule are used in Japan for control of hoppers on rice, lepidopterous larvae on vegetables, and scale insects on fruits. The rat oral LD_{50} is 380 mg/kg and the mouse dermal LD_{50} is >1,500 mg/kg.

Mercaptodimethur

See Mesurol.

Mesurol

In Europe, Mesurol insecticide-acaricide is registered for use on cotton, ornamentals, vegetable crops, fruit trees, and hops at rates from ¼ to 1 lb active/acre.[48-50] Also known by its other trade name, Draza, and by common names methiocarb, mercaptodimethur, and metmercapturon, the contact and stomach poison offers fast cleanup and long residual effectiveness against such pests as thrips, grasshoppers, fruit flies, plum curculio, pear psylla, certain mites, leafhoppers, aphids, and codling moth. The active principle, 4-methylthio-3,5-dimethylphenyl N-methylcarbamate is formulated as a 50% or 75% wettable powder, a 5% dust, and a 4% bait.

The product was discovered by Bayer Leverkusen of West Germany and is an experimental compound of Baychem Corporation in the U.S.[58] In addition to its action on arthropods, Mesurol also provides control of slugs and snails and functions as a bird repellent when used as a seed dressing. The rat oral LD_{50} is about 130 mg/kg and the dermal LD_{50} is more than 2,000 mg/kg.

Metacrate

See Tsumacide.

Metalkamate

See Bux.

Metmercapturon

See Mesurol.

Methiocarb

See Mesurol.

Methomyl

Methomyl (S-methyl N – [(methylcarbamoyl)-oxy]thioacetimidate) was introduced as an experimental insecticide-nematicide by E.I. du Pont de Nemours in 1967 under the trade name Lannate.[48-50,59] At the present time, the product has 29 U.S. registrations for foliar application to a variety of fruit and vegetable crops, cotton, alfalfa, ornamentals, tobacco, and turf. Residue tolerances are 10 ppm for forage and fodder and range from 0.1 to 6 ppm for food crops.

Although there is some systemic movement of methomyl from soil into plants, most insect kill is due to direct contact or ingestion of the oxime carbamate. Effective rates are ¼ to 1 lb active/acre using a 90% water-soluble powder, 25% wettable powder, 24% emulsifiable concentrate, 5% granule, or 2% to 5% dust. Residual persistence of methomyl is short, with a half-life of less than 1 week on most plant foliage. A large number of pest insects are controlled by the insecticide, including alfalfa weevil, various aphids, codling moth, corn earworm, diamondback moth, European corn borer, tobacco hornworm, thrips, and others.

Lannate is compatible with many pesticides, but should not be mixed with alkaline materials. It is one of the few carbamates that is only moderately toxic to honeybees (when sprays are dry). With respect to mammals, it is virtually nontoxic when applied dermally as a water slurry (more than 5,000 mg/kg LD_{50} for rabbits), but its low rat oral LD_{50} of 21 mg/kg requires that considerable caution be observed when applying Lannate. The insecticide is also marketed by Shell Chemical Company under the name Nudrin.

Mexacarbate

See Zectran.

Mipcin

The experimental insecticide 2-isopropylphenyl N-methylcarbamate is trademarked Etrofolan by Bayer A.G. of Germany and Mipcin by Mitsubishi Chemical Industries of Japan.[50,60] In addition, this insecticide, which shows promise for control of leafhoppers, planthoppers, and bugs in rice and cacao, has the common names isoprocarb and MIPC. Reported rat toxicities include an oral and dermal LD_{50} of 485 mg/kg and of more than 500 mg/kg respectively.

Mobam

Discovered in 1964 by Mobil Chemical Company, Mobam (4-benzothienyl N-methylcarbamate) is not under active development at the present time. Past studies have shown it to be a contact residual insecticide effective against a broad range of household, animal, and crop pests at rates of ¼ to 3 lb active/acre.[40,50,61] Its half-life in soil is about 30 days and on plants the $t_{1/2}$ averages 3 to 7 days. Formulations available for testing are 10% granule, 50% or 80% wettable powder, 48% emulsifiable concentrate, and 2%, 5%, or 50% dust. The acute oral LD_{50} has been determined to be 234 mg/kg for rats and the dermal LD_{50} to rabbits is greater than 6,230 mg/kg.

Nudrin

See methomyl.

Osbac

See Bassa.

Oxamyl

This oxime carbamate is registered in the U.S. for early-season control of root knot and lesion nematodes in tobacco and for control of various nematodes on certain ornamentals and nursery crops.[50] Known by the common name oxamyl and by the chemical designation methyl-N′,N′-dimethyl-N-[(methylcarbamoyl)oxy]-1-thioxamimidate, the toxin is the active ingredient of Vydate, a trademark of E.I. du Pont de Nemours.[62] Besides acting as a broad-spectrum contact nematicide, oxamyl has the unique ability to move from the foliage of certain plants down into the roots for nematode control. In addition, foliar sprays at rates of 0.1 to 1 lb active/acre show promise for suppression of Colorado potato beetle, green peach aphid, cotton leafhopper, thrips, fleabeetles, and others on cotton, potato, citrus, peanut, and tobacco, whereas sprays of 0.1 to 0.3 lb active/100 gal on apples and citrus control several mite species. The compound may also be useful as a soil-applied systemic insecticide-miticide.

Formulations available include a 24% emulsifiable concentrate and a 10% granule. Caution must be observed when using Vydate because of the high mammalian toxicity of the active ingredient (rat acute oral LD_{50} is 5.4 mg/kg). However, dermal exposure should cause few problems, since a 24% liquid formulation applied to the skin of rabbits yielded an acute dermal LD_{50} of 2,960 mg/kg.

Pirimicarb

Plant Protection Limited, a subsidiary of ICI of Britain, introduced Pirimor in 1969 as an experimental fast-acting selective aphicide which killed by contact, vapor, or systemic action.[49,50] Subsequently, the principal toxin, 2-dimethylamino-5,6-dimethyl-4-pyrimidinyl N,N-dimethylcarbamate, was given the common name pirimicarb. The product is particularly effective against organophosphorus-resistant aphids infesting fruit trees, cole crops, beans, lettuce, flowers, and certain ornamentals. Available as a 50% wettable powder or 10% smoke generator, pirimicarb possesses an acute oral LD_{50} of 147 mg/kg for rats; dermal application of 500 mg/kg to rabbits caused no mortality.

Pirimor

See pirimicarb.

Promecarb

Promecarb (3-methyl-5-isopropylphenyl N-methylcarbamate) originated from the laboratories of Schering A. G. of West Germany, where it was trademarked Carbamult in 1965.[48-50] Outside of the United States, promecarb is used as a contact and stomach poison for insect pests of potatoes at ½ lb active/acre, and for pests of certain fruit crops at 0.1 to 0.25% active in water. The carbamate gives moderate residual protection against infestations of leafminers, aphids, ticks, sawflies, Colorado beetle, corn rootworm, and others.

Formulated products consist of a 50% and 37.5% wettable powder, a 5% dust, and a 25% emulsifiable concentrate. Ingestion of the active ingredient by rats yielded an acute LD_{50} of 75 mg/kg, whereas rabbit dermal LD_{50} values are over 2,000 mg/kg. Promecarb is under development in the U.S. by NOR-AM Agricultural Products.

Propoxur

See Baygon.

Sendran

See Baygon.

Sevin

See carbaryl.

Suncide

See Baygon.

Temik

See aldicarb.

Tsumacide

Two Japanese companies have trademarked 3-methylphenyl N-methylcarbamate as an active ingredient of an insecticidal product.[49,50] Tsumacide is produced by Nihon Nohyaku Company, and Metacrate belongs to Sumitomo Chemical Company. Also known by the common name MTMC, this methylcarbamate is intended for use against leafhoppers and planthoppers on rice, but it is also effective via contact, stomach, or systemic action for control of codling moth, citrus mealybug, onion thrips, Mediterranean fruit fly, pink bollworm, and cotton aphid. The active ingredient in 2% dusts, 50% wettable powders, and 30% emulsifiable concentrates has an acute mouse oral LD_{50} of 268 mg/kg and a rat dermal LD_{50} of 6,000 mg/kg.

UC 10854

The compound 3-isopropylphenyl N-methylcarbamate was introduced in 1961 as a herbicide by Union Carbide Corporation under the code number UC 10854, and by Hercules Incorporated as experimental chemical H 5727.[63] It was later discovered to have considerable potential as a broad-spectrum insecticide, but the product never progressed beyond experimental status. Part of the reason may be due to its erratic mammalian toxicity; for example, it yields a rat acute oral LD_{50} of 50 mg/kg, but it is lethal to guinea pigs at only 10 mg/kg. There are also indications that the compound is more toxic to humans than the rat oral LD_{50} suggests (see Chapter 3). The acute dermal LD_{50} to rats is 113 mg/kg.

Unden

See Baygon.

Vydate

See Oxamyl.

Zectran

One of the oldest methylcarbamate insecticide trademarks (1961), Zectran, belongs to the Dow Chemical Company.[49,50,63] The toxin, 4-dimethylamino-3,5-dimethylphenyl N-methylcarbamate has the common name mexacarbate and has shown promise as an insecticide, acaricide, and molluscicide. Its history is rather unusual; the compound has been in and out of production on several occasions. At the present time Zectran is not being manufactured by Dow, but it is registered in the U.S. for use as a 2% meal bait for slugs and snails and as a 25% wettable powder or 23% emulsifiable concentrate for use on turf, shade trees, shrubs, flowers, and other ornamentals. For a while, Zectran's greatest potential appeared to lie in its use as a forest insecticide at rates of 0.03 to 1 lb active/acre. Part of the reason this never resulted in wide-scale commercialization may have been mexacarbate's high oral mammalian toxicity (LD_{50} to rat is 30 mg/kg), although it is essentially nontoxic dermally (LD_{50} to rabbit is more than 8,000 mg/kg).

REFERENCES

1. **Shepard, H. H.,** *The Chemistry and Toxicology of Insecticides,* Burgess Publishing Co., Minneapolis, 1939.
2. **Brooks, G. T.,** *Chlorinated Insecticides,* Vol. 1, CRC Press, Cleveland, 1974.
3. **Eto, M.,** *Organophosphorus Pesticides: Organic and Biological Chemistry,* CRC Press, Cleveland, 1974.
4. **Carson, R. L.,** *Silent Spring,* Houghton Mifflin Co., Boston, 1962.
5. *Pest Control Strategies for the Future,* National Academy of Sciences, Washington, D.C., 1972.
6. **Holmstedt, B.,** The ordeal of Old Calabar: the pageant of *Physostigma venenosum* in medicine, in *Plants in the Development of Modern Medicine,* Swain, T., Ed., Harvard University Press, Cambridge, Mass., 1972.
7. **Balfour, J. H.,** Description of the plant which produces the ordeal bean of Calabar, *Trans. R. Soc. Edinb.,* 22, 305, 1861.
8. **Fraser, T. R.,** On the characters, actions and therapeutic uses of the bean of Calabar, *Edinb. Med. J.,* 9, 36, 1863.
9. **Jobst, J. and Hesse, O.,** Über die Bohne von Calabar, *Ann. Chem. Pharm.,* 129, 115, 1864.
10. **Stedman, E. and Barger, G.,** Physostigmine (eserine). Part III, *J. Chem. Soc.,* 127, 247, 1925.
11. **Julian, P. L. and Pikl, J.,** Studies in indole series. V. The complete synthesis of physostigmine (eserine), *J. Am. Chem. Soc.,* 57, 755, 1935.
12. **Christison, R.,** On the properties of the ordeal bean of Old Calabar, Western Africa, *Mon. J. Med.* (London), 20, 193, 1855.
13. **Fraser, T. R.,** On atropia as a physiological antidote to the poisonous effects of physostigma, *Practitioner* (London), 4, 65, 1870.
14. **Pal, J.,** Physostigmin ein Gegengift des Curare, *Zentralbl. Physiol.,* 14, 255, 1900.
15. **Robertson, D. A.,** The Calabar bean as a new agent in ophthalmic medicine, *Edinb. Med. J.,* 8, 815, 1863.
16. **Laqueur, L.,** Ueber Atropin und Physostigmin und ihre Wirkung auf den intraocularen Druck. Ein Beitrag zur Therapie des Glaucoms, *Arch. Ophthalmol.,* 23, 149, 1877.
17. **Walker, M. B.,** Treatment of myasthenia gravis with physostigmine, *Lancet,* 226, 1200, 1934.
18. **Loewi, O.,** uebertragbarkeit der Herznervenwirkung. I. Mitteilung, *Pflüger's Arch. Gesamte Physiol.,* 189, 237, 1921.
19. **Loewi, O. and Navratil, E.,** Über humorale Ubertragbarkeit der Herznervenwirkung. XI. Mitteilung. Über den Mechanismus der Vaguswirkung von Physostigmin und Ergotamin, *Pflüger's Arch. Gesamte Physiol.,* 214, 689, 1926.
20. **Feldberg, W. and Gaddum, J. H.,** The chemical transmitter at synapses in a sympathetic ganglion, *J. Physiol.,* 81, 305, 1934.
21. **Stedman, E. and Stedman, E.,** Studies on the relationship between chemical constitution and physiological action. III. The inhibitory action of certain synthetic urethanes on activity of liver esterases, *Biochem. J.,* 25, 1147, 1931.
22. **Stempel, A. and Aeschlimann, J. A.,** Synthetic analogs of physostigmine, in *Medicinal Chemistry,* Vol. 3, Blickse, F. F. and Cox, R. H., Eds., John Wiley & Sons, New York, 1956.
23. **Tisdale, W. H. and Flenver, A. L.,** Derivatives of dithiocarbamic acid as pesticides, *Ind. Eng. Chem.,* 34, 501, 1942.
24. **Stoddard, E. M., Gries, G. A., and Plumb, G. H.,** Red spider control with disodium ethylene bisdithiocarbamate, *Phytopathology,* 35, 657, 1945.

25. **Rich, S.**, Miticidal action of barium and manganese ethylene bisdithiocarbamates, *Phytopathology*, 44, 387, 1954.
26. **McMullen, R. D.**, Insecticidal activity of ethylene bisdithiocarbamates, *Nature*, 184, 1338, 1959.
27. **Gysin, H.**, Über einige neue Insektizide, *Chimia*, 8, 205, 1954.
28. **Geigy, J. R.**, Swiss Patent 279,553, March 1, 1952.
29. **Geigy, J. R.**, Swiss Patent 282,655, August 16, 1952.
30. **Ferguson, G. R. and Alexander, C. C.**, Heterocyclic carbamates having systemic insecticidal action, *J. Agric. Food Chem.*, 1, 888, 1953.
31. **Gubler, K., Margot, A., and Gysin, H.**, Alicyclic and heterocyclic carbamates, *J. Sci. Food Agric. Suppl.*, p. 13, 1968.
32. **Lambrech, J. A.** (Union Carbide Corp.), U.S. Patent 2,903,478, September 8, 1959.
33. **Weiden, M. H. J.**, Toxicity of carbamates to insects, *Bull. W.H.O.*, 44, 203, 1971.
34. Union Carbide Corporation, *Station to Station Research News* New York, 2, 1, 1956.
35. **Haynes, H. L., Lambrech, J. A., and Moorefield, H. H.**, Insecticidal properties and characteristics of 1-naphthyl N-methylcarbamate, *Contrib. Boyce Thompson Inst.*, 18, 507, 1957.
36. **Back, R. C.**, Significant developments in eight years with Sevin insecticide, *J. Agric. Food Chem.*, 13, 198, 1965.
37. **O'Brien, R. D. and Matthysse, J. G.**, Carbamate insecticides in the U.S. and abroad, *Agric. Chem.*, 16(11), 27, 1961.
38. **Matsumura, F.**, *Toxicology of Insecticides*, Plenum Press, New York, London, 1975, p. 6.
39. **Kolbezen, M. J., Metcalf, R. L., and Fukuto, T. R.**, Insecticidal activity of carbamate cholinesterase inhibitors, *J. Agric. Food Chem.*, 2, 864, 1954.
40. **Payne, L. K., Jr., Stansburg, H. A., Jr., and Weiden, M. H. J.**, The synthesis and insecticidal properties of some cholinergic trisubstituted acetaldehyde O-(methylcarbamoyl)oximes, *J. Agric. Food Chem.*, 14, 356, 1966.
41. **Weiden, M. H. J., Moorefield, H. H., and Payne, L. K.**, O-(Methylcarbamoyl)oximes: a new class of carbamate insecticide-acaricides, *J. Econ. Entomol.*, 58, 154, 1965.
42. **Kilsheimer, J. R. and Manning, D. T.** (Union Carbide Corp.), South African Patent Application 62/3821, 1962.
43. **Kilsheimer, J. R. and Manning, D. T.** (Union Carbide Corp.), French Patent 1,343,654, October 14, 1962.
44. **Dorough, H. W., Leeling, N. C., and Casida, J. E.**, Nonhydrolytic pathway in metabolism of N-methylcarbamate insecticides, *Science*, 140, 170, 1963.
45. **Dorough, H. W. and Casida, J. E.**, Nature of certain carbamate metabolites of the insecticide Sevin, *J. Agric. Food Chem.*, 12, 294, 1964.
46. **Abdel-Wahab, A. M., Kuhr, R. J., and Casida, J. E.**, Fate of C^{14}-carbonyl-labeled aryl methylcarbamate insecticide chemicals in and on bean plants, *J. Agric. Food Chem.*, 14, 290, 1966.
47. **Kuhr, R. J. and Casida, J. E.**, Persistent glycosides of metabolites of methylcarbamate insecticide chemicals formed by hydroxylation in bean plants, *J. Agric. Food Chem.*, 15, 814, 1967.
48. **Thompson, W. T.**, *Agricultural Chemicals. Book I - Insecticides*, Thompson Publications, Fresno, California, 1970.
49. **Martin, H., Ed.**, *Pesticide Manual*, 3rd ed., British Crop Protection Council, London, 1972.
50. **Shepard, H. H., Ed.**, *Pesticide Dictionary, Farm Chemicals Handbook*, Meister Publishing Co., Willoughby, Ohio, 1975.
51. Union Carbide Corporation, *Temik Aldicarb Pesticide, a Systemic Pesticide for Control of Insects, Mites and Nematodes*, Technical Information Bulletin, 1975.
52. Chemagro Division of Baychem Corporation, *Baygon Insecticide*, Technical Information Bulletin, 1973.
53. Chevron Chemical Company, *Bux Insecticide (Ortho 5353)*, Technical Information, Experimental Data Sheet, 1973.
54. Union Carbide Corporation, *Sevin Carbaryl Insecticide for Abundant Food and Fiber*, Technical Information Bulletin, 1974.
55. Agricultural Chemical Division of FMC Corporation, *Furadan Insecticide/Nematicide*, Technical Information Bulletin, 1974.
56. NOR-AM Agricultural Products, Inc., *Carzol SP Acaracide/Insecticide*, Technical Information Bulletin.
57. Agricultural Chemicals Division of Shell Chemical Company, *Summary of Basic Data for Technical Landrin Insecticide*, Technical Data Bulletin, 1969.
58. Chemagro Division of Baychem Corporation, *Mesurol Insecticide*, Technical Information Bulletin, 1973.
59. E.I. du Pont de Nemours & Co., *Lannate L Methomyl Insecticide, Lannate Methomyl Insecticide*, Information Bulletin.
60. **Kenaga, E. E. and End, C. S.**, *Commercial and Experimental Organic Insecticides*, 1974 revision, Entomological Society of America, College Park, Maryland, 1974.
61. Mobil Chemical Company, *Mobam*, Technical Bulletin.
62. E.I. du Pont de Nemours & Co., *Vydate Oxamyl Insecticide Nematicide*, Information Bulletin, 1972.
63. **Martin, H., Ed.**, *Pesticide Manual*, 1st Ed., British Crop Protection Council, London, 1968.
64. Dow Chemical Company, *Zectran Insecticide*, Technical Data Sheets.

Chapter 2

CHEMISTRY AND SYNTHESIS

INTRODUCTION

The general chemistry associated with synthetic carbamate insecticides is relatively easy to understand, especially when compared to that accompanying other classes of insecticides, such as pyrethroids, cyclodienes, or organophosphorus compounds. Thus, for most carbamate compounds, it is a simple task to transcribe a written structural formula into a molecular figure, and metabolic pathways, though elaborate, are not terribly complicated. On the following pages is a brief introduction to the principles of carbamate insecticide chemistry for those not familiar with this interesting group of compounds.

The backbone of all carbamate structures is carbamic acid, the monoamide of carbon dioxide:

$$\mathrm{HO\overset{\overset{O}{\|}}{C}NH_2}$$

The acid does not exist in a free form, however, since it spontaneously decomposes into carbon dioxide and ammonia. On the other hand, salts of carbamic acid, called carbaminates or carbamates, are much more stable. Indeed, one such salt, namely ammonium carbamate,

$$\mathrm{NH_4^+ \; {}^-O\overset{\overset{O}{\|}}{C}NH_2}$$

has been mixed with aluminum phosphide and used as a commercial insecticide and rodenticide.[1] The acid salts are readily soluble in water, but hydrolyze at higher temperatures.

Carbamic acid can also be stabilized by formation of a simple alkyl ester, such as ethyl carbamate, which is commonly referred to as urethane:

$$\mathrm{C_2H_5O\overset{\overset{O}{\|}}{C}NH_2}$$

This material has a wide variety of uses, including applications as a veterinary anesthetic, topical bactericide, and solubilizer or cosolvent for pesticides, fumigants, and cosmetics.[2] Unfortunately, certain analogs of urethane are carcinogenic to lungs and to a few other mammalian tissues,[3,4] and this has caused considerable concern regarding the use of carbamate chemicals as insecticides. However, the structures and properties of insecticide molecules differ markedly from carcinogenic urethanes, as will be shown shortly. Also, the biological and physical breakdown of carbamate insecticides following application does not include the formation of urethanes (see Chapters 7 and 8).

In addition to alkyl esters, aryl esters of carbamic acid are also possible. A typical example is phenyl carbamate:

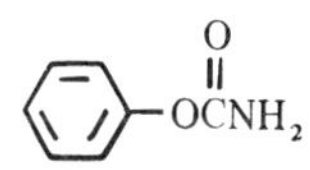

If one of the protons attached to the nitrogen atom is replaced by a methyl group, the resulting compound, phenyl N-methylcarbamate, is mildly toxic to a few insect species. Going one step further by adding substituents to the phenyl ring yields the majority of commercial and experimental methylcarbamate insecticide chemicals; a few of these are pictured in Figure 2.1. The chemical name for Banol is 2-chloro-4,5-dimethylphenyl N-methyl-carbamate, Baygon is 2-isopropoxyphenyl N-methylcarbamate, and so forth. Many more compounds and the effects of structural alterations on biological activity will be thoroughly examined in Chapter 4.

Other aryl methylcarbamate esters with insecticidal activity can be prepared from aromatics such as 1-naphthol, 4-hydroxybenzothiophene, or 2,3-dihydro-2,2-dimethyl-7-hydroxybenzofuran. The resulting insecticides, carbaryl (1-naphthyl N-methylcarbamate), Mobam (4-benzothienyl N-methylcarbamate), and carbofuran (2,3-dihydro-2,2-dimethyl-7-benzofuranyl N-methylcarbamate), are shown in Figure 2.2. Two of the chemicals, carbaryl and carbofuran, are the most widely used commercial carbamate insecticides. Of course, nomenclature becomes a bit more complicated and less familiar when discussing heterocyclic derivatives.

Nomenclature is also somewhat difficult for the newest group of successful methylcarbamate in-

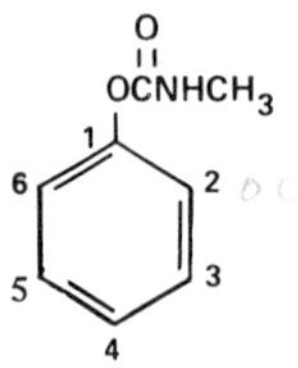

Banol	2–Cl–4,5–diCH_3
Baygon	2–$OCH(CH_3)_2$
Butacarb	3,5–di$C(CH_3)_3$
Bux	3–$CH(C_3H_7)(CH_3)$ and 3–$CH(C_2H_5)_2$
Formetanate	3–N=CH $N(CH_3)_2$
Landrin	3,4,5–triCH_3 and 2,3,5–triCH_3
Mesurol	4–SCH_3–3,5–diCH_3
Meobal	3,4–diCH_3
Tsumacide	3–CH_3
UC10854	3–$CH(CH_3)_2$
Zectran	4–$N(CH_3)_2$–3,5–diCH_3

FIGURE 2.1. Structures of substituted phenyl N-methylcarbamate insecticide chemicals.

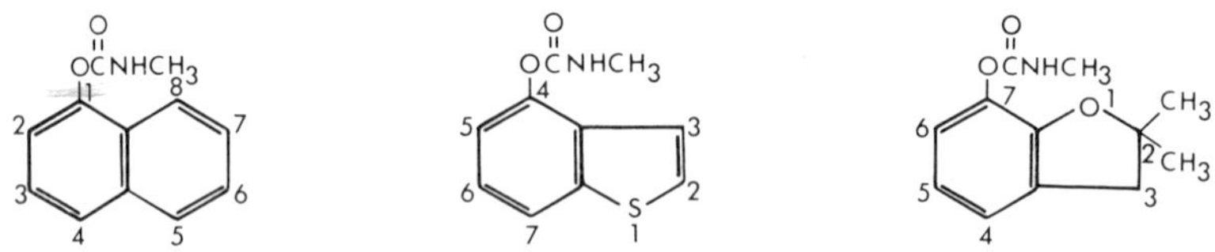

carbaryl Mobam carbofuran

FIGURE 2.2. Structures of various aromatic N-methylcarbamate insecticides.

secticides, which are derived from aliphatic oximes:

$$\begin{matrix} R \searrow \\ (R)H \nearrow \end{matrix} C{=}NOH$$

The latter are compounds resembling ketones or aldehydes, as shown above. When coupled with methylcarbamic acid, insect poisons such as 2-methyl-2-(methylthio)propionaldehyde O-(methylcarbamoyl)oxime (aldicarb), S-methyl-N-[(methylcarbamoyl)oxy]thioacetimidate (methomyl), and S-2-cyanethyl-N-[(methylcarbamoyl)oxy]thioacetimidate (thiocarboxime) are possible (Figure 2.3).

All of the above groups of insecticides are N-methylcarbamates (also called monomethylcarbamates or simply methylcarbamates), and together they comprise the bulk of carbamate insecticide chemicals. Thus, the greatest share of this book will be devoted to their chemistry, biochemistry, and toxicology. However, another group of chemicals, namely the N,N-dimethylcarbamates, have also been studied for insecticidal activity, particularly in Europe. Most of these materials originate from heterocyclic enols and dimethylcarbamic acid. Examples of the more important compounds (Figure 2.4) are dimetilan (2-dimethylcarbamoyl-3-methyl-5-pyrazoyl N,N-dimethylcarbamate), Isolan (l-isopropyl-3-methyl-5-pyrazolyl N,N-dimethylcarbamate), and Pyramat (2-*n*-propyl-4-methyl-6-pyrimidinyl N,N-dimethylcarbamate). Dimethylcarbamates will be discussed periodically on the following pages, but their coverage will be less thorough.

Not included in this book are the many carbamates which have been developed because of their ability to function as herbicides and fungicides. Most of these compounds differ structurally from carbamate insecticides in that they are primarily thiocarbamates or dithiocarbamates. Where carbamates are active, they tend to be aliphatic and are usually N-substituted with groups other than methyl. A few examples appear in Figure 2.5. Thiram (tetramethylthiuram disulfide) and nabam (disodium ethylene-bis-dithiocarbamate) are commercial fungicides, and IPC (isopropyl N-phenylcarbamate), EPTC (ethyl N,N-di-*n*-propylthiocarbamate), and CDEC (2-chloroallyl N,N-diethyldithiocarbamate) are sold as herbicides.

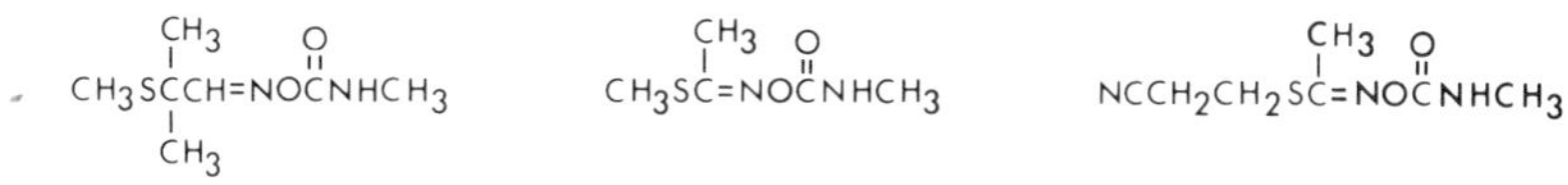

aldicarb methomyl thiocarboxime

FIGURE 2.3. Structures of O-(methylcarbamoyl)oxime insecticide chemicals.

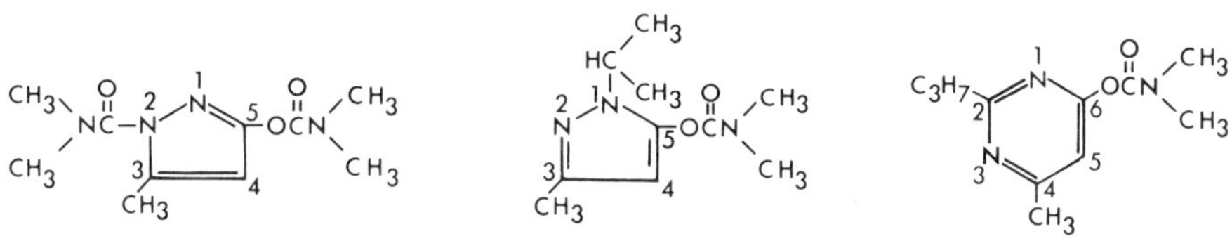

dimetilan Isolan Pyramat

FIGURE 2.4. Structures of substituted heterocyclic N,N-dimethylcarbamate insecticide chemicals.

thiram nabam

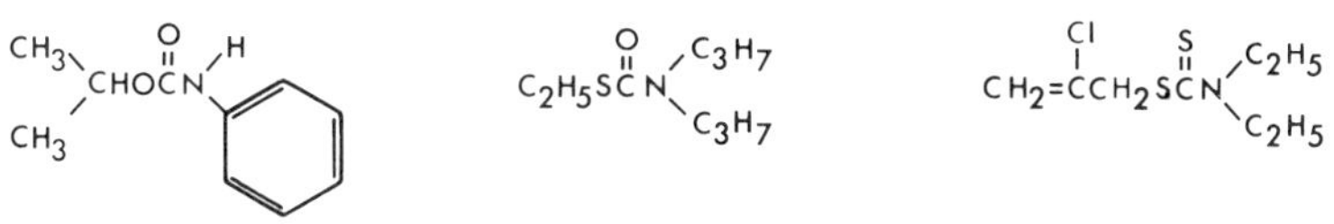

IPC EPTC CDEC

FIGURE 2.5. Structures of various carbamate fungicides and herbicides.

CHEMICAL PROPERTIES AND REACTIONS

The physical and chemical properties of methylcarbamate insecticides make them ideal compounds to work with in the laboratory. As shown in Table 2.1, most of the commercial materials available in the United States can be obtained in pure form as slightly odorous white crystalline solids. Their high melting points and low vapor pressure contribute to excellent shelf stability, though they will decompose very slowly in aqueous solution, a process that is hastened by elevated temperatures and alkalinity. They are generally noncorrosive and mix well with a variety of diluents, adjuvants, solubilizers, extenders, and other formulation materials. Most dissolve readily in a variety of organic solvents, but are only slightly soluble in water, a factor which probably limits their use as systemic insecticides; the exceptions are the hydrochloride salt of formetanate and the methylcarbamoyloximes aldicarb and methomyl.

Following application to environmental surfaces, certain methylcarbamates are susceptible to photodegradation and to air oxidation (see Chapter 8), but these processes do not alter the packaged compounds tc any significant degree. The characteristic pungent odor of methyl isocyanate is easily recognized as a thermal decomposition product of aromatic N-methylcarbamates when the crystalline compounds are subjected to

TABLE 2.1

Chemical and Physical Properties of the Active Ingredients in Commercial Methylcarbamate Insecticides Available in the United States[a]

Preferred name[b]	Empirical formula	Molecular weight	Appearance	Odor	Melting point (°C)	Vapor pressure (mmHg)	Density (g/ml)	Water-solubility (%)
Aldicarb	$C_7H_{14}N_2O_2S$	190	White crystals	Slight sulfur	98–100	1×10^{-4} (25°)	1.195 (25/20°)	0.9 (30°)
Baygon	$C_{11}H_{15}NO_3$	209	White crystals	None	91	–	–	0.1 (25°)
Bux	$C_{13}H_{19}NO_2$	221	Amber solid	Faint sweet	26–39	3×10^{-5} (30°)	1.024 (26/26°)	<0.005 (25°)
Carbaryl	$C_{12}H_{11}NO_2$	201	White crystals	None	142	$<5 \times 10^{-3}$ (26°)	1.232 (20/20°)	0.004 (30°)
Carbofuran	$C_{12}H_{15}NO_3$	221	White crystals	Slight phenol	153–154	2×10^{-5} (33°)	1.180 (20/25°)	0.07 (25°)
Formetanate·HCl	$C_{11}H_{15}N_3O_2 \cdot HCl$	258	White crystals	–	200–202	Nonvolatile	–	>50 (25°)
Landrin	$C_{11}H_{15}NO_2$	193	Buff crystals	Mild ester	105–114	5×10^{-5} (23°)	–	0.006 (23°)
Mesurol	$C_{11}H_{15}NO_2S$	225	White crystals	Mild	121	–	–	–
Methomyl	$C_5H_{10}N_2O_2S$	162	White crystals	Slight sulfur	78–79	5×10^{-5} (25°)	1.295 (25/40°)	5.8 (25°)
Mobam	$C_{10}H_9NO_2S$	207	White crystals	None	128	1×10^{-8} (25°)	–	<0.1 (25°)

[a]The data for this table have been accumulated from Technical Information Bulletins available from the manufacturers.

[b]The preferred name is the one chosen for use throughout this book and may be either an accepted common name or a commercial trade name, whichever appeared to be used most often in the literature.

high temperatures.[5] In addition, the corresponding naphthol or phenol is released, as shown below:

$$\mathrm{Ar O\overset{\overset{O}{\|}}{C}NHCH_3} \xrightarrow{\Delta} \mathrm{ArOH + CH_3NCO}\uparrow$$

The same reaction occurs when methylcarbamates are placed in boiling water.[5-7]

Although the conditions required for destruction by air, light, and heat are quite drastic, hydrolysis of carbamate esters in alkaline media occurs with relative ease, and it is this degradative pathway that most seriously limits the chemical stability of carbamate insecticides. Hence, considerable attention has been given to elucidation of the kinetics and mechanisms regulating base-catalyzed hydrolysis of carbamates.

One of the first studies demonstrated that susceptibility to hydrolysis is determined by the chemical structure of the insecticide, especially by substituents attached to either the phenyl ring or the carbamate nitrogen.[8] Thus, holding the N-methyl end of the molecule constant yielded second-order hydrolysis rate constants for a series of ring-substituted analogs ranging from 3.4×10^6 $mole^{-1}$ min^{-1} for 2-nitrophenyl N-methylcarbamate to 4 $mole^{-1}$ min^{-1} for 3-*t*-butylphenyl N-methylcarbamate (Table 2.2). Similarly, changing the N-substituent illustrated that resistance to attack followed the order N-phenyl < N-benzyl < N-ethyl < N-methyl: that is, N-methylcarbamates were the most stable (Table 2.2).

A later study with several different compounds confirmed the superiority of the N-methyl derivatives by showing that the ability of N-alkyl groups to stabilize the carbamate ester bond was in the order N-isopropyl < N-ethyl < N-*n*-propyl < N-methyl.[9] Furthermore, the corresponding N,N-disubstituted carbamates were even more resistant to hydrolysis than their mono counterparts (Table 2.2). However, the relative ranking was different, particularly with respect to the dimethyl analog, as indicated by the following series: N,N-dimethyl < N,N-diethyl < N,N-di-*n*-propyl < N,N-di-*n*-butyl < N,N-diisopropyl. Greater protection from alkaline hydrolysis by doubly substituting the carbamate nitrogen atom has also been demonstrated for various physostigmine derivatives.[6,7,10,11] In addition, several papers have reported that a wide array of aromatic N-methylcarbamates hydrolyze from 10^3 to 10^7 times faster than corresponding N,N-dimethylcarbamates.[9,12-18]

TABLE 2.2

Second-order Rate Constants for Hydrolysis of Substituted Phenyl N-Methylcarbamates and Phenyl N-Substituted Carbamates[8,9]

Substituted phenyl N-methylcarbamate[a]	K_{hyd} ($mole^{-1}$ min^{-1})
2-NO_2	3.4×10^6
4-NO_2	3.5×10^5
2-Cl	2.0×10^3
3-Cl	1.7×10^3
4-Cl	1.0×10^3
3-CH_3	3.0×10^2
2-CH_3	2.6×10^2
2-*i*-C_3H_7	5.5×10^1
2-*t*-C_4H_9	2.8×10^1
3-$N(CH_3)_2$	2.0×10^1
3-*t*-C_4H_9	0.4×10^1
Phenyl N-substituted carbamates[a]	
NHC_6H_5	5.8×10^3
$NHCH_2C_6H_5$	8.2×10^2
NHC_2H_5	5.0×10^2
$NHCH_3$	2.5×10^2
Phenyl N,N-disubstituted carbamates[b]	
N,N-diCH_3	3.9×10^{-3}
N,N-diC_2H_5	6.2×10^{-5}

[a]Barbital buffer, 0.1*M*, pH 9.5, 37.5°C.
[b]Sodium phosphate buffer, 0.05*M*, pH 7.8, 22°C.

Returning to variations in the phenyl nucleus, the data in Table 2.2, together with those in Table 2.3 for some of the more common experimental and commercial N-methylcarbamate insecticides,[19] emphasize the tremendous contribution of ring substituents to hydrolytic stability. As might be expected for attack by OH^-, groups that withdraw electrons from the phenyl ring, and hence reduce the electronegativity around the carbamate moiety, tend to enhance basic decomposition. In fact, a good correlation between Hammett sigma values (a measure of electron withdrawing power) for phenyl substituents and hydrolysis rates was obtained with a series of aryl N-methyl- and N,N-dimethylcarbamates.[15,20] Later, an even higher degree of correlation appeared between first-order hydrolysis constants and sigma values for a group of O-(methylcarbamoyl)oximes of substituted benzaldehydes ($R{-}C_6H_4CH{=}NOC(O)NHCH_3$).[21]

TABLE 2.3

Hydrolysis of Aryl N-Methylcarbamate Insecticides in 0.1*M* Barbital Buffer, pH 9.3, 25°C

Carbamate	Half-life (hr)
Banol	0.1
Mesurol	0.4
Carbaryl	0.5
Zectran	2.3
UC 10854	2.8
Baygon	3.1
Matacil	4.0
HRS-1422	6.8

Adapted from Reference 19, with permission.

However, when the proton on the α-carbon was exchanged for another grouping, hydrolysis rates were mostly governed by field and resonance effects associated with ring substituents.[22] One final change in structure which favors alkaline stability is replacement of the aromatic phenyl or naphthyl portion of the molecule with an aliphatic group.[12,13,23]

Factors other than molecular configuration which contribute to carbamate decomposition in alkaline media include temperature, pH, and hydroxyl ion concentration. For example, with each 10°C rise in temperature, hydrolysis rates for 4-nitrophenyl N-methylcarbamate, Baygon, carbaryl, dimetilan, and Pyrolan increase from two- to threefold.[16,24] Although the latter two dimethylcarbamates are fairly stable in phosphate-buffered solutions ranging in pH from 6 to 10 for up to 100 days, the monomethyl compounds are readily degraded as the pH increases. For example, carbaryl has a half-life of 10 days at pH 7 (0.01*M* phosphate buffer, 20°C), but at pH 10 its half-life is only 15 min.[24] A similar trend appeared for carbaryl hydrolysis in tris-HCl buffer[25] and for 4-nitrophenyl N-methylcarbamate breakdown in phosphate buffer.[16]

Varying the concentration of hydroxyl ion or carbamate indicates that cleavage of the carbamate ester bond is first order with respect to both OH^- and substrate.[10,12,13,18,24] Based on results with physostigmine and related compounds, early predictions of mechanistic pathways included a hydroxyl ion-catalyzed reaction, with methylisocyanate as one of the end products.[6,7] However, this did not explain the large differences in hydrolysis rate between corresponding N-methyl- and N,N-dimethylcarbamates discussed above. To account for such variations, comparative studies were conducted with aliphatic and aromatic esters of carbamic, N-methylcarbamic, and N,N-dimethylcarbamic acid.[12,13] As a result, the two distinct pathways shown in Figure 2.6 were proposed to explain OH^--mediated decomposition of carbamate esters. Subsequent research generally supports this proposal.[16,18,26,27]

Mechanism I for aliphatic N-methyl- and N,N-dimethylcarbamates and aromatic N,N-dimethylcarbamates is no different from the normal expected reaction of OH^- with an ester whereby the hydroxyl ion approaches the carbonyl carbon atom to yield an unstable tetrahedral intermediate which collapses to an alcohol (or phenol) and dimethylcarbamate ion. The latter rapidly picks up a proton from water and then spontaneously decomposes into dimethylamine and carbon dioxide. Therefore, any structural group which pulls electrons away from the carbonyl center will hasten hydrolysis by facilitating OH^- attack. This has been shown to be the case for substituents attached to the ring of phenyl N,N-dimethylcarbamates.[15]

Mechanism II for hydrolysis of aromatic unsubstituted carbamates and aromatic N-methylcarbamates involves initial removal of an N-proton by OH^- to produce an unstable intermediate which in turn forms methylisocyanate plus a phenoxide ion. The methylisocyanate reacts instantaneously with water to give N-methylcarbamic acid, an unstable arrangement that degrades to methylamine and carbon dioxide. Hence, aryl substituents which support a stable phenoxide ion and/or draw electrons away from the N atom enhance hydrolysis rates.

A slight variation in the general scheme for Mechanism II has been postulated whereby a water molecule is involved in the transition state (Figure 2.7).[16] Note that methylisocyanate does not appear as a transitory product under these conditions. The alternate pathway is kinetically indistinguishable from Mechanism II, but has the advantage that the carbamate never becomes a charged ion at any stage.

Results concerning hydrolysis of carbamates by bases other than OH^- do not always agree. For example, one report states that imidazole and phosphate dianion are able to catalyze carbamate

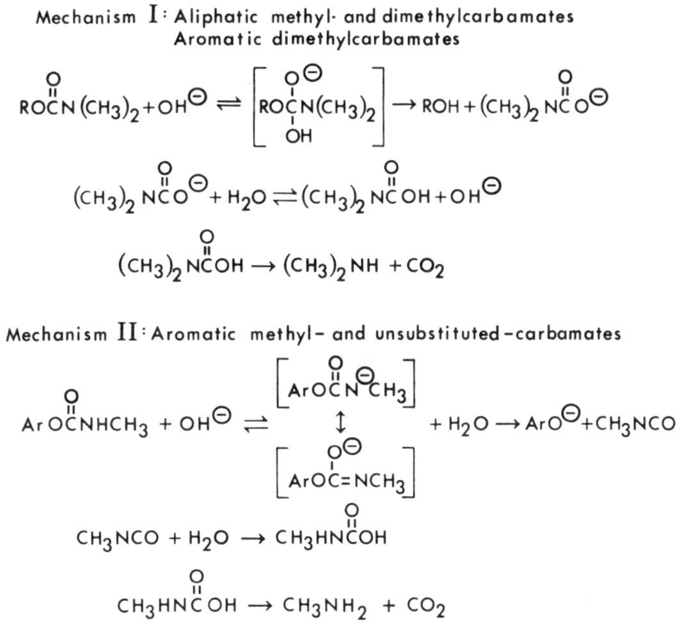

FIGURE 2.6. Mechanisms for alkaline hydrolysis of carbamate chemicals.[12,13]

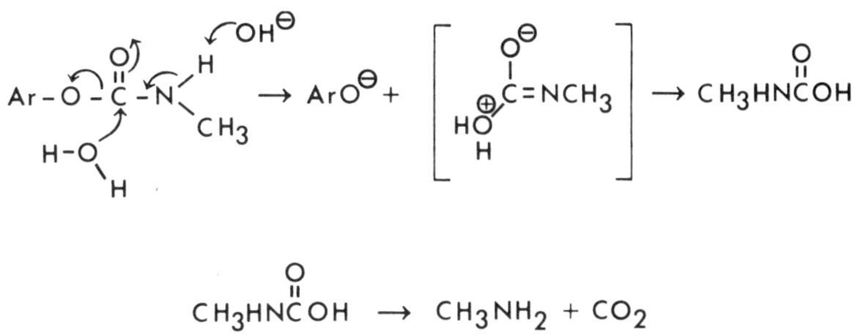

FIGURE 2.7. Alternate mechanism for hydrolysis of aromatic N-methylcarbamate chemicals. (Adapted from Reference 16, with permission.)

hydrolysis.[27] Another study clearly demonstrates that phosphate anions are not the cause of hydrolysis in phosphate buffers, cleavage being due to hydroxide ions in solution.[16] Also, solvolysis by water apparently contributes little to hydrolysis rate, and the weak bases, imidazole and pyridine, do not catalyze the hydrolytic reaction in aqueous media. However, under acid conditions, solvolysis by water apparently proceeds.[18]

Decomposition of carbamates in acid solution generally occurs at very slow rates, and the responsible pathways have not been completely defined. A limited amount of evidence favors mechanisms slightly apart from typical ester $A_{AC}1$ and $A_{AC}2$ cleavage by H^+, the prevailing mechanism dependent on acid concentration[28] and on whether the carbamate is monomethyl or dimethyl.[18] However, significant breakdown does not result under concentrations of acid likely to be encountered during normal handling, application, and weathering conditions.

SYNTHESIS OF CARBAMATES

N-Methylcarbamates

Almost all of the substituted phenyl N-methylcarbamates prepared for evaluation as insecticides have been synthesized by one of the three reaction sequences depicted in Figure 2.8. Each method is a very simple one- or two-step process, and each depends on having the parent phenol available as a starting material. In fact, should a compound look promising, it is the cost and supply of phenol which primarily determine the commercial

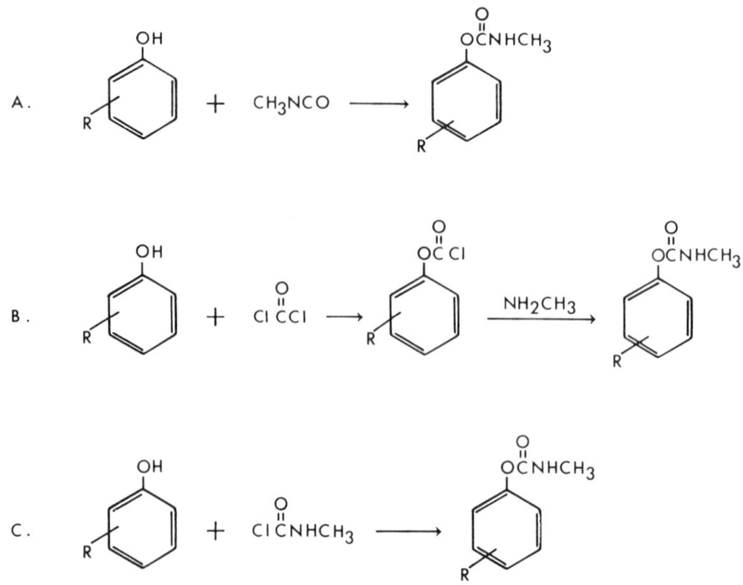

FIGURE 2.8. Synthesis of substituted phenyl N-methylcarbamate insecticides via methylisocyanate (A), phosgene and methylamine (B), or methylcarbamoyl chloride (C).

feasibility of development. The synthesis of substituted phenols will not be described, however, though their preparation may be quite elaborate and elegant. On the other hand, many are only a few steps away from petroleum distillates.

Method A, consisting of phenol esterification with methylisocyanate, is generally the easiest and most convenient route for synthesis of both aliphatic and aromatic N-methylcarbamates. Yields can be very high, whether preparation is on a microscale or industrial-production level. The reaction was first described for insecticidal carbamates as a mixing of a substituted phenol with a slight molar excess of CH_3NCO in anhydrous isopropylether.[8] After storing the ingredients in a stoppered flask or tube for several days at room temperature, the product appeared as crystals, or precipitation was forced by cold and various solvents.

Later, the conditions were slightly modified by reaction of the starting materials in dry diethylether containing a catalytic amount of triethylamine.[29] After standing stoppered for a week, volatiles were evaporated, leaving crystalline methylcarbamate product. For more difficult reactions where yields were low, methylisocyanate was added in $> 2M$ excess to the phenol and triethylamine in a pressure bottle. The temperature was raised to 100°C and held there for 4 hr. These procedures were used routinely to synthesize a large variety of substituted-phenyl N-methylcarbamates during early structure–activity studies.[30–32]

Since that time, an almost endless number of minor modifications to method A have been made, but they are too numerous to accurately and fairly document. Examples of some of the experimental changes follow.

Reaction solvents include ethers, acetone, chloroform, methylene chloride, carbon tetrachloride, acetonitrile, dioxane, benzene, xylene, hexane, petroleum ether, and ligroin. In some cases no solvent is used, and neat CH_3NCO is added directly to phenol plus catalyst. The latter has also changed from triethylamine to other bases, such as trimethylamine, triethylenediamine, and pyridine, or to organometal catalysts, such as dibutyltin dilaurate, dibutyltin diacetate, and tin octoate. At least one preparation included both an amine and a metal catalyst.[33] Reaction time can be speeded up by refluxing for 3 or 4 hr at 40 to 80°C, or for more stubborn esterifications, pressure bottles can be maintained at 100°C for longer periods. Most of these changes have been incorporated to improve yields, which often run higher than 90%.

The ease with which reaction A proceeds has made the preparation of candidate N-methylcarbamate insecticides an easy task for even a novice chemist. However, some precautions are worthy of note. First, it is important to keep all reagents, particularly solvents, free from water because of its ready reaction with methylisocyanate. Second, the addition of CH_3NCO to phenols is usually exothermic, so that cooling of the reaction vessel may be required. In one instance where esterification of 2-*s*-butenylphenol with methylisocyanate was initiated by addition of catalyst, a violent release of heat was reported.[34] This is one reason why some chemists have turned to use of pressure bottles for carbamate synthesis. Third, CH_3NCO is a very potent "tear gas", and thus reactions should be conducted in a hood. The authors recall some of their early attempts at snythesis in a laboratory without proper ventilation. When the volatile methylisocyanate was transferred from its container into the reaction vessel, escaping gas made it extremely difficult to stopper the appropriate bottles.

Of course, it is also possible to make other N-alkyl carbamates, or even N-aryl carbamates, by substituting the required isocyanate for methylisocyanate in reaction A.[8] Or, instead of starting with the phenol, one could mix CH_3NCO with a benzyl alcohol, cyclohexanol, or other aliphatic "hydroxy" compound including an oxime,[23] the latter giving insecticides such as aldicarb.[35] In fact, esterification of oximes often occurs quite readily and does not always require a catalyst.

Method B (Figure 2.8) also begins with a phenol, but in this case it is mixed with an excess of phosgene to yield the phenyl chloroformate, which in turn is reacted with methylamine for production of the desired N-methylcarbamate. Since this amounts to a two-step reaction, yields based on initial phenol may be lower than from method A, but for some phenols route B is preferred.

The first step, production of chloroformate, is performed in the cold in aromatic solvents, e.g., benzene or toluene. Usually, the reaction is complete in only a few hours, but sometimes the mixture is allowed to stand overnight. Often an agent to soak up liberated HCl is added, such as quinoline, triethylamine, dimethyl aniline, or diethyl aniline. Alternatively, the sodium or potassium salt of the desired phenol can be placed together with phosgene, thus avoiding release of hydrochloric acid. Once again, great care must be taken with this reaction to prevent escape of phosgene, a very poisonous gas.

The chloroformate is then separated and purified by distillation, or the crude product is allowed to react directly with methylamine in cold toluene, dioxane, or aliphatic hydrocarbon. Another procedure calls for slow addition of crude product to aqueous CH_3NH_2 at room temperature. In any event, care is usually taken to insure that approximately stoichiometric amounts of chloroformate and methylamine are present for the final reaction.

Substituted phenols are not the only suitable starting materials for synthesis via method B. However, caution must be observed with oximes, since at temperatures above ambient the intermediate chloroformates rapidly form their corresponding nitriles, as shown in Figure 2.9.[36] Reaction B is particularly useful for the preparation of esters other than N-methylcarbamates. Thus, unsubstituted ($-NH_2$) carbamates, which are often synthesized as suspected metabolites, can be made by replacing methylamine with ammonia in step two. Different monoalkylcarbamates result when the chloroformate is mixed with other

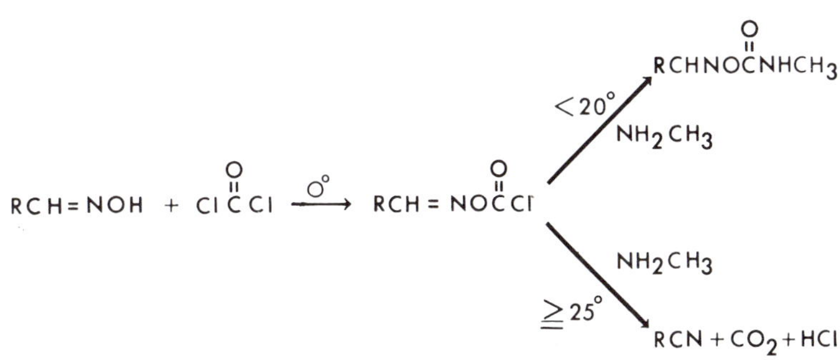

FIGURE 2.9. Reaction of oxime chloroformates with methylamine at different temperatures. (Adapted from Reference 36, with permission.)

primary amines, whereas secondary amines form the basis for preparation of N,N-dialkylcarbamates, including dimethylcarbamates. As will be seen, however, another procedure is favored for synthesis of the dimethyl compounds. More exotic amines, such as CD_3NH_2, $Si(CH_3)_3NH_2$, $\begin{matrix}CH_2\\CH_2\end{matrix}\!>\!NH$, CH_3NHOH, $(CH_3)_2NNH_2$, and NH_2OH, have also been used in accordance with reaction B.[37-39]

Method C (Figure 2.8) is only rarely employed for N-methylcarbamate synthesis. It consists of adding a substituted phenol or salt of phenol to methylcarbamoyl chloride in solvents such as toluene, with a weak base present for absorption of released acid. The reaction may go to completion in only a short time without heat, or it may be necessary to reflux for several hours. Again, by utilizing other carbamoyl chlorides, a variety of mono- or dialkylcarbamates can be produced. Indeed, this is how the majority of N,N-dimethylcarbamate insecticides are made (see below).

The fact that the last step in the preparation of substituted-phenyl N-methylcarbamates involves addition of the carbamate moiety is largely due to the sensitivity of the ester bond to hydrolysis. However, there are a few instances where ring substituents have been placed on the molecule in the presence of the carbamate group. For example, Banol can be made by chlorinating 3,4-dimethylphenyl N-methylcarbamate.[40] Also, certain N-methyl- and N,N-dimethylcarbamates of hydroxybenzaldehyde acetals are synthesized by first making the benzaldehyde carbamates, which are then cyclized as pictured in Figure 2.10.[41]

The methods in Figure 2.8 have also been employed for carbamylation of numerous heterocyclic aromatics, such as benzofuranols, especially 2,3-dihydro-2,2-dimethyl-7-benzofuranol.[42-45] Thus, when carbofuran is synthesized according to route A, the yield can be as great as 99%, without formation of undesirable side products.[45] Many benzothienyl[46,47] and benzotriazolyl[48] carbamate pesticides, including Mobam, are a result of similar production methods.

Although not heterocyclic, the double-ringed 1-naphthol reacts in identical manner to form the best-known carbamate insecticide, carbaryl. Commercial synthesis often occurs through the naphthyl chloroformate route with yields of 80 to 95%,[49-51] though even greater returns have been claimed by using a variation of method A.[52] In addition, several adaptations of method C are known to result in near-quantitative recovery of carbaryl.[53-55]

One method for synthesis of N-methylcarbamates not shown in Figure 2.8 is worthy of mention. It is simply an exchange reaction wherein phenyl N-methylcarbamate is heated with a substituted phenol for several hours.[56] During this time, free phenol is liberated and removed by pressure distillation while the substituted phenol becomes carbamylated (by reaction with liberated methyl isocyanate). This type of reaction has also been utilized as a low-cost radiosynthesis procedure, though resulting specific-activity levels were quite low.[5]

N,N-Dimethylcarbamates

Synthesis of N,N-dimethylcarbamates, particularly insecticidal dimethylcarbamates of heterocyclic enols,[56a] is accomplished primarily by method C in Figure 2.8. For example, Pyramat crystals appear after refluxing the sodium salt of 2-propyl-4-methyl-6-hydroxypyrimidine with dimethylcarbamoyl chloride in benzene for 12 hr

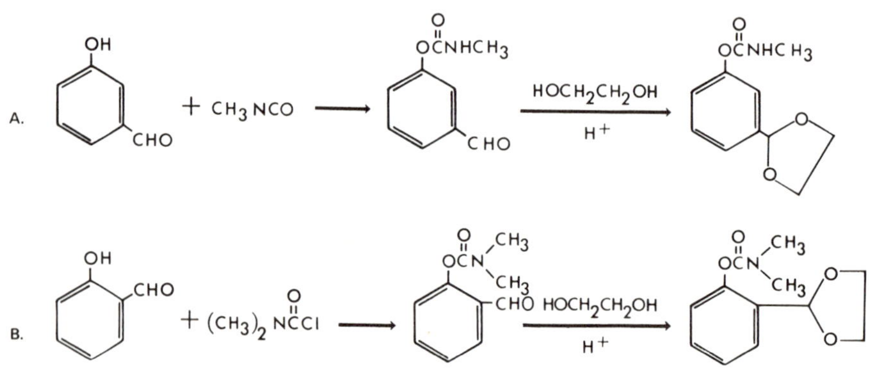

FIGURE 2.10. Synthesis of N-methyl- (A) and N,N-dimethylcarbamates (B) of hydroxybenzaldehydes. (Adapted from Reference 41, with permission.)

(Figure 2.11, A).[57] The same reaction is utilized for industrial synthesis of Isolan from 1-isopropyl-3-methyl-5-pyrazolone-5-enolate.[1] Numerous other Na or K salts of cyclic enols have been converted to dimethylcarbamates via this route, though other reaction solvents (tetrahydrofuran, methyl ethyl ketone, or acetonitrile) and different reflux times were needed. If the enol is treated in place of a salt, then an HCl scavenger, such as pyridine, is included with the solvent.

Interestingly, dimethylcarbamates can also be prepared by direct action of phosgene on ketone analogs, depicted as reaction B in Figure 2.11. Thus, 2-propyl-4-methyl-6-pyrimidone plus phosgene in cold benzene forms the chloroformic acid ester which is separated and subsequently treated with dimethylamine at foom temperature to yield Pyramat after several hours.[57] The original synthesis of dimetilan was similar; that is, 3-methylpyrazolone-5 was mixed with at least two equivalents of dimethylcarbamoyl chloride in benzene containing K_2CO_3 as acid acceptor (Figure 2.12).[17] Two isomers of the product resulted, a situation not uncommon during synthesis of heterocylic enol-derived dimethylcarbamates, but relatively rare with phenyl N-methylcarbamate preparations. Conversion (98%) of the minor isomer into dimetilan was accomplished by treating it first with phosgene and then with dimethylamine, as shown (Figure 2.12).

Synthesis of most N,N-dimethylcarbamates depends, of course, on having the starting enol or other hydroxy derivative, the last step then being carbamylation. However, just as with phenyl methylcarbamates, there are occasions when substituents can be added to structures containing the dimethylcarbamate group. Consider the sequences pictured in Figure 2.13, where 3-methyl-5-pyrazolyl N,N-dimethylcarbamate reacts with chloromethylpropyl sulfide in the presence of triethylamine (A), or where ethyl acrylic ester is attached via a Michael addition (B).[17]

N-Acyl-N-methylcarbamates, N-Sulfenyl-N-methylcarbamates, and Others

In an effort to improve the selective toxicity of

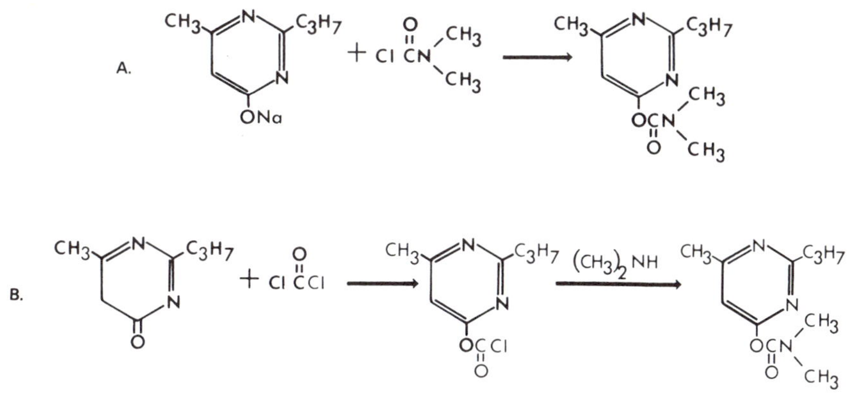

FIGURE 2.11. Synthesis of Pyramat via dimethylcarbamoyl chloride (A) or phosgene and dimethylamine (B).[57]

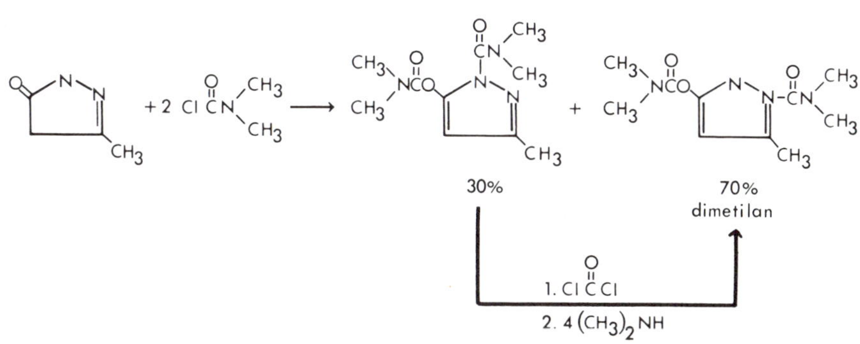

FIGURE 2.12. Synthesis of dimetilan.[17]

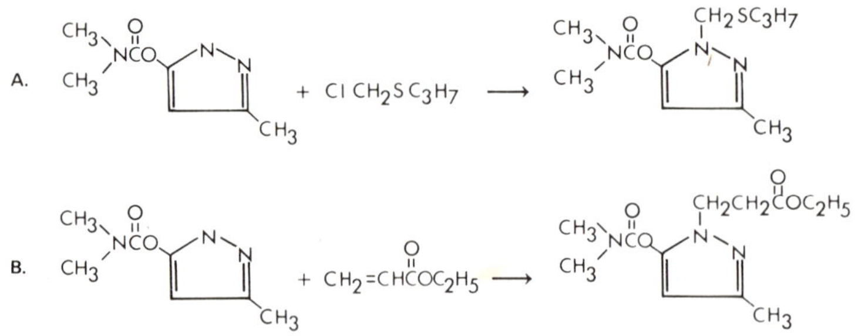

FIGURE 2.13. Addition of substituents to pyrazolyl N,N-dimethylcarbamates.[17]

N-methylcarbamates and their degree of persistence, several derivatives have been synthesized with a labile protective group in place of the N-proton; the resulting compounds are considered to be insecticide precursors. The most widely investigated are the N-acyl-N-methylcarbamates, which can be prepared by the routes summarized in Figure 2.14.[58-61] Methods A and B are used most often and begin with the available aryl N-methylcarbamate. Acylation is accomplished with an acid anhydride catalyzed by sulfuric acid (A) or with an acid chloride in the presence of pyridine or dimethylaniline (B), a less facile combination. Reflux for various time periods in solvents such as xylene or toluene is necessary, and generally an excess of anhydride is employed whereas equimolar quantities of carbamate and acyl chloride are usually the rule.

When the parent N-methylcarbamate cannot be acylated directly, one of the three alternate methods shown in Figure 2.14 (C to E) is used. For example, the oxime carbamate Tirpate (2,4-dimethyl-1,3-dithiolane-2-carboxaldehyde O-methylcarbamoyloxime) could not be acetylated, but when the sodium salt of the oxime was treated with N-acetyl-N-methylcarbamoyl chloride (shown as reaction E in Figure 2.14), the desired N-acetyl Tirpate resulted.[62]

The next important group of precursors includes the aryl N-sulfenyl-N-methylcarbamates. The only synthesis published thus far is illustrated by reaction A at the top of Figure 2.15 and it is responsible for the creation of a wide variety of insecticidally active structures.[63-65] Starting materials in a 1:1 ratio include the available aromatic N-methylcarbamate plus an alkyl-, phenyl-, or substituted phenylsulfenyl chloride. The reactants are stirred in anhydrous pyridine for several hours to give high product yields.

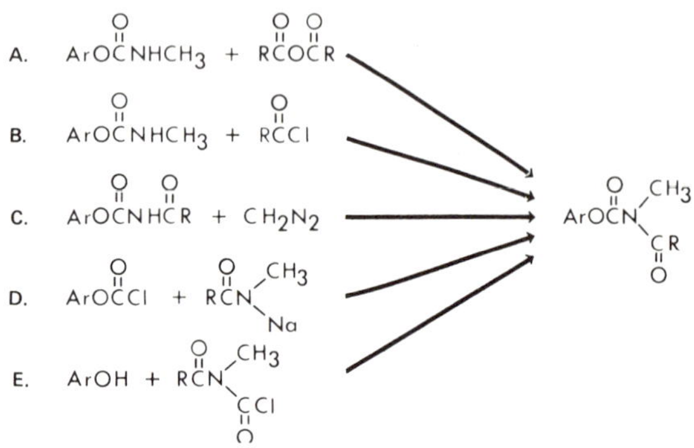

FIGURE 2.14. Synthesis of aromatic N-acyl-N-methylcarbamate insecticides.

Two other modifications of the carbamate N-proton are also illustrated in Figure 2.15. The first of these, shown as reaction B, consists of adding an aryl chloroformate to O,O-dialkyl N-methyl lithiumphosphoramidothioate in a dry ice-acetone bath.[66] Stirring overnight at room temperature yields the anticipated N-dialkylphosphorothioyl-N-methylcarbamate. In this way, the organophosphorus derivatives of carbaryl, UC 10854, Baygon, and carbofuran were made. However, due to instability of the chloroformate of the oxime of aldicarb, an exchange reaction was carried out whereby the phosphorus grouping on 4-nitrophenyl N-dimethylphosphorothioyl-N-methylcarbamate was transferred to the sodium salt of aldicarb oxime. Reaction C in Figure 2.15 describes the synthesis of the N-nitro derivative of UC 10854, namely, 3-isopropylphenyl N-nitroso-N-methylcarbamate.[67] The compound does have insecticidal activity, but its toxicity to mammals is not as great as that of the parent carbamate. In fact, this is one advantage common to most of these types of structures, though none have as yet reached commercial production.

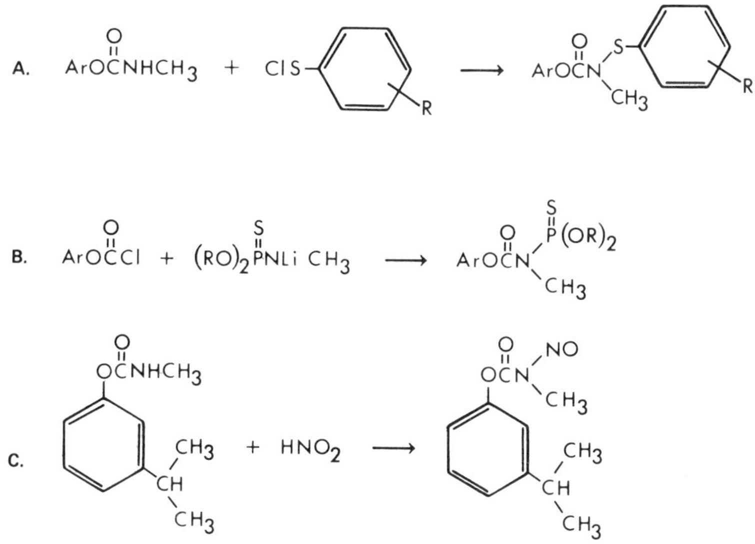

FIGURE 2.15. Synthesis of N-sulfenyl-N-methylcarbamates (A),[63-65] N-phosphorothioyl-N-methylcarbamates (B),[66] and N-nitroso-N-methylcarbamates (C).[67]

SYNTHESIS OF METABOLITES

Evaluation of the safety, selectivity, and degradation of insecticide chemicals requires a complete understanding of the primary metabolic pathways which develop following application. The metabolic routes for carbamate insecticides have been widely researched, and many suspected metabolites have been isolated in the course of these studies (see Chapter 7). Identification of the minute quantities of isolated products and examination of their toxicity require that they be synthesized in large amounts. Unlike the parent insecticide, however, low cost of starting materials, one- or two-step reactions, and quantitative yields are not mandatory as long as a reasonable quantity of pure metabolite can be recovered. Since the types of reactions involved are numerous and varied, it is impossible to thoroughly cover all synthetic routes. However, the major groups of carbamate metabolites will be considered.

The most common degradative reactions involving methylcarbamates are hydrolysis, oxidation, and conjugation. The first yields parent phenols, oximes, enols, naphthols, etc., plus methylamine and carbon dioxide as described on previous pages. These metabolites are readily available because of their use in the synthesis of the original insecticides.

Oxidative Metabolites

Oxidative detoxication results in formation of N-hydroxymethylcarbamates, hydroxyaryl N-methylcarbamates, and hydroxyalkylaryl N-methylcarbamates. Hydroxylation of the carbamate N-methyl group is, of course, a potential reaction common to all methylcarbamate insecticides, and considerable effort has been spent preparing these derivatives. Initial attempts with carbaryl consisted of prolonged refluxing of 1-naphthyl carbamate with excess paraformaldehyde in glacial acetic acid;[68] the product yield was almost negligible, however. Only a modest improvement was achieved with another method at a later date, but the procedure was found to be very useful for synthesis of N-hydroxymethyl analogs of Baygon, UC 10854, HRS-1422, Landrin, Matacil, Zectran, Banol, and Bux.[69]

The method, illustrated by sequence A at the top of Figure 2.16, calls for a mixture of equal amounts of benzyloxymethylisocyanate and substituted phenol in dry benzene with triethylamine as catalyst. After 18 hr of reflux, the benzylated product is recovered, dissolved in isopropanol, and placed in a Parr hydrogenator for quantitative hydrogenation with palladium-charcoal catalyst at 20°C under atmospheric pressure. The same procedure has been used for preparation of N-hydroxymethyl carbofuran,[70] and again for N-hydroxymethyl Bux.[71]

Continued efforts with carbaryl produced scheme B shown in Figure 2.16. Here, 1-naphthol plus 2-(pyranyloxy)methylisocyanate in methylene chloride containing a catalytic amount of dibutyltin diacetate is stirred at room temperature for 12 hr.[72] The resulting stable intermediate is then recovered and boiled in acetone spiked with 1% HCl for 30 min to give 1-naphthyl N-hydroxymethylcarbamate. The corresponding aldicarb metabolite is made the same way, although slightly more drastic conditions are needed to remove the pyranyl group.

Perhaps the best method of all is pictured as synthesis C in Figure 2.16. High yields of N-hydroxymethyl analogs of carbaryl (59%), carbofuran (61%), and phenyl N-methylcarbamate (35%) were accomplished by mixing the appropriate aryl unsubstituted carbamate with a 25% excess of paraformaldehyde in THF:H_2O (4:1) containing hydrochloric acid as a catalyst.[73] The mixture was stirred at 50 to 60°C until clear, then 1 hr more at the same temperature, and finally 1 hr at room temperature. The method has been used for preparation of other hydroxymethyl derivatives, such as those of Meobal and Tsumacide,[74] though it failed to synthesize N-hydroxymethyl aldicarb.[73]

Further oxidation of the carbamate N-CH_2OH moiety could eventually result in loss of the group from the molecule, leaving unsubstituted carbamate as a metabolite, or hydroxylation could possibly occur directly on the N atom, producing the N-hydroxy-N-methyl derivative. Either of these products can be synthesized by the standard methylcarbamate method B shown in Figure 2.8. The aromatic chloroformate is allowed to react with ammonia to give $ArOC(O)NH_2$, or with methyl hydroxylamine to give $ArOC(O)N{<}^{OH}_{CH_3}$.

Another group of important oxidative metabolites includes aryl ring-hydroxylated structures. Once again, pioneer efforts at synthesis centered around carbaryl.[68] Anticipated products were prepared in small amounts by holding respective dihydroxynaphthalenes with methylisocyanate at ambient temperature for 4 hr (Figure 2.17). Unreacted diol was separated from mono-

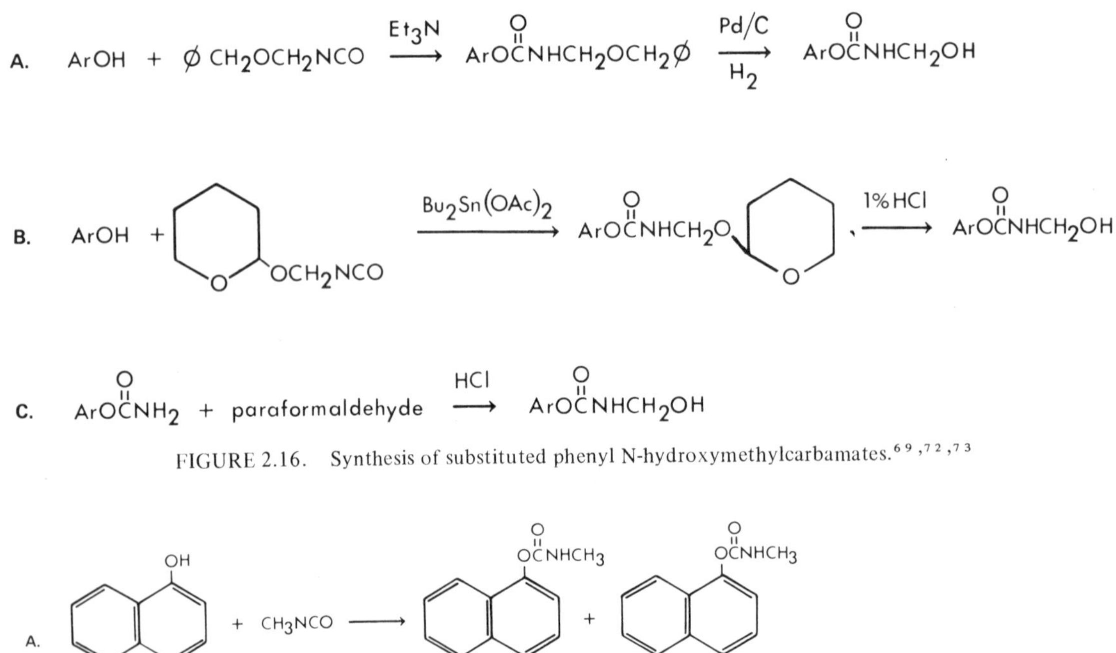

FIGURE 2.16. Synthesis of substituted phenyl N-hydroxymethylcarbamates.[69,72,73]

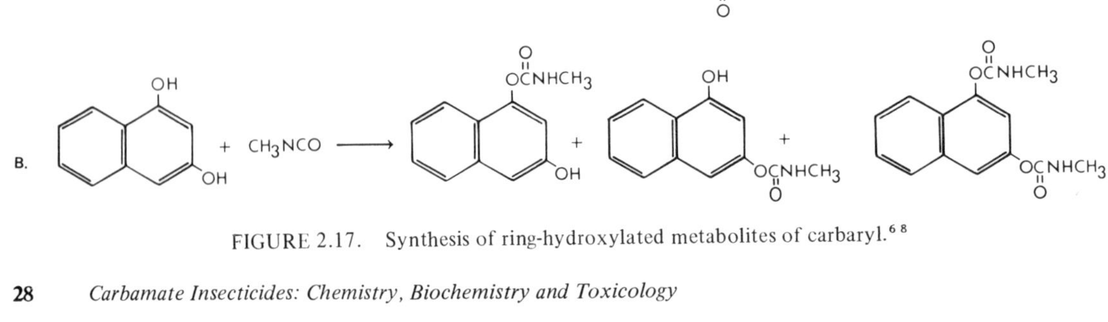

FIGURE 2.17. Synthesis of ring-hydroxylated metabolites of carbaryl.[68]

and biscarbamates by thin-layer chromatography. Symmetrically substituted naphthalenes, such as 1,4-hydroxynaphthalene (Figure 2.17, A), formed only one monocarbamate, i.e., the desired product, 4-hydroxy-1-naphthyl N-methylcarbamate. But when 1,2-, 1,3- (Figure 2.17, B), or 1,7-dihydroxy analogs were used, separation of products was more difficult because of the presence of two different monocarbamates.

More recent improved procedures for preparation of 4- and 5-hydroxy carbaryl require that the respective diol and CH_3NCO be kept in a pressure bottle overnight with dibutyltin diacetate.[75] However, synthesis of 3-, 6-, or 7-hydroxy-1-naphthyl N-methylcarbamate free from unwanted carbamates is much more difficult. One approach illustrating the elaborate steps required is outlined in Figure 2.18 for 7-hydroxy carbaryl.[76]

A carbaryl metabolite which proved to be very difficult to synthesize is 5,6-dihydro-5,6-dihydroxy-1-naphthyl N-methylcarbamate. It has recently been prepared, however, by the pathway given in Figure 2.19.[77] Fremy's salt (moist potassium nitrosodisulfonate) dissolved in aqueous $KHSO_4$ plus KH_2PO_4 was treated with 1,5-dihydroxynaphthalene in methanol. After stirring for 1.5 hr, the quinone was recovered and reduced with lithium aluminum hydride. Following workup, a low yield (8%) of the *trans*-1,2-dihydro-1,2,5-trihydroxynaphthalene resulted, which was selectively carbamylated by normal procedures to give *trans*-5,6-dihydro-5,6-dihydroxy carbaryl together with trace impurities of biscarbamates.

Synthesis of ring-hydroxylated phenyl N-methylcarbamates is not unlike synthesis of hydroxy carbaryl metabolites. For example, substituted hydroquinone is held in a sealed bottle with an equimolar amount of methylisocyanate dissolved in ether containing a few drops of triethylamine.[78] After 2 or 3 days, a mixture of one biscarbamate and two monocarbamates is formed (Figure 2.20, A), and these are separated by column chromatography. Even though the hydroxyl groups are symmetrically attached to the phenyl ring, substituted hydroquinones can yield two monocarbamates, though one may be favored sterically over another. Nevertheless, this procedure has been used to synthesize 4-hydroxy metabolites of Baygon, UC 10854, HRS-1422,[78] Landrin,[79] and Bux,[71] among others. Steric hindrance is also used to advantage during the production of some hydroxy analogs by selective decarbamylation, as shown in pathway B (Figure 2.20) for 4-hydroxy Baygon.[78] But just as with

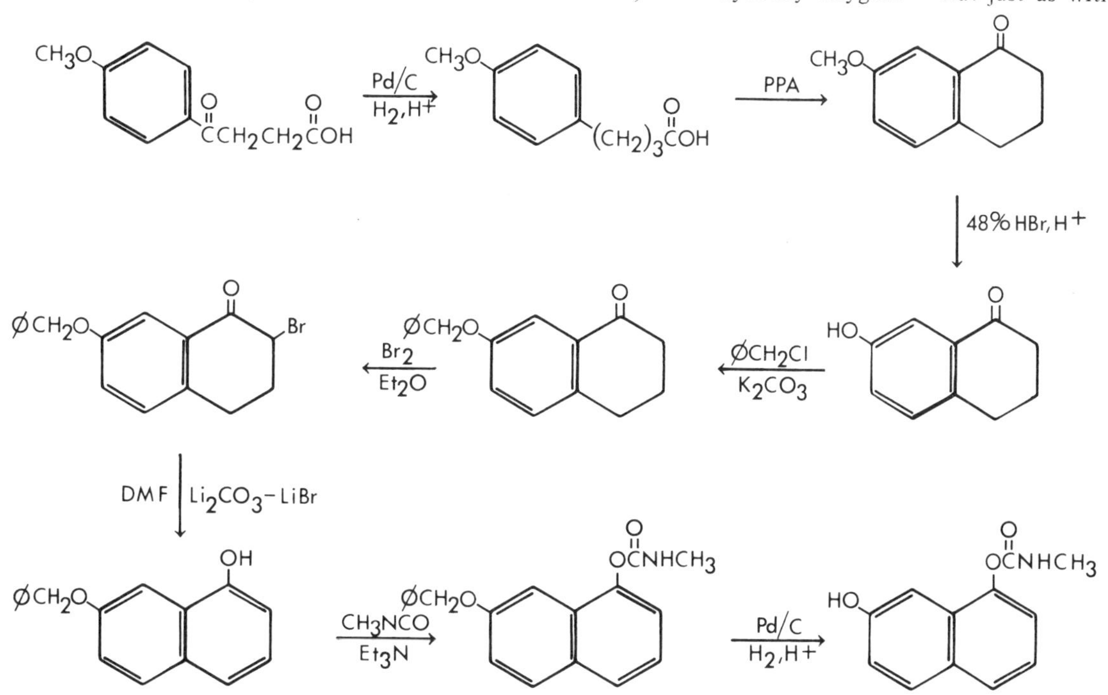

FIGURE 2.18. Synthesis of 7-hydroxy carbaryl. (Adapted from Reference 76, with permission.)

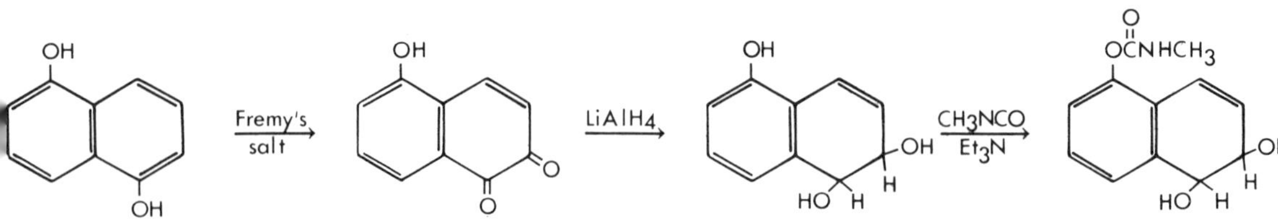

FIGURE 2.19. Synthesis of 5,6-dihydro-5,6-dihydroxy carbaryl. (Adapted from Reference 77, with permission.)

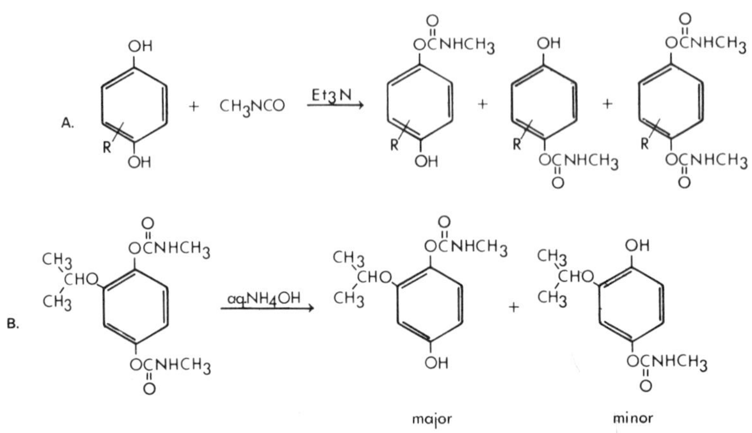

FIGURE 2.20. Synthesis of ring-hydroxylated metabolites of substituted-phenyl N-methylcarbamates by selective carbamylation (A) or decarbamylation (B).[78]

carbaryl, a multistep assembly of pure hydroxyphenyl methylcarbamate metabolites may be required.

Hydroxyalkylphenyl N-methylcarbamates also appear as metabolites in living organisms, and their synthesis has received some attention. Selected examples are outlined in Figure 2.21. The first reaction sequence (A) consists of mixing methyl magnesium iodide with 3-hydroxyacetophenone in cold ether, a holding period of 18 hr, and finally refluxing for 2 hr to give 3-(1-hydroxy-1-methylethyl)phenol.[78] This plus an equimolar quantity of methylisocyanate provides selective carbamylation of the phenol OH and production of the anticipated hydroxypropyl UC 10854. In similar fashion, the 3-(1-hydroxy-1-methylbutyl)phenol of Bux was prepared from the Grignard reagent C_3H_7MgBr.[71] Selective carbamylation gave the desired Bux metabolite.

Hydroxyl attack on the benzylic carbon of carbofuran yields the familiar metabolite 3-hydroxy carbofuran. This product has been synthesized by refluxing a solution of parent insecticide together with a fourfold molar excess of potassium permanganate in 10:1 glacial acetic acid: acetic anhydride for 5 hr (Figure 2.21, B).[70] The separated 3-acetoxy carbofuran was hydrolyzed almost quantitatively to 2,2-dimethyl-3-hydroxy-2,3-dihydro-benzofuranyl-7 N-methylcarbamate by refluxing in dilute sulfuric acid. The more highly oxidized 3-keto carbofuran was prepared in one step by treating carbofuran with four equivalents of chromium trioxide in glacial acetic acid at 25 to 35°C for 16 hr.

The number of mono-, di-, and trihydroxymethylphenyl analogs of Landrin which can be considered potential metabolites is very large. Several of the mono derivatives have been prepared by a variety of methods; two of these are depicted in the lower half of Figure 2.21.[79] In one procedure (Figure 2.21,C), a solution of 3,5-xylenol and 37% aqueous formaldehyde in 5% sodium hydroxide was kept for 5 days without heating. Cleanup of the resulting 4-hydroxymethyl-3,5-dimethylphenol preceded the usual carbamylation in dry ether with methylisocyanate and triethylamine. A more involved synthesis was needed to obtain 3-hydroxymethyl-4,5-dimethylphenyl N-methylcarbamate, and this is briefly summarized by route D in Figure 2.21. Similar multistep reactions were devised for the pro-

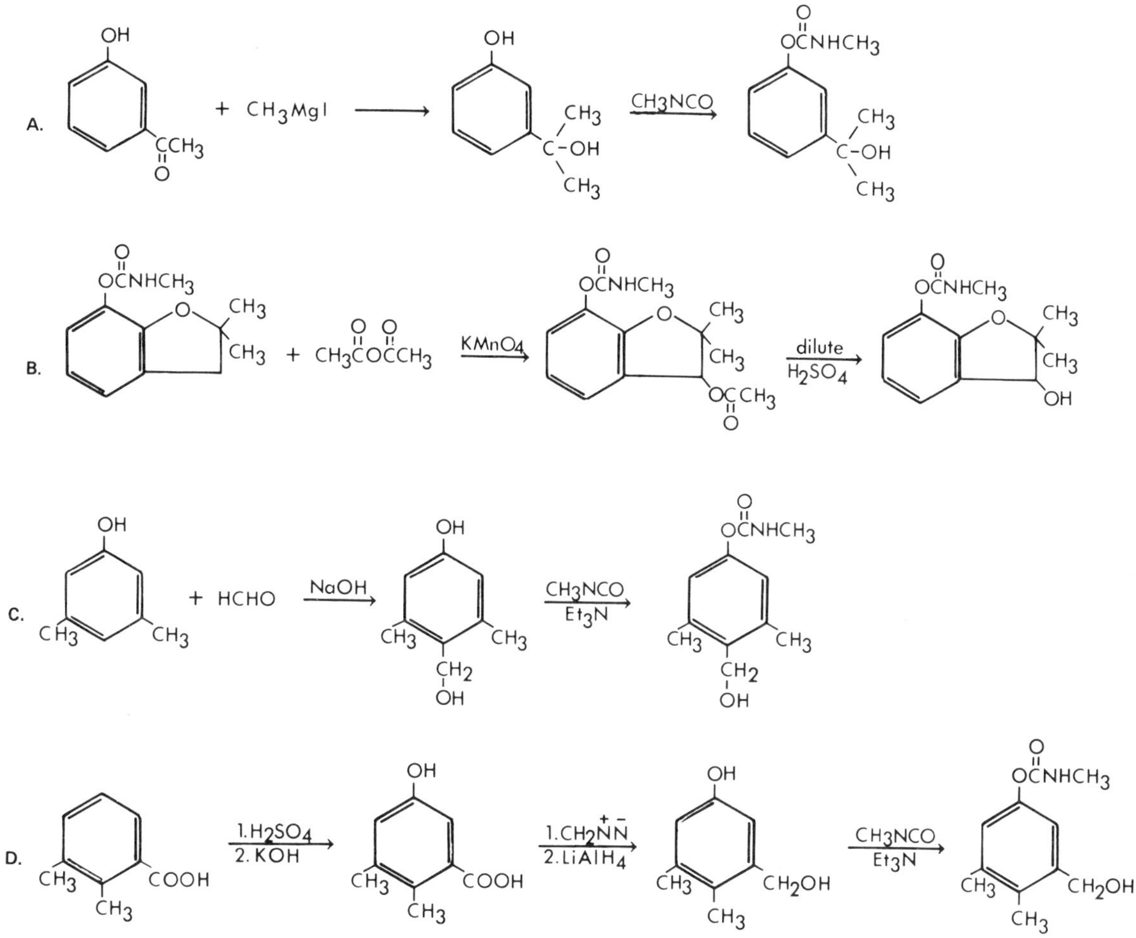

FIGURE 2.21. Synthesis of hydroxypropyl UC 10854 (A),[78] 3-hydroxycarbofuran (B),[70] 4-hydroxymethyl Landrin (C),[79] and 3-hydroxymethyl Landrin (D).[79]

duction of hydroxymethylphenyl metabolites of Meobal.[80] Some of the intermediates isolated during the above procedures are substituted hydroxybenzoic acids. It is possible to carbamylate these intermediates to give another group of oxidative metabolites, including, for example, 3-methyl-4-carboxyphenyl N-methylcarbamate.[80]

Potential breakdown products which demand synthesis are the sulfoxides and sulfones of carbamates containing sulfur atoms, since these compounds often possess toxicity equal to that of the parent compounds, e.g., the oxime carbamates. Formation of the sulfoxide (sulfinyl, SO) analog of aldicarb or any of its decarbamylated metabolites is an easy process, as illustrated by reaction A at the top of Figure 2.22. The conditions vary slightly with each chemist, but usually aldicarb is oxidized by slight warming in ethyl acetate while an equivalent amount of ~20% peracetic acid is slowly added.[36,81,82] The contents are then permitted to steep overnight before recovery of the product. By doubling the amount of acid and extending the holding time, the sulfone (sulfonyl, SO_2) metabolite can be made. Methomyl sulfoxidation products have been synthesized in similar fashion, though 40% peracetic acid was used, with chloroform as the solvent.[83] Aldicarb sulfoxide has also been prepared by dissolving aldicarb in cold ethanol and titrating with 5% hydrogen peroxide until 90% conversion to 2-methyl-2-(methylsulfinyl)propionaldehyde O-(methylcarbamoyl)oxime is complete, as determined by periodic thin-layer chromatography of the reaction mixture.[84]

A different oxidation procedure has been developed for making the mono- and disulfoxide and -sulfone metabolites of 2,4-dimethyl-1,3-dithiolane-2-carboxaldehyde O-(methylcarbamoyl) oxime (Tirpate).[62] Single oxidation on one or both of the cyclic sulfur atoms is achieved by

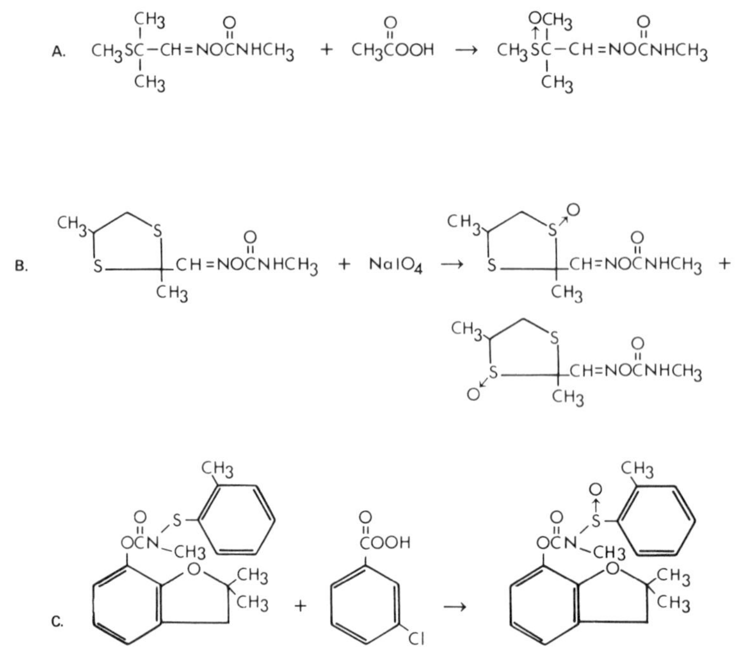

FIGURE 2.22. Synthesis of sulfoxide analogs of aldicarb (A),[36,81,82] Tirpate (B),[62] and N-toluenesulfenyl carbofuran (C).[86]

stirring Tirpate with one or two equivalents, respectively, of sodium metaperiodate in methanol at 0 to 5°C for 4 hr (Figure 2.22, B). Obviously, two isomers of the monosulfoxide product are possible, as shown. To get the disulfone, the carbamate is placed in acetic acid, and a large excess of 30% H_2O_2 is added. The mixture is heated slowly and with great care, since the reaction is exothermic near reflux. A few minutes of boiling yields disulfonyl Tirpate.

The same heterocyclic ring system present in Tirpate has been attached to the nucleus of phenyl N-methylcarbamate. However, in this instance, the sulfoxide and sulfone derivatives were prepared in a manner analogous to that described for peracetic acid oxidation of aldicarb.[85] Sulfur-containing substituents placed on the carbamate N-methyl group are also subject to sulfoxidation, so that appropriate sulfinyl and sulfonyl analogs have been synthesized. An example is reaction C at the bottom of Figure 2.22. A 2:1 mixture of N-toluenesulfenyl carbofuran and 3-chloroperbenzoic acid was heated to 40°C in methylene chloride for 30 min to obtain the sulfoxide, whereas excess acid was employed as usual to prepare the sulfone.[86] An alternative method is to add the already oxidized aryl sulfinyl group to the N-methylcarbamate.[87]

Conjugates

After a carbamate has been hydrolyzed or hydroxylated by a living organism, it is usually conjugated very quickly as a water-soluble sulfate, glucuronide, glycoside, or amino acid analog (see Chapters 6 and 7). Since such types of metabolites are generally assumed to be toxicologically safe and are difficult to isolate in a pure state, only a handful have been synthesized for identification purposes. The simplest to make are the sulfates of parent phenols.[88] In the case of Mobam phenol (Figure 2.23,A), for example, a slight excess of chlorosulfonic acid is added to 4-hydroxythiophene and dimethylaniline at 0°C to yield the sulfate, which is actually recovered by making the potassium salt with 50% KOH.[89] The same procedure has been used to prepare 1-naphthyl sulfate[75] and 4-chlorophenyl sulfate,[90] and undoubtedly could be used for synthesis of sulfates of many phenols derived from methylcarbamate insecticides.

Also relatively easy to prepare are the glucuronides and glucosides of phenols; one method is represented by reaction B in Figure 2.23.[89] Thus, fourfold excess of 4-chlorophenol is mixed with methyl-1,2,3,4-tetra-O-acetyl-β-D-glucuronate in the presence of a catalytic amount of 4-toluenesulfonic acid. Continuous heating on a steam bath

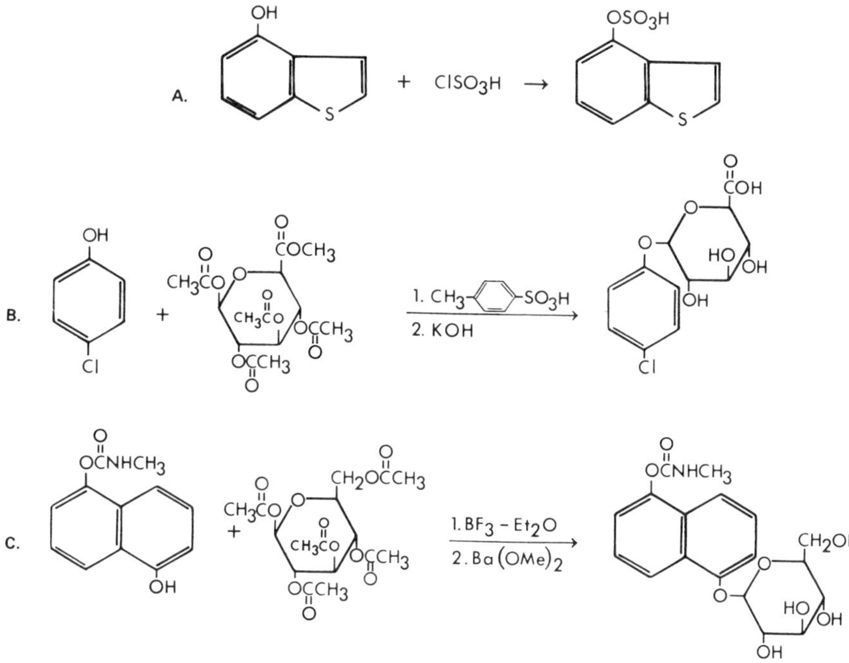

FIGURE 2.23. Synthesis of the sulfate of Mobam phenol (A),[8,9] the glucuronide of 4-chlorophenol (B),[90] and the glucoside of 5-hydroxy carbaryl (C).[92]

with intermittent evacuation for 2.5 hr gives the methyltriacetyl intermediate, which is cleaned up and deacetylated by exposure to mild KOH in ethanol at 65°C for 1 hr.[90] The final product is 4-chlorophenyl glucuronide. The same reaction with some modifications, including deacetylation by ammonia gas, yields 1-naphthyl glucoside from 1-naphthol plus penta-O-acetyl-β-D-glucopyranose.[91]

Although the above reactions proceed smoothly with carbamate hydrolysis products, great difficulty has been encountered in attempting to add conjugate groups to metabolites still possessing the intact carbamate moiety. Usually the conditions are such that the carbamate ester bond is cleaved in the course of preparation. This problem has been surmounted by the recent synthesis of the β-D-glucosides of 4- and 5-hydroxy carbaryl.[92] For example, a slight excess of 5-hydroxyl-1-naphthyl N-methylcarbamate was coupled to β-D-glucose pentaacetate by refluxing for 1 hr in anhydrous benzene spiked with a boron trifluoride-ether complex (Figure 2.23, C). Recrystallized 5-hydroxy carbaryl tetraacetylglucoside was deacetylated without loss of carbamate by dissolution in cold dry methanol and treatment with barium methylate solution for 3 hr. Attempts at making the glucosides of 3-hydroxy carbofuran or N-hydroxymethyl carbaryl by this method failed, however.

One other type of conjugate which may enter the metabolic pathway of a carbamate insecticide is an amino acid conjugate. Although evidence for their formation is limited mostly to speculation, in one study with carbaryl, S-(4-hydroxy-1-naphthyl) cysteine was implicated as a metabolite.[93] Synthesis of the suspected product was achieved by refluxing 4-methoxy-1-naphthylmercapturic acid with HBr in glacial acetic acid for 3 hr under nitrogen.

RADIOSYNTHESIS

All of the commercially available and many experimental carbamate insecticides have been synthesized with a carbon-14 atom in at least one position of the molecule. Such radiolabeling has been instrumental in accumulating the majority of metabolic information concerning this group of toxins. In fact, the use of isotope-tagged insecticides for defining degradation pathways is a critical part of the EPA-registration process.

The principal advantage of working with a radiolabeled compound is that treatment, transport, metabolism, and excretion can be monitored quantitatively at the nanogram level. In addition, a total-balance sheet accounting for all of the applied material can usually be constructed. By strategically inserting the radioatom in the toxophoric group, a labile leaving group, or a stable ring nucleus, much can be learned about the fate of the insecticide, even without actual product indentification.

Thus, a very important aspect of the chemistry and synthesis of carbamates is the specialized area of radiosynthesis. Here the chemist trades liter flasks and beakers for thimble-sized glassware and preparations on a mmole scale. Because the cost of radiolabeled starting materials tends to be very high, good yields are important, although a considerable amount of research can be done with only a fraction of a mmole of ^{14}C insecticide. Tracers other than carbon-14 are occasionally used in carbamate studies, but in most cases it is advantageous to prepare the radiocarbon analog.

The first carbamate insecticide to be labeled was carbaryl. Following the usual recipe for N-methylcarbamate synthesis (Figure 2.8, A), methylisocyanate (1.71 mmole) and 1-^{14}C-naphthol (1.62 mmole) dissolved in benzene containing a trace of pyridine were sealed in an ampule (Figure 2.24, A).[94] After 5 hr at 80°C, 1-^{14}C-naphthyl N-methylcarbamate was recovered in 82% yield after cleanup. Later, the same product was synthesized by combining 1-^{14}C-naphthol and CH_3NCO in the absence of solvent or catalyst.[68] Since then, numerous other researchers have prepared ring-labeled carbaryl by variations of this general method. The desirability for carbaryl labeled on the other end of the molecule led to the development of a special two-compartment glass tube reaction for synthesis of carbonyl-^{14}C and N-methyl-^{14}C carbamates.[5] In one end of the tube, sodium azide was heated with acetyl-1-^{14}C chloride or acetyl-2-^{14}C chloride to give $CH_3N^{14}CO$ or $^{14}CH_3NCO$ respectively. The tagged methylisocyanate was then forced into the other end of the tube, which contained triethylamine catalyst and 1-naphthol or substituted phenol. After 2 hr at 100°C and overnight cooling, yields ranging from 38 to 74% were reported. The routes for ^{14}C-carbaryl synthesis are

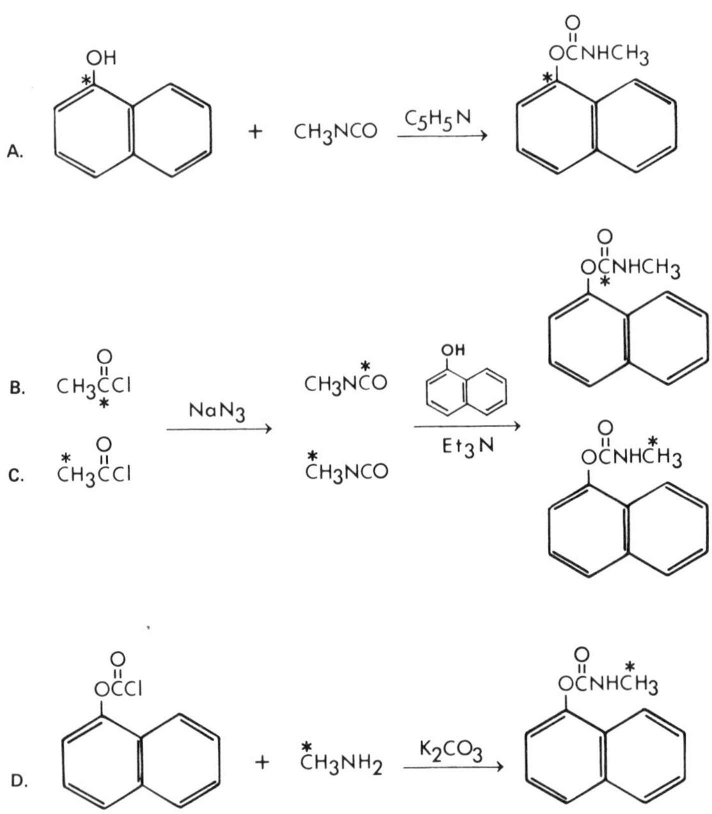

FIGURE 2.24. Radiosynthesis of carbaryl with a ^{14}C atom in the ring (A),[94] in the carbonyl position (B),[5] or in the N-methyl position (C, D).[5,75]

pictured as methods B and C in the middle portion of Figure 2.24. A different procedure for making N-methyl-^{14}C carbaryl, involving reaction of 1-naphthyl chloroformate with methyl-^{14}C-amine·HCl, has been used by other investigators (Figure 2.24, D).[75,95]

The reaction-tube synthesis of N-methyl-^{14}C and carbonyl-^{14}C Banol, Baygon, HRS-1422, Matacil, Mesurol, UC 10854, and Zectran[5] paved the way for uncovering the principal reaction mechanisms regulating the metabolic fate of phenyl methylcarbamate insecticides in plants and animals.[96] Subsequent investigations with newer carbamates, such as aldicarb,[84] carbofuran,[97] Bux,[71] 3,5-dimethyl N-methylcarbamate,[98] and 3-(2-butyl)phenyl N-methylcarbamate,[87] have also been facilitated by synthesis of carbonyl-^{14}C preparations via tagged methylisocyanate. However, the latter did not always originate from labeled acetyl-1-^{14}C chloride, but could be just as easily obtained from sodium azide plus acetyl-1-^{14}C bromide[97] or acetic-^{14}C anhydride.[99] Samples of methyl-^{14}C carbofuran,[70] aldicarb,[100] and tolyl N-methylcarbamate[101] have also been synthesized from the appropriate ^{14}C-methylisocyanate; while Baygon, UC 10854, and carbofuran N-$^{14}CH_3$ preparations became available after reacting corresponding chloroformates with $^{14}CH_3NH_2$.[95] Finally, the dimethylcarbamate dimetilan was produced with a carbonyl-^{14}C marker through standard synthesis procedures, using $(CH_3)_2N^{14}C(O)Cl$ as the carbamylation agent.[102]

All of the above syntheses are, of course, scaled-down versions of the established procedures used to prepare unlabeled carbamates, discussed earlier in the chapter. However, for placing ^{14}C in the aromatic ring or oxime portion of a carbamate, more complicated, and sometimes new, pathways are required. The only exceptions are carbaryl made from readily available 1-^{14}C-naphthol (Figure 2.24, A) and unsubstituted phenyl N-methylcarbamate from phenol-U-^{14}C.[103] The difficulty encountered in preparing labeled substituted phenols from low-cost available ^{14}C-starting materials can be appreciated by examining the shorthand scheme in Figure 2.25 for Zectran (A).[99]

An alternative to ring-labeling *per se* is tagging an alkyl ring substituent. One example is given in Figure 2.25, B for marking the butyl group of RE 5365.[87] Other investigators have published methods for preparing ring-$^{14}CH_3$ derivatives of Zectran,[99] Banol,[104] and Meobal.[105]

Isotopes have been synthetically inserted into various positions of the oxime portion of aldicarb,

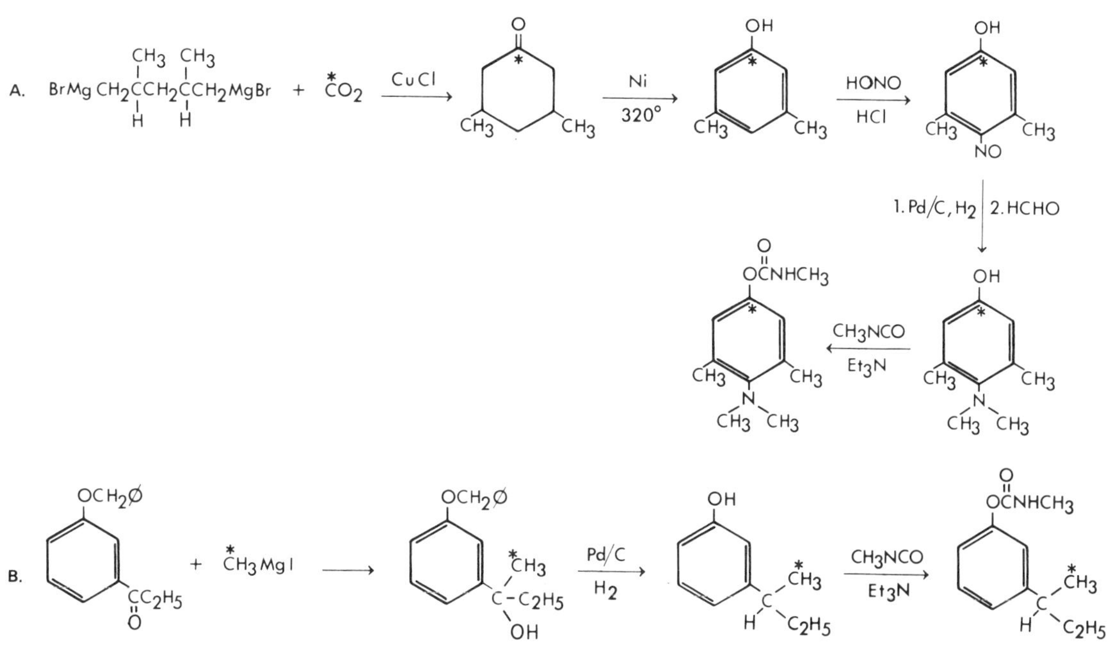

FIGURE 2.25. Radiosynthesis of ring-^{14}C-labeled Zectran (A)[99] and ring-alkyl-^{14}C-labeled RE 5365 (B).[87]

as illustrated in Figure 2.26. Thus, ^{14}C can be placed in the tertiary-carbon slot (A) or S-methyl position (B),[100] or a ^{35}S atom can substitute for normal sulfur (C).[84] In analogous fashion, the pseudo-tertiary carbon of methomyl becomes labeled following various reactions on $^{14}CH_3CN$.[83]

Carbon tracers have also been utilized in studies with certain of the N-methyl-N-substituted carbamate precursors discussed earlier in the chapter. For N-sulfenyl derivatives of carbofuran[85] and RE 5365[87] the label has been restricted to the parent carbamate portion of the molecule, but for N-acetyl Zectran a method was devised to place the radiocarbon on the labile acetyl group.[106]

On a few occasions, tritium-labeled carbamates have been prepared. The method used is a general procedure for aromatic insecticides and involves soaking the compound in cyclohexane containing $^3H_3PO_4 \cdot BF_3$ for 1 day.[107] In this way, 58 to 76% yield of tritiated carbaryl, Baygon, Mesurol, UC 10854, and HRS-1422 was obtained, with specific activities varying from 0.9 to 9.75 mCi/mmole. It is also possible to tritiate the parent phenol by this process and then carbamylate with methylisocyanate.[108] Alternatively, the desired phenol can be prepared from a readily available labeled compound, e.g., ^3H-carbofuran phenol from ^3H-catechol.[70]

Many other labeled and unlabeled carbamates have been synthesized by chemists in pesticide companies or radiochemical supply organizations by methods which have not been published. Undoubtedly, the procedures do not differ substantially from those described on the previous pages. Despite the thousands of structures already prepared and tested, the creation of new carbamates continues in chemistry laboratories throughout the world. Although the number of possible molecular configurations is virtually unlimitied, the relative ease of synthesis and favorable chemical and physical properties of most carbamates make "unlimited" appear as an attainable goal.

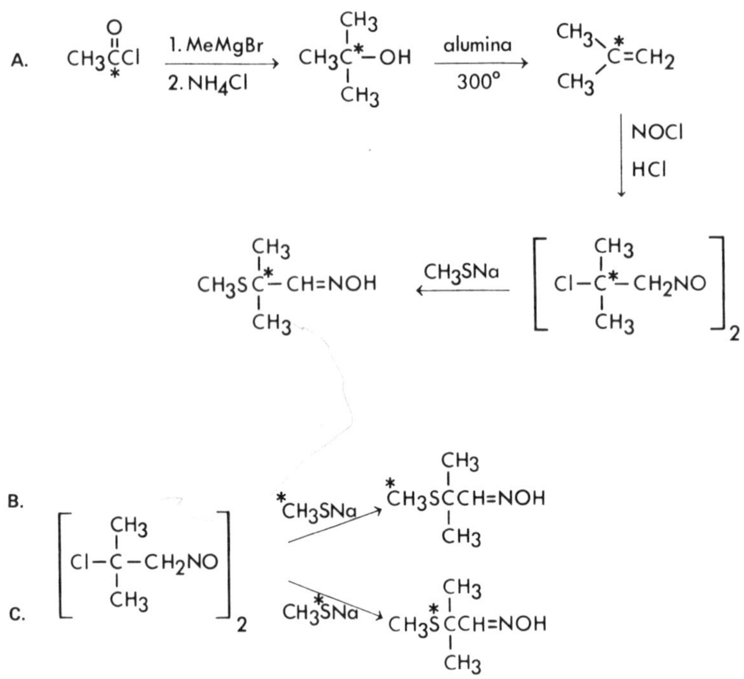

FIGURE 2.26. Radiosynthesis of aldicarb oxime with a ^{14}C atom in the tertiary carbon position (A) or in the S-methyl position (B),[100] or with a ^{35}S atom in the S-methyl position (C).[84]

REFERENCES

1. **Schlagbauer, B. G. L. and Schlagbauer, A. W. J.,** The metabolism of carbamate pesticides – a literature analysis. Part I, *Residue Rev.*, 42, 1, 1972.
2. **Stecher, P. G., Ed.,** *The Merck Index,* 8th ed., Merck and Co., Rahway, N. J., 1968.
3. **Berenblum, I., Ben-Ishai, D., Haran-Ghera, N., Lapidot, A., Simon, E., and Trainin, N.,** Skin-initiating action and lung carcinogenesis by derivatives of urethane (ethyl carbamate) and related compounds, *Biochem. Pharmacol.*, 2, 168, 1958.
4. **Casida, J. E.,** Mode of action of carbamates, *Annu. Rev. Entomol.*, 8, 39, 1963.
5. **Krishna, J. G., Dorough H. W., and Casida, J. E.,** Synthesis of N-methylcarbamates via methyl isocyanate-C^{14} and chromatographic purification, *J. Agric. Food Chem.*, 10, 462, 1962.
6. **Stedman, E.,** Studies on the relationship between chemical constitution and physiological action. Part I. Position isomerism in relation to miotic activity of some synthetic urethanes, *Biochem. J.*, 20, 719, 1926.
7. **Aeschlimann, J. A. and Reinert, M.,** The pharmacological action of some analogues of physostigmine, *J. Pharmacol. Exp. Ther.*, 43, 413, 1931.
8. **Kolbezen, M. J., Metcalf, R. L., and Fukuto, T. R.,** Insecticidal activity of carbamate cholinesterase inhibitors, *J. Agric. Food Chem.*, 2, 864, 1954.
9. **Casida, J. E., Augustinsson, K.-B., and Jonsson, G.,** Stability, toxicity, and reaction mechanisms with esterases of certain carbamate insecticides, *J. Econ. Entomol.*, 53, 205, 1960.
10. **Ellis, S., Plachte, F. L., and Straus, O. H.,** Studies on physostigmine and related substances. II. The destruction of physostigmine in buffered solutions and in serum, *J. Pharmacol. Exp. Ther.*, 79, 295, 1943.
11. **Chaikin, S. W.,** Study of the hydrolysis of several physostigmine analogs, *J. Am. Chem. Soc.*, 69, 1266, 1947.
12. **Dittert, L. W.,** The Kinetics and Mechanisms of the Base-catalyzed Hydrolysis of Organic Carbamates and Carbonates, Ph.D. thesis, University of Wisconsin, Madison, 1961.
13. **Dittert, L. W. and Higuchi, T.,** Rates of hydrolysis of carbamate and carbonate esters in alkaline solution, *J. Pharmacol. Sci.*, 52, 852, 1963.
14. **Metcalf, R. L., Fukuto, T. R., and Winton, M. Y.,** Insecticidal carbamates: comparison of the activities of N-methyl- and N,N-dimethylcarbamates of various phenols, *J. Econ. Entomol.*, 55, 345, 1962.
15. **O'Brien, R. D., Hilton, B. D., and Gilmour, L.,** The reaction of carbamates with cholinesterase, *Mol. Pharmacol.*, 2, 593, 1966.
16. **Fukuto, T. R., Fahmy, M. A. H., and Metcalf, R. L.,** Alkaline hydrolysis, anticholinesterase, and insecticidal properties of some nitro-substituted phenyl carbamates, *J. Agric. Food Chem.*, 15, 273, 1967.
17. **Gubler, K., Margot, A., and Gysin, H.,** Alicyclic and heterocyclic carbamates, *J. Sci. Food Agric. Suppl.*, p. 13, 1968.
18. **Vontor, T., Socha, J., and Vecera, M.,** Kinetics and mechanism of hydrolysis of 1-naphthyl N-methyl- and N,N-dimethylcarbamates, *Collect. Czech. Chem. Commun.*, 37, 2183, 1972.
19. **Abdel-Wahab, A. M., Kuhr, R. J., and Casida, J. E.,** Fate of C^{14}-carbonyl-labeled aryl methylcarbamate insecticide chemicals in and on bean plants, *J. Agric. Food Chem.*, 14, 290, 1966.
20. **Furdik, M., Drabek, J., Locigova, I., and Ondrejka, J.,** Effect of substituents and their position in the benzene nucleus on the insecticide activity of the phenyl-substituted O-phenyl thiophosphates, and N-methyl phenyl carbamates, *Chem. Zvesti*, 20, 650, 1966.
21. **Jones, R. L., Fukuto, T. R., and Metcalf, R. L.,** Structure, reactivity, and biological activity of O-(methylcarbamoyl)oximes of substituted benzaldehydes, *J. Econ. Entomol.*, 65, 28, 1972.
22. **Fukuto, T. R., Metcalf, R. L., Jones, R. L., and Myers, R. O.,** Structure, reactivity, and biological activity of O-(diethylphosphoryl)oximes and O-(methylcarbamoyl)oximes of substituted acetonphenones and α-substituted benzaldehydes, *J. Agric. Food Chem.*, 17, 923, 1969.
23. **Metcalf, R. L. and Fukuto, T. R.,** Some effects of molecular structure upon anticholinesterase and insecticidal activity of substituted phenyl N-methylcarbamates, *J. Agric. Food Chem.*, 15, 1022, 1967.
24. **Aly, O. M. and El-Dib, M. A.,** Studies of the persistence of some carbamate insecticides in the aquatic environment, in *Fate of Organic Pesticides in the Environment,* Advances in Chemistry Series III, American Chemical Society, Washington, D.C., 1972.
25. **Chin, B. H., Eldridge, J. M., and Sullivan, L. J.,** Metabolism of carbaryl by selected human tissues using an organ-maintenance technique, *Clin. Toxicol.*, 7, 37, 1974.
26. **Christenson, I.,** Alkaline hydrolysis of carbamic acid esters, *Acta Chem. Scand.*, 18, 904, 1964.
27. **Bender, M. L. and Homer, R. B.,** The mechanism of the alkaline hydrolysis of *p*-nitrophenyl N-methylcarbamate, *J. Org. Chem.*, 30, 3975, 1965.
28. **Armstrong, V. C. and Moodie, R. B.,** Hydrolysis of amides, esters, and related compounds in acid solution. II. carbamates, *J. Chem. Soc. Sect. B,* p. 934, 1969.
29. **Metcalf, R. L., Fukuto, T. R., and Winton, M. Y.,** Alkoxyphenyl N-methylcarbamates as insecticides, *J. Econ. Entomol.*, 53, 828, 1960.
30. **Metcalf, R. L., Fukuto, T. R., and Winton, M. Y.,** Insecticidal carbamates: position isomerism in relation to activity of substituted-phenyl N-methylcarbamates, *J. Econ. Entomol.*, 55, 889, 1962.

31. **Metcalf, R. L., Fuertes-Polo, C., and Fukuto, T. R.,** Carbamate insecticides: multisubstituted chloro- and methylphenyl N-methylcarbamates, *J. Econ. Entomol.*, 56, 862, 1963.
32. **Metcalf, R. L., Fukuto, T. R., and Frederickson, M.,** *Para*-Substituted *meta*-xylenyl diethyl phosphates and N-methylcarbamates as anticholinesterases and insecticides, *J. Agric. Food Chem.*, 12, 231, 1964.
33. **Szabo, K. and Broadbent, D. J.** (Stauffer Chemical Co.), British Patent 1,143,726, February 26, 1969.
34. **Hess, R., Boecker, E., and Unterstenhoefer, G.** (Farbenfabriken Bayer A.-G.), German Patent 1,142, 599, January 24, 1963.
35. **Payne, L. K. and Weiden, M. H. J.** (Union Carbide Corp.), U.S. Patent 3,217,037, November 9, 1965.
36. **Payne, L. K., Jr. and Stansbury, H. A., Jr.,** The synthesis and insecticidal properties of some cholinergic trisubstituted acetaldehyde O-(carbamoyl)oximes, *J. Agric. Food Chem.*, 14, 356, 1966.
37. **Fahmy, M. A. H., Metcalf, R. L., Fukuto, T. R., and Hennessey, D. J.,** Effects of deuteration, fluorination, and other structural modifications of the carbamyl moiety on the anticholinesterase and insecticidal activities of phenyl N-methylcarbamates, *J. Agric. Food Chem.*, 14, 79, 1966.
38. **Scharpf, W. G.** (FMC Corp.), U.S. Patent 3,547,955, December 15, 1970.
39. **Donninger, C. and Schofield, J. A.** (Shell Internationale Research Maatschappij N. V.), British Patent 1,216,312, December 16, 1970.
40. **Szabo, K.** (Syracuse University Research Corp.), U.S. Patent 3,523,968, August 11, 1970.
41. **Nickles, E. F.,** N-Methyl- and N,N-dimethylcarbamates of hydroxybenzaldehyde acetals and mercaptals, *J. Agric. Food Chem.*, 17, 939, 1969.
42. FMC Corp., Dutch Patent 6,500,340, July 26, 1965.
43. **Scharpf, W. G.** (FMC Corp.), U.S. Patent 3,564,605, February 16, 1971.
44. **Wilson, K. R. and Kennedy, R. M.** (FMC Corp.), U.S. Patent 3,655,696, April 11, 1972.
45. **Pelley, R. L. and Orwall, E. F.** (FMC Corp.), South African Patent 68 00,794, March 13, 1969.
46. **Kilsheimer, J. R. and Kaufman, H. A.** (Mobil Oil Corp.), U.S. Patent 3,288,673, November 29, 1966.
47. **Kilsheimer, J. R., Kaufman, H. A., Foster, H. M., Driscoll, P. R., Glick, L. A., and Napier, R. P.,** Benzo[b]thienyl carbamate insecticides, *J. Agric. Food Chem.*, 17, 91, 1969.
48. **Sacher, R. M., Alt, G. H., and Darlington, W. A.,** Insecticidal and anticholinesterase activity of benzotriazolyl methyl and dimethylcarbamates, *J. Agric. Food Chem.*, 21, 132, 1973.
49. **Lambreck, J. A.** (Union Carbide Corp.), U.S. Patent 2,903,478, September 8, 1959.
50. **Wildgrube, W., Schelle, R., and Heidenreich, S.,** East German Patent 36, 946, March 5, 1965.
51. **Moldoran-Velniceriu, A. and Mihaiu, T.** (Romania, Ministry of the Chemical Industry), Romanian Patent 51,249, September 12, 1968.
52. **Lehureau, J. and Pierrot, F.** (Progil S. A.), French Patent 1,344,236, November 29, 1963.
53. **Lehureau, J. and Gras, G.** (Progil S. A.), French Patent 1,344,934, December 6, 1963.
54. **Pelah, Z.** (Makhteshim Chemical Works Ltd.), German Patent 1,910, 295, October 9, 1969.
55. **Naumov, Y. A., Bazhanova, L. G., and Knyazeva, A. P.** (All-Union Scientific Research Institute of Chemicals for Plant Protection), U.S.S.R. Patent 186,438, October 3, 1966.
56. **Heiss, R., Boecker, E., and Morschel, H.** (Farbenfabriken Bayer A.-G.), German Patent 1,166,180, March 26, 1964.
56a. **Geigy, J. R., A.-G.,** Swiss Patents 279,553, March 1, 1952; 282,655, August 16, 1952.
57. **Gysin, H., Margot, A., and Simon, C. (J. R. Geigy A.-G.),** U.S. Patent 2,694,712, November 16, 1954.
58. **Fraser, J., Harrison, I. R., and Wakerley, S. B.,** Synthesis and insecticidal activity of N-acyl-N-methyl carbamates, *J. Sci. Food Agric. Suppl.*, p.8, 1968.
59. **Fraser, J., Clinch, P. G., and Reay, R. C.,** N-Acylation of N-methylcarbamate insecticides and its effect on biological activity, *J. Sci. Food Agric.*, 16, 615, 1965.
60. **Robertson, W. A. H., Fraser, J., and Clinch, P. G.** (Boots Pure Drug Co. Ltd.), British Patent 982,235, February 3, 1965.
61. Boots Pure Drug Co. Ltd., Belgian Patent 654,121, April 8, 1965.
62. **Fridinger, T. L., Mutsch, E. L., Bushong, J. W., and Matteson, J. W.,** Synthesis and biological activity of α-mercaptol O-(methylcarbamoyl)oximes, *J. Agric. Food Chem.*, 19, 422, 1971.
63. **Brown, M. S. and Kohn, G. K.** (Chevron Research Co.), U.S. Patent 3,663,594, May 16, 1972.
64. **Farbenfabriken Bayer A.-G.,** Belgian Patent 772,634, March 15, 1972.
65. **Black, A. L., Chiu, L., Fahmy, M. A. H., and Fukuto, T. R.,** Selective toxicity of N-sulfenylated derivatives of insecticidal methylcarbamate esters, *J. Agric. Food Chem.*, 21, 747, 1973.
66. **Fahmy, M. A. H., Fukuto, T. R., Myers, R. O., and March, R. B.,** The selective toxicity of new N-phosphorothioylcarbamate esters, *J. Agric. Food Chem.*, 18, 793, 1970.
67. **Haubein, A. H.** (Hercules Inc.), U.S. Patent 3,345,256, October 3, 1967.
68. **Dorough, H. W. and Casida, J. E.,** Nature of certain carbamate metabolites of the insecticide Sevin, *J. Agric. Food Chem.*, 12, 294, 1964.
69. **Balba, M. H., Singer, M. S., Slade, M., and Casida, J. E.,** Synthesis of possible metabolites of methylcarbamate insecticide chemicals. Substituted-aryl N-hydroxymethylcarbamates, *J. Agric. Food Chem.*, 16, 821, 1968.
70. **Metcalf, R. L., Fukuto, T. R., Collins, C., Borack, K., El-Aziz, A., Monoz, R., and Cassil, C. C.,** Metabolism of 2,2-dimethyl-2,3-dihydrobenzofuranyl-7 N-methylcarbamate (Furadan) in plants, insects, and mammals, *J. Agric. Food Chem.*, 16, 300, 1968.

71. **Tucker, B. V. and Pack, D. E.,** Bux insecticide soil metabolism, *J. Agric. Food Chem.*, 20, 412, 1972.
72. **Durden, J. A., Stollings, H. W., Casida, J. E., and Slade, M.,** The synthesis of hydroxymethylcarbamates, *J. Agric. Food Chem.*, 18, 459, 1970.
73. **Fahmy, M. A. H. and Fukuto, T. R.,** Convenient method for synthesis of hydroxymethylcarbamates, *J. Agric. Food Chem.*, 20, 168, 1972.
74. **Ohkawa, H., Yoshihara, R., Kahara, T., and Miyamoto, J.,** Metabolism of *m*-tolyl N-methylcarbamate (Tsumacide) in rats, houseflies and bean plants, *Agric. Biol. Chem.*, 38, 1035, 1974.
75. **Knaak, J. B., Tallant, M. J., Bartley, W. J., and Sullivan, L. J.,** The metabolism of carbaryl in the rat, guinea pig and man, *J. Agric. Food Chem.*, 13, 537, 1965.
76. **Durden, J. A.,** The synthesis of nuclearly hydroxylated 1-naphthyl methlcarbamates. The 6- and 7-isomers, *J. Agric. Food Chem.*, 19, 432, 1971.
77. **Richey, F. A., Bartley, W. J., Fitzpatrick, J. T., and Kurtz, A. P.,** Chemical synthesis of the carbaryl metabolite *trans*-5,6-dihydro-5,6-dihydroxy-1-naphthyl methylcarbamate, *J. Agric. Food Chem.*, 20, 825, 1972.
78. **Balba, H. M. and Casida, J. E.,** Synthesis of possible metabolites of methylcarbamate insecticide chemicals: hydroxyaryl and hydroxyalkylphenyl methylcarbamates, *J. Agric. Food Chem.*, 16, 561, 1968.
79. **Slade, M. and Casida, J. E.,** Metabolic fate of 3,4,5- and 2,3,5-trimethylphenyl methylcarbamates, the major constituents in Landrin insecticide, *J. Agric. Food Chem.*, 18, 467, 1970.
80. **Miyamoto, J., Yamamoto, K., and Matsumoto, T.,** Metabolism of 3,4-dimethylphenyl N-methylcarbamate in white rats, *Agric. Biol. Chem.*, 33, 1060, 1969.
81. **Knaak, J. B., Tallant, M. J., and Sullivan, L. J.,** The metabolism of 2-methyl-2-(methylthio)propionaldehyde O-(methylcarbamoyl)oxime in the rat, *J. Agric. Food Chem.*, 14, 573, 1966.
82. **Durden, J. A., Jr., Bartley, W. J., and Stephen, J. F.,** The preparation of standards for aldicarb (Temik) metabolism study, *J. Agric. Food Chem.*, 18, 454, 1970.
83. **Harvey, J., Jr., Jelinek, A. G., and Sherman, H.,** Metabolism of methomyl in the rat, *J. Agric. Food Chem.*, 21, 769, 1973.
84. **Andrawes, N. R., Dorough, H. W., and Lindquist, D. A.,** Degradation and elimination of Temik in rats, *J. Econ. Entomol.*, 60, 979, 1967.
85. **Durden, J. A., Jr. and Weiden, M. H. J.,** Insecticidal 2-(methylcarbamoyloxyphenyl)-1,3-dioxolanes, -oxathiolanes, and -dithiolanes, *J. Agric. Food Chem.*, 17, 94, 1969.
86. **Black, A. L., Chiu,. Y. C., Fukuto, T. R., and Miller, T. A.,** Metabolism of 2,2-dimethyl-2,3-dihydrobenzofuranyl-7 N-methyl-N-(2-toluenesulfenyl) carbamate in the housefly and white mouse, *Pestic. Biochem Physiol.*, 3, 435, 1973.
87. **Cheng, H.-M. and Casida, J. E.,** Metabolites and photoproducts of 3-(2-butyl)phenyl N-methylcarbamate and N-benzenesulfenyl-N-methylcarbamate, *J. Agric. Food Chem.*, 21, 1037, 1973.
88. **Feigenbaum, J. and Neuberg, C. A.,** Simplified method for the preparation of aromatic sulfuric acid esters, *J. Am. Chem. Soc.*, 63, 3529, 1941.
89. **Robbins, J. E., Bakke, J. E., and Feil, V. J.,** Metabolism of 4-benzothienyl N-methylcarbamate (Mobam) in rats: balance study of urinary-metabolite separation, *J. Agric. Food Chem.*, 17, 236, 1969.
90. **Paulson, G. D. and Zehr, M. V.,** Metabolism of *p*-chlorophenyl N-methylcarbamate in the chicken, *J. Agric. Food Chem.*, 19, 471, 1971.
91. **Dorough, H. W., McManus, J. P., Kumar, S. S., and Cardona, R. A.,** Chemical and metabolic characteristics of 1-naphthyl β-D-glucoside, *J. Agric. Food Chem.*, 22, 643, 1974.
92. **Cardona, R. A. and Dorough, H. W.,** Syntheses of the β-D-glucosides of 4- and 5-hydroxy-1-naphthyl N-methylcarbamate, *J. Agric. Food Chem.*, 21, 1065, 1973.
93. **Bend, J. R., Holder, G. M., Protos, E., and Ryan, A. J.,** Water-soluble metabolites of carbaryl (1-naphthyl N-methylcarbamate) in mouse liver preparations and in the rat, *Aust. J. Biol. Sci.*, 24, 535, 1971.
94. **Skraba, W. J. and Young, F. G.,** Radioactive Sevin (1-naphthyl-1-carbon-14 N-methylcarbamate), a convenient synthesis, *J. Agric. Food Chem.*, 7, 612, 1959.
95. **Metcalf, R. L., Osman, M. F., and Fukuto, T. R.,** Metabolism of ^{14}C-labeled carbamate insecticides to $^{14}CO_2$ in the housefly, *J. Econ. Entomol.*, 60, 445, 1967.
96. **Casida, J. E.,** Radiotracer studies on metabolism, degradation and mode of action of insecticidal chemicals, *Residue Rev.*, 25, 149, 1969.
97. **Dorough, H. W.,** Metabolism of Furadan (NIA-10242) in rats and houseflies, *J. Agric. Food Chem.*, 16, 319, 1968.
98. **Kazano, H., Kearney, P. C., and Kaufman, D. D.,** Metabolism of methylcarbamate insecticides in soil, *J. Agric. Food Chem.*, 20, 975, 1972.
99. **Williams, E. A., Meikle, R. W., and Redemann, C. T.,** Radiosynthesis of carbon-14-labeled 4-dimethylamino-3,5-xylyl methylcarbamate, *J. Agric. Food Chem.*, 13, 210, 1965.
100. **Bartley, W. J., Heywood, D. L., Steele, T. E. N., and Skraba, W. J.,** Synthesis of C^{14}-labeled 2-methyl-2-(methylthio)propionaldehyde *O*-(methylcarbamoyl)oxime, *J. Agric. Food Chem.*, 14, 604, 1966.
101. **Svindlerman, G. S., Sivokhin, P. A.,Filator, G. V., and Baskakov, Yu. A.,** *Tr. Vses. Nauchno-Issled. Inst. Vet. Sanit.*, 31, 319, 1968.
102. **Casida, J. E.,** Radiotracer approaches to carbamate insecticide toxicity, in *Radiation and Radioisotopes Applied to Insects of Agricultural Importance*, International Atomic Energy Agency, Vienna, 1963.

103. **Kuhr, R. J.,** Possible role of tyrosinase and cytochrome P-450 in the metabolism of 1-naphthyl methylcarbamate (carbaryl) and phenyl methylcarbamate by houseflies, *J. Agric. Food Chem.,* 17, 112, 1969.
104. **Friedman, A. R. and Lemin, A. J.,** Metabolism of 2-chloro-4,5-dimethylphenyl N-methylcarbamate in bean plants, *J. Agric. Food Chem.,* 15, 642, 1967.
105. **Hazue, M. and Miyake, K.,** Synthesis of 3,4-dimethyl-4-$^{14}CH_3$ methylcarbamate (Meobal-^{14}C), *Botyu-Kagaku,* 34, 120, 1969.
106. **Look, M.,** Synthesis of N-acetyl-1'-^{14}C-Zectran (4-dimethylamino-3,5-xylyl N-acetyl-1'-^{14}C methylcarbamate) by microacetylation, *J. Agric. Food Chem.,* 16, 893, 1968.
107. **Hilton, B. D. and O'Brien, R. D.,** A simple technique for titration of aromatic insecticides, *J. Agric. Food Chem.,* 12, 236, 1964.
108. **Douch, P. G. C. and Smith, J. N.,** Metabolism of *m-tert*-butylphenyl N-methylcarbamate in insects and mice, *Biochem. J.,* 125, 385, 1971.

Chapter 3

MODE OF ACTION

INTRODUCTION

Almost all insecticides are toxic by virtue of their ability to disrupt the transmission of impulses along the nervous system of animals. This, of course, makes them very effective poisons, since any gross tampering with a system that regulates such vital functions as breathing and heartbeat can only be tolerated for the briefest period of time without fatal consequences. Though the mechanism of impulse transmission is apparently identical in the target insect and in other nontarget animals, fortunately there exist biochemical, physiological, and morphological differences in the gross anatomy of nerves and other tissues which permit selective toxicity. That is, a chemical that is lethal to insects can be relatively nontoxic to other organisms.

Understanding the causes for this selectivity and, more precisely, how and why insecticides kill is important for several reasons. First of all, the more learned about why one particular molecule is more toxic than a closely related analog, or how the same cluster of atoms can be lethal to one species but not another, the more effectively can new safer insecticides be developed. Certainly, there are a multitude of other factors which must also be considered, but the discovery of an exploitable difference between the nerves of insects and other animals could prove extremely useful in the chemist's search for a structure that is environmentally and toxicologically more acceptable.

Secondly, information on mode of action can be instrumental in understanding why compounds with certain structures are biologically inactive. This is particularly important for commercially successful products that have become less effective due to resistance of insects to the toxic action of the compounds. Again, there are a number of parameters that may cause resistance, but subtle physiological or biochemical alterations at the site of action could prove to be the deciding factor.

Finally, mode-of-action studies contribute to our overall comprehension of the insect. If man is to successfully compete with insects for food and fiber, then the more is known about deleterious species, the better they can be controlled.

Despite the almost countless number of experiments conducted in the search for the mode of action of DDT and other organochlorine insecticides (see Reference 1), the complete mechanism remains a mystery. There seems to be little doubt, however, that interruption of impulse transmission along the nerve is involved. With carbamate insecticides, the mode of action was known before their inception as insecticides. This was mostly because of prior knowledge of carbamate drug pharmacology and because of similarities to the known mode of action of organophosphorus insecticides. Put simply, carbamate toxicity is due to inhibition of acetylcholinesterase at certain synaptic junctions in the nervous system. All of the manifestations from this action, and the exact biochemical mechanism of inhibition, are not completely understood, but it is very likely that cholinesterase inhibition is the prima-facie cause of death.

Obviously, this mechanism does not explain why some carbamate insecticides, particularly when applied in large doses, are toxic to certain plants or defoliate others. Nor does it tell why carbaryl is effective as a thinning agent for apple trees.[1a] On the other hand, cholinesterase inhibition may explain why some carbamate chemicals are active as selective rodenticides,[2] molluscicides,[3] anthelmintics,[4] and bird immobilizers.[5]

In order to fully appreciate carbamate mode of action, a cursory familiarity with the nervous system of animals is necessary. This has been described, in relation to organophosphorus poisoning, by other monographs,[6-8a] therefore, only a brief account will be given here.

A schematic representation of the nervous system of vertebrates is given in Figure 3.1. It is divided into a central coordinating unit consisting of the brain and spinal cord (central nervous system or CNS) and a peripheral organization composed of the autonomic and somatic systems. Impulses travel along the nerve fibers in the direction of the arrows, and at certain points the impulse must cross a gap or synapse. There are two known chemicals that mediate transfer of the impulse across these junctions, and hence there are two types of synapses. The significance of this will be made clear later.

The somatic system is primarily responsible for

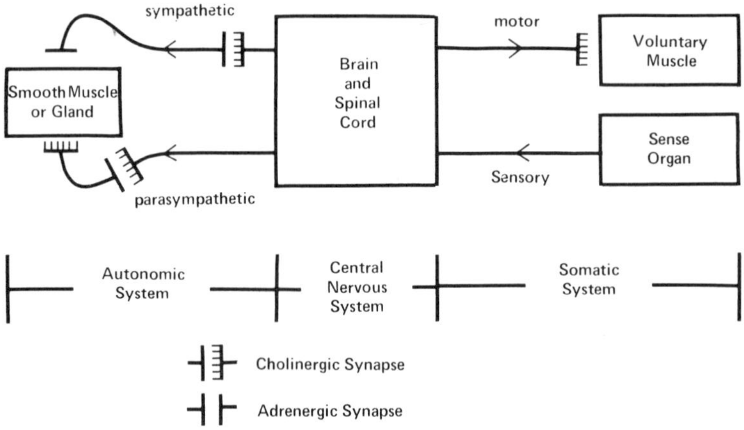

FIGURE 3.1. Schematic representation of vertebrate nervous system. (Adapted from Reference 6, with permission.)

an animal's awareness of its surroundings and of its internal well-being. Incoming stimuli, such as light, pain, hunger, and cold, are picked up by receptor cells or sense organs and transmitted via afferent nerves, without synapsing, to the CNS. After a decision is made, efferent nerves carry instructions back to the voluntary muscles across a synapse, where a response is elicited.

The autonomic system, perhaps more appropriately called the "automatic" system, regulates the involuntary muscles of the heart, lungs, alimentary canal, pupil, bladder, glands, etc. All impulses travel from the CNS to the muscles or glands. Sympathetic fibers contain a cholinergic synapse near the central nervous system and an adrenergic synapse at the terminal junction (exceptions are the adrenal medulla and sweat gland junctions, which are cholinergic). On the other hand, the parasympathetic nerve has a distal cholinergic synapse and another cholinergic gap at the muscle or gland. The two systems act in opposition to each other; for example, one contracts the pupil and the other dilates it, one accelerates heartbeat and the other retards it.

There are certain similarities and very basic differences in the anatomical construction of insect and vertebrate nervous systems. The more primitive CNS of insects is composed of a less complex brain and a ventral chain of ganglia, which, like the central nervous system in other animals, is protected from exogenous chemicals by a protective sheath. In mammals, the sheath is myelinated and extends to include the peripheral nervous system (except for postganglionic sympathetic fibers), but insect peripheral nerves are nonmyelinated. Although both possess autonomic systems, the construction is much less complex in insects; it is composed only of a sympathetic (or visceral) branch, with no counteracting parasympathetic branch. In addition, sympathetic nerve–nerve synapses in insects are not peripheral, but occur within the central nervous system. Therefore, the very susceptible autonomic nerve–nerve junctions in vertebrates are absent in insects.

Within the somatic system, the most significant difference occurs at the neuromuscular junction of the motor fibers. This synapse is cholinergic in vertebrates, but the transmitting substance in insects is unknown. Hence, there are no cholinergic synapses in the insect peripheral nervous system. However, all of the afferent and efferent fibers end with a synaptic junction within the CNS, and there are numerous other synapses between nerve cells inside the CNS. A number of these central junctions are undoubtedly cholinergic.

The transmission of an impulse along a nerve cell appears to be identical in vertebrates and insects. The cell, or neuron, is composed of a cell body with short dendrite extensions and a long axon fiber (Figure 3.2). Impulses travel from the dendrites through the cell body and down the axon, as shown by the arrow in Figure 3.2. This movement is thought of as electrical movement, since it is propagated by a change in charge across the outer cell membrane. When a message passes a particular point on the axon, for example, the cell membrane at that point allows sodium ions to rush into the nerve. Then, the membrane becomes suddenly permeable to potassium ions, which in

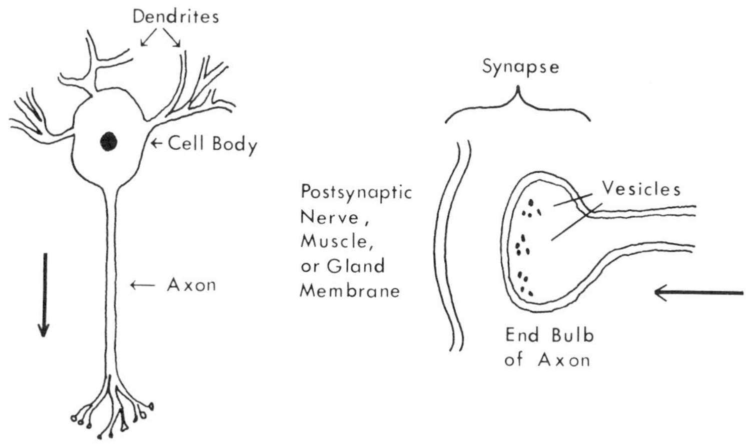

FIGURE 3.2. Schematic drawing of nerve cell and synapse between cells.

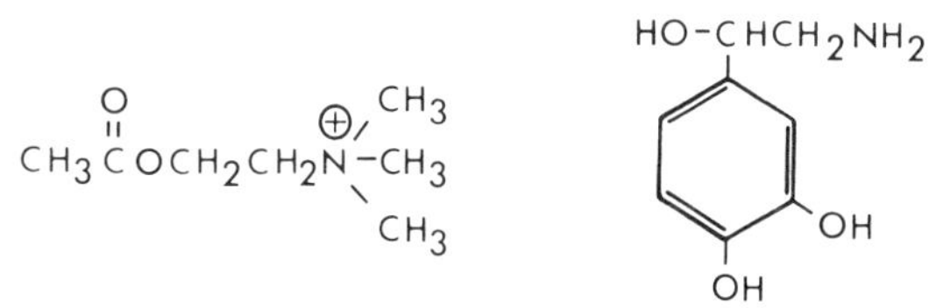

acetylcholine   norepinephrine

FIGURE 3.3. Structures of nerve transmission chemicals.

turn burst out of the axon to restore the status quo. The precise mechanism for this extremely rapid wave of charge transfer down the nerve is unknown. Considerable circumstantial evidence suggests that DDT acts by interfering with conduction along the axon of the nerve.

When the impulse reaches the end of a nerve, it must be transferred from the presynaptic axon membrane to the postsynaptic membrane of another nerve, muscle, or gland. This transfer, which covers a distance of only a few hundred angstroms, is made chemically. Upon reaching the end of an axon, the wave of depolarization triggers the expulsion of a cloud of transmitter substance from the many vesicles clustered in the end bulb (Figure 3.2). The cloud diffuses through the presynaptic membrane, across the synapse, and contacts the postsynaptic membrane which is depolarized and thus instigates an electrical impulse that continues down the next neuron. If the end bulb is adjacent to a muscle or gland, then the transfer of impulse results in a response by the effector tissue.

Once depolarization has been achieved, it is necessary to "shut off" the stimuli. This is accomplished by an enzyme present in the synapse which rapidly inactivates the transmitter substance so that the gap is ready for the next impulse. Synaptic transmission is considerably slower than axonic transmission, but nerve impulses generally move along at speeds near 30 m/sec, depending on fiber size and a few other parameters.

As indicated in Figure 3.1, there are two known types of synapses with two corresponding chemical mediators. One of these is norepinephrine or noradrenaline, and the other is acetylcholine (Figure 3.3). The former is responsible for transmission across synapses between axons and muscles or glands in the sympathetic fibers of vertebrates. The shut-off enzyme in these adrenergic synapses is monoamine oxidase.

All other peripheral synapses are cholinergic; that is, transmission is mediated by acetylcholine (ACh). This is probably also true for many of the synapses within the central nervous system of all animals. Destruction of the mediator is catalyzed by the enzyme acetylcholinesterase (AChE) to give acetic acid and choline (Figure 3.4).

It is this cholinergic transmission with which carbamate insecticides interfere. More precisely, the carbamate insecticide enters the synapse and reacts with or inhibits the AChE so that the enzyme cannot cleave the acetylcholine. As a result, the mediator continues to depolarize the postsynaptic membrane, causing prolonged stimulation and eventual failure of the nerve or effector tissue. Thus, in vertebrates, where peripheral cholinergic synapses control such vital functions as heartbeat and respiration, inhibition of acetylcholinesterase can cause rapid death. In insects, where peripheral cholinergic junctions are not

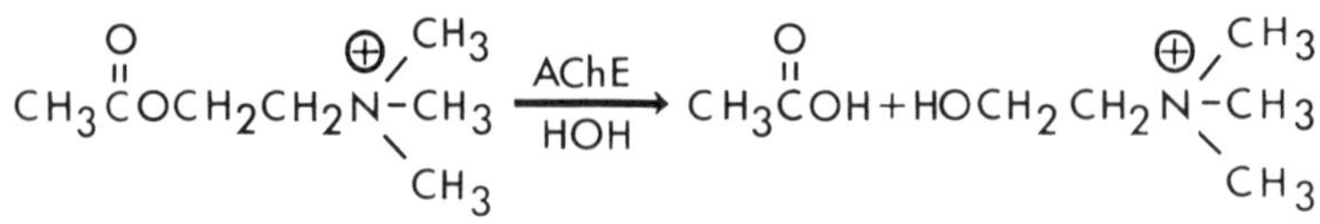

FIGURE 3.4. Hydrolysis of acetylcholine by acetylcholinesterase.

present, inhibition of AChE takes place in the CNS synapses, and hence death is often not as rapid; indeed, it may be due to secondary or indirect responses resulting from interruption of nerve impulse transmission.

Some of the literature pertaining to acetylcholinesterase inhibition by carbamates is confusing because of the different types of cholinesterases that exist in animals.[9] Acetylcholinesterase, also called "true cholinesterase" or acetycholine acetylhydrolase (3.1.1.7),[10] has been partially purified from bovine erythrocytes and from the electric organ of electric eels; it is believed to be most like the principal cholinesterase of vertebrate nervous systems and of the central nervous system of insects. This is not to say that the AChE from each of these sources is biochemically identical, but they all appear to have acetylcholine as their natural substrate.

Carbamates also inhibit cholinesterases (3.1.1.8) present in the blood plasma of man and other vertebrates. The biofunction of these pseudocholinesterases or plasma cholinesterases is not known, but total inhibition of the enzymes does not result in any recognized ill effects. Cholinesterases have also been found in liver, pancreas, intestine, heart, and muscle of mammals. These cholinesterases can be distinguished from each other, and from AChE, by their substrate and inhibitor specificity.

Thus, the reason carbamate insecticides are toxic lies in their ability to inhibit the nerve enzyme acetycholinesterase. Blockage of this enzyme results in a failure of the nervous system due to accumulation of acetylcholine in the nerve synapse. Without proper innervation of respiration, blood circulation, and other vital functions, the animal dies. It should be recognized that there have been some indications that other modes of action may be involved, especially with insects, but unless more evidence is uncovered, it seems very probable that carbamate poisoning is due solely to the ramifications of acetylcholinesterase inhibition.

CHOLINESTERASE INHIBITION

General Reaction

Although, as indicated above, there is general agreement that carbamate insecticides are toxic by virtue of their ability to inhibit acetylcholinesterase, the precise biochemical mechanism for this interaction is not completely understood. Over the past 25 years, a reaction mechanism has evolved which explains most of the available observations. Although the experimental methods used to obtain the data were fairly straightforward, the mathematical derivations and enzyme kinetics used to interpret results often varied from one laboratory to the next. The authors have chosen not to delve into the mathematical details of these investigations, but have attempted to present a simple evaluation of the observed results. Those interested in a more thorough description of the kinetics should refer to the references cited.

The equation sequence presented in Figure 3.5 is the most recent explanation for the reaction of N-methylcarbamate insecticides with acetylcholinesterase. It is the same mechanism that occurs with dimethylcarbamate and organophosphorus insecticides and with the natural substrate, acetylcholine.[7,11-13a] The enzyme of importance with respect to carbamate poisoning is acetylcholinesterase, but other cholinesterases react in a similar fashion.

The enzyme HAChE couples with the insecticide to form an intermediate complex (step +1), which may dissociate to enzyme and carbamate (step −1) or decompose into a stable carbamylated enzyme plus a leaving group (step 2). For methylcarbamate insecticides, the leaving group, HOX, is the parent phenol, naphthol, or oxime, and for most dimethylcarbomate insecticides it is a heterocyclic enol. Finally, the carbamylated enzyme hydrolyzes to generate free enzyme and methylcarbamic acid (step 3). The overall reaction, then, does not result in the destruction of any acetylcholinesterase, but it does yield hydrolysis of the insecticide molecule.

Within a nerve synapse, AChE cleaves the

$$HAChE + XO\overset{O}{\overset{\|}{C}}NHCH_3 \underset{k_{-1}}{\overset{k_{+1}}{\rightleftharpoons}} HAChE \cdot XO\overset{O}{\overset{\|}{C}}NHCH_3 \xrightarrow{k_2} HOX + AChE\overset{O}{\overset{\|}{C}}NHCH_3 \underset{HOH}{\xrightarrow{k_3}} HAChE + HO\overset{O}{\overset{\|}{C}}NHCH_3 + HOX$$

FIGURE 3.5. Reaction of carbamate insecticides with acetylcholinesterase.

transmitter acetylcholine according to this scheme, where movement through the reaction sequence from left to right is extremely rapid. When a carbamate insecticide enters the synapse, which occurs during poisoning, it "competes" with ACh for the active site on the enzyme. If successful, the insecticide ties up or inhibits the enzyme, since steps 2 and particularly 3 are thousands of times slower with carbamates than with ACh. The more enzyme inhibited, the less enzyme available for destruction of transmitter and the more severe the poisoning.

However, if the pressure is relieved by preventing the influx of new carbamate into the synapse, the AChE "recovers" due to dissociation of the insecticide-enzyme complex via step −1 and, more importantly, decarbamylation through step 3. The half-life for decarbamylation has been calculated to be 30 to 40 min;[7,14] thus, if all of the acetylcholinesterase in a synapse were carbamylated and no further insecticide entered the synapse, the enzyme would begin to recover in only a few minutes and would be almost fully restored by several hours. This restoration of AChE explains why animals showing symptoms of carbamate poisoning often recover within a matter of hours after exposure to carbamate insecticides.

In the case of most organophosphorus insecticides, the phosphorylated enzyme is considerably less susceptible to hydrolysis, so that step 3 is very slow. Also, whereas carbamate inhibition frequently involves a small but measurable quantity of insecticide-enzyme complex at any one point in time, organophosphorus kinetics suggest that virtually all inhibition is due to phosphorylated AChE. Thus, recovery from organophosphorus poisoning is a much slower process.

These differences between reaction of the two classes of insecticide with AChE have led to coinage of the terms "reversible" and "irreversible" inhibitors for carbamate and organophosphorus insecticides respectively. The terms are not entirely correct, for two reasons. First, a reversible inhibitor is not destroyed in the course of reaction, but carbamates are hydrolyzed, as discussed above. Second, most phosphorylated acetylcholinesterase is eventually hydrolyzed, so that the enzyme is able to recover or "reverse." However, these differences tend to give carbamates an added margin of safety over the organophosphorus materials, particularly in reference to inhibition accumulation due to repeated exposure to low levels of insecticide at closely timed intervals.

The series of reactions in Figure 3.5 grew from earlier proposals, which began to appear in the late 1940's. The fact that carbamate-inhibited cholinesterase was freed by washing, dialysis, or simple dilution, as well as the sensitivity of the reaction to substrate concentration, led to the conclusion that inhibition was due to the formation of a reversible complex.[15-20] Consequently, carbamate mode of action was explained simply by reaction constants k_{+1} and k_{-1} (Figure 3.5).

This, of course, did not account for destruction of the carbamate during the reaction. That is to say, the carbamate was not a substrate for cholinesterase. But in the early 1950's, evidence emerged indicating that certain carbamate structures did indeed act as substrates with extremely low turnover rates.[18,21,22] The importance of the carbamylation reaction, i.e., step 2, was not fully appreciated from these studies, however.

Later experiments, based mostly on analogy with phosphorylation studies, implied that carbamates did in fact react with the active site of the enzyme.[23,24] Further research culminated in the publication of a reaction sequence which was the immediate precursor of the one shown in Figure 3.5.[25,26] Using acetylcholinesterase from the electric organ of an eel and various noninsecticidal methyl- and dimethylcarbamate substrates, the investigators uncovered convincing evidence for the existence of constants k_2 and k_3. In addition, it was found that the ease with which the enzyme-substrate complex would form a carbamylated esterase was partially dependent on the size of the carbamate grouping on the substrate. Thus, the most potent AChE inhibitors had acyl groups as follows:

$$-C(O)NH_2 > -C(O)NHCH_3 > -C(O)N(CH_3)_2 > -C(O)N(C_2H_5)_2$$

Once chemical coupling with the enzyme had occurred, the ease with which the enzyme was regenerated (step 3) also depended on the structure of the coupler, so that enzyme reactivation rate followed the order

$$-C(O)CH_3 >> -C(O)NH_2 > -C(O)N(CH_3)_2 > -P(O)(OR)_2$$

These results instigated many of the structure activity studies which have been conducted with potential carbamate insecticides (see Chapter 4).

Since all N-methylcarbamates produced the same methylcarbamyl acetylcholinesterase, the rate constant, k_3, was identical for all monomethyl compounds (0.018 min^{-1}). Also, all dimethylcarbamates had similar k_3 values of 0.026 min^{-1}. More recent studies with bovine erythrocyte AChE and electric eel AChE gave a range of 0.011 to 0.036 min^{-1} for N-methylcarbamates, and of 0.012 to 0.052 min^{-1} for N,N-dimethylcarbamates.[14,27-32] Thus, it appears that the rate of decarbamylation is more or less the same for mono- and for dimethylcarbamate insecticides.

The last piece of information needed to confirm the mechanism of action in Figure 3.5 was positive proof for production of a free leaving group, XOH. This was clearly demonstrated in the mid 1960's by isolating the parent phenols following incubation of carbamates with acetylcholinesterase.[33,34] What was particularly significant in these studies was the use of carbamate insecticides as substrates. Prior to this, almost all of the reaction mechanisms were based on research with eserine, prostigmine, carbamyl fluorides, carbamyl cholines, and other noninsecticidal compounds. Some of the information resulting from the studies with insecticides is listed in Table 3.1. Interpretation of these data provides considerable insight into carbamate insecticide interactions with acetylcholinesterase.

Consider first the equilibrium constant or affinity constant, K_a, which has been defined as follows, using the notation pictured in Figure 3.5:

$$K_a = \frac{k_{-1}}{k_{+1}}$$

This is analogous to the well-known Michaelis dissociation constant, K_m, which is based on the assumption that the concentration of substrate is much greater than the concentration of enzyme. If the methylcarbamate and acetylcholinesterase are easily complexed, that is, if the substrate and enzyme have a great affinity for each other, then the velocity constant k_{+1} in Figure 3.5 will be much larger than k_{-1} and the equilibrium will be shifted to the right. In other words, the greater the affinity, the smaller is the constant K_a.

Of course, this definition of K_a does not take into account the effect of step k_2 on the affinity of carbamate for AChE. The constant for the total reaction, which is the constant that is actually

TABLE 3.1

Reaction Constants for Carbamates and Bovine Erythrocyte Acetylcholinesterase at 38°C and pH 7.0–7.4

Compound	K_a (M)	k_2 (min^{-1})	k_i (min^{-1} l $mole^{-1}$)
4-Fluorophenyl N-methylcarbamate	1.8×10^{-2}	1.9	1.0×10^{2}
4-Methylphenyl N-methylcarbamate	1.3×10^{-3}	1.5	11.6×10^{2}
Phenyl N-methylcarbamate	2.9×10^{-3}	1.6	5.4×10^{2}
4-Fluorophenyl N,N-dimethylcarbamate	7.3×10^{-2}	2.0	3.0×10^{1}
4-Methylphenyl N,N-dimethylcarbamate	6.4×10^{-3}	0.5	8.0×10^{1}
Phenyl N,N-dimethylcarbamate	1.5×10^{-3}	0.2	1.5×10^{2}
HRS-1422	3.4×10^{-6}	1.4	4.1×10^{5}
Mesurol	6.7×10^{-6}	1.2	1.9×10^{5}
Baygon	1.0×10^{-5}	1.1	1.1×10^{5}
Carbaryl	1.1×10^{-5}	1.3	1.3×10^{5}
Isolan	8.0×10^{-6}	1.3	1.6×10^{5}

Adapted from Reference 34, with permission.

measured in many kinetic studies, is defined as $K_a = (k_{-1} + k_2)/k_{+1}$.[35] However, if $k_2 << k_{-1}$, then it is clear that K_a is essentially equal to k_{-1}/k_{+1} as defined above. It has also been shown that this relation holds under certain conditions if k_2 is approximately equal to k_{-1}.[35a] From the available data, then, it appears that carbamate-cholinesterase affinity constants can be compared from study to study with some degree of reliability.

As shown in Table 3.1, the chemical structure of the carbamate can drastically alter affinity and thus produce a wide range of K_a values. In general, the carbamate insecticides possess affinity constants of $10^{-4}M$ or lower, but there are exceptions. It is also apparent that differences in K_a values for N-methylcarbamates and N,N-dimethylcarbamates from the same phenolic parent were not found to be very large, suggesting that the change from a monomethyl to a dimethyl group had a limited influence on the rate of enzyme-carbamate complex formation. Thus, differences in affinity were primarily a result of the molecular makeup of the aromatic portion of the insecticide and how well it fit the active center of acetylcholinesterase (this will be examined more carefully in the next sections of this chapter and in Chapter 4). By way of comparison, acetylcholine, the normal substrate for AChE, has a dissociation constant for bovine erythrocyte acetylcholinesterase of about 2×10^{-5}.[36]

The velocity constant k_2 (carbamylation constant), which represents the rate of formation of carbamylated acetylcholinesterase from the insecticide-esterase complex, differs by only one order of magnitude for all of the chemicals listed in Table 3.1. Thus, it appeared that the carbamate's structure played a minor role in regulating carbamylation of AChE. However, more recent work with bovine erythrocyte AChE and with other aryl N-methylcarbamates and N,N-dimethylcarbamates showed that k_2 values varied from 81 min^{-1} for UC 10854 to 0.14 min^{-1} for 4-nitrophenyl N,N-dimethylcarbamate.[36]

There was a slight but consistent indication in both of these studies[34,36] that methylcarbamates had higher carbamylation constants than did the corresponding dimethyl analogs. Once again for comparative purposes, the k_2 for acetylation of AChE from bovine erythrocytes by acetylcholine has been calculated to be 2.95×10^5 min^{-1}, and even higher values have been reported for electric eel AChE.[14] Such tremendous differences in k_2 separate a good substrate from a poor substrate or inhibitor.

Finally, the bimolecular reaction constant, k_i, which describes the overall potential of a compound to inhibit acetylcholinesterase, is defined by the ratio shown below:

$$k_i = \frac{k_2}{K_a}$$

This is simply the carbamylation ability of the insecticide divided by its affinity for AChE. Therefore, if two carbamates have identical k_2 values, and many of them are very similar, then the compound with the better enzyme affinity (smaller K_a) will have the larger k_i and will be most likely to carbamylate the enzyme. In the case of ACh, the bimolecular reaction constant has been reported as 1.5×10^9 for bovine erythrocyte acetylcholinesterase.[14] Bimolecular reaction constants for carbamate insecticides are considerably lower than this, as shown in Table 3.1.

The last step in the overall reaction of a carbamate insecticide with acetylcholinesterase, the decarbamylation of the enzyme, is defined by the velocity constant k_3 (see Figure 3.5). As discussed above, k_3 is identical for all methylcarbamates and is very similar to the k_3 reported for dimethylcarbamates. Therefore, assuming that a carbamate insecticide has reached the nerve synapse, its effectiveness as a poison depends almost entirely on its k_i value. However, k_3 is important when comparing carbamates to phosphates or acetylcholine. Obviously, k_3 must be extremely high for ACh, so that none of the enzyme is tied up by acetylation during nerve transmission. With carbamates, k_3 is about 0.02 min^{-1}, which allows for recovery of enzyme with time, but k_3 values for phosphorylated AChE can be much lower, giving more "permanent" inhibition.

The preceding brief description of the kinetics of cholinesterase inhibition by carbamate chemicals is believed to account for the mode of action of this group of insecticides. The assumption is made that the insecticide has surmounted the problems of penetration, dispersal, and metabolism to arrive at the nerve synapse in sufficient quantity to block nerve transmission. More detailed information on the interactions of carbamate insecticides with the cholinesterases of insects and other animals is presented on the following pages.

Kinetics of Animal Cholinesterase Inhibition

Interest in the kinetics associated with the inhibition of cholinesterase by carbamates began with the realization that physostigmine was a potent anticholinesterase agent.[37,38] But it was not until several years later that a series of papers appeared concerning the kinetic interactions of physostigmine, carbamyl cholines, carbamyl fluorides, and other drug-related carbamates with various esterases.[17,18,20-26,39-43] The first study that reported the reaction of insecticidally active carbamates with esterases was conducted in 1960.[44] Since that time, many more carbamate insecticides have been used for elucidation of the mechanism of cholinesterase inhibition.

The following review will consider primarily results obtained from more recent investigations using carbamate insecticides (and closely related analogs) in reaction with acetylcholinesterase. Most of these studies have utilized either bovine erythrocytes or the electric organ of the electric eel as the enzyme source. These preparations are rich in acetylcholinesterase and are commercially available in fairly pure form. However, reaction conditions, such as temperature, pH, concentration of reactants, incubation time, and method of enzyme analysis, varied from laboratory to laboratory, so that sometimes rather gross discrepancies resulted. In addition, the mathematical and graphical calculation of equilibrium and velocity constants also differed from one study to the next. Nevertheless, the limited amount of data describing the hydrolysis of carbamates by acetylcholinesterase has contributed very significantly to our understanding of the kinetics of carbamate insecticide inhibition of this critical enzyme.

Correlation of anticholinesterase activity of a particular carbamate with toxicity to mammals is generally good, although there are other limiting factors that enter the picture (see Chapter 5). However, for the present time, only those parameters that influence the reaction of the carbamate with the enzyme will be considered. A compilation of most of the published reaction constants for aryl N-methyl- and N,N-dimethylcarbamates with bovine erythrocyte and electric eel AChE is presented in Table 3.2.

Omitted from the tabulation are aryl carbamates with a formal positive charge, since these materials are not useful as insecticides. Generally, they have been characterized by better affinity (lower K_a value) and by better carbamylation ability (higher k_2 value) than their corresponding uncharged isosteres. For example, the ratios of K_a for 3-*t*-butyl-, 3-isopropyl-, and 3-dimethylaminophenyl N-methylcarbamate to the K_a for 3-trimethylaminophenyl N-methylcarbamate (methylneostigmine) were shown to be 15, 18, and 214 respectively.[45] The increased affinity by the quaternary compounds is presumably due to an enhanced binding energy resulting from coulombic attraction between the charged nitrogen and the anionic site of the enzyme.

A much less dramatic change was apparent when the rate of carbamylation of AChE by methylneostigmine was compared to k_2 for the corresponding uncharged analogs. However, in the dimethylcarbamate series of compounds, neostigmine (3-trimethylaminophenyl N,N-dimethylcarbamate) carbamylated at a rate about 50 times faster than did analogous uncharged chemicals. It is likely, therefore, that a formal positive charge on the molecule not only assists in binding to AChE, but also aids in carbamylation of the enzyme.

Drawing correlations between N-methyl- and N,N-dimethylcarbamates synthesized from identical phenols, be they charged or uncharged, led to the conclusion that N-methylcarbamates are more effective inhibitors of acetylcholinesterase. In every case, bimolecular reaction constants (k_i) for the monomethyl analogs were higher, albeit the degree of difference varied with structure of the phenol (Table 3.2). The reason for the increased k_i is primarily due to a higher carbamylation rate by the methylcarbamates, since the effect of the number of methyl groups on affinity, K_a, was mixed. Thus, the greater inhibitory power of methylcarbamates appears to be due principally to an enhanced ability to carbamylate the enzyme.

Within the methyl or dimethyl series of compounds, altering the position and type of substituent on the aromatic portion of the molecule also changed inhibition characteristics. Thus, K_a values for aryldimethylcarbamates varied by almost 10^4 from 0.008 mM for Isolan to 73 mM for 4-fluorophenyl dimethylcarbamate (Table 3.2). In the case of monomethyl compounds, the K_a for unsubstituted phenyl N-methylcarbamate was around 25 mM in most studies. By adding substituents to the ring, the affinity constant was reduced to as low as 0.0034 mM for HRS-1422, which again represents a range of about 10^4.

TABLE 3.2

Reaction Constants for Carbamates and Acetylcholinesterase

Compound	AChE source,[a] °C, pH	K_a (mM)	k_2 (min^{-1})	k_i (mM^{-1} min^{-1})	Reference
Aryl N,N-dimethylcarbamates					
3-Trifluorophenyl	BE, 38, 7.2	0.44	0.63	1.43	34
4-Chlorophenyl	BE, 38, 7.2	1.9	0.60	0.32	34
4-Fluorophenyl	BE, 38, 7.2	73	2.0	0.03	34
4-Methylphenyl	BE, 38, 7.2	6.4	0.51	0.08	34
4-Methoxyphenyl	BE, 38, 7.2	0.36	0.13	0.35	34
1-Naphthyl	BE, 38, 7.2	0.093	2.4	26	34
3,5-Diisopropylphenyl	BE, 38, 7.2	0.058	0.75	13	34
	BE, 25, 8.0	–	–	2.9	14
4-Nitrophenyl	BE, 38, 7.2	0.32	0.25	0.75	34
	BE, 25, 7.6	3.2	0.14	0.045	36
1-Nitrophenyl	BE, 25, 7.6	1.4	7.3	5.1	36
3-Nitrophenyl	BE, 25, 7.6	0.92	0.15	0.17	36
3-Isopropylphenyl	BE, 25, 7.6	2.5	6.3	2.6	36
	EE, 25, 7.0	0.73	2.5	3.6	45
Phenyl	BE, 25, 7.6	10	0.24	0.023	36
	BE, 38, 7.2	1.5	0.22	0.15	34
	BE, 25, 7.0	17	0.38	0.024	45
	EE, 25, 7.0	8.9	0.37	0.039	45
3-Dimethylaminophenyl	EE, 25, 7.0	0.52	1.5	3.2	45
Isolan	BE, 38, 7.2	0.008	1.3	164	34
Aryl N-methylcarbamates					
3-Trifluorophenyl	BE, 38, 7.2	1.3	3.0	2.3	34
4-Chlorophenyl	BE, 38, 7.2	3.3	1.3	0.41	34
4-Fluorophenyl	BE, 38, 7.2	18	1.9	0.10	34
4-Methylphenyl	BE, 38, 7.2	1.3	1.5	1.2	34
4-Methoxyphenyl	BE, 38, 7.2	2.7	1.4	0.50	34
3-Nitrophenyl	BE, 25, 7.6	0.68	0.45	0.66	36
3-Dimethylaminophenyl	EE, 25, 7.0	1.4	48	34	45
3-*t*-Butylphenyl	BE, 25, 7.0	0.055	21	370	45
	BE, 25, 7.6	0.067	23	350	36
	EE, 25, 7.0	0.097	64	660	45

TABLE 3.2 (continued)

Reaction Constants for Carbamates and Acetylcholinesterase

Compound	AChE source,[a] °C, pH	K_a (mM)	k_2 (min^{-1})	k_i (mM^{-1} min^{-1})	Reference
Aryl N-methylcarbamates (cont.)					
Phenyl	BE, 38, 7.2	2.9	1.6	0.54	34
	BE, 25, 7.6	24	6.8	0.29	36
	BE, 25, 7.6	26	7.1	0.27	122
	BE, 25, 7.0	24	8.1	0.26	45
	EE, 25, 7.0	23	9.1	0.37	45
	BE, 25, 7.4	–	–	0.27	30
HRS 1422	BE, 38, 7.2	0.0034	1.4	407	34
Baygon	BE, 25, 7.4	–	–	40	30
	BE, 25, 8.0	–	–	49	14
	BE, 38, 7.2	0.0099	1.1	110	34
Carbofuran	BE, 25, 7.8	0.21	22	100	69
Mesurol	BE, 38, 7.2	0.0067	1.2	186	34
Carbaryl	BE, 25, 7.4	–	–	18	30
	BE, 25, 7.6	>0.5	>20	22	36
	BE, 38, 7.2	0.011	1.3	125	34
	BE, 37, 7.4	–	–	47	28
	BE, 25, 7.4	–	–	21	28
UC 10854	BE, 25, 7.0	0.16	76	480	45
	BE, 25, 7.6	0.18	81	460	36
	EE, 25, 7.0	0.11	97	840	45
	BE, 25, 8.0	–	–	540	14
O-methylcarbamoyl oximes					
Aldicarb	BE, 25, 7.6	10	150	16	36
Aldicarb sulfone	BE, 25, 7.6	17	74	4.4	36

[a]BE = bovine erythrocytes; EE = electric eel.

Less pronounced effects were found for k_2, ranging from 0.13 to 7.3 min^{-1} for dimethylcarbamates and from 0.45 to 97 min^{-1} for methylcarbamates (Table 3.2). Putting affinity and carbamylation factors together yielded k_i values from 0.023 to 26 mM^{-1} min^{-1} for all dimethylcarbamates except Isolan. None of these compounds have been used as insecticides, but Isolan, having a k_i of 164 mM^{-1} min^{-1}, has enjoyed some success as an insect toxin. Methylcarbamate bimolecular reaction constants ranged from 0.1 to 840 mM^{-1} min^{-1}, with insecticidal structures generally possessing k_i values above 50. However, this is not meant to suggest that all that is needed for insecticidal success by a methylcarbamate is a k_i value above 50. Obviously, many other conditions must be met.

Perhaps the most important fact to extract from the data in Table 3.2 is that both carbamylation (k_2) and affinity (K_a) for acetylcholinesterase contribute to inhibition of AChE by carbamate insecticides. For example, the affinity of compounds such as Baygon and HRS-1422 for the active center of AChE approaches that of acetylcholine itself, but their k_2 values are not appreciably great. On the other hand, the oxime carbamates, and probably carbaryl as well, owe their anticholinesterase activity primarily to a very good carbamylation efficiency. Finally, insecticides UC 10854 and carbofuran have passable affinity and carbamylation constants, which, acting together, produce good cholinesterase inhibition. Results such as these negate the often-stated phrase that the major factor to consider in the design of carbamate insecticides is the fit on the enzyme surface.[34,46]

There are variables that must be weighed when comparing the numbers in Table 3.2. Obviously, enzyme source, pH, and temperature are three of these variables. Consider first the results obtained from bovine erythrocyte acetylcholinesterase and electric eel acetylcholinesterase under similar reaction conditions in the same laboratory, using the same carbamate.[36,45] The two enzymes are generally considered to be very similar to each other and to the AChE in nerve synapses.

From Table 3.3 it can be seen that K_a and k_2 values found for individual methyl- and dimethylcarbamates did not differ by more than threefold for the two enzymes. Even more noteworthy, the k_i, representing the overall inhibition, always varied by less than twofold for all of the inhibitors tested. Considering that the range of K_a and k_i for the carbamates used was several hundredfold, the differences between bovine erythrocyte and eel AChE inhibition by the same compound appear to be sufficiently small to assume that the binding and carbamylation sites on the two enzymes are virtually identical, especially after recalling that the enzymes originated from different tissues of different animals.

Although the effect of temperature has not been evaluated extensively, it may be a crucial factor in determining the degree of inhibition. As shown in Table 3.2 for carbaryl, the k_i for bovine erythrocyte AChE at pH 7.4 increased from 21 mM^{-1} min^{-1} at 25°C to 47 mM^{-1} min^{-1} at 37°C.[28] Temperature differences could also explain some of the disparity in the k_i values reported from different studies with the same carbamate, e.g., 4-nitrophenyl N,N-dimethylcarbamate, phenyl N,N-dimethylcarbamate, phenyl N-methylcarbamate, and Baygon. It is worth noting that in all of these instances an increase in temperature from 25°C to 37 or 38°C always yielded an increase in k_i value.

Raising the temperature also results in an accelerated rate of decarbamylation (k_3) of acetylcholinesterase.[12] In fact, the activation energy for regeneration of enzyme was found to be 10 to 20 kcal/mol, which is similar to that reported for some organophosphorus compounds. Such dependence on temperature is not surprising, however, since the rate of most enzyme–substrate reactions varies with temperature. Until more is learned about this interaction, it would be well for researchers to remember that the natural environment for erythrocyte AChE includes a temperature regime near 38°C, whereas that of the eel may be closer to 25°C.

The pH of the reaction medium seems to be less important, as judged by differences produced with the same substrate, enzyme, and temperature, but with different pH values (for example, phenyl N-methyl- and N,N-dimethylcarbamate, 3-*t*-butylphenyl N-methylcarbamate, Baygon, and UC 10854). The reason for this becomes apparent from the results shown in Figure 3.6. With Baygon and with 3-isopropylphenyl N,N-dimethylcarbamate, there is very little change in k_i from pH 5.5 to pH 8.5; and in the region from pH 7.0 to pH 8.0, where all of the experiments in Table 3.2 fall, even UC 10854 yielded little k_i variation.[14] The part of the curve beyond pH 8.0 for each substrate is, for all purposes, identical with a curve calculated from a reactive center with a dissociation constant pK equal to 10.25. The same response was found with charged carbamates, such

TABLE 3.3

Comparison of Reaction Constants for Bovine Erythrocyte and Eel Acetylcholinesterase[36,45]

Carbamate	K_a eel/ K_a bov. ery.	k_2 eel/ k_2 bov. ery.	k_i eel/ k_i bov. ery.
3-*t*-Butylphenyl N-methyl	1.8	3.0	1.8
3-Isopropylphenyl N-methyl	0.7	1.3	1.7
3-Isopropylphenyl N,N-dimethyl	0.3	0.4	1.4
Phenyl N-methyl	0.9	1.1	1.4
Phenyl N,N-dimethyl	0.5	1.0	1.7

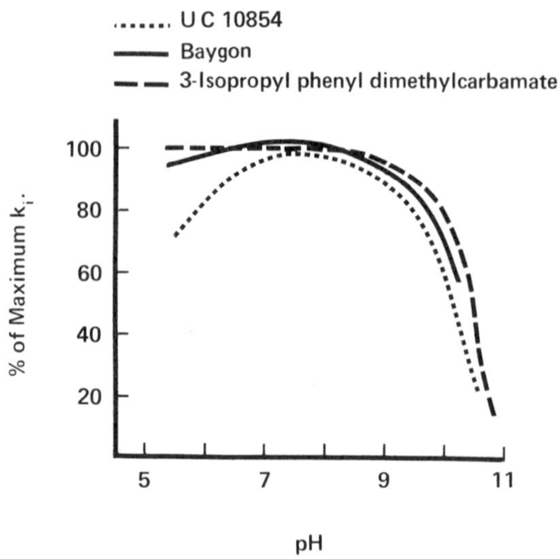

FIGURE 3.6. Effect of pH on inhibition of bovine erythrocyte acetylcholinesterase by carbamates. (Adapted from Reference 14, with permission.)

as neostigmine, but in addition, k_i was also depressed in the acidic range, so that a smooth bell-shaped curve resulted which fit a theoretical plot based on pK groups of 6.2 and 10.25. This suggests that pK 6.2 is derived from a site on acetylcholinesterase which is required for binding of quaternary nitrogen, but which is much less involved in binding of uncharged carbamates.

Hydrolysis of methyl- or dimethylcarbamyl acetylcholinesterase also depends on pH. Values of k_3 for carbamates and bovine erythrocyte AChE were shown to maximize between pH 8 and 9.[14,47] An explanation for this rather sharp response to hydrogen ion concentration is that changes in pH influence the ionization of areas near the catalytic center of acetylcholinesterase, which in turn alters the reactivity of the enzyme–carbamate bond.

One other parameter that could contribute significantly to the value of reaction constants is the presence or absence of ions. Reports to date deal almost exclusively with effects on k_3 and include many contradictions. For example, one experiment[47] showed that various inorganic salts had no effect on the reactivation of bovine erythrocyte AChE which had been carbamylated with methyl- or dimethylcarbamates, whereas other investigators[31] reported that both NaCl and $MgCl_2$ accelerated decarbamylation of dimethylcarbamyl acetylcholinesterase four- to sixfold. A third study revealed that NaF, even at concentrations as high as $10^{-2} M$, had no effect on the rate of reactivation of dimethyl- or methylcarbamylated AChE from electric eel, although the salt did decrease the rate of carbamylation by N-methyl- and N,N-dimethylcarbamyl choline and N,N-dimethylcarbamyl fluoride.[32]

Large organic cations influence spontaneous reactivation of AChE, although the results here also disagree. Tetraalkylammonium salts did not alter the k_3 for erythrocyte AChE and 2-nitrophenyl N,N-dimethylcarbamate, except for the tetraethyl analog, which caused a fourfold increase in the decarbamylation rate.[31] However, in another case, tetramethylammonium and pentyltrimethylammonium ions inhibited dimethylcarbamylation,[48] although the reaction conditions, particularly with respect to the presence of inorganic ions, differed between the two experiments. Presumably, these large ions either interfere with the reaction of water and carbamylated enzyme, or they cause unfavorable geometrical changes of the carbamate-enzyme complex, so that decarbamylation is inhibited.

Reaction constants for carbamates with other cholinesterases, notably human and horse serum cholinesterase, have been determined for only a few compounds. The results point out kinetic differences between pseudocholinesterase and true cholinesterase (AChE). For example, the rate of reactivation of carbamylated serum ChE was about one order of magnitude slower than the k_3 for AChE.[49] Also, the marked effect of pH on the k_3 for acetylcholinesterase is not apparent with human serum cholinesterase, the latter appearing to be independent of pH.[12] However, the k_3 for both enzymes increased with increased temperatures. Bimolecular reaction constants were also three- to tenfold lower for the interaction of Baygon, UC 10854, and Mesurol with serum ChE than for that with AChE.[49,50]

It has also been shown that inhibition of serum cholinesterases responds quite differently to pH changes than does AChE. With human serum, plots of k_i against pH are similar for charged and uncharged compounds and for methyl and dimethyl compounds.[12,49] The shape of the curve, with a maximum value near 9.5, follows a theoretical calculation consistent with a change in ionization of a single group of pK 7.5 on the enzyme surface. This, of course, is not in agreement with the pH response of AChE described above. There were also differences in some of the kinetic values for human serum and horse serum cholinesterase.[49]

From the above brief description of reaction kinetics it can be seen that our knowledge of carbamate insecticide interaction with acetylcholinesterase is more than rudimentary. However, a comparison of constants obtained for the same carbamate compound in different experiments is often disappointing. Obviously, variation of in-vitro incubation medium and conditions played a part in the production of incongruous results. Perhaps the most significant variable has been the "condition" of the enzyme. Many workers purchased an erythrocyte or eel preparation commercially and used it either with or without further purification. Others isolated the cholinesterase from a crude homogenate. In any event, it seems that some standardization of the enzyme source and purity is needed, so that those working on carbamate–cholinesterase kinetics can successfully compare their results.

The questions which enzyme to use and how pure should it be are difficult to answer. There are those who strive for the isolation and use of only crystalline AChE, which certainly would produce the best biochemical kinetics. Others believe that a crystalline enzyme is too far from AChE as it exists in the living animal. For example, there is evidence that acetylcholinesterase is membrane-bound in vivo. An attempt at simulating this condition in vitro included the use of AChE covalently bonded to sepharose.[51] It was found that reaction kinetics were quite different under these circumstances when compared to free AChE. Even if no membrane were involved, the in-vivo existence of acetylcholinesterase often occurs in the presence of various proportions of other cholinesterases. Perhaps the most important of these are isoacetylcholinesterases which may vary significantly in response to substrates.[35a,45] Thus, work with a crude homogenate may more closely mimic reality.

Another aspect deserving consideration is the choice of enzyme used in kinetic studies. The presumption that erythrocyte and/or eel AChE is essentially identical to nerve AChE is based on research with drugs and organophosphorus insecticides and has not been confirmed by comparative studies with carbamate insecticide structures. Certainly, there is ample evidence to show that carbamates inhibit synaptic AChE, but there are not enough data to show that the kinetics of inhibition are of the same order with nerve AChE and erythrocyte AChE. The use of brain, spinal cord, or even nerves outside of the CNS would not only assist in our understanding of the kinetics of acetylcholinesterase inhibition by carbamate insecticides, but it could provide valuable clues for the design of more selective carbamate structures.

Kinetics of Insect Cholinesterase Inhibition

A considerable bank of knowledge exists on the anatomy, physiology, and biochemistry of the insect nervous system.[52–58] There is no doubt that acetylcholine is present in insect nerves as well as in various other tissues. In addition, the accompanying enzyme, acetylcholinesterase, has been demonstrated to be present in numerous insect species, particularly in the nervous system and, more precisely, in the neuropil close to synaptic membranes. Although there is fragmentary evidence which does not entirely support the role of acetylcholine as the synaptic transmitter (most notably the relative insensitivity to acetylcholine of insect nerves with the sheaths removed), the majority of results point toward a viable ACh-AChE system.

In a study of the kinetics of carbamate–acetylcholinesterase interactions it is presumed that the mode of action is inhibition of AChE within the central nervous system of insects. Although considerable circumstantial evidence supports this plea, there also exists a growing number of observations which suggest that other modes are involved or that variations may be possible. Thus, before considering kinetics, it seems appropriate to bring some of these studies to the attention of the reader.

One report which is invariably cited in any review of carbamate mode of action suggested that Zectran killed certain lepidopterous larvae by means other than cholinesterase inhibition.[59] The reason for this conclusion was a marked darkening and shriveling of Zectran-killed insects, a phenomenon that did not occur with carbaryl-treated larvae. Also, the insecticide was very rapid in action; all insects were dead within one half hour after topical application. No attempt has ever been made to rationalize these results, but it is unlikely that such rapid death would occur unless the nervous system was affected by the carbamate.

An often neglected "alternate-mode" hypothesis originated from studies which demonstrated that the neural synapses of the insect heart are cholinergic.[60] When American cockroach hearts were exposed to very low concentrations (5 X

$10^{-7}M$) of UC 10854 or 2-isopropylthiophenyl N-methylcarbamate, heartbeat rate was substantially increased. This was often followed by erratic beating with incomplete systolic and diastolic movements and eventual cessation of activity. In this respect, the carbamates appeared to direct their effects on the receptor protein of the heart rather than on AChE, causing a build-up of acetylcholine and failure of the heart. However, the function of the insect heart in vivo is not nearly as critical as it is in vertebrates, since much of the insect's circulation is dependent on body movement. Thus, a relatively long-term interruption of heartbeat could possibly be tolerated by an insect without any ill effect.

Many investigators have questioned whether cholinesterase inhibition is the true mode because they have been unable to show correlation between in-vivo AChE inhibition by carbamates or organophosphates and death of insects. It was believed at one time that death should be concomitant with total generalized inhibition of acetylcholinesterase in the entire insect nervous system. However, autopsies of fatally poisoned cockroaches revealed that, although the AChE of thoracic ganglia was mostly inhibited, abdominal ganglionic synapses were normal.[61] Also, decapitated houseflies (which can live for a considerable time) were killed by the same level of insecticide as were normal insects, suggesting that inhibition of brain acetylcholinesterase was not necessary for death.[62] Thus, total-body analysis of either of these insects at the time of death would have revealed far less than total AChE inhibition.

Such localization of inhibition within the thorax of poisoned insects has been confirmed by more recent histochemical studies, where cholinesterase inhibition in the thoracic ganglia of the housefly was demonstrated to be much more extensive than inhibition in the brain.[58,63] However, with most insecticides, including the two carbamates UC 10854 and aldicarb, ChE inhibition in the ganglia was confined mostly to the peripheral regions, occurring to a lesser degree in the neuropil, where synaptic transmission takes place. This agrees with several other reports, in which the link between organophosphorus insecticide toxicity and thoracic neuropil AChE inhibition was very weak.[64-66] Thus, it appears that cholinesterase inhibition in areas outside the synapse may be most critical.

The function of the "peripheral" cholinesterase is unknown, but it could serve to hydrolyze endogenous ACh, which might otherwise enter synapses within the neuropil. Presumably, flooding the ganglia with a carbamate would not only result in inhibition of synaptic AChE, but inactivation of the peripheral enzyme would permit an influx of acetylcholine, thus further antagonizing the lethal effects of the insecticide. It is also possible that cholinergic synapses exist in the cell-body area of the ganglion as well as within the neuropil, but then the mode of action would still be acetylcholinesterase inhibition.

One final example concerns the nature of cholinesterase in the nerve synapses. Aldicarb and its sulfoxide analogs were discovered to be nontoxic to certain lepidopterous larvae, possibly because of their water-soluble properties, which hindered penetration into the CNS of these particular insects.[67] However, there were some indications that the ChE present in the head and body of the insects was not acetylcholinesterase. Confirmation of this finding awaits further study.

Returning to the assumption that acetylcholinesterase inhibition is the principal mode of action, there is great difficulty in comparing kinetic constants from one experiment with those from another. Perhaps the main reason for this is the diversity in the enzyme preparations used for determination of the constants. Table 3.4 presents a list of kinetic information obtained from four different laboratories. Two of these utilized partially pure insect-head AChE,[68,69] another used whole-head homogenates containing mixed acetylcholinesterase and pseudocholinesterase,[70] and a fourth used homogenates of whole spider mites.[71] These reports are tabulated despite such differences, because they are among the few that have included kinetic constants other than k_i.

Even with variations in the nature of the enzyme, some general trends are evident. As expected, the insecticidally active structures elicited higher k_i values than did the unsubstituted phenyl N-methyl- and N,N-dimethylcarbamates. This was a result of better affinity for the enzyme (Isolan), better carbamylation ability (carbofuran), or a combination of both (carbaryl). Affinity and carbamylation constants varied by about 100-fold for this very limited group of compounds. Thus, just as with bovine and eel acetylcholinesterase, both carbamylation rate and affinity contribute to the overall insect anticholinesterase activity of carbamate chemicals.

TABLE 3.4

Kinetic Constants for Reaction of Carbamates with Insect Cholinesterases

Compound	Enzyme[a] °C, pH	K_a (mM)	k_2 (min^{-1})	k_i (mM^{-1} min^{-1})	k_3 (min^{-1})	Reference
Phenyl N,N-dimethylcarbamate	HF, 35, 7.5	0.0033	0.50	150	0.028	68
Isolan	HF, 35, 7.5	0.00025	0.60	2400	0.028	68
Phenyl N-methylcarbamate	HF, 35, 7.5	0.0067	0.28	42	0.048	68
Carbofuran	HF, 25, 7.8	0.009	10	1100	–	69
	HB, 25, 7.8	0.03	28	930	–	69
Carbaryl	HF, 35, 7.5	0.0005	0.80	1600	0.048	68
	HF, 30, 8.0	–	–	740	0.029	70
	HB, 30, 8.0	–	–	4000	0.027	70
N-Hydroxy Baygon	SM, 25, 7.5	0.018	0.80	43	0.3	71

[a]Enzyme sources: HF = housefly heads; SM = whole spider mites; HB = honeybee heads.

The rate (k_3) reported for regeneration of methylcarbamylated cholinesterase from housefly heads was considerably higher for partially pure preparations[68] than for crude homogenates[70] (Table 3.4). However, temperature and pH were not the same in the two studies. With the whole-head homogenates, indications were that the k_3 was identical for acetylcholinesterase and pseudocholinesterase, and that housefly and honeybee breis had similar k_3 values. Reactivation of methylcarbamyl acetylcholinesterase (0.048 min^{-1}) was faster than that of the dimethylcarbamyl enzyme (0.028 min^{-1}) under identical in-vitro conditions, although this was not substantiated by in-vivo studies.[68] Results of the same order were obtained in another in-vitro study with housefly head AChE, where the reactivation rate for the dimethyl-esterase was 0.024 min^{-1} and for the methyl-esterase 0.04 min^{-1}.[47] The effect of altering the carbamate group by placing a hydroxyl group on the nitrogen resulted in an almost tenfold increase in k_3, as shown for N-hydroxy Baygon and mite cholinesterase.[71]

The effect of temperature and pH on reaction constants of carbamates and insect cholinesterases has only been touched upon. A slightly higher pH optimum (8.3 to 8.8) was found for regeneration of fly head dimethylcarbamyl acetylcholinesterase than for methylcarbamyl acetylcholinesterase (8.0 to 8.5).[47,70] In addition, higher temperatures appeared to yield higher regeneration rates.[68]

A value not shown in Table 3.4, but often used in conjunction with insect cholinesterase studies, is the I_{50} value.[6] This is related to k_i by the following equation:

$$I_{50} = \frac{0.695}{k_i\, t}$$

Kinetically, I_{50} is the molar concentration of inhibitor needed to reduce the rate of acetylcholine hydrolysis by 50% of the normal level. It has also been described as the amount of inhibitor necessary to produce 50% inhibition of cholinesterase. I_{50} values have been determined for more than 1,000 carbamate chemicals, mostly with housefly head AChE as the esterase. One of the major problems when comparing results from paper to paper is the presence of the variable t (time) in the equation. When this is held constant, I_{50} is satisfactorily reproducible from experiment to experiment and from laboratory to laboratory.[72] Thus, it is I_{50} that is calculated in most structure–activity studies (cf Chapter 4).

A listing of I_{50} values for a few carbamate insecticides is given in Table 3.5.[72] A decrease of at least two orders of magnitude separates the I_{50} values of insecticides from the I_{50} value of unsubstituted phenyl N-methylcarbamate. Differences of similar magnitude have been found between monomethyl- and dimethylcarbamates of identical phenols. However, certain carbamates

TABLE 3.5

Values of I_{50} for Various Methylcarbamate Insecticides and Housefly Acetylcholinesterase[72]

Compound	I_{50} (M)
Phenyl N-methylcarbamate	2.0×10^{-4}
Carbaryl	9.0×10^{-7}
UC 10854	3.4×10^{-7}
HRS-1422	3.3×10^{-8}
Baygon	6.7×10^{-7}
Mesurol	1.2×10^{-6}
Zectran	3.3×10^{-6}
Carbofuran	2.5×10^{-7}

possessing positively charged groupings yielded I_{50} values about two orders of magnitude lower than corresponding uncharged analogs, thus producing some of the most potent anticholinesterases. Yet these materials, as well as the drug-related carbamates, are ineffective insect toxicants. The reason for their lack of activity against living insects is physiological. As indicated previously, the central nervous system of insects containing cholinergic synapses is surrounded by a fibrous sheath, the neural lamella.[73,74] It is this protective lipid barrier, together with the lack of peripheral cholinergic synapses, that is primarily responsible for the total ineffectiveness of charged carbamates as insecticides.[74-81] Of course, it is likely that the sheath can also have some effect on the penetration of uncharged, more polar carbamates. Such an effect could partially explain differences in toxicity and differences in in-vivo cholinesterase inhibition found at various portions of the CNS.

Comparison of insect AChE with that from bovine erythrocytes and electric eel tissues involves an evaluation of substrate specificity and inhibition kinetics. Because of the variability associated with the determination of reaction constants for insects and other animals, comparisons are not too meaningful. The most useful information arises from self-contained comparative studies, but results are confusing. For example, it has been reported that the substrate specificity of acetylcholinesterase from 23 insect species toward various choline analogs was of the same order as that for bovine erythrocyte or electric eel AChE.[46] On the other hand, studies have indicated that there is a definite difference in the response of mammalian and insect AChE to the substrates acetylcholine and butyrylcholine.

It has been known for a long time that acetylcholinesterase from the electric eel and from mammalian erythrocytes and brain tissue hydrolyzes acetylcholine, but not butyrylcholine.[82,83] The finding that honeybee and housefly head breis hydrolyzed both substrates was at first thought to be due to the presence of at least two enzymes, each responsible for hydrolysis of one of the cholines.[84] However, later work provided support for the theory that a single AChE in fly brain and thoracic nerve tissue was able to hydrolyze both acetylcholine and butyrylcholine.[66,85,86] Thus, it appears that housefly acetylcholinesterase is less specific than erythrocyte or eel AChE.

Other differences become apparent when one examines results for carbamate inhibitors. Generally, comparison of the kinetic constants in Table 3.2 with those in Table 3.4 shows that carbamylation rates (k_2) are lower for bovine erythrocyte AChE, but affinity constants are lower for housefly head AChE. The combined effect usually gives bimolecular reaction constants (k_i) indicating higher anticholinesterase activity toward insect acetylcholinesterase. A specific example of this relationship is given in Table 3.6 for carbofuran.[69] As seen, affinity between the insecticide and esterase is much greater in the insects, so that, with similar k_2 values, a pronounced difference in k_i is realized. Similar but sometimes less dramatic differences were found for numerous other heterocyclic methylcarbamates[69a] Interestingly, when the carbamate N-alkyl group was changed to propyl or butyl, these correlations were not as good.[69b]

Regeneration of free enzyme from methyl- and dimethylcarbamylated housefly acetylcholinesterase occurred twice as fast as reactivation of carbamylated bovine erythrocyte AChE.[47] With substituted phenyl biscarbamates or other N-alkyl carbamates, the reverse was true; that is, inhibited erythrocyte acetylcholinesterase recovered much faster than inhibited insect AChE.[69b,87] There were also marked deviations in specificity of the esterases from honeybee, housefly, and erythrocyte toward the biscarbamates.[87] Effects of pH on decarbamylation also varied with enzyme; the pH maximum for removing the methyl- or dimethylcarbamyl group from erythrocyte AChE was 8.6 to 8.9, but for housefly preparations, the pH maximum was 8.0 to 8.5 and 8.3 to 8.8 for methylcarbamyl and dimethylcarbamyl groups respectively.[47] However, activation energies of 27

TABLE 3.6

Kinetic Constants for Inhibition of Bovine Erythrocyte and Insect Acetylcholinesterase by Carbofuran[69]

Enzyme source	K_a (mM)	k_2 (min^{-1})	k_i (mM^{-1} min^{-1})
Bovine erythrocyte	0.21	22	100
Housefly head	0.009	10	1,100
Honeybee head	0.03	28	930

Adapted from Reference 69, with permission.

kcal/mole for fly and 18 kcal/mole for erythrocyte support carbamylation of both enzymes at a similar reactive center.

Further modifications prevail among acetylcholinesterase in various species of insects. As early as 1950 it was demonstrated that head AChE of honeybee and housefly responded differently to acetylcholine, acetyl-β-methylcholine, and an organophosphorus insecticide.[88] Substrate specificity was also shown to vary among brain acetylcholinesterase of American cockroaches, honeybees, and houseflies.[89] More recently, differences in the kinetics of acetylthiocholine reaction with AChE from housefly heads or cricket heads was clearly defined.[58] Response of insect AChE to carbamate insecticides also changes with species, and this has been suggested as a cause of selective toxicity. For instance, the k_i values for acetylcholinesterase inactivation in housefly heads and honeybee heads was 7.4×10^5 and 4.0×10^6 respectively.[70] This difference could partially explain the sensitivity of honeybees to carbaryl, whereas most houseflies are not killed by this carbamate.

It is difficult to draw any conclusions from the limited comparative kinetic data for insects, but it may be that some minor variations in the properties of nerve acetylcholinesterase among different species coincides with taxonomic order. For example, reaction constants of AChE from brains of Mediterranean fruitfly, oriental fruitfly, and melon fly with acetylcholine, acetylmethylcholine, and numerous organophosphorus and carbamate compounds were remarkably similar.[90] Also, the specificity of acetylcholinesterase for acetylcholine in the three species was very much like that reported for the housefly.[88]

Insect nervous tissue apparently contains several other esterases that react with carbamate compounds. One of these, an aliesterase, has been identified from houseflies and from fireflies.[91,92] This enzyme was inhibited by carbaryl and by Isolan in a manner analogous to acetylcholinesterase, although it appeared to be even more sensitive to carbaryl; it was not inhibited by eserine. The aliesterase is probably not universally present in all species, since very little was found in mosquitoes. There is also evidence for the presence of AChE isoenzymes in insect brains.[58,63]

It should be clear from the above discussion that the simple phrase, "the mode of action of carbamate insecticides in insects is acetylcholinesterase inhibition," is not necessarily a statement of fact. Histochemistry, enzymology, and reaction kinetics have provided considerable evidence for alternate or secondary modes that cannot be ignored. Indeed, the biochemical definition of insect acetylcholinesterase needs re-evaluation based on response of the enzyme(s) to choline analogs, carbamate insecticides, and organophosphorus compounds. Obviously, more research and less assumption is needed to unravel the interactions of carbamate insecticides with the vital processes of insects. Comparative studies among different insect species and between insects and other animals could prove to be extremely valuable in our understanding of the mode of action of carbamate toxicants in insects.

A few attempts have been made to incorporate the carbamate structure into compounds possessing a different mode of action. For example, various derivatives of insect juvenile hormones containing carbamate moieties have been synthesized, as shown in Figure 3.7.[93,94] Compound I, phenyl citronellylcarbamate, exhibited considerable juvenile hormone activity toward yellow mealworm pupae, and many other derivatives, such as compounds II and III, were also active. Although it might seem appropriate to utilize carbamates such as Baygon, UC 10854, or carbaryl

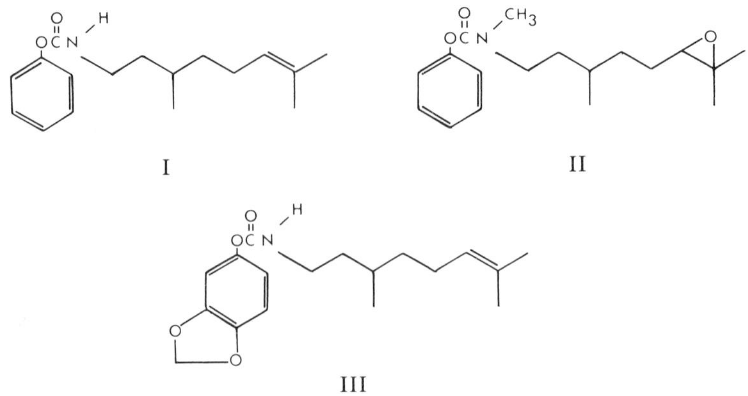

FIGURE 3.7 Structures of carbamate-insect juvenile hormone compounds.[93,94]

for the aryl portion of the hormone, this double-punch approach could negate the proclaimed unique insect specificity of this promising class of chemicals.

Two compounds that have shown significant chemosterilant activity toward houseflies bear a remarkable similarity to insecticidal carbamates[95] (Figure 3.8). Both the 4-chloro-3-methyl (I) and the 2-chloro-4,5-dimethyl (II) phenyl esters of aziridine-1-carboxylic acid were found to be relatively nontoxic to the flies, but the methylcarbamate counterparts were highly toxic. Presumably, in-vivo metabolism of the chemosterilants could yield active insecticide molecules that might restrict the usefulness of these compounds.

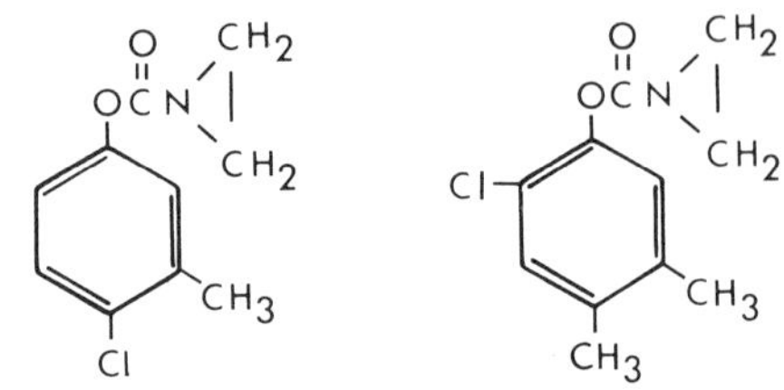

I II

FIGURE 3.8. Structures of insect chemosterilants.[95]

SIGNS AND SYMPTOMS OF POISONING

Mammals

Recognition of the symptoms associated with sublethal insecticide poisoning is particularly important for prevention of accidental contamination of humans with these chemicals. Because of obvious ethical problems, most research on symptomology is conducted with laboratory animals and only occasionally with human volunteers. For the most part, the cholinergic symptoms attributed to carbamate insecticides have been based on analogy with medicinal carbamates and with organophosphorus insecticides. More recent studies of detailed symptomology accompanying carbamate insecticide administration have confirmed the existence of typical anticholinesterase symptoms.

An animal receiving a toxic dose of an anticholinesterase agent responds by salivation, lacrimation (tearing), urination, defecation, miosis (constriction of pupils), muscle spasms, general muscular weakness, prostration, and convulsions.[6] The latter sometimes occur with stiff rigid outstretched limbs (tonic convulsions), but generally death is preceded by very rapid frenzied body movements (clonic convulsions). Respiratory failure and asphyxiation are the primary cause of death, but the precise mechanism for this process is uncertain. More probably than not it is a combination of bronchial constriction, lowered blood pressure, neuromuscular block of respiratory muscles, and failure of the respiratory center in the brain. Other possible causes include cardiac arrest and pulmonary edema.

The rapidity with which these symptoms appear and their severity are dependent on the amount of toxicant administered, chemical structure of the toxicant, mode of entry into the body, and biochemistry and physiology of the animal treated. An acute dose of carbamate insecticide will produce most of the symptoms listed. Exposure of humans to sublethal doses usually causes headache, salivation, nausea, vomiting, muscular weakness, and diarrhea. Recovery from these symptoms is often complete in only a matter of hours after initial carbamate contact.

Oral ingestion of a toxic quantity of an anticholinesterase insecticide is followed by a relatively slow appearance of symptoms. It is this method of treatment that is used to obtain the mammalian LD_{50} (lethal dose needed to kill 50% of the treated animals) of a compound. Chronic feeding of low levels of insecticide generally does not elicit cholinergic symptoms; these studies are conducted for an evaluation of other pharmacological and physiological responses. The oral LD_{50} of an insecticide is most representative of potential hazards from accidental ingestion by man or other animals (see Chapter 5).

Exposure of shaved animal skin to toxic doses of insecticides results in an even slower production of signs of poisoning. Determination of dermal LD_{50} values and the accompanying symptomology is necessary to ensure minimal risk in handling these chemicals. Perhaps the greatest potential for accidental poisoning by insecticides lies in dermal contact through manufacture, distribution, application, and harvest. Inhalation studies to simulate respiratory poisoning have not been extensive, partially because of procedural problems. However, results can be used as guidelines for potential hazards. The two techniques most often utilized for examination of symptomology are intravenous and intraperitoneal injection. The former places the toxicant directly into the bloodstream, resulting in rapid transport to target nerves and almost immediate symptoms. The other injects the compound into the abdominal cavity beyond the serous peritoneum membrane, which also results in a fast neurological response.

Neither of these techniques should be confused with subcutaneous injection, which places the chemical just under the skin. Signs of poisoning appear much more slowly with this method of administration. Of course, none of the injections are related to exposure from normal insecticide usage, but intravenous and intraperitoneal treatment moves the poison to the site of action with a minimum of metabolism and dilution, so that symptoms can be readily observed.

Up until the last few years, most toxicity determinations were obtained with chemically pure insecticides. However, recent efforts have expanded to include examination of oral and dermal LD_{50} values for formulated compounds. In some cases, the adjuvants appear to alter intoxication by the insecticide. It is probable that this results from changes in penetration of the compound through the skin or gut lining. Whether or not formulation can also alter poisoning symptoms has not been thoroughly researched.

The goal of many symptom-toxicity investigations is to compare measurable physiological changes with the degree of poisoning. If the physiological response is visible, and if it occurs a considerable length of time prior to intake of a toxic dose, then antidotal or other measures can be taken. For cases where chronic exposure to low levels of toxicant cannot be avoided, it is necessary to have a method for assessing overexposure that is sufficiently reliable to prevent sickness to those working with the chemicals.

For carbamate insecticides, poisoning is frequently assessed by measuring the degree of cholinesterase inhibition in blood plasma and in erythrocytes. With laboratory animals, brain cholinesterases and other tissue cholinesterase levels may also be obtained. Fortunately, changes in erythrocyte AChE are usually noticeable following exposure to levels of carbamate far below the toxic dose.

A variety of methods are employed for cholinesterase determinations with various advantages and disadvantages.[96] Results with organophosphorus insecticides are fairly reliable because of the "irreversible" nature of cholinesterase inhibition. In the case of carbamates, however, the ease with which carbamylated cholinesterases spontaneously reactivate requires that blood or tissue analysis be performed within a few minutes of sampling, or that precautions be taken to prevent decarbamylation prior to analysis. In many studies this has not been accomplished satisfactorily. In addition, daily variations in blood cholinesterase levels and possible changes due to illness or other reasons complicate the picture. Certainly this is one area of carbamate insecticide symptomology that needs more research.

The antidote for carbamate poisoning is atropine (Figure 3.9). This compound antagonizes the effect of excess acetylcholine within synapses by blocking ACh receptor sites, thus preventing continuous nerve stimulation. The primary site of action appears to be on the parasympathetic nerve fibers controlling salivation, urination, heartbeat, etc.[6] Secondary relief also occurs within the central nervous system.

Because of its action, atropine is a very good

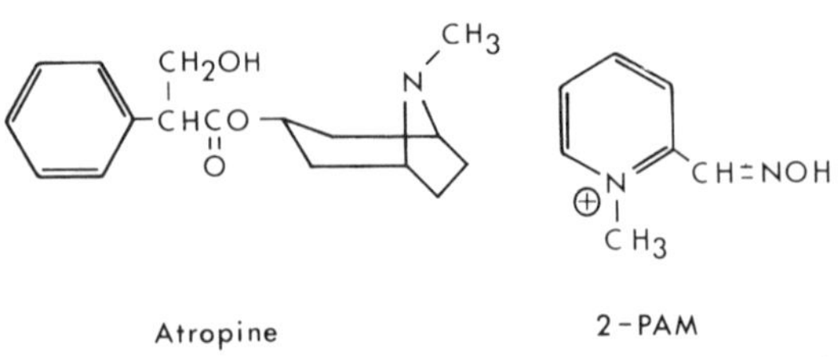

FIGURE 3.9. Structures of atropine and pyridine-2-aldoxime methiodide.

therapeutic agent, producing visible signs that can be correlated to recovery from anticholinesterase poisoning. Thus, atropinization of a severely poisoned individual is continued at regular intervals until pupil constriction, salivation, and other visible symptoms show definite improvement. Normally this requires an initial intravenous injection of about 2 mg of atropine sulfate for an adult (slightly higher doses for extreme poisoning), followed by repeated 2-mg doses at 10- to 15-min intervals until recovery is obvious.[8] If respiration is distressed, oxygen is administered before atropine treatment. It is important to remember that atropine in large doses is toxic *per se,* so that use of this antidotal procedure should be under the supervision of a trained medical person if at all possible.

Another antidote commonly used for treatment of poisoning by organophosphorus insecticides is 2-PAM or pyridine-2-aldoxime methiodide (Figure 3.9). However, this compound is *not* an antidote for carbaryl poisoning,[97] and it should probably not be used against other carbamates (cf Chapter 5). The reason 2-PAM and various other oxime derivatives effectively reverse organophosphorus symptoms is that the hydroxyl group on the oximes is a much better nucleophilic reagent than the hydroxyl of water. Therefore, hydrolysis of phosphorylated AChE within the nerve synapse occurs much more rapidly in the presence of 2-PAM, so that recovery of the nervous system is facilitated. On the other hand, it has been shown that decarbamylation of AChE is not accelerated by 2-PAM or by hydroxylamine.[25,27] There are compounds, such as diethyldi(2-hydroxyethyl) ammonium iodide, which have increased the rate of hydrolysis of carbamylated AChE in vitro 70-fold,[25] but these have not been given a great deal of attention as potential carbamate antidotes.

Case histories of carbamate insecticide poisoning, especially of humans, appear rarely in the scientific literature. One of the first was published in 1961, shortly after the registration of carbaryl.[97] This compound was given orally to dogs at rates of 250, 375, or 500 mg/kg. No cholinergic symptoms were observed at the low-dosage treatment, but the high dosage caused severe poisoning, although the dogs survived. The 375-mg level produced typical carbamate symptoms as follows. For the first 15 to 30 min the dogs were quiet; then excess salivation began, and the respiration rate increased. From 30 min to 90 min lacrimation, muscular twitching, and uncontrolled urination and defecation occurred. By 2½ hr all dogs had constricted pupils, poor coordination, diarrhea, increased respiration rate, loss of bladder control, and profuse salivation. Within 3 hr vomiting of mucus and of thick saliva, violent intestinal movements, general body weakness, and considerable muscle spasms became evident. However, 5 hr after ingestion symptoms began to subside, although lacrimation, salivation, slight miosis, muscle twitching, and occasional vomiting continued. By 7 hr the dogs appeared to be nearly normal, and the next day no symptoms were apparent.

During this poisoning, there was no measurable effect on dog plasma cholinesterase. Erythrocyte acetylcholinesterase, on the other hand, was depressed to a maximum of 33% after 2 to 3 hr, when symptoms were most pronounced. After 7 hr and after 24 hr blood cell AChE activity was reduced by only 20% and 5% respectively.

In another experiment, a dog treated with 375 mg/kg of carbaryl was injected intramuscularly with 10 mg of atropine sulfate at a time when symptoms were obvious (1½ hr). The poisoning was reversed and controlled by two successive injections of 5 mg atropine at 2.5 and 7 hr. Similar results occurred with a carbaryl dose of 795 mg/kg and antidote treatment with doses of 20, 10, and 10 mg. On ingestion of a 1,000-mg/kg dose of insecticide, one dog died, at which time 70 to 90% of blood cholinesterases and 100% of brain AChE was inhibited.

Rats injected intraperitoneally with an LD_{50} dose (4.7 mg/kg) of formetanate also exhibited the anticipated symptoms, but at a much faster rate than that observed with the oral administration described above.[98] Within only 3 min head tremors were noticeable, and by 10 min the tremors had become quite intense; convulsions and paralysis followed, and death of half the rats was evident by 35 to 40 min. Depression of brain acetylcholinesterase correlated fairly well with external symptoms; there was 41% inhibition 5 min after injection, and 70% or more was depressed from 10 to 40 min after injection. Return of AChE to 85% of normal was achieved by 5 hr in the rats that survived the experiment.

Comparison of brain and plasma cholinesterase following administration of large doses of Baygon to rats showed similar inhibition of both enzymes and close correlation with symptoms.[99] However, after intramuscular injection of small amounts of insecticide (0.25 to 1.0 mg/kg), as much as 60% of cholinesterase was inactivated in samples from rats showing no visible signs of poisoning. Increasing the dose to 2 mg/kg provided the first detectable tremors, which corresponded to an average brain and plasma ChE depression of 47% and 49% respectively. At levels from 10 to 50 mg/kg, brain esterases were blocked about 10 to 15% more than plasma esterases. The onset of symptoms after infusion of Baygon into the jugular vein of rats also occurred at the time of about 50% depression of brain cholinesterase.[100]

Other studies have demonstrated that variation of insecticide dosage is generally accompanied by changes in degree and extensiveness of outward signs. This has been effectively illustrated in an experiment with rats and the carbamate insecticide UC 10854, as described in Table 3.7.[101] Mild tremors and salivation were clearly evident at 1/30 of the intramuscular dose needed to kill one of the animals. Symptoms became progressively more intense and recovery was much slower as the amount of injected toxicant increased.

The nature of symptoms also depends somewhat on the structure of the carbamate. This has been proved by intravenous injection of rats with LD_{50} quantities of fifteen different carbamate insecticides.[102] Recordable signs of poisoning persisted for 45 to 180 min with varying degrees of intensity. Compounds that induced minor symptoms of short duration (Baygon, UC 10854, Banol, promecarb) were detected in the blood for only 1 to 2 hr after treatment. Those producing long-lasting, more intense symptoms (carbaryl, Landrin) persisted in the bloodstream for more than 6 hr. Nearly identical differences in symptomology have been found for carbaryl and Baygon after oral ingestion of equitoxic doses.[100]

In those cases where laboratory animals die almost instantaneously upon injection of a carbamate, the cause of death may not be due to cholinesterase inhibition, although some of the symptoms may appear. For instance, intravenous application of lethal levels of carbaryl, UC 10854,

TABLE 3.7

Correlation of Symptoms with Dose Following Intramuscular Injection of UC 10854 into Rats

Dose (mg/kg)	Symptoms
0.1	None
0.4	Moderate tremors, salivation, symptoms gone in 2 hr
1.6	Pronounced tremors, fasciculations, nausea, rapid recovery after 2 hr
6.5	Very pronounced tremors, excessive salivation, difficult breathing, slow recovery, symptoms gone in 6 hr
12.5	Same symptoms as above, plus convulsions, prostration, unconsciousness, vomiting, and very slow recovery or death
25.0	Extensive intense symptoms, with death in 3 to 5 hr

Adapted from Reference 101, with permission.

HRS-1422, or Landrin to rats produced a very pronounced anesthetic response, with severe respiratory difficulty and respiratory failure within a few minutes after injection.[102] Administration of artificial respiration revived the animals, and spontaneous breathing resumed, together with slow recovery from anesthesia. Then, typical cholinergic symptoms gradually developed. This phenomenon was not observed after oral treatment, and only moderate indications followed intraperitoneal injections. The reason for this narcotic response is not known, but it has been suggested that similar effects elicited by certain organophosphorus compounds were due to a blockage of sodium-ion transport across axon membranes in the nervous system.[103]

Development of cholinergic symptoms in humans exposed to large amounts of carbamate insecticides are very much like those found for other animals. A 42-year-old male volunteer weighing 90 kg ate 1.5 mg/kg of Baygon 2 hr after his breakfast.[100] Within 15 min his blood erythrocyte AChE level was reduced in activity by 73% (Figure 3.10), but no signs of poisoning were apparent except for moderate discomfort described as "pressure in the head." By 18 min blurred vision and nausea had developed, and after 20 min the subject was pale and his face perspired profusely. At this time the volunteer's pulse rate was 140/min (76/min prior to treatment), and blood pressure was 175/95 mm Hg (up from 135/90). At 30 min post-treatment pronounced nausea with repeated vomiting and excessive sweating occurred, which continued for 15 min. As seen in Figure 3.10, blood cell AChE activity during this period was around 50% of normal. After 1 hr the man was still tired and nauseous, but he felt much better. A few minutes later his pulse and blood pressure returned to pretreatment levels, and by 2 hr after ingestion the subject had recovered. It should be noted that erythrocyte AChE had also been fully reactivated by this time (Figure 3.10) and that plasma ChE was not depressed throughout the experiment. This is consistent with differences in the reaction of Baygon with the two esterases in vitro; I_{50} values of $4.6 \times 10^{-7} M$ (AChE) and $2.3 \times 10^{-5} M$ (ChE) have been reported.[100]

Exposure to a lower amount of Baygon (0.36 mg/kg) gave less AChE depression (57%) after 10 min.[104] The symptoms were also moderate; they included stomach upset, blurred vision, some facial

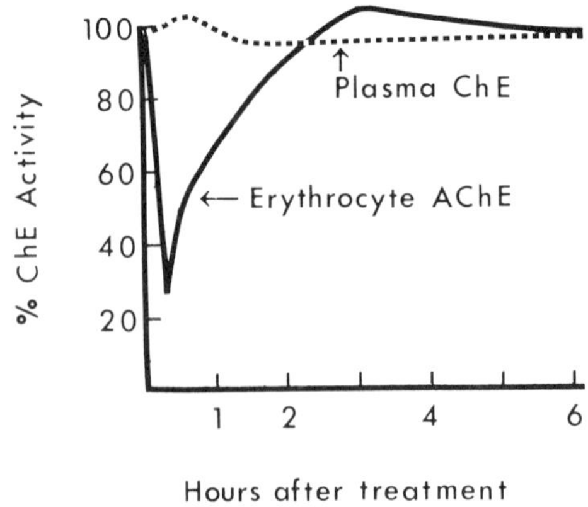

FIGURE 3.10. Changes in cholinesterases of the blood in a man following ingestion of a 1.5-mg/kg dose of Baygon.[100]

redness, and sweating, all of which lasted only a few minutes. Production of urinary phenolic materials also increased during the first few hours following treatment. By 3 hr after ingestion, however, AChE was back to normal.

Swallowing a higher dose of Baygon (1.0 mg/kg) over a longer period of time, i.e., 0.2 mg/kg at five ½-hr intervals, resulted in no signs of poisoning, despite AChE inhibition of about 60% in erythrocytes at the time of the last ingestion. The volunteer's blood esterase recovered completely in only 3 hr.

It is not surprising, as indicated above, that ingestion of carbamate insecticides by humans should be followed by excretion of hydrolysis products in the urine (see Chapter 7). The use of this concept as a means of detecting exposure to carbamates has been attempted with mixed results. Findings from two studies in which volunteers ingested Baygon are given in Table 3.8.[100,105,106] There is a rough correlation between the amount of insecticide ingested and the amount of hydrolysis products excreted. Severe cholinergic symptoms appeared following the 135-mg dose, but levels of 116 mg and lower caused no signs of illness in the volunteers. This does not agree with results from field experiments, where urinary residues of 80 ppm Baygon were uniformly associated with illness of spraymen.[105] It was also shown that the amount of interference by natural urinary phenols, together with variability of excretion volumes, mode of exposure, length of exposure, differences in metabolic rate, etc., yielded considerable discrepancies among exposed individuals. But, even with these prob-

TABLE 3.8

Urinary Excretion of Baygon Following Oral Ingestion by Human Volunteers[100,105]

Dose (mg/person)	Baygon in urine (ppm)	Analytical method
50	25	Gas-chromatographic
110	140	Colorimetric
116	140	Colorimetric
135	256	Colorimetric

lems, it was concluded that, provided urine was collected within 8 hr after exposure, the method would probably identify any adult who had received more than 5% of the LD_{50} dose of Baygon.

Measurement of urinary metabolites of carbaryl in workers exposed to this carbamate has been used with some success.[107] However, if this indicator of carbamate exposure is to be of value as a warning device, the analytical methods and sensitivity will have to be improved considerably. At present, the detection of carbamates or their phenols in urine is not reliable because of the many variables and inconsistencies inherent in the technique.

Large-scale field trials with a few carbamate insecticides for control of insect-borne diseases have been conducted in cooperation with the World Health Organization. Part of this program encompasses careful examination of applicators and village inhabitants for ill effects due to the use of the insecticides as surface treatments on and in dwellings.[108] Under these conditions, carbaryl has proved to be a very safe compound. When sprayed as a 5% suspension on walls and roofs of mud huts, there were no apparent adverse effects on spraymen or villagers.[109] The day after application the spraymen showed a slight depression (15%) of total blood cholinesterase, and after one week some village inhabitants exhibited about 8% ChE inhibition with a slight increase in urinary levels of 1-naphthol.

Symptomology was much more evident in tests with Baygon, although conditions and location were different.[110] During the 6-week trial, 240 spray operators worked 807 man-days applying 5% suspensions of Baygon to village dwellings. Throughout the total period there were only 20 reports of headache, 11 of nausea, and 7 of excess sweating. These symptoms mostly disappeared 1 to 3 hr after the spraymen left the job. Similar transient signs appeared in less than 1% of the 11,000 villagers. Very little correlation was found between complaints of illness and depression of blood cholinesterase. The most reliable preventative measure was the appearance of symptoms, which automatically resulted in termination of spraying, thus preventing further exposure long before a dangerous dose was absorbed. Similar findings were reported from field trials with Baygon involving 4,000 man-days of spraying.[104]

A final example concerns village trials with UC 10854, a carbamate that appeared to be harmless on the basis of laboratory animal studies.[108] When spraymen began applying 5% suspensions of UC 10854, rather severe symptoms appeared within a few hours, and the experiment had to be terminated.[109] From the eight operators and two supervisors, the following complaints (with the number of cases given in parentheses) were registered: weakness (10); dizziness (9); tightness in the chest (8); blurred vision (7); headache (6); pinpoint pupils (6); nausea (5); profuse sweating (5); abdominal cramps (4); and vomiting (2). At that time, whole-blood ChE was inhibited by more than 50%. One day later, cholinesterase levels were back to 80% of normal, but weakness (7), chest tightness (8), and headache (5) continued to persist. All symptoms except weakness (4) were gone by 4 days, and blood enzymes were back to normal activity. Among the 70 inhabitants of the village, three showed mild symptoms, and five progressed to the vomiting stage. Four days after spraying, the average blood cholinesterase was depressed by 23%, and urinary phenols were high for as long as 8 days after exposure. A very pronounced, persistent rash broke from the exposed skin of 17 of the 34 males. The only serious poisoning occurred with a 14-month-old child. By the time the baby reached the hospital (1:30 a.m.), he was semiconscious, groaning, rolling his eyes upwards, breathing rapidly, sweating profusely, and wheezing audibly. Examination revealed a pulse rate of 200/min, pinpoint pupils, and excessive rales and crepitations in the lungs. A 0.5-mg dose of atropine sulfate was injected intramuscularly, and within 1 hr marked improvement was evident: perspiration had mostly ceased, and the lungs were clearing. In the morning (8:00 a.m.) the lungs had recovered, and the child appeared near normal. The following day the patient was discharged from the hospital.

The observation in many of these village trials that cholinergic symptoms developed long before serious poisoning occurred prompted a comparative study between LD_{50} and ED_{50} (effective dose needed to produce first sign of tremors).[101] Rats were used as the test animals, and five carbamates plus two organophosphorus compounds served as the toxicants. Intravenous or intramuscular injection of the insecticides yielded the results shown in Table 3.9. The ratios of LD_{50} over ED_{50} were much larger with all of the carbamates than with the organophosphates, regardless of application method. As little as 1/40 to 1/19 of the lethal carbamate dose produced tremors in the rats. Ratios of the same order were found in the case of dogs injected with Baygon, UC 10854, and promecarb.[100] Of interest was the fact that UC 10854, the compound discontinued in village field trials, had the lowest ratios among the carbamate insecticides.

Summarizing all of these studies on carbamate symptomology in mammals provides several conclusions. First, the manifestation of very obvious adverse symptoms occurs long before a lethal dose of carbamate insecticide is absorbed. Second, the rapid rate of reactivation of carbamate-inhibited acetylcholinesterase, together with efficient metabolic detoxication, makes it extremely difficult to obtain severe, prolonged blockage of nerve transmission by a carbamate insecticide. Third, there is a rough parallelism between erythrocyte AChE depression, urinary excretion of phenolic metabolites, and degree of exposure, though none of these measurements is as reliable an indicator of exposure as is the appearance of early symptoms. Taken together, these factors supply carbamate insecticides with a very wide margin of safety when compared with other classes of insecticides.

Insects

The symptoms produced by poisoning insects with carbamate esters are neither as complex nor as dramatic as those evidenced by mammals. However, there exists an unmistakable carbamate intoxication syndrome of hyperactivity, incoordination (ataxia), clonic and tonic convulsions, paralysis, and death. The remarkable aspect of insect poisoning is the ability of these animals to withstand extended periods of paralysis without dying. Complete recovery can take place after hours and hours of motionless prostration (knockdown). This is partly because of their infusion-type tracheal respiratory system and primitive unstructured circulatory system, both of which are relatively unaffected by AChE inhibition. Presumably, with time, the carbamate is detoxified and/or the acetylcholinesterase is spontaneously reactivated, allowing recovery.

However, the actual cause of death remains a mystery. One report of Pyrolan-poisoned American cockroaches suggested that death was due partly to autointoxication.[111] Whether carbamates induce production of toxic substances *in situ*, similar to that observed with certain organochlorine-poisoned insects,[79] is not known. It is probable that death is simply due to exhaustion following general failure of critical areas within the central nervous system of the insect.

The appearance, intensity, and duration of

TABLE 3.9

Comparison of Lethal Dose (LD_{59}) with Dose (ED_{50}) Needed to Elicit First Signs of Poisoning

	Intravenous injection		Intramuscular injection	
Compound	LD_{50}	LD_{50}/ED_{50}	LD_{50}	LD_{50}/ED_{50}
Baygon	11	32	53	32
UC 10854	3.2	19	14	24
Banol	3.0	24	25	29
Promecarb	5.3	28	44	31
Landrin	32	22	280	40
Paraoxon-methyl	0.46	6	1.7	4
Paraoxon	0.25	3	0.45	2.5

Adapted from Reference 101, with permission.

symptoms are dependent on dose, chemical structure of the toxicant, length of exposure, mode of administration, and site of application. Repeated failures through the years to relate insect toxicity of carbamate insecticides with in-vitro cholinesterase inhibition[44,112-117] have been attributed to differences in cuticular penetration, metabolism (both detoxication and oxidation to toxic metabolites), and membrane permeability in the central nervous system. All of these factors, particularly biotransformations to toxic metabolites, also affect symptomology. Finally, the ever-present suggestions of alternate modes of action may explain the production of variable signs of poisoning in insects, but this has received little attention.

The majority of structure–activity and comparative toxicity studies have been conducted with houseflies, and thus the symptoms of carbamate poisoning are well documented for this insect. A recently reported example involved a dose of UC 10854 near the LD_{50}, topically applied.[118] About 5 to 10 min after treatment the flies began to exhibit hyperactivity, with body jerks and wing fanning being the most obvious symptoms. Next the insects fell over on their backs and buzzed around in circles, or they flew wildly around the test chamber. Within 30 min paralysis set in, together with periodic tremors. After several hours a few flies began kicking their legs, buzzed around for a while, and then righted themselves. By 10 hr the flies appeared perfectly normal.

With lepidopterous larvae, symptoms are a little different.[113,118] Regurgitation and defecation are usually readily apparent shortly after treatment. There is also a rapid rolling up and opening of the body before flaccid paralysis takes over. Recovery is not as successful, however, as it is with adult flies.

The more hydrophilic oxime carbamates, such as aldicarb, often elicit symptoms in insects much more slowly.[119] Houseflies treated topically with aldicarb sulfoxide, sulfone, and one other analog resisted knockdown for 2 to 24 hr. The slower production of symptoms was blamed on the decreased rate of penetration through the more lipophilic insect cuticle. On the other hand, another oxime carbamate, methomyl, produced cholinergic symptoms only a few minutes after application to various lepidopterous larvae (Kuhr, R. J., unpublished observations).

Two species of calliphorid flies treated with median lethal doses of seven methylcarbamates displayed typical symptoms.[112] Within 10 to 20 min of treatment the flies dropped onto their backs. Shortly thereafter, hyperactivity in the form of intermittent wing movement and twirling, twitching, and stretching of the legs took place. These symptoms became less intense with time until paralysis occurred. Paralyzed flies generally recovered in 8 to 12 hr, although several were down for as long as 24 hr before recovery.

Another "symptom" that has been noticed by the authors as well as by other investigators is inhibition of feeding. For example, Mexican bean beetle larvae refused to eat foliage treated with aldicarb, and it was suggested that this response was reliable enough to serve as a bioassay method.[119] In experiments with cabbage looper larvae, the antifeeding response was found to be dependent on dose, i.e., foliage covered with lower levels of carbaryl was eaten more rapidly and with less regurgitation (unpublished data by R. J. Kuhr). It was also discovered that, if carbaryl treatment was spotty, larvae chewed through the uncontaminated areas of leaves without ingesting any significant quantity of insecticide. Finally, if last-instar larvae were given only food dipped in high concentrations of carbaryl, the insects would refuse to eat and eventually pupate. On the other hand, fourth instars would crawl off the food and die (starve) rather than consume the treated leaves. This phenomenon has also been observed with other larvae and with other carbamate insecticides.

Just as in mammalian studies, several investigators have attempted to correlate the degree of poisoning and symptomology with cholinesterase inhibition within the insect. A very elaborate experiment with two species of flies and several carbamate insecticides indicated no consistent relationship between in-vivo head cholinesterase inhibition and toxicity.[112] However, there was a general trend of increased inhibition with time until paralysis. Cholinesterase activity remained constant for awhile, and then decreased when the flies recovered from the symptoms. Maximum inhibition in affected living flies ranged from about 25% to 42%. Death usually occurred in 8 to 12 hr, with 36% to 56% brain ChE depression.

Inhibition of acetylcholinesterase in housefly heads had to reach almost 60% before knockdown was effected by various aryl N-methyl- and N,N-dimethylcarbamates.[68] When reactivation returned esterase activity to 60% of normal, recovery began. However, enzyme reactivation and recovery took much longer with carbaryl and phenyl N-methylcarbamate than with Isolan. It was felt that the longer paralysis allowed more secondary damage to occur following monomethyl carbamate poisoning.

Contrary results were found in an earlier study with the same insect and two of the same carbamates.[120] Analysis of brain cholinesterase activity from flies treated topically with LD_{50} doses of carbaryl or Isolan revealed similar depression (60% maximum) and recovery profiles. However, 21 hr after treatment the enzyme levels in the carbaryl-treated flies had returned to normal, whereas the Isolan-treated insects still possessed 50%-depressed brain ChE. In both cases hyperactivity, paralysis, and recovery closely paralleled cholinesterase inhibition. Still another laboratory showed that Isolan caused greater housefly head ChE inhibition (84% maximum) and more severe anticholinesterase symptoms than did carbaryl (75% maximum, with only slight paralysis), even though both insecticides were used in LD_{50} amounts.[91] In addition, flies from the two experiments had completed over 80% enzyme reactivation by 24 hr.

Greater than 60% inhibition of housefly head AChE has also been found in flies knocked down by Baygon, formetanate, and other N-methylcarbamate insecticides.[98,121] But all these projects where fly head enzyme levels have been measured may not be too meaningful. As discussed in some detail above, there are convincing bits of evidence implying that inhibition of acetylcholinesterase in the brain of flies is only of secondary importance with respect to toxicity of carbamates. Peripheral regions of the thoracic ganglia may prove to be the primary site of action of carbamate toxicants in insects.[58,63] But generalizations must be made cautiously, since very few insect species have been utilized in insect mode-of-action studies.

REFERENCES

1. **Brooks, G. T.,** *Chlorinated Insecticides,* Vols. 1 and 2, CRC Press, Cleveland, 1974.
1a. **Batjer, L. P. and Thompson, B. J.,** Effect of 1-naphthyl N-methylcarbamate (Sevin) on thinning apples, *Proc. Am. Soc. Hortic. Sci.,* 77, 1, 1961.
2. **Peardon, D. L., Kilbourn, E. E., and Ware, J. E.,** New selective rodenticides, *Soap Cosmet. Chem. Spec.,* 48, 70, 1972.
3. **Judge, F. D. and Kuhr, R. J.,** Laboratory and field screening of granular formulations of candidate molluscicides, *J. Econ. Entomol.,* 65, 242, 1972.
4. **Sanz, F. and Jurado, R.,** Comparative study of organic phosphates and carbamates in experimental helminthiasis, *Arch. Inst. Farmacol. Exp. Madrid,* 18, 91, 1966.
5. **Schafer, E. W., Starr, R. I., Cunningham, D. J., and Decino, T. J.,** Substituted phenyl N-methylcarbamates as temporary immobilizing agents for birds, *J. Agric. Food Chem.,* 15, 287, 1967.
6. **O'Brien, R. D.,** *Toxic Phosphorus Esters,* Academic Press, New York, London, 1960.
7. **O'Brien, R. D.,** *Insecticides Action and Metabolism,* Academic Press, New York, London, 1967.
8. **Fest, C. and Schmidt, K.-J.,** *The Chemistry of Organophosphorus Pesticides,* Springer-Verlag, New York, Heidelberg, Berlin, 1973.
8a. **Eto, M.,** *Organophosphorus Pesticides: Organic and Biological Chemistry,* CRC Press, Cleveland, 1974.
9. **Augustinsson, K.-B.,** Comparative aspects of the purification and properties of cholinesterases, *Bull. W. H. O.,* 44, 81, 1971.
10. **Florkin, M. and Stolz, E. H. (Eds.),** *Comprehensive Biochemistry,* Vol. 1, Elsevier Publishing Co., Amsterdam, London, New York, 1964.
11. **Aldridge, W. N.,** The nature of the reaction of organophosphorus compounds and carbamates with esterases, *Bull. W. H. O.,* 44, 25, 1971.
12. **Aldridge, W. N. and Reiner, A.,** *Enzyme Inhibitors as Substrates,* North-Holland Publishing Co., Amsterdam, London, 1972.
13. **Casida, J. E.,** Insecticide biochemistry, *Annu. Rev. Biochem.,* 42, 259, 1973.
13a. **O'Brien, R. D.,** The design of organophosphate and carbamate inhibitors of cholinesterase, in *Drug Design,* Vol. 2, Ariens, E. J., Ed., Academic Press, New York, London, 1971.

14. **Reiner, E. and Aldridge, W. N.,** Effects of pH on inhibition and spontaneous reactivation of acetylcholinesterase treated with esters of phosphorus acids and of carbamic acids, *Biochem. J.*, 105, 171, 1967.
15. **Aldridge, W. N.,** Some properties of specific cholinesterase with particular reference to the mechanism of inhibition by diethyl *p*-nitrophenyl thiophosphate (E605) and analogues, *Biochem. J.*, 46, 451, 1950.
16. **Aldridge, W. N.,** The inhibition of erythrocyte cholinesterase by triesters of phosphoric acid. 3. The nature of the inhibitory process, *Biochem. J.*, 54, 442, 1953.
17. **Goldstein, A.,** The mechanism of enzyme-inhibitor-substrate reactions as illustrated by the cholinesterase-physostigmine-acetylcholine system, *J. Gen. Physiol.*, 27, 529, 1944.
18. **Goldstein, A.,** Properties and behavior of purified human plasma cholinesterase. III. Competitive inhibition by physostigmine and other alkaloids with special reference to differences in kinetic behavior, *Arch. Biochem. Biophys.*, 34, 169, 1951.
19. **Burgen, A. S. V.,** The mechanism of action and anticholinesterase drugs, *Br. J. Pharmacol.*, 4, 219, 1949.
20. **Augustinsson, K. B. and Nachmansohn, D.,** Cholinesterase. VI. Kinetics of the inhibition of acetylcholinesterase, *J. Biol. Chem.*, 179, 543, 1949.
21. **Myers, D. K.,** Studies on cholinesterase. 8. Determination of reaction velocity constants with a reversible inhibitor of pseudocholinesterase, *Biochem. J.*, 52, 46, 1952.
22. **Goldstein, A. and Hamlisch, R. E.,** Properties and behavior of purified human plasma cholinesterase. IV. Enzymatic destruction of the inhibitors prostigmine and physostigmine, *Arch. Biochem. Biophys.*, 35, 12, 1952.
23. **Myers, D. K.,** Studies on cholinesterase. 10. Return of cholinesterase activity in the rat after inhibition by carbamoyl fluorides, *Biochem. J.*, 62, 556, 1956.
24. **Myers, D. K. and Kemp. A., Jr.,** Inhibition of esterases by the fluorides of organic acids, *Nature*, 173, 33, 1954.
25. **Wilson, I. B., Hatch, M. A., and Ginsburg, S.,** Carbamylation of acetylcholinesterase, *J. Biol. Chem.*, 235, 2312, 1960.
26. **Wilson, J. B., Harrison, M. A., and Ginsburg, S.,** Carbamyl derivatives of acetylcholinesterase, *J. Biol. Chem.*, 236, 1498, 1961.
27. **O'Brien, R. D.,** Kinetics of the carbamylation of cholinesterase, *Mol. Pharmacol.*, 4, 121, 1968.
28. **Winteringham, F. P. W. and Fowler, K. S.,** Substrate and dilution effects on the inhibition of acetylcholinesterase by carbamates, *Biochem. J.*, 101, 127, 1966.
29. **Kitz, R. J., Braswell, L. M. and Ginsburg, S.,** On the question: is acetylcholinesterase an allosteric protein, *Mol. Pharmacol.*, 6, 108, 1970.
30. **Reiner, E. and Simeon-Rudolf, T. V.,** The kinetics of inhibition of erythrocyte cholinesterase by monomethylcarbamates, *Biochem. J.*, 98, 501, 1966.
31. **Roufogalis, B. D. and Thomas, J.,** The effect of organic and inorganic cations on the decarbamylation of dimethylcarbamylacetylcholinesterase: a comparison with deacetylation, *Mol. Pharamacol.*, 5, 286, 1969.
32. **Greenspan, C. M. and Wilson, J. B.,** The effect of fluoride on the reactivation of acetylcholinesterase with carbamates, *Mol. Pharmacol.*, 6, 266, 1970.
33. **O'Brien, R. D.,** Mechanism of inhibition of cholinesterase by carbamates, *Exp. Med. Surg.*, Suppl. 23, 117, 1965.
34. **O'Brien, R. D., Hilton, B. D., and Gilmour, L.,** The reaction of carbamates with cholinesterase, *Mol. Pharmacol.*, 2, 593, 1966.
35. **Main, A. R.,** Affinity and phosphorylation constants for the inhibition of esterases by organophosphates, *Science*, 144, 992, 1964.
35a. **Main, A. R.,** Kinetics of cholinesterase inhibition by organophosphates and carbamate insecticides, *Can. Med. Assoc. J.*, 100, 161, 1969.
36. **Hastings, F. L., Main, A. R., and Iverson, F.,** Carbamylation and affinity constants of some carbamate inhibitors of acetylcholinesterase and their relation to analogous substrate constants, *J. Agric. Food Chem.*, 18, 497, 1970.
37. **Loewi, O. and Navoratil, E.,** Über den Mechanismus der Vaguswirkung von Physostigmin und Ergotamin, *Arch. Gesamte Physiol.*, 214, 689, 1926.
38. **Engelhart, E. and Loewi, O.,** Fermentative Acetylcholinspaltung im Blut und ihre Hemmung durch Physostigmin, *Arch. Exp. Pathol. Pharmakol.*, 150, 1, 1930.
39. **Stedman, E. and Stedman, E.,** Studies on the relationship between chemical constitution and physiological action. III. The inhibitory action of certain synthetic urethanes on the activity of liver esterases, *Biochem. J.*, 25, 1147, 1931.
40. **Stedman, E. and Stedman, E.,** Studies on the relationship between chemical constitution and physiological action. IV. The inhibitory action of certain synthetic urethanes on the activity of esterases, *Biochem. J.*, 26, 1214, 1932.
41. **Easson, L. H. and Stedman, E.,** The absolute activity of cholinesterase, *Proc. R. Soc. London*, B121, 142, 1936.
42. **Ellis, S., Plachte, F. L., and Straus, O. H.,** Studies on physostigmine and related substances. II. The destruction of physostigmine in buffered solutions and in serum, *J. Pharmacol. Exp. Ther.*, 79, 295, 1943.
43. **Casida, J. E.,** Mode of action of carbamates, *Annu. Rev. Entomol.*, 8, 39, 1963.
44. **Casida, J. E., Augustinsson, K.-B., and Jonsson, G.,** Stability, toxicity, and reaction mechanism with esterases of certain carbamate insecticides, *J. Econ. Entomol.*, 53, 205, 1960.
45. **Iverson, F. and Main, A. R.,** Effect of charge on the carbamoylation and binding constants of eel acetylcholinesterase in reaction with neostigmine and related carbamates, *Biochemistry*, 8, 1889, 1969.

46. **Metcalf, R. L. and Fukuto, T. R.,** Effect of chemical structure on intoxication and detoxication of phenyl N-methylcarbamates in insects, *J. Agric. Food Chem.*, 13, 220, 1965.
47. **Reed, W. D. and Fukuto, T. R.,** The reactivation of carbamate-inhibited cholinesterase, kinetic parameters, *Pestic. Biochem. Physiol.*, 3, 120, 1973.
48. **Wilson, J. B. and Alexander, J.,** Acetylcholinesterase: reversible inhibition, substrate inhibition, *J. Biol. Chem.*, 237, 1323, 1962.
49. **Simeon, V., Reiner, E., and Vernon, C. A.,** Effect of temperature and pH on carbamoylation and phosphorylation of serum cholinesterases, *Biochem. J.*, 130, 515, 1972.
50. **Voytenko, G. A. and Podlinyayeva, M. Y.,** Toxicity and anticholinesterase activity of some N-methyl- and N-phenylcarbamic acid-derived pesticides, *Farmakol. Toksikol.*, 35, 356, 1972.
51. **Axen, R., Heilbronn, E., and Winter, A.,** Preparation and properties of cholinesterases covalently bound to sepharose, *Biochem. Biophys. Acta,* 191, 478, 1969.
52. **Schmitt, J. B.,** The comparative anatomy of the insect nervous system, *Annu. Rev. Entomol.*, 7, 137, 1962.
53. **Narashashi, T.,** Properties of insect axons, *Adv. Insect Physiol.*, 1, 175, 1963.
54. **Smith, D. S. and Treherne, J. E.,** Functional aspects of the organization of the insect nervous system, *Adv. Insect Physiol.*, 1, 401, 1963.
55. **Colhoun, E. H.,** The physiological significance of acetylcholine in insects and observations upon other pharmacologically active substances, *Adv. Insect Physiol.*, 1, 1, 1963.
56. **Davey, K. G.,** The control of visceral muscles in insects, *Adv. Insect Physiol.*, 2, 219, 1964.
57. **Boistel, J.,** The synaptic transmission and related phenomena in insects, *Adv. Insect Physiol.*, 5, 1, 1968.
58. **Booth, G. M. and Lee, A.-H.,** The distribution of cholinesterases in insects, *Bull. W. H. O.*, 44, 91, 1971.
59. **Shorey, H. H.,** Synergism of carbamate insecticides against cabbage and beet armyworm larvae, *J. Econ. Entomol.*, 54, 1243, 1961.
60. **Metcalf, R. L., Winton, M. Y., and Fukuto, T. R.,** The effects of cholinergic substances upon the isolated heart of *Periplaneta americana, J. Insect Physiol.*, 10, 353, 1963.
61. **Burt, P. E., Gregory, G. E., and Molloy, F. M.,** A histochemical and electrophysiological study of the action of diazoxon on cholinesterase activity and nerve conduction in ganglia of the cockroach *Periplaneta americana* L., *Ann. Appl. Biol.*, 58, 341, 1966.
62. **Mengle, D. C. and Casida, J. E.,** Biochemical factors in the acquired resistance of houseflies to organophosphate insecticides, *J. Agric. Food Chem.*, 8, 431, 1960.
63. **Booth, G. M. and Metcalf, R. L.,** Histochemical evidence for localized inhibition of cholinesterase in the house fly, *Ann. Entomol. Soc. Am.*, 63, 197, 1970.
64. **Molloy, F. M.,** The histochemistry of the cholinesterase in the central nervous system of susceptible and resistant strains of the housefly, *Musca domestica* L., in relation to diazinon poisoning, *Bull. Entomol. Res.*, 52, 667, 1961.
65. **Farnham, A. W., Gregory, G. E., and Sawicki, R. M.,** Bioassay and histochemical studies of the poisoning and recovery of houseflies (*Musca domestica* L.) treated with diazinon and diazoxon, *Bull. Entomol. Res.*, 57, 108, 1966. their toxicities toward these animals, *J. Agric. Food Chem.*, 20, 923, 1972.
66. **Brady, U. E.,** Localization of cholinesterase activity in housefly thoraces: inhibition of cholinesterase with organophosphate compounds, *Entomol. Exp. Appl.*, 13, 423, 1970.
67. **Bull. D. L., Lindquist, D. A., and Coppedge, J. R.,** Metabolism of 2-methyl-2-(methylthio)propionaldehyde O-(methylcarbamoyl) oxime (Temik, UC-21149) in insects, *J. Agric. Food Chem.*, 15, 610, 1967.
68. **Hellenbrand, K.,** Inhibition of housefly acetylcholinesterase by carbamates, *J. Agric. Food Chem.*, 15, 825, 1967.
69. **Yu, C.-C., Metcalf, R. L., and Booth, G. M.,** Inhibition of acetylcholinesterase from mammals and insects by carbofuran and its related compounds and their toxicities toward these animals, *J. Agric. Food Chem.*, 20, 923, 1972.
69a. **Yu, C. C., Park, K. S., and Metcalf, R. L.,** Correlation of toxicity and acetylcholinesterase (AChE) inhibition in 2-alkyl-substituted 1,3-benzodioxolyl-4 N-methylcarbamates and related compounds, *Pestic. Biochem. Physiol.*, 4, 178, 1974.
69b. **Yu, C. C., Kearns, C. W., and Metcalf, R. L.,** Acetylcholinesterase inhibition by substituted phenyl N-alkyl carbamates, *J. Agric. Food Chem.*, 20, 537, 1972.
70. **Kunkee, R. E. and Zweig, G.,** Inactivation and reactivation rates of fly and bee cholinesterases inhibited by Sevin, *Biochem. Pharamacol.*, 14, 1011, 1965.
71. **Post, L. C.,** Inhibition of cholinesterase by carbamates: a new kinetic approach, *Biochim. Biophys. Acta,* 250, 121, 1971.
72. **Metcalf, R. L.,** Structure–activity relationships for insecticidal carbamates, *Bull. W. H. O.*, 44, 43, 1971.
73. **Twarog, B. M. and Roeder, K. D.,** Properties of the connective-tissue sheath of the cockroach nerve cord, *Biol. Bull. Woods Hole,* 111, 278, 1956.
74. **Winton, M. Y., Metcalf, R. L., and Fukuto, T. R.,** The use of acetylthiocholine in the histochemical study of the action of organophosphorus insecticides, *Ann. Entomol. Soc. Am.*, 51, 436, 1958.
75. **Kolbezen, M. J., Metcalf, R. L., and Fukuto, T. R.,** Insecticidal activity of carbamate cholinesterase inhibitors, *J. Agric. Food Chem.*, 2, 864, 1954.
76. **Roeder, K. D.,** The effect of anticholinesterases and related substances on nervous activity in the cockroach, *Bull. Johns Hopkins Hosp.*, 83, 587, 1948.

77. **Harlow, P. A.,** The action of drugs on the nervous system of the locust (*Locusta migratoria*), *Ann. Appl. Biol.,* 46, 55, 1958.
78. **O'Brien, R. D. and Fisher, R. W.,** The relation between ionization and toxicity to insects for some neuropharmacological compounds, *J. Econ. Entomol.,* 51, 169, 1958.
79. **Colhoun, E. H.,** Approaches to mechanisms of insecticidal action, *J. Agric. Food Chem.,* 8, 252, 1960.
80. **Winteringham, F. P. W.,** Metabolism and significance of acetyl choline in the brain of the adult house fly *Musca domestica, J. Insect Physiol.,* 12, 909, 1966.
81. **Metcalf, R. L., Gruhn, W. B., and Fukuto, T. R.,** Electrophysiological action of carbamate insecticides in the central nervous system of the American cockroach, *Ann. Entomol. Soc. Am.,* 61, 618, 1968.
82. **Augustinsson, K. B. and Nachmansohn, D.,** Distinction between acetylcholine esterase and other choline-splitting enzymes, *Science,* 110, 98, 1949.
83. **Augustinsson, K. B.,** Acetylcholine esterase and cholinesterase, *Enzymes,* 1, 443, 1950.
84. **Metcalf, R. L., March, R. B., and Maxon, M. G.,** Substrate preferences of insect cholinesterases, *Ann. Entomol. Soc. Am.,* 48, 222, 1955.
85. **Van Asperen, K.,** Distribution and substrate specificity of esterases in the housefly, *Musca domestica* L., *J. Insect Physiol.,* 3, 306, 1959.
86. **Dauterman, W. C., Talens, A., and Van Asperen, K.,** Partial purification and properties of fly head cholinesterase, *J. Insect Physiol.,* 8, 1, 1962.
87. **Yu, C. C., Kearns, C. W., Metcalf, R. L., and Davies, J. D.,** Acetylcholinesterase inhibition by biscarbamates, *Pestic. Biochem. Physiol.,* 1, 241, 1971.
88. **Metcalf, R. L. and March, R. B.,** The comparative properties of acetylcholinesterases from the bee, the fly, and the mouse, and their relation to insecticidal action, *J. Econ. Entomol.,* 43, 670, 1950.
89. **Babers, F. F. and Pratt, J. J.,** A comparison of the cholinesterase in the heads of the housefly, the cockroach, and the honeybee, *Physiol. Zool.,* 24, 127, 1951.
90. **Roan, C. C. and Maeda, S.,** The cholinesterase systems of three species of fruit flies and the effects of certain insecticidal compounds on these enzymes, *J. Econ. Entomol.,* 47, 507, 1954.
91. **Plapp, F. W. and Bigley, W. S.,** Carbamate insecticides and aliesterase activity in insects, *J. Econ. Entomol.,* 54, 793, 1961.
92. **Erdos, E. G.,** Esterases in fireflies and their inhibition, *Arch. Biochem. Biophys.,* 93, 302, 1961.
93. **Sonnet, P. E., Redfern, R. E., Schwarz, M., Wakaboyashi, N., and Waters, R. M.,** Compounds related to insect juvenile hormones. VII. Activity of aryl citronellylcarbamates on the yellow mealworm and the large milkweed bug, *J. Econ. Entomol.,* 64, 1378, 1971.
94. **Sonnet, P. E., Waters, R. M., Redfern, R. E., Schwarz, M., and Wakaboyashi, N.,** Compounds related to juvenile hormones. IX. Activity of citronellylamine and citronellol derivatives on the yellow mealworm and the large milkweed bug, *J. Agric. Food Chem.,* 20, 65, 1972.
95. **Geering, Q. A., Brooker, P. J., and Parsons, J. H.,** The chemosterilant activity of some substituted phenyl esters of 1-aziridinecarboxylic acid, *J. Econ. Entomol.,* 58, 574, 1965.
96. **Witter, R. F.,** Measurement of blood cholinesterase, *Arch. Environ. Health,* 6, 537, 1963.
97. **Carpenter, C. P., Weil, C. S., Palm, P. E., Woodside, M. W., Nair, J. H., III, and Smyth, H. F., Jr.,** Mammalian toxicity of 1-naphthyl-N-methylcarbamate (Sevin insecticide), *J. Agric. Food Chem.,* 9, 30, 1961.
98. **Knowles, C. O. and Ahmad, S.,** Mode-of-action studies with formetanate and formparanate acaricides, *Pestic. Biochem. Physiol.,* 1, 445, 1971.
99. **Plestina, R. and Vandekar, M.,** Studies in the Toxicology of N-Methylcarbamates. IV. Symptoms as Related to Cholinesterase Activity, 15th International Congress on Occupational Health, Vienna, 1966, Vol. 2, p. 525.
100. **Vandekar, M., Plestina, R., and Wilhelm, K.,** Toxicity of carbamates for mammals, *Bull. W. H. O.,* 44, 241, 1971.
101. **Vandekar, M., Reiner, E., Svetlicic, B., and Fajdetic, T.,** Value of ED_{50} testing in assessing hazards of acute poisoning by carbamates and organophosphates, *Br. J. Ind. Med.,* 22, 317, 1965.
102. **Wilhelm, K. and Vandekar, M.,** Studies in the Toxicology of N-Methylcarbamates. I. Comparative Toxicity Tests and Estimation of Persistence of Inhibitor in the Body, 15th International Congress on Occupational Health, Vienna, 1966, Vol. 2, p. 517.
103. **Heath, D. F.,** *Organophosphorous Poisons,* Pergamon Press, Oxford, England, 1961.
104. **Wright, J. W., Fritz, R. F., Hocking, K. S., Babione, R., Gratz, N. G., Pal, R., Stiles, A. R., and Vandekar, M.,** *ortho*-Isopropoxyphenyl methylcarbamate (OMS-33) as a residual spray for control of anopheline mosquitoes, *Bull. W. H. O.,* 40, 67, 1969.
105. **Dawson, J. A., Heath, D. F., Rose, J. A., Thain, E. M., and Ward, J. B.,** The excretion by humans of the phenol derived in vivo from 2-isopropoxyphenyl N-methylcarbamate, *Bull. W. H. O.,* 30, 127, 1964.
106. **Hayes, W. J., Jr.,** Studies on exposure during the use of anticholinesterase pesticides, *Bull. W. H. O.,* 44, 277, 1971.
107. **Best, E. M., Jr. and Murray, B. L.,** Observations on workers exposed to Sevin insecticide: a preliminary report, *J. Occup. Med.,* 4, 507, 1962.
108. **Arnan, A.,** Experience in the WHO field programme for evaluating the safety of new insecticides, *Bull. W. H. O.,* 44, 273, 1971.
109. **Vandekar, M.,** Observations on the toxicity of carbaryl, Folithion and 3-isopropylphenyl N-methylcarbamate in a village-scale trial in Southern Nigeria, *Bull. W. H. O.,* 33, 107, 1965.

110. **Vandekar, M., Hedayat, S., Plestina, R., and Ahmady, G.,** A study of the safety of *o*-isopropoxyphenylmethylcarbamate in an operational field trial in Iran, *Bull. W. H. O.,* 38, 609, 1968.
111. **Weismann, R. and Kocher, C.,** Untersuchungen über ein neues, gegen resistente *Musca domestica* L. wirksames Insektizid, *Z. Angew. Entomol.,* 33, 297, 1951.
112. **Respicio, N. C. and Sherman, M.,** Toxicity and anticholinesterase activity of phenyl methylcarbamate insecticides to *Chrysomya megacephala* and *Phoenicia cuprina, J. Econ. Entomol.,* 65, 1535, 1972.
113. **Metcalf, R. L., Fukuto, T. R., and Winton, M. Y.,** Alkoxyphenyl N-methylcarbamates as insecticides, *J. Econ. Entomol.,* 53, 828, 1960.
114. **Fukuto, T. R., Metcalf, R. L., Winton, M. Y., and Roberts, D. A.,** The synergism of substituted phenyl N-methylcarbamates by piperonyl butoxide, *J. Econ. Entomol.,* 55, 341, 1962.
115. **Weiden, M. H. J. and Moorefield, H. H.,** Synergism and species specificity of carbamate insecticides, *J. Agric. Food Chem.,* 13, 200, 1965.
116. **Reay, R. C. and Lewis, D. K.,** Evidence for differences in mode of action between two related carbamate insecticides, *J. Sci. Food Agric.,* 17, 17, 1966.
117. **Green, L. R. and Dorough, H. W.,** Housefly age as a factor in their response to certain carbamates, *J. Econ. Entomol.,* 61, 88, 1968.
118. **Weiden, M. H. J.,** Toxicity of carbamates to insects, *Bull. W. H. O.,* 44, 203, 1971.
119. **Weiden, M. H. J.,** Insecticidal carbamoyloximes, *J. Sci. Food Agric.,* Suppl., p. 19, 1968.
120. **Mengle, D. C. and Casida, J. E.,** Inhibition and recovery of brain cholinesterase activity in houseflies poisoned with organophosphate and carbamate compounds, *J. Econ. Entomol.,* 51, 750, 1958.
121. **Goszczynska, K. and Styczynska, B.,** Inhibition of acetylcholinesterase activity in *Musca domestica* by N-methylcarbamate insecticides, *Rocz. Panstw. Zakl. Hig.,* 23, 611, 1972.
122. **Main, A. R.,** in *Structure and Reactions of DFP-Sensitive Enzymes,* Research Institute for National Defense, Stockholm, 1967.

Chapter 4

STRUCTURE–ACTIVITY RELATIONSHIPS

INTRODUCTION

The discovery of new insecticide chemicals usually occurs during the random biological screening of the thousands of compounds synthesized each year by industrial chemists. In some instances the compounds are extensions of known bioactive structures with minor modifications or additions, although new classes of toxicants generally appear by chance. However, the methylcarbamate insecticides, more than any other group of insecticides, owe their existence to a preconceived biochemical plan of attack which began with some experiments in the late 1940's and early 1950's.

It was known that organophosphorus insecticides were active by virtue of their ability to inhibit acetylcholinesterase (AChE), and that pharmacologically active carbamate esters also blocked cholinesterases in mammals. When some of these esters, such as physostigmine (Figure 4.1), were incubated with housefly and honeybee brain preparations, they were found to be very effective inhibitors of insect AChE.[1] However, there were variations of as much as 1,000-fold in the response of the two insect tissues and of mouse brain to certain of the carbamates. Further distinct biochemical differences in the acetylcholinesterase of fly, bee, and mouse brains with respect to organophosphorus compounds and choline substrates left the definite impression that carbamate structures might be synthesized which were not only selective in toxicity between insects and mammals, but were selective even among different insect species.

At about the same time it was also discovered that prostigmine, methylprostigmine (Figure 4.1), and other quaternary-nitrogen aryl N-methyl- and N,N-dimethylcarbamates were essentially nontoxic when applied in large amounts to the cuticle of various insect species.[1-3] This was contrary to their very good potential as toxicants anticipated from in-vitro housefly head AChE inhibition (I_{50} values near $5 \times 10^{-8} M$). On injection into locusts, which discounted cuticle penetration difficulties, prostigmine had very marginal insecticidal activity, but physostigmine was quite toxic.[4] The toxicity was not reduced by injection of the antidote, atropine.

Thus, it appeared that the charge on prostigmine (and probably atropine) prevented the compound from reaching the site of action even after injection into the insect's body. Presumably, physostigmine was not ionized at physiological pH and therefore was able to effect its action. This action was undoubtedly AChE inhibition, since it had been demonstrated that physostigmine applied to the nerve cord of American cockroach produced a rise in acetylcholine levels, which in turn led to disruption of synaptic transmission in the nervous tissue.[5-7] In this instance, when atropine was introduced into the system, the action of physostigmine was blocked.

Because of the difference in specificity of insect and vertebrate acetylcholinesterases, and because of the variable activity of some carbamates (uncharged) in comparison to others (charged), a series of methylcarbamic acid esters of substituted phenols similar to methylprostigmine, but without quaternary groups, was synthesized with the idea that the more lipophilic compounds would more readily penetrate insect cuticle and nerve sheaths to yield good insect poisons.[8] Simultaneously, a group of pyrazolyl N,N-dimethylcarbamates, including Pyrolan and Isolan, was prepared and tested for insecticidal activity.[9,10] These reports prompted a series of structure-activity studies which eventually provided today's commercially successful carbamate insecticides.

Before examining the structural requirements of a good carbamate anticholinesterase, a picture of the reactive center on the surface of acetylcholinesterase and the chemical interactions thereupon should be fixed in the reader's mind. To assist in this exercise, a simplified pictorial representation of the sequence described kinetically in Chapter 3 (Figure 3.5) is shown in Figure 4.2.[11-13] The substrate in this case is unsubstituted phenyl N-methylcarbamate, but the same reaction occurs with other aromatic methylcarbamates, dimethylcarbamates, and oxime derivatives. The mechanism is based primarily on the proposed scheme for hydrolysis of the natural substrate, acetylcholine.[14-16]

The process is basically a three-step reaction, that is, formation of the enzyme-substrate complex governed by the affinity constant $K_a = k_{-1}/k_{+1}$, carbamylation of the enzyme (k_2), and reactivation of the enzyme (k_3). All of this appears to take place at a single reactive center on acetylcholinesterase consisting of an esteratic site and an anionic site, separated by a distance of approximately 5 Å [17,18] (Figure 4.2). The anionic site is often visualized as a negatively charged pocket that is responsible for coulombically attracting the positively charged trimethylammonium moiety of acetylcholine.[19] Much of the area in and around the pocket is hydrophobic in nature and binds the methyl groups of ACh or the phenolic portion of aryl N-methylcarbamates. The source of the anionic character is probably a carboxylic group of glutamic or aspartic acid.[20] However, more recent studies imply that the negative part of the site is very weak and that binding at the "anionic" site is due mostly to hydrophobic and van der Waal forces.[21]

The nature of the esteratic site is more complex, consisting of a basic unit (B), which is probably a histidine imidazole nitrogen atom, and of an acidic unit (HA), thought to be the aromatic hydroxyl group of tyrosine (Figure 4.2). Between these two units lies the true catalytic area of the site (OH), which appears to be the hydroxyl of serine incorporated into a peptide sequence, glycyl–aspartyl–seryl–glycine.[13-16,22,23] It is carbamylation of this aliphatic serine hydroxyl

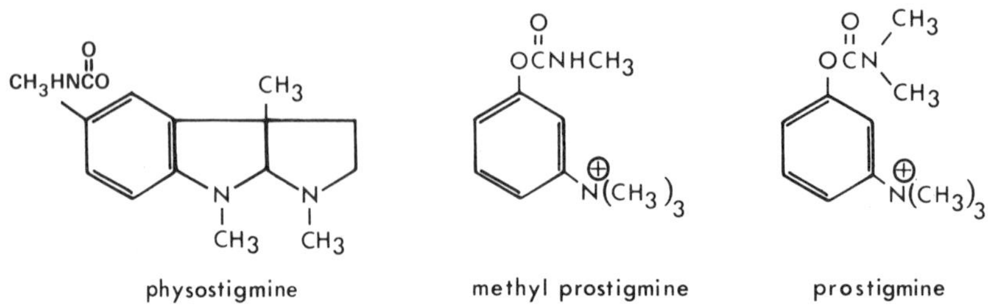

FIGURE 4.1. Structures of pharmacologically active carbamates.

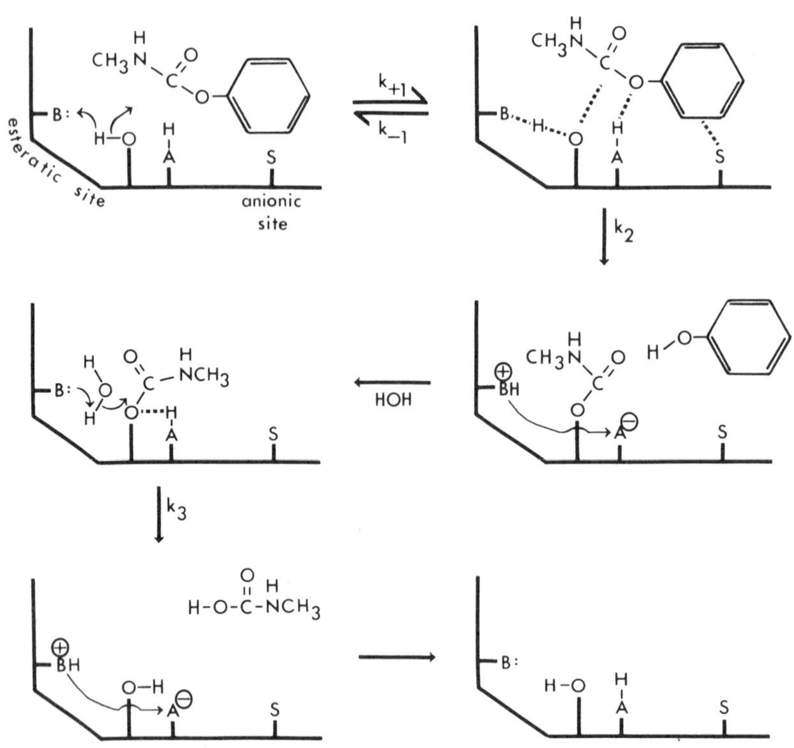

FIGURE 4.2. Pictorial representation of the reaction of phenyl N-methylcarbamate with acetylcholinesterase.[11-13]

group that prevents the esteratic site from performing its normal function of hydrolyzing acetylcholine. Note that carbamylated serine forms an aliphatic carbamate, which is more resistant to alkaline hydrolysis than an aromatic carbamate.[11]

An approaching substrate molecule is probably drawn to AChE primarily by opposite charge attraction and/or hydrophobic forces at the anionic site. Once there, the stability of the enzyme-substrate complex is determined by the configurational fit of the substrate onto both the esteratic and the anionic site, and by the strength of the accompanying hydrogen bonds, van der Waal forces, hydrophobic attractions, and charge transfers [24] (Figure 4.2). Many believe that the formation of this complex is the most critical step in cholinesterase inhibition by carbamates.

Carbamylation of the enzyme occurs via electrophilic attack by the substrate's carbonyl carbon on the serine hydroxyl of AChE, resulting in the formation of a covalent bond between carbamate moiety and enzyme (Figure 4.2). A consequence of this is, of course, hydrolysis of the substrate molecule, releasing a phenolic leaving group. Since the success of carbamylation is partially dependent on the electrophilicity of the carbonyl center, the overall structure of the carbamate can affect the carbamylation rate (k_2). This has been verified kinetically (cf Chapter 3). There exists some limited information implying that carbamylation may involve a more complex interaction with various subsites within the esteratic site.[21]

Finally, the carbamylated enzyme is reactivated (k_3) in the presence of water to yield free AChE and methylcarbamic acid (Figure 4.2). This acid-base-catalyzed hydrolysis has been confirmed by recent investigations using deuterium oxide, different pH conditions, and various inorganic ions.[13] Obviously, the phenolic portion of a carbamate does not alter enzyme reactivation, but the nature of the substituents on the carbamate nitrogen atom can be quite important. Although the rates of reactivation (k_3) for methyl- and dimethylcarbamyl acetylcholinesterase are similar (cf Chapter 3), large differences can be found when other groups are attached to the nitrogen atom.[25-29]

The above rather simple description of substrate–enzyme interaction has been complicated by more current investigations. For example, one project provided evidence for the existence of two anionic sites on the AChE of houseflies.[30] Thus, two substrate molecules could be bound to one molecule of enzyme at the same time, and it was proposed that, while one site was occupied by ACh, the other might be filled by an inhibitor (noncompetitive inhibition). On the contrary, a stoichiometric relationship was found between substrate and bovine erythrocyte or electric eel AChE.[30,31] However, there were some curare-like inhibitors that did not complement the normal reactive center of AChE and were possibly bound to an alternate site on the enzyme.[32]

Other reports hypothesize that there are three sites clustered about the esteratic site, all contributing significantly to the complexing of AChE and carbamate.[21,33,34] One of these, the α-site, is very much like the anionic site in Figure 4.2 and is responsible for binding quaternary ammonium compounds and certain alkylating agents. The β-site occupies most of the area around the catalytic center, and part of it is very hydrophobic in character; this is where methylcarbamate, organophosphorus, and certain acetate molecules are bound. Finally, an ill-defined γ-site is responsible for binding other acylating agents. When the α-site is filled, a configurational shift in the catalytic center (serine OH) occurs, which affects carbamylation rate, but not affinity or reactivation.

Experiments showed that such compounds as physostigmine, carbaryl, and HRS-1422 produced slightly depressed k_2 values when reacted with alkylated AChE and therefore were said to be bound at the β-site. But such carbamates as prostigmine and 2-isopropyl N-methyl-N-phenylcarbamate seemed to bind primarily at the α-site.[33] Evidence for even further divisions and alternate descriptions of the reactive center of AChE also exist.[21]

Regardless of these minor variations in the proposed mechanisms of carbamate-AChE reaction, there is little doubt that certain steric and electronic features must be present if a carbamate molecule is to successfully mimic acetylcholine. Bioassay for active structures generally involves a Warburg manometric measurement of inhibition of housefly head AChE at various concentrations of carbamate in order to obtain the 50% inhibition level (I_{50}).[1,35] Such an in-vitro assay does not account for problems that may arise with living insects, such as cuticular penetration and metabolic destruction, but the procedure is very reliable for biochemical definition of the requirements needed for maximum anticholinesterase activity. Structure–toxicity studies, where the candidate insecticide is applied to the living insect and an LD_{50} is determined, do take these factors into account, although the co-administration of synergists may rule out metabolic differences. Finally, compounds that are potent inhibitors of insect AChE must be tested for their effects on vertebrate AChE and for mammalian toxicity.

FACTORS DETERMINING INSECTICIDAL ACTIVITY OF AROMATIC CARBAMATES

Basic Structure and Effects of Carbamate N-Substituents

At this point it should be obvious that the most successful carbamate insecticide structures consist of a relatively simple aromatic portion and either a dimethyl- or monomethylcarbamate portion. This basic format was established very early and, with the exception of the oxime carbamates, has differed little over the past 25 years. Just as the first group of pharmacologically active carbamates was patterned after the naturally occurring physostigmine and after derivatives of prostigmine and methylprostigmine[36,37] (Figure 4.1), so were initial insecticidal carbamates which resulted in the familiar substituted phenyl N-methylcarbamate structure of today.[8]

Interestingly, during the early development of carbamate structure–activity correlations, only minor attention appears to have been given to the importance of aromaticity and to the presence of the oxygen atoms in the ester part of the molecule. Later work revealed that removal of the aromatic character by replacing the planar benzene ring with a chair-shaped cyclohexane ring destroyed all insecticidal activity.[38,39] For example, the housefly head AChE I_{50} value, housefly toxicity (synergized), and mosquito larval LC_{50} changed from 6×10^{-6} *M*, 24 μg/g, and 0.56 ppm respectively to $> 10^{-3}$ *M*, > 500 μg/g, and > 10 ppm in a comparison of 2-isopropylphenyl N-methylcarbamate to 2-isopropylcyclohexyl N-methylcarbamate. Very similar changes took place when a methylene group was inserted between the ring and carbamic portions of Baygon to give 2-isopropoxybenzyl N-methylcarbamate.

This separation or removal of the aromatic nucleus from the carbamate moiety reduces the reactivity of the OC=O center of the molecule. Thus, interaction and reaction of the carbamate with acetylcholinesterase should be expected to decline. In addition, the changes in configuration alter the fit of the compound to the enzyme. Lastly, there is probably considerable π–π hydrophobic attraction between the ring of phenyl methylcarbamates and the anionic site of AChE, but this is eliminated when benzene is exchanged for cyclohexane. These reasons also partially explain the inactivity of straight-chained aliphatic carbamates, though some insect toxicity has been obtained with various chlorinated ethyl and propyl N,N-dimethyl- and N,N-diethylcarbamates.[40]

Changing the carbamate ester linkage to thiol (SC=O), thiono (OC=S), or dithio (SC=S) reduced the anticholinesterase activity of 2-isopropylphenyl N,N-dimethylcarbamate by 12 to 40 times.[12] None of the thiocarbamates were toxic to houseflies or *Culex* larvae, except for the thiono analog, which gave mosquito kill about equal to that of the parent compound. It is possible that the living insect oxidized the thiono sulfur to produce an active carbamate, similar to the P=S to P=O activation which occurs with many thionophosphates.

The reduced in-vitro anticholinesterase activity of thiono N,N-dimethylcarbamates results from poorer affinity and less enzyme carbamylation, brought about by a decrease in the electrophilic character of the C=S carbon as compared to the C=O carbon.[12] This, of course, would also make in-vivo desulfuration of C=S to C=O very difficult. Synthesis of thiono methylcarbamates was not achieved, but a number of thiol N-methylcarbamates have been shown to be nontoxic to three species of mosquitoes.[12,41] The thiol counterpart of Mobam also possessed considerably reduced anticholinesterase activity and was nontoxic.[42]

The discovery of insecticidally active Isolan-type dimethylcarbamates has often been compared to the almost simultaneous appearance of phenyl N-methylcarbamates, with the implication that both groups of chemicals resulted from modifications of medicinal carbamates as described above. However, although the activity of such compounds as prostigmine probably influenced the decision to synthesize and evaluate various dimethylcarbamates as insecticides, the original compounds appeared during a search for useful insect-repellent agents.[10,43]

Compounds such as phenyl and dimethyldihydroresorcinol N,N-diethylcarbamate (Figure 4.3) had good repellent action, and when the N,N-dimethyl analog of the resorcinol was made (dimetan), it was found to be not only repellent, but also toxic to houseflies. Alteration of the carbamate N-alkyl groups or substitution of one or two sulfur atoms for carbamate oxygen atoms canceled activity. These results set the stage for commercial development of the more toxic Isolan, a systemic aphicide, and of dimetilan, a fly stomach poison (Figure 4.4). More recent alteration of the ring substituent of Isolan (Figure 4.4, I) created a broad-spectrum insecticide with much more favorable mammalian toxicity.[44]

A question that is asked early in any examination of carbamate structure–activity relationships is, "Why are heterocyclic enol carbamate insecticides always dimethylcarbamates, whereas substituted phenyl carbamates are invariably monomethylcarbamates?" Some examples comparing toxicity, anticholinesterase activity, and hydrolytic stability of various N-methyl- and N,N-dimethylcarbamates make the answer very apparent. For example, when the monomethyl derivatives of Isolan-type insecticides were studied, it was found that these structures were less toxic

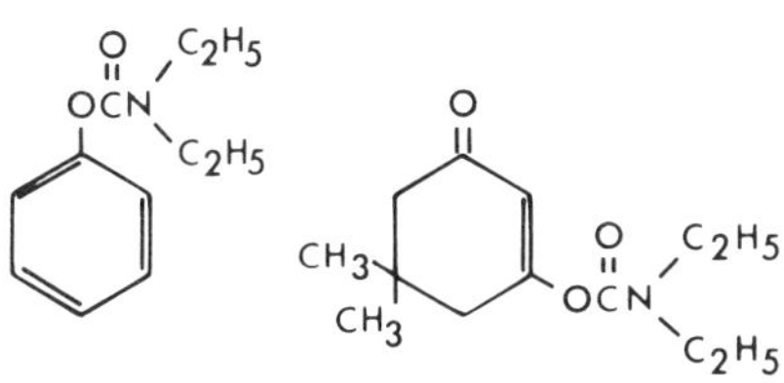

FIGURE 4.3. Structures of carbamate insect repellents.

to warm-blooded animals without significant loss in insecticidal properties.[44] However, an extreme sensitivity to alkaline hydrolysis (half-life of 0.5 min at pH 7.8) prohibited their use as commercial insecticides.

With phenyl esters of carbamates, anticholinesterase activity of dimethyl structures was shown to be from 4 to 150 times poorer than that of corresponding monomethyl compounds.[45-48] (Table 4.1), though there are examples where both structures inhibited AChE equally.[49] Generally, effectiveness as insecticides decreased accordingly, though relative toxicity within each series was similar, suggesting that variation in specificity was due mostly to changes in the phenol moiety. When flies were pretreated with piperonyl butoxide to block metabolism, a better relationship between anticholinesterase activity and topical toxicity was obtained, except for 3-isopropylamidophenyl carbamates, which were not very toxic as dimethyl or methyl analogs. Dimethylation of bendiocarb (2,2-dimethyl-1,3-benzodioxyl-4-yl N-methyl-carmate) reduced its toxicity to aphids, mosquitoes, and cabbage white catepillars.[50]

Other studies have shown that, under certain structural circumstances, phenyl dimethylcarbamates can be quite toxic. Thus, houseflies, bean weevils, rice stem borers, and green peach aphids were more susceptible to some cyano-substituted phenyl N,N-dimethylcarbamates than to corresponding N-methylcarbamates.[51] Dichlorophenyl and naphthyl N,N-dimethylcarbamates were also insecticidal to houseflies and mosquito larvae,[52] and 2-(1,3-dithiolan-2-yl)phenyl N,N-dimethylcarbamate (Figure 4.4, II)

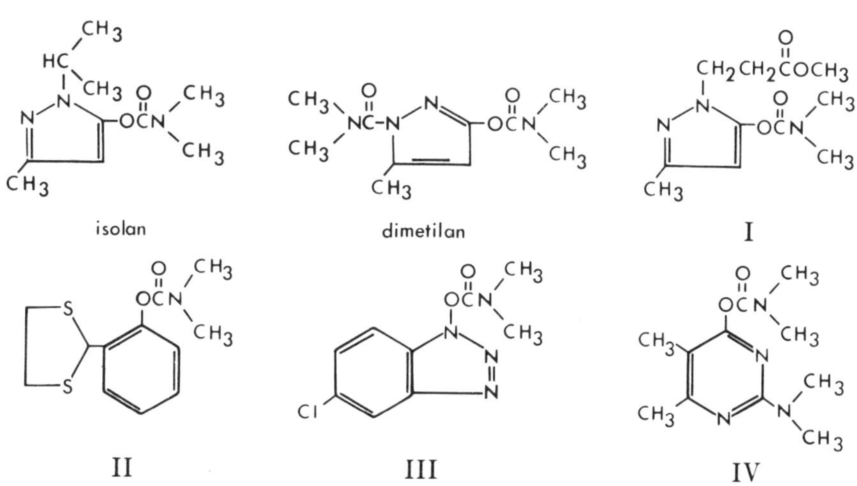

FIGURE 4.4. Insecticidally active N,N-dimethylcarbamates.

TABLE 4.1

Inhibition of Housefly Head AChE by Various Aryl N-Methyl- and N,N,-Dimethylcarbamates[4 5-4 7]

Aryl group	Molar I_{50} for carbamate group: $OC(O)NHCH_3$	$OC(O)N(CH_3)_2$	$N(CH_3)_2/NHCH_3$
Phenyl (Ø)	2×10^{-4}	8×10^{-4}	4
Naphthyl	9×10^{-7}	9.5×10^{-6}	9
2-*i*-C_3H_7 Ø	6.9×10^{-7}	1.3×10^{-5}	19
3-*i*-$C_3H_7C(O)NH$ Ø	5×10^{-6}	1.5×10^{-4}	30
3-*t*-C_4H_9 Ø	4×10^{-7}	1.8×10^{-5}	45
3-*i*-C_3H_7 Ø	3.4×10^{-7}	5×10^{-5}	150

was found to be systematically as toxic as Isolan to various aphid species.[53] The structurally different benzotriazolyl N,N-dimethylcarbamates were not active against mosquitoes or houseflies, but compound III was poisonous to some insect species.[54] Very selective diptera and aphid mortality was found for pirimicarb[55] (Figure 4.4, IV). Finally, when analyzed with bovine erythrocyte AChE, a number of dimethyl analogs of II and III inhibited activity as well as did their monomethyl counterparts,[49,54] though this is generally not the rule with phenyl carbamate esters (cf Chapter 3).

The much greater tendency for better anticholinesterase activity in phenyl methylcarbamates, compared to that in dimethylcarbamates, implies that this structure is more complementary to the esteratic site of insect AChE, both sterically and electronically, and/or that it is more efficient at carbamylation of the serine hydroxyl subsite. Although the latter is of primary importance with vertebrate AChE (Chapter 3), there are insufficient kinetic data to determine which is most critical for insect AChE. However, the existing evidence favors an enhanced affinity between the monomethylcarbamates and the reactive center of housefly acetylcholinesterase.

Differences in hydrolytic stability of phenyl methyl- and dimethylcarbamates might also explain variable anticholinesterase and insecticidal activity. Consider the fact that 4-nitrophenyl N-methylcarbamate ($k_{hyd} = 5.4 \times 10^5 M^{-1}$ min^{-1}) hydrolyzed about 10^7 times faster than its dimethyl counterpart. At that rate, the compound was almost completely destroyed during assay for AChE inhibition, and it was not very persistent in vivo during toxicity determinations. Thus, most of the nitrophenyl esters were found to be poor insecticides. Other methylcarbamates were not as readily hydrolyzed, but under basic conditions they all were 10^4 to 10^7 times more susceptible to hydroxyl attack than were dimethyl analogs.[45,56,57]

Such tremendous differences in hydrolysis rate cannot be rationalized by electronic or steric effects caused by addition of a second methyl group to the carbamate nitrogen atom. It has been suggested that a different mechanism is operating on the two structures, i.e., dimethylcarbamates hydrolyze through nucleophilic attack on the carbonyl carbon atom, whereas methylcarbamates react through initial removal of the hydrogen atom attached to nitrogen.[11,58] Perhaps a combination of these mechanisms occurs at the esteratic site of AChE during methylcarbamylation, with the N-hydrogen atom reacting with the basic subsite (B) and the carbonyl carbon reacting with the serine hydroxyl subsite (OH) (Figure 4.2). This could possibly result in higher k_2 constants, which in turn would yield lower I_{50} values and greater inhibition by monomethyl compounds.

Allowing for the activity of methyl and dimethylcarbamates, the next question is, "What happens when other alkyl groups are attached to the carbamate nitrogen atom?" Again, the inactivity of these structures was determined early in carbamate insecticide history. A change from N-methyl to N-ethyl, -phenyl, or -benzyl resulted in a dramatic loss of anticholinesterase activity for a variety of phenolic carbamate esters (Table 4.2).[8,12] Even the small change from an N-methyl to an N-ethyl group severely reduced enzyme inhibition with most compounds. These findings have been substantiated by numerous other studies in which anticholinesterase activity and/or insect toxicity were reduced when the N-methyl group was replaced by a wide range of alkyl and aryl substituents.[8,12,42,59,60]

The order of activity of the N-alkyl substituents

TABLE 4.2

Inhibition of Housefly Head AChE by Various Phenyl N-Substituted Carbamates[8]

	Molar I_{50} for N-substituent			
Aryl group	CH_3	C_2H_5	ϕCH_2	ϕ
Phenyl (ϕ)	2×10^{-4}	7×10^{-4}	8×10^{-3}	$>1 \times 10^{-3}$
3-CH_3 ϕ	8×10^{-6}	4.6×10^{-4}	$>1 \times 10^{-3}$	$>1 \times 10^{-3}$
3-*t*-C_4H_9 ϕ	4×10^{-7}	2×10^{-5}	1×10^{-3}	$>1 \times 10^{-3}$
2-*i*-C_3H_7, 5-CH_3 ϕ	1.4×10^{-6}	2×10^{-5}	3×10^{-3}	$>1 \times 10^{-3}$

($CH_3 > C_2H_5 > C_6H_5CH_2 > C_6H_5$) was explained by a decrease in favorable fit on the esteratic site of AChE, suggesting that steric requirements for the carbamate end of the molecule were very rigid.[8] It was also proposed that the groups may have altered the reactivity of each compound's carbonyl center, since the order of activity followed the ease with which the substituents could release their electrons. In other words, hydrolytic stability and reaction with the active center of AChE might be affected.

This hypothesis was tested years later by synthesizing numerous other alkylcarbamate structures. The deuterated derivative ($-NHCD_3$) of UC 10854 was prepared and found to have similar insecticidal and anticholinesterase activity; this was interpreted to mean that the N-methyl group did not react chemically with AChE although it did interact sterically.[61] When the $-NHCF_3$ isostere was synthesized and evaluated, it was found to be inactive, presumably because the fluorinated carbamate hydrolyzed about 2,000 times faster. Finally, a number of $-NH_2$ carbamates were also shown to be noninsecticidal (I_{50} = $>1 \times 10^{-3}$ to 3.7×10^{-4}*M*), with enzyme inhibition abilities about 0.001 of those of their monomethyl counterparts. More rapid hydrolysis (100 times) of $-NH_2$ compounds, compared to that of $-NHCH_3$ compounds, and lack of a methyl group needed for site interaction can explain inactivity of the unsubstituted carbamates.

This lack of activity of alkylcarbamates other than methylcarbamates was re-examined more critically using kinetics of alkyl group interactions with bovine, housefly, and honeybee acetylcholinesterase.[29] A sharp decrease in bimolecular rate constants (k_i), which is inversely related to I_{50}, occurred with all enzymes when 3-dimethylaminophenyl or 2-dithiolanephenyl N-methylcarbamates were changed to N-ethylcarbamates. Interestingly, the reason for the drop in activity varied with the AChE; that is, for bovine and bee AChE it was due mainly to a decrease in carbamylation rate (k_2), but for fly AChE it was due to a lowered affinity (K_a). The ethylcarbamate was a much less potent inhibitor of erythrocyte AChE than of insect AChE.

Certain other differences in K_a and k_2 were discovered for N-propyl-, N-butyl-, and N-hexylcarbamates, but the overall effect on k_i caused a general decrease in anticholinesterase activity with an increase in alkyl chain length. Altering the alkyl group of bendiocarb-type methylcarbamates to propargyl ($CH_2C{\equiv}CH$) or allyl ($CH_2CH{=}CH_2$) lowered toxicity to several insect species.[50]

With respect to dialkylcarbamates other than dimethylcarbamates, little research has been conducted for obvious stereochemical reasons. An example of their inactivity can be seen by comparing the fly AChE I_{50} value for UC 10854 (4.5×10^{-7}*M*) with that of its N,N-diethyl sister ($>1 \times 10^{-3}$*M*).[61] The latter compound was not toxic to *Culex* mosquito larvae or to adult houseflies, even in the presence of a synergist.

A very innovative alteration of the carbamate moiety of insecticides involves replacing the N-proton with an acyl group.[62] Originally, it was hoped that attaching such a labile handle to the carbamate nitrogen would provide an insecticide precursor with more desirable environmental properties without sacrificing toxicity to insects. Following the synthesis of more than 250 acylcarbamates, no obvious relationship between insect toxicity and the nature of the acyl group was apparent, although the usual rule was that acylated compounds were less toxic than their parents.[62,63] Within a narrow range of acyl substituents, however, some correlation of structure with insect toxicity has been obtained.[64]

One fortunate property of many of these carbamates is their pronounced reduction in activity towards mammals without significant loss of topical or ingested effects on insects, as illustrated in Table 4.3.[65] This dramatic decrease in mammalian toxicity and maintenance of insecticidal activity has been demonstrated with several other acetylated carbamates.[12,66,67] For example, the mouse oral LD_{50} of Zectran, 3-methyl-4-aminophenyl N-methylcarbamate, and 3-methyl-4-methylthiophenyl N-methylcarbamate ranged from 20 to 50 mg/kg, but after acetylation the LD_{50} was > 1,000 mg/kg for all compounds.[66] At the same time, toxicity to the spruce budworm remained about the same (1 to 15 μg/g). N-acetyl Zectran has also been found to be one of the few carbamates that are equally toxic to both the tobacco budworm and the cotton bollworm.[68]

In some instances, acetylation of certain methylcarbamates did not result in retention of sufficient toxicity to insects, for example, sheep blowfly larvae,[69] bollworms,[68] cattle tick, poultry red mite,[65] aphids, mosquitoes, and cabbage white caterpillars.[50] In other cases, even though toxicity of such carbamates as Baygon to adult mosquitoes was maintained after N-acetylation, the increase in volatility of the acyl derivatives rendered the compounds useless as residual insecticides.[70] Finally, acylation of oxime carbamates, such as methomyl and thiocarboxime [1-(2′-cyanoethylthio) ethanaldoxime N-methylcarbamate], resulted in loss of toxicity to both mammals and insects.[67]

TABLE 4.3

Toxicity of Various N-Acyl Derivatives of Baygon

N-Acyl group	LC_{50} to vetch aphid (ppm) Contact	Systemic	Oral LD_{50} to mice (mg/kg)
None (Baygon)	7	10	100
$C(O)CH_3$	13	55	370
$C(O)C_2H_5$	150	>1,000	350
$C(O)CH_2OCH_3$	14	11	60–250
$C(O)CH_2Cl$	8	11	>1,000
$C(O)CHCl_2$	5	12	250–1,000

Adapted from Reference 65, with permission.

Unlike most structure–activity studies with carbamates, toxicity of acylated derivatives cannot be measured by the degree of cholinesterase inhibition. Quite often, acetylation reduces enzyme inhibitory powers by two to three orders of magnitude, as illustrated in Table 4.4 with honeybee head AChE.[63,71] But with increasing acyl alkyl chain length, anticholinesterase activity was found to increase, though it never reached a par with unacylated parent compounds. Lowered AChE inhibition with short-chained N-acylated UC 10854 derivatives was similar with housefly head enzymes.[12,61]

It is surprising that the N-acyl-N-methylcarbamates inhibit AChE as well as depicted in Table 4.4, considering the extra burden placed on conformational fit by the bulky acyl moiety. Although there is some indication that the acyl compounds might act by an alternate mechanism, it seems more likely that AChE inhibition measured in vitro may have been due to cleavage of the acyl group, releasing small amounts of parent methylcarbamate, which in turn inhibited the enzyme.[67] Thus, some other explanation is needed to account for these large differences in bee and fly I_{50} values between acylated and parent methylcarbamates when in-vivo toxicity only varies by 2 to 10 times.[61,63,72]

The most reasonable hypothesis is that the N-acyl-N-methylcarbamates are deacylated in vivo to produce toxic methylcarbamate insecticides. This would explain their lethality to insects, but why the acyl compounds are generally less toxic to mammals is of interest. Apparently, the

TABLE 4.4

Inhibition of Honeybee Head AChE by Various N-Acyl-N-Methylcarbamate Insecticides[63,71]

N-Methylcarbamate	Molar I_{50} for N-acyl group: H	$CH_3C(O)$	$C_3H_7C(O)$	$C_4H_9C(O)$	$C_6H_{13}C(O)$
Carbaryl	4.8×10^{-8}	4.0×10^{-6}	8.5×10^{-7}	6.0×10^{-7}	–
UC 10854	2.2×10^{-8}	2.0×10^{-5}	–	8.4×10^{-6}	8.0×10^{-6}
Baygon	7.5×10^{-7}	6.5×10^{-4}	–	3.0×10^{-4}	2.3×10^{-6}
Zectran	1.25×10^{-8}	4.0×10^{-6}	–	–	–

acylmethylcarbamate is hydrolyzed in mammals much more readily than it is deacylated, so that the acyl group prevents AChE inhibition while enhancing metabolic detoxication. Studies using kinetics of N-acetyl-N-methylcarbamate inhibition, base-catalyzed solvolysis, metabolite characterization, and in-vivo enzyme inhibition have provided substantial evidence for the above proposals.[67,72,73]

A few other alterations of the carbamate portion of the molecule are worthy of brief mention. Some success has been achieved by substituting the carbamate proton with a thiophenyl group, producing compounds such as RE 11775, 3-(2-butyl)phenyl N-methyl-N-phenythiomethylcarbamate (Figure 4.5, I), which is a potent mosquito larvicide,[12,74,75] and the toluenesulfenyl derivative of carbofuran, which is twice as toxic to houseflies as the parent carbamate.[76] Addition of an O,O-dimethylphosphorothioyl group to various insecticidal N-methylcarbamates (Figure 4.5, II) was shown to decrease mammalian toxicity and anticholinesterase activity, but to increase toxicity to flies.[77] Doubling the carbamate end of the molecule by making biscarbamates, such as III in Figure 4.5, resulted in a loss of toxicity to houseflies and to *Culex* mosquito larvae, though the charged trimethylammonium compounds were active inhibitors of insect and vertebrate AChE.[28] Lastly, evaluation of a few N-hydroxy- and N-methoxy-N-methylcarbamates showed them to be true reversible inhibitors of bovine erythrocyte AChE with high affinity constants (K_a), though their toxicity was not described.[78]

Effects of Charge

Even though carbamate compounds containing charged nitrogen atoms are not toxic to insects, some of these materials have been prepared for various structure–activity studies. As an example, the activity of certain phenyl N-methylcarbamates against housefly AChE is listed in Table 4.5.[12,38,79] Quaternization of the dimethylamino nitrogen in the 3- or *meta*-position and in the 4- or *para*-position on the phenyl ring enhanced anticholinesterase activity 450- and 70-fold respectively when compared to their uncharged analogs. However, if the 3,5-dimethylaminophenyl N-methylcarbamate was quaternized in only one position [3-$N(CH_3)_2$, 5-$N^+(CH_3)_3$], cholinesterase inhibition increased only 70-fold, whereas diquaternization resulted in only a 20-fold increase. Presumably, the unexpected smaller change in activity of the multicharged structure was due to a higher degree of inherent hydrolytic instability.

Among the uncharged monoaminophenyl compounds, substitution in the 2- or *ortho*-position on the ring gave the most active structure. This implies that the *ortho*-structure provided optimal complementarity to the anionic site of AChE. Forming the quaternary amino derivative

TABLE 4.5

Effects of Quaternization on the Anticholinesterase Activity of Substituted Phenyl N-Methylcarbamates[12,38,79]

Phenyl substituent	Molar I_{50} for housefly head AChE
2-$N(CH_3)_2$	2.0×10^{-6}
2-$N^+(CH_3)_3$	1.0×10^{-5}
3-$N(CH_3)_2$	8.0×10^{-6}
3-$N^+(CH_3)_3$	1.8×10^{-8}
4-$N(CH_3)_2$	2.4×10^{-4}
4-$N^+(CH_3)_3$	3.5×10^{-6}
3,5-di-$N(CH_3)_2$	2.6×10^{-6}
3-$N(CH_3)_2$, 5-$N^+(CH_3)_3$	3.7×10^{-8}
3,5-di-$N^+(CH_3)_3$	1.2×10^{-7}
2-S CH_3	9.0×10^{-7}
2-$S^+(CH_3)_2$	1.5×10^{-5}
3-S CH_3	7.0×10^{-6}
3-$S^+(CH_3)_2$	6.5×10^{-7}
3-$P(C_2H_5)_2$	7.4×10^{-7}
3-$P^+CH_3(C_2H_5)_2$	3.6×10^{-8}

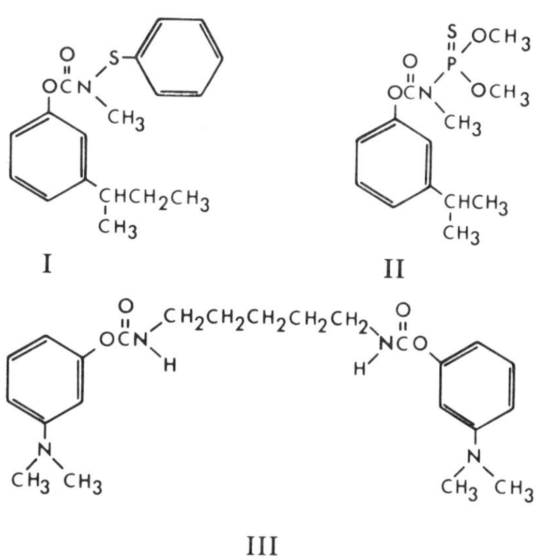

FIGURE 4.5. Insecticidally active N-substituted carbamates.

of 2-dimethylaminophenyl N-methylcarbamate did not enhance activity, however; in fact, the charged molecule was less effective. It is probable that quaternization altered orientation of the carbamate onto the enzyme surface, so that in one case inhibition was decreased (*ortho*), whereas in others it was increased.

Studies with other compounds confirmed that the better the fit of the uncharged carbamate to the anionic site, the less was affinity improved by formation of the charged analog. Also, additional optimally placed alkyl groups on the phenyl ring increased attraction through interactions with the hydrophobic area surrounding the anionic site. Thus, trimethylammonium anticholinesterases were synthesized with very potent I_{50} values, from 2×10^{-8} to $6 \times 10^{-9} M$. Similar structures have also proved to be extremely successful inhibitors of bovine erythrocyte and electric eel acetylcholinesterase.[28,29,80]

Conversion of an appropriate ring methylthio substituent to a charged dimethylsulfonium group increased anti-AChE activity in an analogous manner[12] (Table 4.5). As with nitrogen, alteration of the sulfur atom in the *ortho*-position was not found to enhance I_{50}, but a charge in the *meta*- or *para*-positions increased inhibition over parent methylthiophenyl N-methylcarbamates. However, the differences were not as pronounced with sulfur as with nitrogen. Finally, quaternization of a *meta*-diethylphosphoro group also lowered the concentration of carbamate needed to inhibit 50% of AChE. Thus, even though all of the charged compounds are nontoxic to insects, they are extremely potent inhibitors of acetylcholinesterase. The easiest explanation for this is coulombic attraction between opposite charges on inhibitor and enzyme; these are forces that act over a longer distance than van der Waal or hydrophobic forces. However, the hydrophobic effect of adding another methyl group to obtain quaternization should not be overlooked and may prove to be equally important.[21]

Effects of Asymmetry

Once at the site, spatial fit onto the enzyme surface is necessary. With substituted phenyl N-methylcarbamates it is the size and shape of the ring substituents that determine complementarity. In most cases these moieties do not contain asymmetric centers, so that optical isomerism is not a factor. However, a study with the *d*- and *l*-isomers of 2-*s*-butylphenyl N-methylcarbamate revealed that the *l*-form ($I_{50} = 10^{-6} M$) was six times more active than the *d*-form as an insect anticholinesterase.[81] But when the asymmetric carbon was removed from the phenyl nucleus by interspacing a sulfur atom to give the corresponding 2-*s*-butylthiophenyl isomers, there was no difference in activity between the enantiomorphs, though both were better inhibitors ($I_{50} = 10^{-7} M$) than the nonsulfur compounds. This suggested that insertion of the sulfur atom, with consequent free rotation between the ring and side chain, canceled the effects of asymmetry. It is also possible that the presence of the polar sulfur atom may have changed interactions with the negatively charged anionic site.

Variation of AChE inhibition with side chain asymmetry can only be effected through the anionic site, since there should be no difference in intermolecular attraction or carbamylation ability of *d*- and *l*-isomers (see Figure 4.2). Thus, the greater anticholinesterase activity of the *l*-form of 2-*s*-butylphenyl N-methylcarbamate was probably due to greater affinity (K_a) for the enzyme, but this has not been confirmed kinetically. Results with optical isomers of other substituted phenyl N-methylcarbamates,[49] organophosphorus insecticides,[82] and choline esters[23] also point to stereospecific requirements at the reactive center of acetylcholinesterase. With both carbamate and organophosphorus insecticides, in-vitro anticholinesterase activity of enantiomorphs was consistent with in-vivo toxicity to houseflies and to mosquito larvae.

Effects of Phenyl Substitution

The position, type, and number of substituents attached to the aromatic nucleus of phenyl N-methylcarbamates can greatly alter anticholinesterase properties as well as insect toxicity. Hence, it is compounds with structural modifications of the phenyl ring that are compared in the vast majority of carbamate structure–activity studies, and the data generated from these projects will be briefly summarized on the following pages. More detailed information concerning I_{50} values and toxicity determinations can be found in the referenced articles.

Alkyl, Alkoxy, and Alkylthio Groups

Because of the anticholinesterase potency of methylprostigmine (3-trimethylammoniumphenyl

N-methylcarbamate), one of the first uncharged carbamates to be synthesized and evaluated as an insecticide was the corresponding carbon isostere, i.e., 3-*t*-butylphenyl N-methylcarbamate (Figure 4.6).[8] The resulting molar I_{50}, when measured against housefly AChE, was shown to be 4 X 10^{-7}, which was not quite as good as the I_{50} of methylprostigmine (1.8 X 10^{-8}*M*). Synthesis and examination of other alkyl-substituted candidates, a small sample of which are listed in Table 4.6, revealed that 3-*s*-butylphenyl N-methylcarbamate was an even better enzyme inhibitor than the *t*-butyl analog.[12,83] In fact, the general trend of anticholinesterase activity and toxicity to houseflies, *Culex* mosquitoes, and salt marsh caterpillars for alkyl group substitution was in the order $s\text{-}C_4H_9 > t\text{-}C_4H_9 = i\text{-}C_3H_7 > C_2H_5 > CH_3$.

Placement of the substituent on the ring relative to the carbamate moiety was also important. As the size of the alkyl group increased, substitution in the 3-position yielded the most active compounds, so that the usual order of effectiveness was 3 > 2 > 4 (Table 4.6). This was to be expected, since molecular models showed that the distance between the carbamate group and the center of a *meta*-substituent was approximately 5 Å,[83] the same distance between the functional groups of acetylcholine and the same distance that

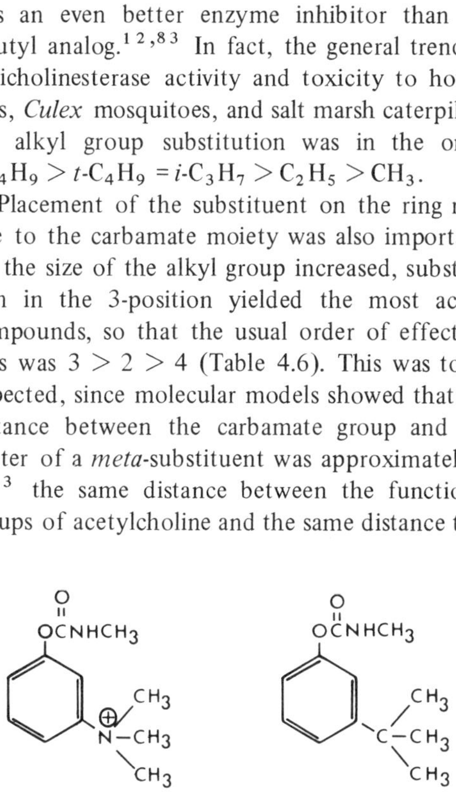

FIGURE 4.6. Carbon-nitrogen isosteres of *meta*-substituted phenyl N-methylcarbamates.

TABLE 4.6

Inhibition of Housefly Head AChE by Various Alkylphenyl N-Methylcarbamates[12,83]

Alkyl substituent	Molar I_{50}	Alkyl substituent	Molar I_{50}
None	2 X 10^{-4}	3-i-C_3H_7	3.4 X 10^{-7}
2-CH_3	1.4 X 10^{-4}	3-t-C_4H_9	4 X 10^{-7}
3-CH_3	1.4 X 10^{-5}	3-s-C_4H_9	1.6 X 10^{-7}
4-CH_3	1 X 10^{-4}	2-cyclo C_5H_9	1.1 X 10^{-6}
2-C_2H_5	1.3 X 10^{-5}	3-cyclo C_5H_9	1.5 X 10^{-6}
3-C_2H_5	4.8 X 10^{-6}	2-cyclo C_6H_{11}	1.4 X 10^{-6}
4-C_2H_5	3.8 X 10^{-5}	3-cyclo C_6H_{11}	2.0 X 10^{-6}

proposedly separates the anionic site and the esteratic site of AChE. Thus, *meta*-substitution should provide greater affinity of substrate for enzyme; this has been documented by kinetic studies with 2- and 3-isopropylphenyl N-methylcarbamate and electric eel AChE, where the carbamylation rate (k_2) was found to be the same for the two compounds, but the affinity (K_a) was about eight times better with 3-substitution.[80]

Data for the cyclopentyl and cyclohexyl groups demonstrated that the *ortho*-position was slightly preferred over the *para*-position (Table 4.6). This is probably due to the bulkiness of the substituents, so that the 2-additions were able to mimic placement in the 3-position. The fact that inhibition decreased with the larger groups may be due to the rigidity imposed by cyclization, producing a more limited van der Waal and hydrophobic interaction at the anionic site.

The effects of substitution of longer-chained alkyl groups was examined, using bovine erythrocyte AChE as the activity screen.[84] First of all, with the 2-, 3-, and 4-*t*-butylphenyl N-methylcarbamates, I_{50} values very similar to those reported for fly AChE were obtained. There were some differences among *n*-, *s*-, and *t*-butyl when placed *meta* to the carbamate grouping, but the *s*-configuration proved to be the best inhibitor (I_{50} = 7 X 10^{-8}*M*). Moving the alkyl branching about a bit did alter activity somewhat, and with a $\text{-}CH(C_2H_5)_2$ substituent in the 3-position, the most potent anticholinesterase of any known simple alkylphenyl N-methylcarbamate was uncovered (I_{50} = 4.1 X 10^{-8}*M*).

A homologous series of α-substituted alkyl chains attached to the phenyl ring in the 3-position showed an increase in inhibitory power of the carbamate from *i*-propyl to *s*-butyl to *s*-amyl, then a decrease as the alkyl tail lengthened, though activity was relatively good even with nine carbons (7.1 X 10^{-6}*M*).[84] Thus, maximum inhibition, and presumably maximum interaction at the anionic site, occurred with a four-carbon chain. This was also true when two α-methyl groups were present on the substituent [$\text{-}C(CH_3)_2C_3H_7$, I_{50} = 4.2 X 10^{-8}*M*]. In contrast, β-branched alkyl chains contributed very little toward anticholinesterase activity.

The requirement for α-branching undoubtedly relates to fit at the anionic site and is reminiscent of early studies with aminoethyl acetate.[85] As methylation of the acetate's amino nitrogen progressed toward acetylcholine, the affinity or binding strength to vertebrate AChE increased fourfold with addition of the first methyl group, 14-fold with the second methyl addition, and not at all with the third group. These results imply a definite shape requirement for maximum van der Waal and hydrophobic attraction of substrate and enzyme (presumably the tetrahedral arrangement around the nitrogen atom makes the third methyl group project away from the enzyme surface, so that no additional effect was observed).

Strikingly similar results were obtained when comparing insect AChE inhibition by unsubstituted phenyl N-methylcarbamate relative to 3-methyl-, 3-ethyl-, 3-*i*-propyl-, and 3-*t*-butylphenyl N-methylcarbamate[83] (Table 4.6). Although the near-identical I_{50} values for the last two compounds implies that the additional methyl group did not alter affinity, breaking I_{50} down into component affinity (K_a) and carbamylation constants (k_2) for these carbamates has been accomplished with bovine erythrocyte AChE.[86] In this case, though not necessarily identical to housefly AChE, the affinity of 3-isopropylphenyl N-methylcarbamate was only half that of the 3-*t*-butylphenyl analog. However, the carbamylation ability of the 3-isopropyl compound was three times better, so that the overall inhibition was about the same for the two carbamates.

The toxicity to insects of these chemicals generally follows their anticholinesterase activity, though there is considerable species variation, so that some insects are more easily disposed of with phenyl N-methylcarbamates containing *ortho*-substituents and others are more efficiently killed with *meta*-substituted compounds.[84,87-89] Could this mean that different insects have different acetylcholinesterases? It seems more likely that different insects have different metabolic capacities. The commercial and experimental success of alkyl- and dialkylphenyl N-methylcarbamates is evidenced by compounds such as Bux, UC 10854, HRS-1422, promecarb, Butacarb, and Moebal.

Synthesis and systematic evaluation of structure–activity correlations among alkoxyphenyl N-methylcarbamates actually took place before the analysis of alkyl compounds.[90] The results were of the same order in each instance, however, as shown in Table 4.7. The one major difference was that replacement of the proton in the 2-position of the phenyl ring with alkoxy

TABLE 4.7

Inhibition of Housefly Head AChE by Various Alkoxyphenyl N-Methylcarbamates[83,90]

Alkoxy substituent	Molar I_{50}	Alkoxy substituent	Molar I_{50}
None	2×10^{-4}	2-O-*i*-C_3H_7	6.9×10^{-7}
2-OC_2H_5	1.6×10^{-5}	3-O-*i*-C_3H_7	9.2×10^{-6}
3-OC_2H_5	6×10^{-6}	2-OC_4H_9	1.2×10^{-5}
4-OC_2H_5	7×10^{-5}	2-O-*s*-C_4H_9	3.1×10^{-7}
2-OC_3H_7	8.7×10^{-6}	2-O-cyclo C_5H_9	4×10^{-7}
3-OC_3H_7	1.6×10^{-5}	3-O-cyclo C_5H_9	8×10^{-6}

groups generally caused the greatest anticholinesterase response. It was decided that this difference was due to the presence of the extra bond between the ring and the oxygen atom, which changed the orientation of the alkyl group relative to the enzyme surface. Thus, the critical 5-Å spread between carbamate moiety and alkyl cluster was best approached with *ortho*-substitution of alkoxy groups.

Once again, the advantages of alkyl-branching can be appreciated by comparing the activities of propyl versus *i*-propyl and of butyl versus *s*-butyl substitution (Table 4.7). It should be noted that increasing the size from straight-chained propoxy to butoxy gave only a slight change in cholinesterase inhibition. In addition, substitution of a butoxy group for the 3-proton on the ring yielded a better I_{50} value ($9.4 \times 10^{-6}M$) than that obtained with 2- substitution, indicating the ability of these long-chain groups to bend around and interact at the anionic site regardless of their position of attachment to the phenyl nucleus.

When comparing the anticholinesterase activity of *meta*-substituted alkylphenyl N-methylcarbamates with that of corresponding *ortho*-substituted alkoxy compounds (those containing the same alkyl moiety), it can be seen that the alkylphenyl carbamates are about twice as potent in each instance. This is probably because in orienting the alkoxy group onto the anionic site it is necessary to place the phenyl ring further away from the enzyme surface, thus decreasing π–π hydrophobic bonding between the ring and enzyme, which in turn causes lower affinity and poorer anticholinesterase activity.

Toxicity of alkoxy analogs to houseflies (synergized), mosquito larvae, and salt marsh caterpillars paralleled anticholinesterase activity.[83,90] However, the need for synergism with piperonyl butoxide to obtain maximum fly kill was much greater than for corresponding alkylphenyl N-methylcarbamates, suggesting an enhanced susceptibility to in-vivo metabolic detoxication of the alkoxy carbamates. The most successful alkoxyphenyl N-methylcarbamate has been Baygon.

The first report of a biologically active thiophenyl N-methylcarbamate appeared in 1944, when the charged methylsulfonium salt of 3-methylthiophenyl N-methylcarbamate was found to be nearly as toxic to mammals as prostigmine.[91] Insecticidal activity of carbamate esters of phenols, e.g., 4-methylthiophenol, 3-methyl-4-methylthiophenol, and 3,5-dimethyl-4-methylthiophenol (Mesurol phenol), was not recognized until several years later.[92] The next step was completed with detailed structure-activity studies, and some representative data are tabulated in Table 4.8.

As might be expected, the alkylthio substituents affected inhibition in the same qualitative manner as did alkoxy groups, yielding maximum insect anticholinesterase activity when attached to the phenyl ring in the 2-position and increased activity with α-alkyl-branching and lengthened carbon chains.[46,93] However, the overall effects of position isomerism and size of substituent were not nearly as pronounced. This could be due to the larger bulk of the sulfur atom (van der Waal radius of 1.85 Å) compared to oxygen (radius of 1.4 Å), giving a "looser" accommodation at the anionic site.

Another noticeable difference was variation in inhibition due to alkyl shape. For example, the change in molar I_{50} between 2-propoxy- and 2-isopropoxyphenyl N-methylcarbamate was about tenfold (Table 4.7), whereas the same propoxythioether isomers were almost identical in

TABLE 4.8

Inhibition of Housefly Head AChE by Various Alkylthiophenyl N-Methylcarbamates[46,93]

Alkylthio substituent	Molar I_{50}	Alkylthio substituent	Molar I_{50}
None	2×10^{-4}	2-S-*i*-C_3H_7	1.4×10^{-7}
2-SCH_3	9×10^{-7}	3-S-*i*-C_3H_7	1.8×10^{-6}
3-SCH_3	7×10^{-6}	2-S-*i*-C_5H_{11}	7.4×10^{-7}
4-SCH_3	3.4×10^{-5}	2-SC_4H_9	1.6×10^{-7}
2-SC_3H_7	1.8×10^{-7}	2-SC_6H_{13}	2×10^{-7}
3-SC_3H_7	1.1×10^{-6}	2-SC_9H_{19}	3.6×10^{-7}

anticholinesterase properties (Table 4.8). This, of course, contradicts the proposed preference of the anionic site for branched alkyl groups, but it is believed that the sulfur atom, which can assume a positive nature under certain circumstances, is attracted to the anionic site, so that steric factors are overridden.[46] Note that even with carbon chain lengths of nine the alkylthio substituents did not alter cholinesterase inhibition appreciably. However, there was a peak of activity with *ortho*-substitution at four carbons,[93] which coincides with the *meta* maximum of four carbons reported for alkyl groups.[84] Thus, it is apparent that the hydrophobic pool surrounding the anionic site must extend to accommodate four methylene groups.

Comparison of anticholinesterase potency between phenyl N-methylcarbamates containing thioether groups and alkoxy groups demonstrated that the sulfur derivatives were up to 200 times better enzyme inhibitors.[46] This was particularly true with comparable *para*-substituted isomers, which incidentally were also about 20 times more toxic to insects. However, correlation of alkylthio structure with toxicity was not good. Almost all compounds, when applied topically with a synergist to houseflies, were equally lethal (LD_{50} range of 6 to 19 μg/g), although carbamates containing alkylthio groups more than seven carbons long were nontoxic.[93]

Without co-administration of a synergist, some very unusual relationships of toxicity to isomer became apparent. For instance, placement of methyl-, butyl-, and isoamylthio groups in the *meta*-position of the ring gave the most toxic carbamates, but *ortho*-substitution for propyl- and isopropylthioethers proved to be the most critical orientation, and *para*-placement was optimal for allylthioether. Some compounds were toxic to flies as well as to mosquito larvae, others were not, and some were equally lethal to both species. The most prominent member belonging to this series of compounds is Mesurol.

Other closely related groups which have been attached to phenyl methylcarbamates are alkenyls, such as propargyl ($-CH_2C{\equiv}CH$) and allyl ($-CH_2CH{=}CH_2$), dimethyl acetate [$-CH(OCH_3)_2$], and mercaptol [$-CH(SCH_3)_2$]. A comparison of I_{50} values for 2-allylphenyl N-methylcarbamate and 2-(1-isobutenyl) phenyl N-methylcarbamate with those for their saturated analogs indicated that the presence of the double

TABLE 4.9

Inhibition of Housefly Head AChE by Various Alkenyl, Acetal, and Mercaptal Phenyl N-Methylcarbamates[38,46,49]

Phenyl substituent	Molar I_{50}	Phenyl substituent	Molar I_{50}
2-C_3H_7	6.0×10^{-6}	2-$SCH_2CH{=}CH_2$	2.6×10^{-7}
2-$CH_2CH{=}CH_2$	3.4×10^{-6}	3-$SCH_2CH{=}CH_2$	3.6×10^{-6}
2-$CH_2CH(CH_3)_2$	2.3×10^{-6}	2-$SCH_2C{\equiv}CH$	3.4×10^{-7}
2-$CH{=}C(CH_3)_2$	1.2×10^{-6}	2-$CH(OCH_3)_2$ [a]	1.9×10^{-5}
2-OC_3H_7	8.7×10^{-6}	3-$CH(OCH_3)_2$ [a]	5.0×10^{-4}
2-$OCH_2C{\equiv}CH$	2.9×10^{-6}	2-$CH(SCH_3)_2$ [a]	2.0×10^{-6}
3-$OCH_2C{\equiv}CH$	4.0×10^{-6}	3-$CH(SCH_3)_2$ [a]	8.3×10^{-5}
2-SC_3H_7	1.8×10^{-7}	4-$CH(SCH_3)_2$ [a]	$<10^{-4}$

[a]Molar I_{50} values for bovine erythrocyte AChE.

bond approximately doubled the anticholinesterase activity (Table 4.9) and insect toxicity.[38] Changing to a triple bond from 2-propoxy- to 2-propargyloxyphenyl N-methylcarbamate resulted in a threefold increase in esterase inhibition and toxicity,[46] but the effects of saturation were minor among the isomeric alkylthiophenyl compounds (Table 4.9). Note that the position and presence or absence of oxygen or sulfur atoms between the phenyl ring and the alkenyl group altered activity in a manner analogous to the above descriptions for the corresponding alkyl series.

Synthesis and evaluation of dimethyl acetal [$-CH(OCH_3)_2$] and mercaptol [$-CH(SCH_3)_2$] benzaldehyde N-methylcarbamates showed that maximum bovine AChE inhibition occurred when the phenyl substituent was in the *ortho*-position (Table 4.9).[49] Enzyme activity was blocked most efficiently with mercaptols, but greater insect toxicity was obtained with acetals. Thus, in total, results were very similar to those found for alkoxy- and alkylthiophenyl N-methylcarbamates.

Halogen, Nitro, Amino, and Miscellaneous Groups

Halogens occur infrequently in phenyl N-methylcarbamate insecticide structures, the only widely studied compound being Banol. However, when appropriately placed on an already substituted phenyl ring, they may further enhance the anticholinesterase activity of the carbamate, as will be seen later.

Results from initial studies with monohalogenated phenyl N-methylcarbamates were used primarily to establish the critical distance between esteratic and anionic sites on the surface of acetylcholinesterase.[8,83] The different atoms possessing almost equal electronic characteristics allowed for an examination of the effects of substituent size with a van der Waal radius of 1.35 Å for fluorine to 2.15 Å for iodine. The best inhibition of housefly AChE occurred with iodine, the largest atom, in the *ortho*-position, as seen in Table 4.10. Other halogens were also most effective in this position, which most closely approximated the proposed 5-Å separation of sites.

TABLE 4.10

Inhibition of Housefly Head AChE by Various Halophenyl and Nitrophenyl N-Methylcarbamates[8,11,83]

Phenyl substituent	Molar I_{50}	Phenyl substituent	Molar I_{50}
None	2×10^{-4}	2-Br	2.2×10^{-6}
2-F	1.6×10^{-5}	2-I	8×10^{-7}
3-F	8.5×10^{-4}	2-NO_2	5.0×10^{-3}
4-F	2.3×10^{-4}	3-NO_2	2.0×10^{-3}
2-Cl	5×10^{-6}	4-NO_2	3.0×10^{-3}
3-Cl	5×10^{-5}	3-NO_2,6-CH_3	2.3×10^{-5}
4-Cl	2.4×10^{-4}	4-NO_2,3-*i*-C_3H_7	2.8×10^{-6}

This finding appears to contradict results with alkyl substituents, particularly with the methyl group (van der Waal radius = 2 Å), where placement in the *meta*-position of the phenyl ring was found to be optimal (Table 4.6). The proposed reason for differences between substitution by methyl and chlorine groups, which are approximately the same size, was electronic rather than steric, with the electron-withdrawing halogens causing more rapid hydrolysis, and thus altered inhibition.[94] However, it is also possible that the longer carbon–halide bond lengths make orientation best when halogens are *ortho* to the carbamate moiety. This would also help explain why fluorine, bromine, or chlorine, but not iodine, in the 4-position yielded I_{50} values very near to that of unsubstituted phenyl N-methylcarbamate (Table 4.10). The topical toxicity (synergized) of these carbamates to houseflies paralleled their in-vitro anticholinesterase activity.

Flyhead anticholinesterase data presented in Table 4.10 for nitro-substituted phenyl N-methylcarbamates show them to be poorer inhibitors of AChE than the unsubstituted carbamate.[11] A great deal of this apparent inactivity is undoubtedly due to the extreme sensitivity of the nitro compounds to alkaline hydrolysis, so extreme that the compounds were up to 97% hydrolyzed during analysis for inhibition. Thus, ineffective anticholinesterase activity and lack of insect toxicity may have been largely a result of chemical instability of the compounds.

Addition of an alkyl group in the appropriate position on the nitro-substituted ring increased toxicity (synergized) to flies and markedly improved AChE inhibition (Table 4.10), presumably by increasing the compound's resistance to hydrolysis and also by providing a hydrophobic group for interaction at the anionic site. However, correlation of activity with other alkyl and nitro substitutions at various places on the phenyl ring was not always good, and interpretation is therefore difficult. Of interest were the much better anticholinesterase and lethal properties of corresponding N,N-dimethylcarbamates, which were considerably more stable to hydrolysis. When these compounds were included in kinetic studies with bovine erythrocyte AChE, 3-nitrophenyl N,N-dimethylcarbamate showed the greatest enzyme affinity, whereas the 2-nitrophenyl analog was best at carbamylating the enzyme.[86]

The use of dimethylamino groups on the carbamate's phenyl ring as anionic interactants has been touched upon earlier in the chapter. Maximum housefly acetylcholinesterase inhibition and insect toxicity corresponded to *ortho*-substitution[79] (Table 4.5). Nevertheless, even though this compound is fairly active, the dimethylamino group is generally used only in addition to other suitably placed alkyl groups. For example, placement of a *para*-dimethylamino substituent in 3-*t*-butylphenyl N-methylcarbamate caused a dramatic increase in toxicity and in the spectrum of insects effected.[95] Structure–activity studies with compounds containing other combinations, including Zectran, showed that the nature of the alkyl moieties on the amino group was not critical, but it was important to include ring *meta*-alkyl groups, which would hinder rotation of the amine.[95–97]

At about the same time it was discovered that changing the *para*-amino group to a *para*-imino group also yielded active compounds, such as compound I in Figure 4.7, which is as toxic to Mexican bean beetles (LD_{95} = 4 μg/g) and to Southern armyworm (LD_{95} = 1 μg/g) as Zectran.[98,99] A final modification brought forth azobenzene carbamates, such as compound II, which is very active against the bean beetle (LD_{95}

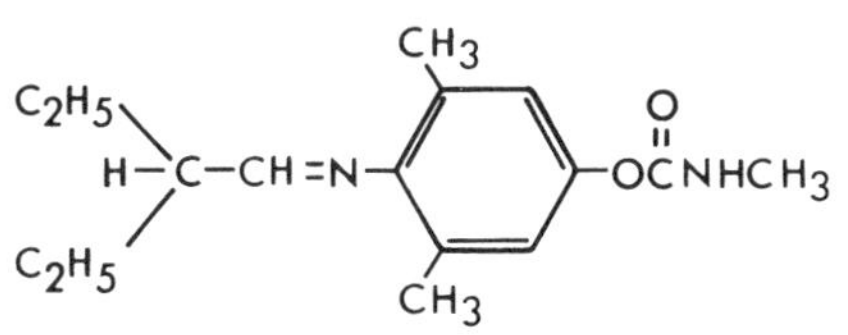

I

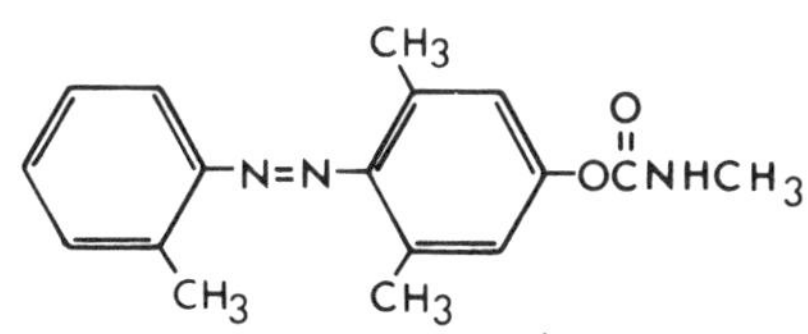

II

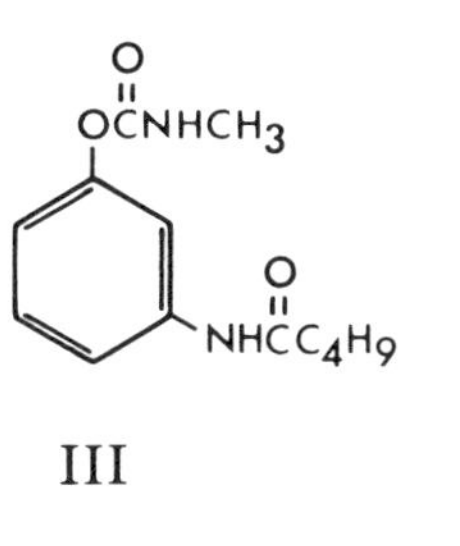

III

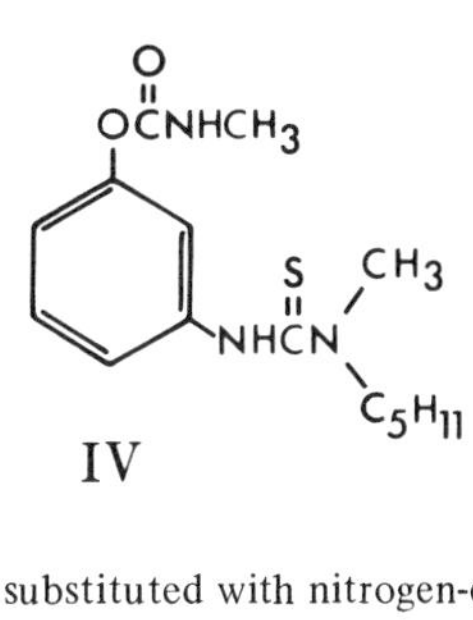

IV

FIGURE 4.7. Structures of phenyl N-methylcarbamates substituted with nitrogen-containing groups.

= 0.35 μg/g), but nontoxic to the armyworm.[100,101] Various substituents in either ring altered activity, suggesting that fit onto acetylcholinesterase was important for activity. The negative character of the nitrogen atom in *para*-amino, -imino, and -azo substitutents also implied that fit could involve a cationic surface on the enzyme quite near to the classical anionic site. However, there are no in-vitro data to support this claim.

Another group of carbamates with nitrogen-containing substitutents are the acylamidophenyl and thioureidophenyl N-methylcarbamates (respective examples are compounds III and IV in Figure 4.7). Even though these compounds were essentially nontoxic to insects, they were quite effective fly anticholinesterase agents[47] (Table 4.11). Note that the acylamido groups follow the pattern of activity for alkyl substituents, that is, 3 > 2 > 4 for placement on the phenyl ring. Also, enzyme inhibition increased with alkyl size and alkyl length up to a maximum of four carbons. Though not shown on the table, branched alkyl groups on the acylamido group enhanced anticholinesterase activity even more. Changing to thioureido groups did not greatly improve insect toxicity, but some very powerful cholinesterase inhibitors were uncovered, such as compound IV in Figure 4.7, which has an I_{50} value of $2.1 \times 10^{-8} M$.

Two final examples include formetanate, where a (dimethylamino)methylene amino group $[-N{=}CHN(CH_3)_2]$ is located *meta* to the carbamate on the phenyl ring, and a series of 4-dimethylaminomethylphenyl N-methylcarbamates. The former compound is an effective broad-spectrum acaricide,[102] and the latter series contains several promising selective anticholinesterase insecticides.[87]

Since the N-methylcarbamate moiety is not too unlike some of the groups that have been used as substituents on the phenyl ring, it is possible that a ring containing two carbamate clusters might be unusually effective as an anticholinesterase. Yet, when bis-N-methylcarbamates, such as compound I in Figure 4.8, were synthesized and evaluated, they were found to be average as enzyme inhibitors (I_{50} for all isomers around $3 \times 10^{-5} M$), and only the 1,3-derivative showed appreciable synergized toxicity to houseflies.[38] In an analogous manner, phenyl N,N-dimethylcarbamate has been substituted with organophosphorus groups to yield compounds such as that designated as II in Figure 4.8.[103] Again these structures were not demonstrated to be extremely effective as anticholinesterases (bovine AChE I_{50} values ranged from 5×10^{-6} to $7 \times 10^{-5} M$) or as insect toxicants. However, position preference was apparent, and the organophosphorus-substituted carbamates were better inhibitors than the unsubstituted carbamate.

Multiple Substitution and Double Rings

There is an almost endless number of substituent combinations that can be put together in a phenyl methylcarbamate structure. The fact that multiple substitution is effective can be judged by

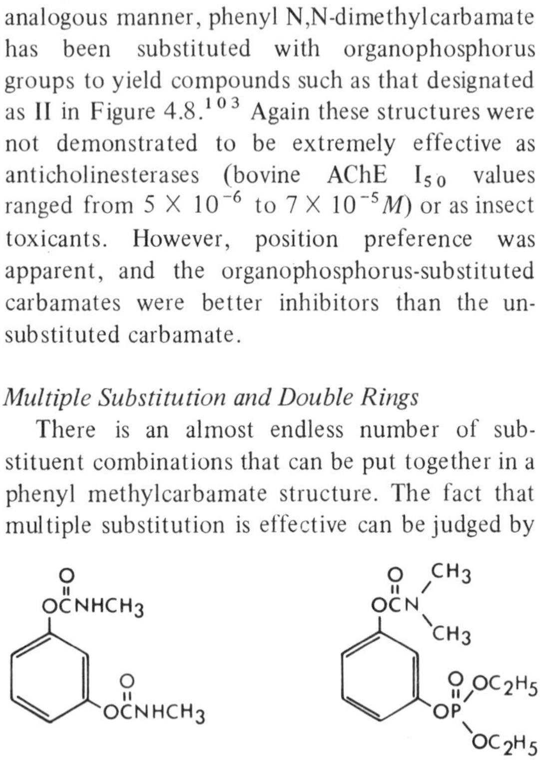

I II

FIGURE 4.8. Structures of compounds containing two ester groups.

TABLE 4.11

Inhibition of Housefly Head AChE by Various Acylamidophenyl N-Methylcarbamates[47]

Acylamido substituent	Molar I_{50}	Acylamido substituent	Molar I_{50}
2-NHC(O)CH_3	5×10^{-5}	3-NHC(O)H	7×10^{-6}
3-NHC(O)CH_3	6×10^{-6}	3-NHC(O)C_3H_7	1.2×10^{-6}
4-NHC(O)CH_3	6×10^{-5}	3-NHC(O)C_4H_9	1.3×10^{-6}
2-NHC(O)C_2H_5	1.5×10^{-5}	3-NHC(O)C_8H_{17}	3×10^{-6}
3-NHC(O)C_2H_5	2×10^{-6}	3-NHC(O)$C_{12}H_{25}$	1×10^{-4}

the large number of experimental and commercial carbamate insecticides possessing two or three substituent groupings. The value of each ring addition depends on its bulk, electronic characteristics, and location on the planar benzene nucleus. Its addition may improve steric fit and affinity for AChE, or it may alter carbamylation rate. Obviously, when all of these factors are interacting simultaneously, comparative I_{50} values do not always agree with the expected activity of a particular structure.

In the unsubstituted carbamate, the flat ring structure rotates freely about the ester oxygen of the carbamate moiety, and when it is lined up with the enzyme surface, it appears that the edge of the ring is in contact with the anionic site (Figure 4.2). Addition of a substituent for better interaction at the site moves the ring further from the enzyme, though the ring maintains its upright orientation. Thus, the 3,5-dimethylphenyl N-methylcarbamate should offer more chances for proper alignment than the 3-methylphenyl structure and should be a better anticholinesterase, which is exactly what has been found[94] (Table 4.12).

Moving the methyl groups to various other positions or addition of a third methyl to the ring altered activity somewhat, but substitution in the 2,6-combination abolished all activity. Similar studies with more bulky alkyl groups and with alkoxy groups produced the same trends.[38,79,83,90] However, double substitution by larger groups, such as isopropyl (Table 4.12), produced a proportionally greater increase in enzyme inhibition, which implied that the multiple methyl groups formed a canopy around the aromatic ring, thereby permitting enzyme interaction at the sides as well as the edges of the ring.

The lack of activity of 2,6-dialkyl-, -dialkoxy-, and -dihalophenyl N-methylcarbamates is due to poor interaction with AChE, probably because of steric interference with the carbamylation step.[83,90,94,104] None of these compounds were found to be toxic to insects, though other disubstituted analogs did exhibit varied degrees of lethality. Steric problems caused by multiple additions to the phenyl ring also lowered insect toxicity by disruption of enzyme–inhibitor interactions at the anionic site.[105] Unsymmetrical disubstituted carbamates often vary considerably in activity, as illustrated by the 2,5-placement of methoxy and isopropoxy groups in the phenyl ring[106] (Table 4.12). Although this variation is certainly due to the different size (and corresponding affinity) of the groups, a similar effect, though less dramatic, can be found by alternating 2,4-attachment of methyl and methylthio substituents.[46] Electronic effects, which could be significantly more important for carbamylation purposes, often negate expected increases in affinity. For example, placing a fluorine atom in position 6 of 3-isopropylphenyl N-methylcarbamate decreased the I_{50} almost tenfold[38] (Table 4.12). However, a chlorine or bromine in this position improved anticholinesterase activity about five times, which is often the case with substitution by larger halides.[84] Perhaps the more bulky *ortho*-chlorine atom forces the carbamate moiety "towards" the 3-alkyl group for maximum interaction with the active center of acetylcholinesterase.

Generally, substitutions that increase enzyme inhibition by the carbamate also increase toxicity to insects, but the presence of two or three attached groups on the aromatic ring often produces considerable species specificity. Just as it is difficult to predict which moieties will yield the highest anticholinesterase properties, so it is with

TABLE 4.12

Inhibition of Housefly Head AChE by Various Multisubstituted Phenyl N-Methylcarbamates[38,46,94,106]

Substituents	Molar I_{50}	Substituents	Molar I_{50}
3-CH_3	1.4×10^{-5}	2-O-*i*-C_3H_7,5-OCH_3	5.6×10^{-7}
3,5-diCH_3	6.0×10^{-6}	2-OCH_3,5-O-*i*-C_3H_7	1.5×10^{-5}
2,6-diCH_3	1×10^{-2}	2-CH_3,4-SCH_3	1.8×10^{-5}
3,4,5-triCH_3	1.9×10^{-6}	2-SCH_3,4-CH_3	1.9×10^{-6}
3-*i*-C_3H_7	3.4×10^{-7}	3-*i*-C_3H_7,6-F	5.8×10^{-6}
3,5-di-*i*-C_3H_7	3.3×10^{-8}	3-*i*-C_3H_7,6-Cl	7.8×10^{-8}

toxicity. Only after testing against the living insect can a compound's effectiveness be assured.

Double-ring carbamates are merely extensions of doubly substituted phenyl N-methylcarbamates. Carbaryl, the best-known of all carbamates, was one of the first structures to deviate from the phenyl format.[107] It has been likened electronically to 2,3-dimethoxyphenyl N-methylcarbamate[106] (Figure 4.9). Numerous other 2,3-ring additions exist, and one of the most interesting of these is carbofuran. As indicated in Figure 4.9, this carbamate is derived from Baygon, where the anionic propoxyl interactant has been fixed in position by formation of a dimethylfuran ring. This fusion increases AChE inhibition about threefold.[108]

Because of the need for interaction of the second ring with the anionic site, most junctions are made on the 2- and 3-carbons of the carbamate's phenyl nucleus. This arrangement, of course, gives maximum interaction of the molecule at both the anionic site and the esteratic site of AChE, as judged by the increased affinity (K_a) of 2,3-placement versus 3,4-placement.[109] An example of the decrease in enzyme activity when moving the second ring to the 3,4-position can be seen by comparing I_{50} values for 1-naphthyl (carbaryl) and 2-naphthyl N-methylcarbamate[110] (Table 4.13). A similar dramatic loss in insect toxicity also occurred. Removing the aromatic character from "ring 2" might be expected to reduce interaction at the anionic site, and this has

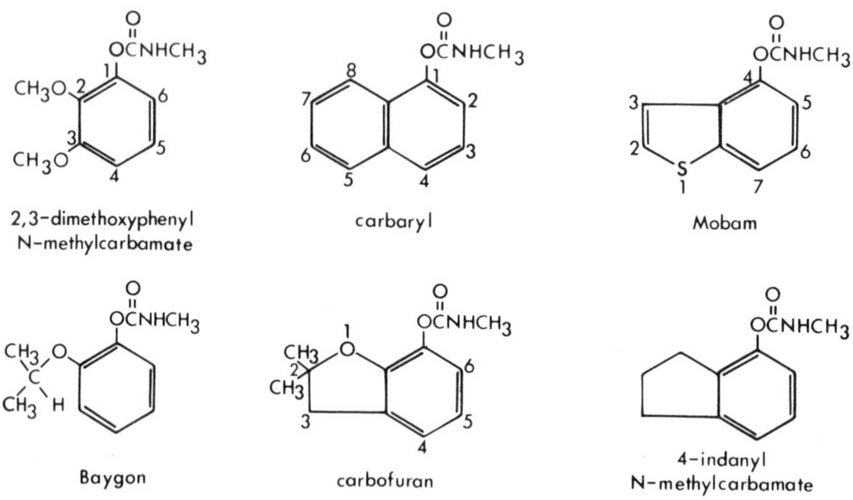

FIGURE 4.9. Comparison of phenyl N-methylcarbamate to double-ring N-methylcarbamates.

TABLE 4.13

Inhibition of Housefly Head AChE by Double-ring N-Methylcarbamates[42,108,110]

N-Methylcarbamate	Molar I_{50}	N-Methylcarbamate	Molar I_{50}
1-Naphthyl (carbaryl)	9×10^{-7}	2,3-Dihydro-2,2-dimethyl-7-benzofuranyl (carbofuran)	2.5×10^{-7}
2-Naphthyl	1.4×10^{-5}	2,3-Dihydro-2-methyl-7-benzofuranyl	2.4×10^{-6}
5,6,7,8-Tetrahydro-1-naphthyl	1.4×10^{-6}	3-OCH_3 carbofuran	6.4×10^{-7}
4-Indanyl	2.4×10^{-6}	4-Cl carbofuran	2.8×10^{-7}
4-Benzothienyl (Mobam)	3×10^{-8}	4-CH_3 carbofuran	1.6×10^{-7}
5-Benzothienyl	9×10^{-7}	4-F carbaryl	1.2×10^{-6}
3-Benzothienyl	1×10^{-7}	5-F carbaryl	1.5×10^{-6}

been observed in the form of lower I_{50} values for 5,6,7,8-tetrahydro-1-naphthyl and 4-indanyl N-methylcarbamates (Table 4.13).

The sulfur atom in 4-benzothienyl N-methylcarbamate (Mobam, Figure 4.9) lends aromaticity to this heterocyclic substituent, so that enzyme activity and toxicity are retained, in contrast to the findings for carbaryl[42,110] (Table 4.13). This, of course, is another example of freezing the phenyl substituent (thioalkyl group) in place for more efficient contact with the enzyme surface. Moving the carbamate moiety to position 5 or 6 (see Mobam, Figure 4.9) yielded a loss in anticholinesterase activity and insect toxicity, analogous, though less pronounced, to the decrease between carbaryl and 2-naphthyl N-methylcarbamate.[42,93] Placing the ester on the pseudoaromatic heterocyclic ring (2- or 3-benzothienyl) also reduced activity, suggesting that loss of inhibition resulted from loss of electronic influence contributed by the fully aromatic benzene ring (which would cause a lower carbamylation rate).

Carbofuran's anionic interactant obviously centers around carbon 2 (Figure 4.9), since removal of one of the methyl groups reduced insect enzyme inhibition tenfold[108] (Table 4.13). Although it is often assumed that such a change in I_{50} is due to a change in affinity of compounds for AChE, there is seldom any proof for this assumption. However, in this case K_a and k_2 constants have been determined for carbofuran analogs.[29] It was found that there was indeed a loss of affinity upon removal of one of the 2,2-dimethyl groups, and a much more dramatic decrease occurred when both methyl groups were replaced by hydrogen atoms. Meanwhile, carbamylation rates varied somewhat, but the changes were minor compared to the alteration of affinity constants. Note that the presence of the oxygen atom in carbofuran's second ring is not contributory, since the I_{50} values for 2-demethyl carbofuran and 4-indanyl N-methylcarbamate are identical.

With the double-ring set up, activity of the molecule could conceivably be affected by substituent additions to either cycle. However, substituents placed in the primary or secondary rings of carbofuran or carbaryl were found to modify enzymic activity only slightly (Table 4.13). On the other hand, additional molecular "handles" did offer more metabolic sites, so that toxicity to insects was unpredictable.

A cross between the above double-cycle carbamates and classical alkyl-type carbamates is the group of structures containing dithiolane or dioxolane groups, such as C-8353 (Figure 4.10). Interestingly, this type of molecule was discovered almost simultaneously by three different research groups.[48,111,112] The possible combinations of structures for the heterocyclic ring attached *ortho* to the carbamate in phenyl N-methylcarbamate are shown in Figure 4.10.

The rules governing the anticholinesterase activity of this series are not unlike those described above for alkylthio and alkoxy substitution. Generally, *ortho* attachment was discovered to be better than *meta* or *para,* and the dithiolanes were about ten times more active than the dioxolanes.[48,49] Consistent with this, the oxathiolanes were intermediate in activity. The contribution of the sulfur atoms to inhibition can be appreciated by comparing the I_{50} for the 2-dithiolane-substituted structure ($2 \times 10^{-8} M$) to that of the 2-cyclohexyl N-methylcarbamate ($2 \times 10^{-7} M$).[48] Anticholinesterase activity of 2-dithiolane N-methylcarbamate was superior to that of 3-*t*-butyl- or 3-dimethylamino N-methylcarbamate when measured with honeybee or cricket AChE.[29]

Substituent effects of alkyl groups added to the *ortho*-attached heterocyclic ring were minor or negative, suggesting that maximum complimentarity with AChE was present in this type of structure.[48,49] However, substitution in the less favorable *meta*-attached ring did improve activity. Only a few appropriately placed bulky substituents in the phenyl nucleus altered anticholinesterase properties.[48] Generally, insect toxicity approximated molar I_{50} values, but some interesting species-specific toxicities were encountered.[48,49,53]

FACTORS DETERMINING INSECTICIDAL ACTIVITY OF OXIME CARBAMATES

The discovery of insecticidal activity associated with an O-(methylcarbamoyl)oxime structure occurred around 1960.[113] Since that time, numerous industrial and academic studies have provided a variety of oxime insecticides with a broad range of insect selectivity. Registered compounds belonging to this series include methomyl and aldicarb; the latter arose from a synthetic attempt to structurally mimic acetylcholine.[114] The resemblance can be seen in Figure 4.11. The critical feature, compared to other alkyl N-methylcarbamates, is the presence of the double bond between carbon and nitrogen, which electronically influences the character of the carbamate ester moiety in a manner analogous to that of the benzene ring in phenyl N-methylcarbamates.

As might be expected, structure–activity relationships among different oxime carbamates are often similar to those for aryl N-methylcarbamates. For example, N-alkyl, N,N-dialkyl, and NH_2 O-(carbamoyl)oximes other than N-methylcarbamates were generally shown to be less potent or even inactive as insecticides.[114-116] In some cases the N,N-dimethyl analogs proved to be better anticholinesterases, but it was believed that the corresponding monomethyl compound was hydrolytically unstable, which may have resulted in a misleading I_{50} determination. Thus, the rule is that O-(dimethylcarbamoyl)-oximes are poorer anticholinesterases than O-(methylcarbamoyl)oximes.

This means that modifications of oxime carbamate structures center around changes in the oxime portion of the molecule, the objective being improvement of the affinity or fit of the candidate to acetylcholinesterase and/or improvement of reactivity or carbamylation ability of the prospective insecticide. Thus, size, shape, and electronic properties of oxime substituents can significantly influence enzyme inhibition. Whether a corresponding change in insect toxicity occurs depends on other factors, such as cuticular penetration, metabolism, etc.

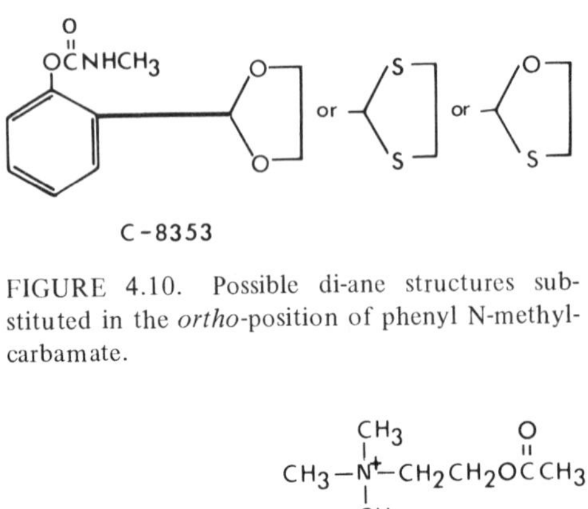

C-8353

FIGURE 4.10. Possible di-ane structures substituted in the *ortho*-position of phenyl N-methylcarbamate.

To date the greatest amount of effort concerning insecticidally active O-(methylcarbamoyl)oximes has revolved around the basic structures pictured in Figure 4.12. For the aliphatic compounds (I), all alterations are essentially composed of substitutions on carbons 1 or 2, or on both. These atoms may be fused into a cycle (II), usually containing four to six carbons, where substitution on positions other than carbon 1 are possible. In some cases the cycle is hetero in nature. Finally, aromaticity can be incorporated by replacing carbon 2 with a benzene ring (III), to which various groups can be attached. In addition, the proton on carbon 1 may also be substituted.

One aspect of oxime carbamate chemistry that does not occur with phenyl carbamates is geometrical isomerism, such as that shown in Figure 4.13. Based on evidence from chemical degradation, the *syn* isomer has been assigned to aldicarb, though it is not certain which isomer is more active.[115] In the case of *syn* and *anti* isomers of isophorone O-(methylcarbamoyl) oxime[38] or of *exo*-3-chloro-*endo*-6-cyano-2-norbornanone O-(methylcarbamoyl)oxime[117]

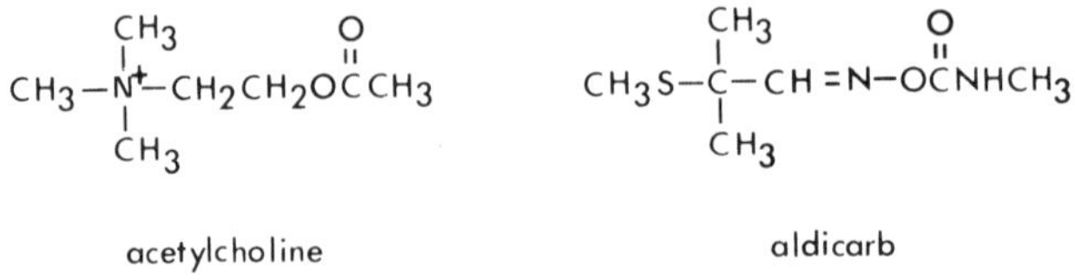

FIGURE 4.11. Structures of acetylcholine and aldicarb.

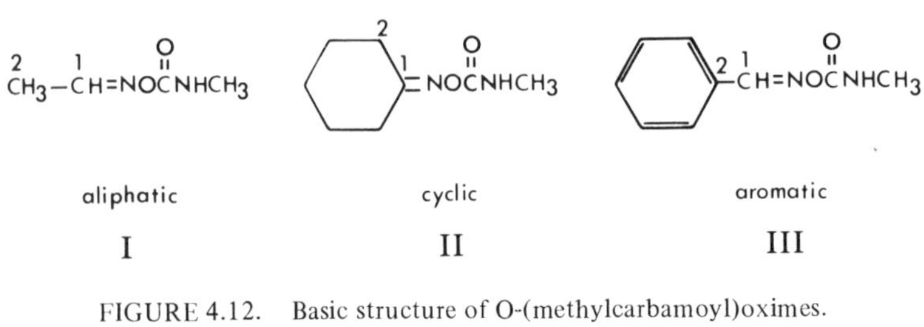

FIGURE 4.12. Basic structure of O-(methylcarbamoyl)oximes.

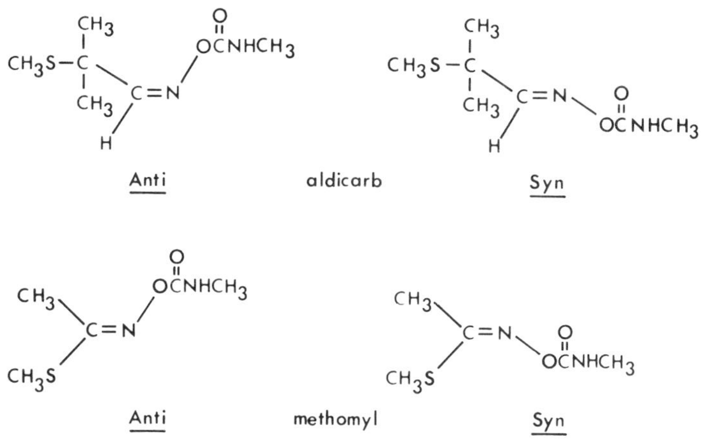

FIGURE 4.13. Stereoisomers of aldicarb and methomyl.

there was little difference in anticholinesterase potency or in insect toxicity between the respective isomers. On the other hand, *syn* and *anti* configurations of methomyl (Figure 4.13) reacted quite differently with acetylcholinesterase, giving I_{50} values of $3.8 \times 10^{-6} M$ and $3.4 \times 10^{-4} M$ respectively.[118] The 1-methoxy derivatives also showed a similar trend, though less pronounced ($3 \times 10^{-5} M$ vs. $1 \times 10^{-4} M$). With methomyl and its oxygen counterpart, insecticidal activity was greatest with the *syn* isomer. Obviously, a thorough understanding of oxime carbamate structure–activity relationships requires more extensive investigation of isomer contribution.

The aliphatic compound shown in Figure 4.12 is acetaldehyde O-(methylcarbamoyl)oxime (I). This chemical is not an effective insect toxicant, but modest activity was obtained by adding a methyl group to carbons 1 and 2,[114] though the anticholinesterase activity was not very high (Table 4.14). Maximum methylation of carbon 2 with no substituents on carbon 1 produced a strong structural similarity to acetylcholine and better cholinesterase inhibition ($1 \times 10^{-5} M$). Addition of alkyl groups onto carbon 1 was not helpful, and moving the tertiary carbon atom further away from the carbamate moiety (more than 5 Å away) decreased inhibition drastically (Table 4.14).

However, exchanging one methyl group of carbon 2 for an electron-withdrawing substituent, such as allyl, methylthio (e.g., aldicarb), or nitro groups, increased enzyme activity and insect toxicity substantially. It was even possible to place two methylthioether linkages on carbon 2 (Figure 4.14, I) without sacrificing significant housefly or mosquito toxicity.[116] Closing the two ether groups into a ring structure and adding a methyl group yielded the broad-spectrum greenhouse insecticide Tirpate, 2,4-dimethyl-1,3-dithiolane-2-carboxyaldehyde O-(methylcarbamoyl)oxime (Figure 4.14). Substituting other alkyl or aromatic groups on the ring, enlarging to a six-membered ring, or replacing the carbon 1 proton with a methyl group narrowed the insect specificity considerably.

Presumably, these electronegative substitutions increased the reactivity of the carbamate molecule, causing a corresponding increase in carbamylation of the esteratic site of acetylcholinesterase. In addition, the electronegative groups may have inductively increased attraction of neighboring methyl groups for the anionic site. Further alterations of methyl groups in aldicarb precipitated a loss of insect toxicity, so that aldicarb's configuration was judged to be optimal for this series of structures.[114]

The next step was to place an electronegative center on carbon 1 of acetaldehyde-type oxime carbamates. A methylthio group so placed in aldicarb enhanced anticholinesterase activity fivefold without any loss of insect toxicity[115] (Table 4.14). In fact, the bulkiness of carbon 2 was not even needed, since methomyl proved to be an even better inhibitor and a better toxicant than aldicarb. However, modest branching of carbon 2 supplied by a single methyl group gave the lowest I_{50} of all, $1 \times 10^{-7} M$, (Table 4.14), though there was no increase in insect toxicity.

If a stronger electron-withdrawing group, such as cyano, was positioned on carbon 1 instead of SCH_3, activity dropped precipitously because of extreme hydrolytic instability.[115] But a change in the S-alkyl moiety to a more suitable anionic interactant, such as S-*i*-propyl, did improve blockage of enzyme action without a significant drop in insecticidal activity[118] (Table 4.14). Lastly, conversion of carbon 1 alkylthio substituents to corresponding alkoxy substituents weakened AChE inhibition in a manner similar to that recorded for alkylthio-alkoxy exchanges among phenyl N-methylcarbamates. With both series of compounds, however, some of the oxygen analogs proved to be better insect toxicants.

TABLE 4.14

Inhibition of Housefly Head AChE by Various Acetaldehyde O-(Methylcarbamoyl)oximes[114,115,118]

Aliphatic substituent	Molar I_{50}
$CH_3CH_2C(CH_3)=$	6×10^{-4}
$(CH_3)_3CCH=$	1×10^{-5}
$(CH_3)_3CCH_2CH=$	3×10^{-4}
$CH_2=CHCH_2C(CH_3)_2-CH=$	5×10^{-6}
$CH_3SC(CH_3)_2-CH=$ (Aldicarb)	1×10^{-5}
$NO_2-C(CH_3)_2-CH=$	5×10^{-7}
$CH_3S-C(CH_3)_2-C(SCH_3)=$	2×10^{-6}
$CH_3C(SCH_3)=$ (Methomyl)	1×10^{-6}
$CH_3CH_2C(SCH_3)=$	1×10^{-7}
$CH_3C(S\text{-}i\text{-}C_3H_7)=$	5×10^{-7}
$CH_3C(O\text{-}i\text{-}C_3H_7)=$	2.5×10^{-6}
$CH_3C(OCH_3)=$	3×10^{-5}

Addition of alkyl groups to the cyclohexanone ring (Figure 4.12, II) increased anticholinesterase activity about tenfold and occasionally produced better insecticides,[115] though no outstanding insecticide candidates have become apparent. Shrinking the ring to cyclopentanone lessened activity ($7 \times 10^{-4} M$), whereas expanding it to cycloheptanone produced better enzyme inhibition ($I_{50} = 6 \times 10^{-5} M$) and higher insect toxicity. A chlorine atom or nitro group in the 2-position of these rings, analogous to placing electronegativity at the 2-position of aliphatic oximes, often improved toxicity and inhibition by several 100-fold. Thus, it seems that electronic effects on the reactivity of the molecule are more important in conferring insecticidal properties than enhanced van der Waal or hydrophobic characteristics which affect enzyme affinity.

Cyclic carbamoyloximes have also been synthesized with other atoms in the ring. Two examples, compounds II and III in Figure 4.14, have been found to be toxic to various insect species, but unfortunately they are also quite toxic to mammals.[119-121]

This brings us to aromatic derivations of benzaldehyde O-(methylcarbamoyl)oxime (Figure 4.12, III). Replacing the carbon 1 proton with alkyl groups yielded a correlation between substituent size and branching (that is, van der Waal dispersion forces) and anticholinesterase activity[122] (Table 4.15). Replacing the proton with electron-withdrawing substituents or a positively charged amino group also contributed to enhanced enzyme inhibition. Thus, once again, structural alterations aided both anionic fit (K_a) and carbamylation (k_2) to give improved I_{50} values.

In three studies where the C-1 substituent was held constant at either H or CH_3, varying additions to the benzene ring of the corresponding oxime carbamates resulted in little alteration of acetylcholinesterase inhibition or insect toxicity. Thus, I_{50} ranges were from 9×10^{-5} to $5 \times 10^{-6} M$,[123] 1.9×10^{-4} to $4.4 \times 10^{-6} M$,[122] and,

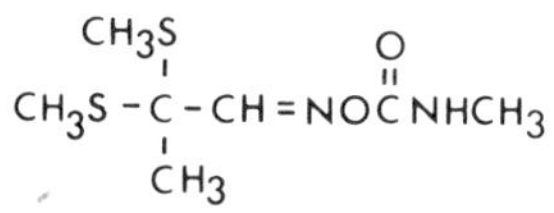

I

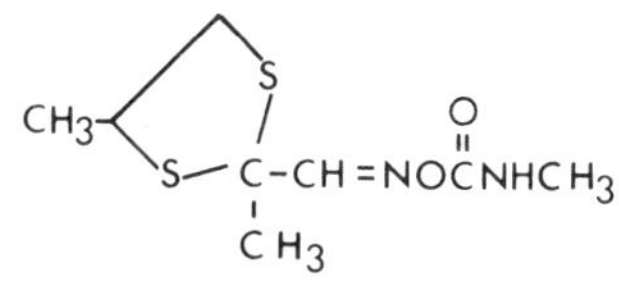

Tirpate

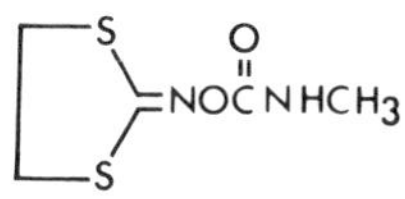

II

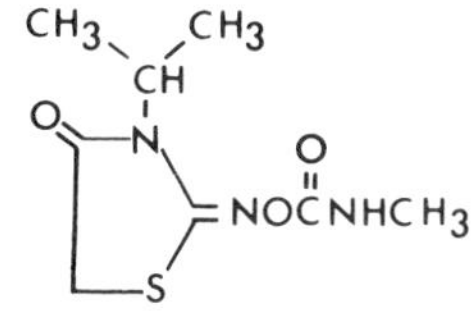

III

FIGURE 4.14. Structures of various insecticidally active oxime carbamates.

TABLE 4.15

Inhibition of Housefly Head AChE by Various Carbon-1-substituted Benzaldehyde O-(Methylcarbamoyl)oximes[122]

Carbon-1-substituent	Molar I_{50}	Carbon-1-substituent	Molar I_{50}
H	9.2×10^{-5}	SCH_3	7.8×10^{-6}
CH_3	7.1×10^{-5}	CF_3	5.8×10^{-7}
C_2H_5	2.8×10^{-5}	$CH_2N(CH_3)_2$	1.5×10^{-5}
$i\text{-}C_3H_7$	5.7×10^{-6}	$CH_2N^{(+)}(CH_3)_3$	1.8×10^{-6}
C_3H_7	1.3×10^{-4}		

except for two compounds, 6×10^{-5} to $1 \times 10^{-6}M$.[115] This amounts to only a 200-fold difference between the best and the poorest of all the benzaldehyde derivatives, compared to the previously discussed phenyl N-methylcarbamates, where ring substituents altered I_{50} values by 10^5 times. The difference is apparently due to the fact that substituents on the phenyl ring of N-methylcarbamates serve as anionic interactants and also influence molecular reactivity, whereas those on benzaldehyde methylcarbamoyloximes serve only to increase carbamylation rate via inductive effects on the carbamate carbonyl carbon.

The importance of induction has been emphasized more recently by taking the most active compound in Table 4.15, the CF_3 analog, and adding electronegative substituents to the benzene ring.[124] In this way, the *ortho*-fluoro-substituted and *ortho*-methoxy-substituted compounds had I_{50} values of $3.4 \times 10^{-8}M$ and $4.3 \times 10^{-9}M$ respectively. Such compounds, though not very toxic to houseflies, appeared to show promise for control of some aphids and Coleopteran pests.

Much of the evidence for the overriding influence of carbamate reactivity versus enzyme affinity in the reaction of O-(methylcarbamoyl)oximes with housefly AChE has been based on indirect approaches. However, interaction of this class of chemicals with bovine erythrocyte acetylcholinesterase has been examined through individual measurements of K_a and k_2. As anticipated from the above structure–activity studies, k_2 was much more important for determining k_i and I_{50} of aldicarb and its sulfoxide than for most phenyl methylcarbamates.[86] In this respect, the oxime carbamates are more like organophosphorus inhibitors than phenyl N-methylcarbamates.

CORRELATIONS OF I_{50}, LD_{50}, AND HYDROLYTIC STABILITY

From the time of the first extensive structure–activity study with phenyl N-methylcarbamates[8] to a very recent study with O-(methylcarbamoyl)oximes,[124] attempts have been made to correlate AChE inhibition data with insect toxicity and/or chemical reactivity of the carbamates. As the amount of available information increased, and as theories explaining the mechanism of carbamate action changed, different interpretations have been attached to the relationships, so that some confusion exists in the literature. We hope that the following pages will clarify most of the confusion.

Consider first the relationship of insecticidal activity to anticholinesterase activity. Assuming that housefly head AChE is representative of the nerve enzyme as it appears in most insects, and that AChE inhibition is the mode of action of carbamates, a relatively good correlation could be expected between insect toxicity and housefly head AChE inhibition. This is exactly what was found when the contact toxicity of various phenyl N-methylcarbamates to greenhouse thrips (expressed as log LC_{50}) was plotted against negative logarithms of fly I_{50} values[8] (Figure 4.15).

However, when a similar plot was drawn comparing topical toxicity to houseflies with I_{50}, a much poorer relationship was observed. Later it was shown that correlation could be vastly improved by synergizing the carbamates with piperonyl butoxide.[12,90,106,125,126] Presumably, the synergist prevented in-vivo metabolic destruction of the insecticides, so that their "true" toxic action could be realized. Those few compounds that did not fit the pattern supposedly had a different mode of action, were not synergized by piperonyl butoxide, or had problems penetrating to the site of action.[126]

This relationship of I_{50} to synergized LD_{50} has not been seriously challenged; in fact, the relationship has been refined. For example, a correlation coefficient of 0.875 was obtained from plots of log synergized housefly LD_{50} versus $-\log K_a$ (affinity constant) for fly AChE of various double-ring phenyl N-methylcarbamates.[109] In contrast, very little correlation ($r = 0.3$) was shown between k_2 (carbamylation constant) and anticholinesterase activity. Thus, it was concluded that the primary structural requirement within this group of carbamates for good insecticidal action

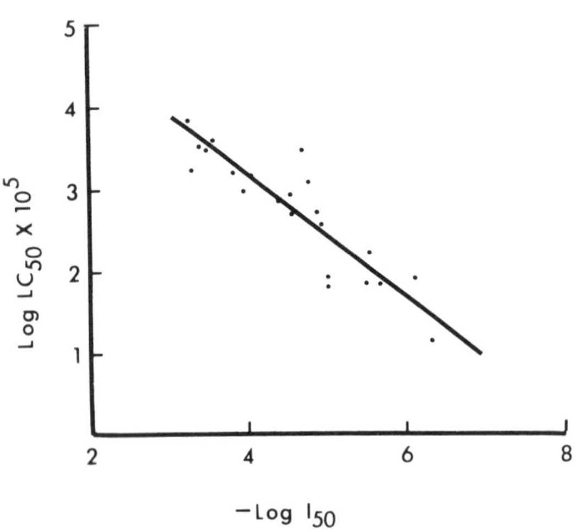

FIGURE 4.15. Correlation of housefly head AChE I_{50} values for various phenyl N-methylcarbamates with greenhouse thrip LC_{50}. (Adapted from Reference 8, with permission.)

was affinity of the carbamate inhibitor for the enzyme. Similar findings have been reported for carbofuran analogs.[29] In these instances then, not only was toxicity correlated with AChE inhibition, but it was significantly correlated with one aspect of the overall inhibition process.

A second relationship discovered early in the history of structure–activity studies was between insect toxicity and hydrolytic stability of the carbamate structure.[8] The explanation for such a correlation was based on the prevailing mechanism-of-inhibition theory, which stated that carbamate inhibitors formed a dissociable complex with acetylcholinesterase and were not hydrolyzed in the course of reaction. Hence, resistance to hydrolysis would be an attribute, and it was not surprising to find that second-order hydrolysis constants of nine substituted phenyl N-methylcarbamates (k_{hyd} ranged from 4 to 5,800 $M^{-1} min^{-1}$) plotted against toxicity to thrips formed a near straight-line relationship with the most toxic compounds possessing the greatest hydrolytic stability.[8]

Hydrolytic stability was then extended to overall chemical reactivity of the carbamate, and a third relationship was drawn between enzyme inhibition and chemical stability of the inhibitor as determined by sigma values.[8] The sigma value of a substituent on a benzene ring is a measure of its electron-attracting or -repelling power.[127,128] Thus, a substituent's effect on carbamate hydrolysis is by supplying or withdrawing electrons from the carbonyl carbon center of the carbamate molecule. For example, strong electron withdrawers, such as $-CN$ or $-NO_2$, in the *para-* or *meta*-positions of phenyl N-methylcarbamate make the carbonyl carbon atom very positive in character (more electrophilic), and thus very susceptible to hydrolysis by hydroxyl ion (see Chapter 2).

When *meta-* and *para*-sigma constants were compared to the negative logarithm of an arbitrary cholinesterase inhibition constant (not k_i or I_{50}), the correlation shown in Figure 4.16 resulted.[8] Although the slopes of the lines were rather "flat", it was concluded that good agreement existed between high cholinesterase activity and low chemical reactivity. Or, stated another way, electron-withdrawing substituents decreased inhibitory power of phenyl N-methylcarbamates.

Later plots of sigma values versus log I_{50} for fly AChE inhibition by *para*-substituted phenyl and xylenyl N-methylcarbamates also provided a flat relationship in the same direction.[129] The fact that the xylenyl carbamates averaged I_{50} values 33-fold lower than the corresponding phenyl carbamates enforced the belief that affinity of the structure for the enzyme (supplied by the methyl groups in the xylenyl structures) was much more important than electron withdrawal or attraction by the *para* ring substituents. Thus, it was essential that a carbamate should be resistant to attack by hydroxide ion and possess sufficient enzyme affinity so that it could function as an effective dissociable inhibitor.[79,83,129,130]

These proposals do not entirely support the accepted interaction of carbamate insecticides with AChE as pictured in Figure 4.2. In this scheme, hydrolytic or chemical stability could work against the compound by lowering its ability to carbamylate the enzyme (k_2). Thus, carbamates with electron-withdrawing substituents would react more readily with the serine hydroxyl of AChE and would be better inhibitors (affinity factors being equal). A plot of this relationship would slope in a direction opposite to that shown in Figure 4.16.

In other words, the present theory does not agree with past results. How can this be explained? First of all, there are some results that do not fit the type of data shown in Figure 4.16. This is particularly true for alkoxy-substituted carbamates, where hydrolytic stability was not found to be essential for high insect toxicity.[106] In addition, highly reactive structures, such as nitro-substituted phenyl N-methylcarbamates (which represent the extreme right-hand points in Figure 4.16, points which contribute considerable bias to the slope of the lines), were found to be so susceptible to alkaline hydrolysis that they degraded spontaneously during the measurement of their I_{50} values.[11] Hence, the reported inhibition constants for these and, perhaps, other carbamates are surely an underestimate of their real anticholinesterase potency, and therefore correlations of inhibition with hydrolytic stability (or σ value) would be misleading.

Secondly, a careful examination of much of the available data using substituent constants and regression analysis revealed that cholinesterase inhibition by phenyl N-methylcarbamates was more or less independent of sigma.[131] Among the constants used in the regression calculations was π, which is a measure of the hydrophobic contribution of a substituent (determined by comparing partition coefficients in a water-octanol system).[131] There was considerable correlation found between I_{50} and π ($r = 0.77$) for 23 *para*- and 30 *meta*-substituted carbamates. Indeed, 60% of all the variability among the compounds was due to π. In contrast, a correlation coefficient of only 0.4 for *para*-σ and of 0.5 for *meta*-σ constants was found. The conclusion was that, the greater the lipophilic character of the benzene substituent, the greater activity it conferred to the parent molecule, regardless of its effect on hydrolysis.

These results appeared contrary to the absolute requirement for hydrolytic stability in a carbamate structure while enforcing, by implication from hydrophobic data, the overriding need for

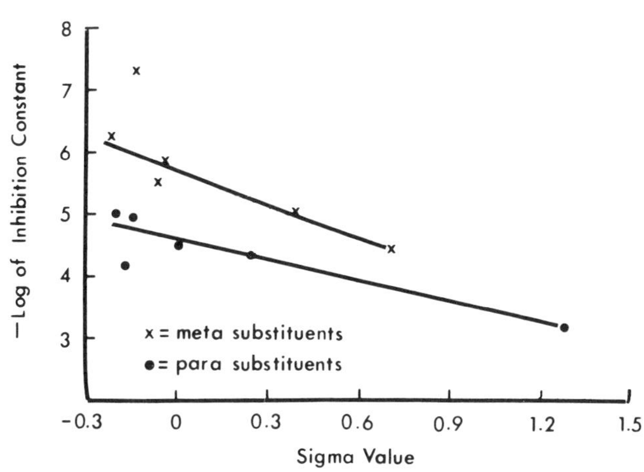

FIGURE 4.16. Relationship of sigma value of ring substituents on phenyl N-methylcarbamates to acetylcholinesterase inhibition. (Adapted from Reference 8, with permission.)

complementary fit to the reactive center of the enzyme. But was this fit necessary for formation of the complex (K_a), carbamylation of the enzyme (k_2), or both? When K_a and k_2 were determined for a number of phenyl N-methyl- and N,N-dimethylcarbamates, it was shown that variation in inhibition of bovine erythrocyte AChE was due primarily to changes in K_a, and that k_2 was similar for all compounds. Since then, data from eight of the N-methylcarbamates have been subjected to regression analysis as described above, and a good correlation was uncovered between π variation and K_a variation.[21] At the same time, little of the difference in inhibition was attributable to σ differences, and there was no correlation of inhibition with k_2. Thus, it seemed that k_2 (or hydrolytic stability) was essentially a constant for all carbamate structures, and that I_{50} was only dependent on K_a or affinity.

However, more recent K_a and k_2 determinations from several other studies with vertebrate and insect acetylcholinesterase revealed that both affinity and carbamylation rate contribute significantly to I_{50} (cf Chapter 3). So, granted that the most valuable function a ring substituent can perform is to fit snugly into the hydrophobic anionic site for maximum enzyme-inhibitor complex formation, what determines carbamylation rate? Obviously, changing ring substituents modified k_2, or the same carbamylation rate would be found for all N-methylcarbamates (which, of course, was the case where σ was found to be independent of k_2).[57] Surely, a good fit would enhance carbamylation by merely bringing the inhibitor's carbonyl carbon atom closer to the catalytic serine hydroxyl, and this may be the dominant factor controlling k_2.[80,86]

But, more important, an electronegative group on the phenyl ring should pull electrons from the C=O and enhance reaction rate (as it enhances alkaline hydrolysis), whereas electron pushers should reduce reactivity. Although this is precisely what happens with aromatic organophosphorus compounds, the phenyl N-methylcarbamates are notorious for disobeying this corollary.[8,57,79,83,129,130,132,133] The only reasonable explanation for this is based on research concerning formation of a charge-transfer complex between tetracyanoethylene and various aryl N-methyl and N,N-dimethylcarbamates.[103] It was demonstrated that the ability of a carbamate to form the complex was directly related to its ability to inhibit AChE (correlation coefficient between association constants and $-\log I_{50}$ was 0.87). Since the evidence favored formation of the complex via donation of electrons from the aromatic nucleus, it was not surprising to find that those ring substituents which push electrons toward the ring (and toward carbonyl carbon) enhanced complex formation.

If the same interaction occurred between carbamate and AChE, then electron-donating groups even though they reduce electrophilicity of C=O, would encourage the formation of the reversible enzyme-carbamate complex, and therefore increase K_a and cholinesterase inhibition, as shown in Figure 4.16, except for the extremely reactive nitro-substituted compounds, which are highly susceptible to alkaline hydrolysis.

One other possible explanation for the negative correlation between inhibition and chemical reactivity is a change in our concept of the esteratic site. If, instead of a serine hydroxyl group, some other catalytic moiety were involved, one that preferred a more nucleophilic carbamic ester setting, hydrolytic and inhibition data would be more consistent. The only support for this hypothesis is that no one has actually isolated the carbamylated serine-hydroxyl combination from the reaction of an aryl carbamate with acetylcholinesterase.

One still unanswered question is "What, exactly, causes changes in k_2?" The answer must await further study, but it is probable that the combination of charge-transfer, hydrophobic bonding, van der Waal forces, steric interactions, electron-donating and -withdrawing effects, hydrolytic instability, and chemical reactivity contribute in such a complex manner that no single or even two factors will correlate with carbamylation rate. At present, the evidence favors maximum enzyme fit over chemical reactivity of the carbonyl carbon atom in phenyl N-methylcarbamates.

With respect to oxime carbamates, regression analysis has shown that, with various ring-substituted α-methyl benzaldehyde O-(methylcarbamoyl)oximes, both alkaline hydrolysis and AChE inhibition were dependent primarily on reactivity of the carbamoyl moiety.[122] In other words, affinity of the carbamate for the active site (within limits) was not very important, and inhibition was largely governed by carbamylation rate.

Similar results have been reported for benzaldehyde-type oximes.[123] This, of course, is opposite to the situation for phenyl N-methylcarbamates, but then the structure of the oxime derivatives is such that the aryl nucleus probably does not critically interact with the anionic site. Thus, when the α-methyl group was exchanged for other substitutents (which do interact with the anionic site), regression analysis indicated that, for this series of structures, AChE inhibition was about equally correlated with steric interactions (affinity) and with carbamate reactivity (carbamylation).

It should be evident from the discussion in this chapter that, even with the extensive amount of research which has been expended on structure–activity relationships over the past 20 years, our understanding of the intricate interplay of the numerous parameters governing inhibition of acetylcholinesterase by carbamate insecticides is only elementary. Further study, particularly in relation to intrinsic differences between reactions of vertebrate versus invertebrate cholinesterases, may help in the design of new, safer insect toxicants.

The fact that chemical companies dealing with pesticides are interested in the carbamate structure as an insecticide is proven by the presence of at least 175 patents or patent applications claiming insecticidal activity for a group of synthetic carbamates. This number does not include multiple patents for the same data, so it is easy to see that literally thousands of derivatives have been prepared and tested. With a few exceptions, the vast majority of structures have not been greatly unlike the phenyl and oxime carbamates discussed throughout this chapter. From this mass of synthetic and biological screening, only a handful of molecular configurations have met the toxicological, environmental, and chemical requirements of a successful commercial insecticide. But the search is by no means over.

REFERENCES

1. **Metcalf, R. L. and March, R. B.,** Properties of acetylcholine esterases from the bee, the fly and the mouse and their relation to insecticide action, *J. Econ. Entomol.,* 43, 670, 1950.
2. **Brown, A. W. A., Robinson, D. B., Hurtig, H., and Wenner, B. J.,** Toxicity of selected organic compounds to insects. I. Tests for general toxicity on larvae of *Musca, Tribolium,* and *Ephestia,* and adults of *Sitophilus, Can. J. Res.,* 26D, 177, 1948.
3. **Brown, A. W. A., Wenner, B. J., and Park, F. E.,** Toxicity of selected organic compounds to insects. II. Tests for contact toxicity on nymphs of *Blattela* and *Oncopeltus,* and adults of *Tribolium, Can. J. Res.,* 26D, 188, 1948.
4. **Hopf, H. S.,** Studies in the mode of action of insecticides. I. Injection experiments on the role of cholinesterase inhibition, *Ann. Appl. Biol.,* 39, 193, 1952.
5. **Tobias, J. M., Kollros, J. J., and Savit, J.,** Acetylcholine and related substances in the cockroach, fly, and crayfish, and the effect of DDT, *J. Cell. Comp. Physiol.,* 28, 1959, 1946.
6. **Chadwick, L. E. and Hill, D. L.,** Inhibition of cholinesterase by diisopropyl fluorophosphate, physostigmine and hexaethyl tetraphosphate in the roach, *J. Neurophysiol.,* 10, 235, 1947.
7. **Roeder, K. D.,** The effect of anticholinesterases and related substances on nervous activity in the cockroach, *Bull. Johns Hopkins Hosp.,* 83, 587, 1948.
8. **Kolbezen, M. J., Metcalf, R. L., and Fukuto, T. R.,** Insecticidal activity of carbamate cholinesterase inhibitors, *J. Agric. Food Chem.,* 2, 864, 1954.
9. **Gysin, H.,** New Group of Insecticidal Substances, 3rd International Congress on Phytopathology, Paris, September 1952.
10. **Gysin, H.,** Über einige neue Insektizide, *Chimia,* 8, 205, 1954.
11. **Fukuto, T. R., Fahmy, M. A., and Metcalf, R. L.,** Alkaline hydrolysis, anticholinesterase, and insecticidal properties of some nitro-substituted phenyl carbamates, *J. Agric. Food Chem.,* 15, 273, 1967.
12. **Metcalf, R. L.,** Structure–activity relationships for insecticidal carbamates, *Bull. W. H. O.,* 44, 43, 1971.
13. **Reed, W. D. and Fukuto, T. R.,** The reactivation of carbamate-inhibited cholinesterase, kinetic parameters, *Pestic. Biochem. Physiol.,* 3, 120, 1973.
14. **Wilson, I. B.,** Acetylcholinesterase, *Enzymes,* 4, 501, 1960.
15. **Krupka, R. M. and Laidler, K. J.,** Molecular mechanisms for hydrolytic enzyme action. IV. The structure of the active center and the reaction mechanism, *J. Am. Chem. Soc.,* 83, 1458, 1961.

16. **Krupka, R. M.,** Acetylcholinesterase, *Can. J. Biochem.*, 42, 677, 1964.
17. **Wilson, I. B. and Quan, C.,** Acetylcholinesterase studies on molecular complementariness, *Arch. Biochem. Biophys.*, 73, 131, 1958.
18. **Foldes, F. F., VanHees, G., Davis, D. L., and Shanor, S. P.,** The structure–action relationship of urethane-type cholinesterase inhibitors, *J. Pharmacol. Exp. Ther.*, 122, 457, 1958.
19. **Wilson, I. B.,** Acetylcholinesterase. XII. Further studies of binding forces, *J. Biol. Chem.*, 197, 215, 1952.
20. **Bergmann, F., Segal, R., Shimoni, A., and Wurzel, M.,** The pH dependence of enzymic ester hydrolysis, *Biochem. J.*, 63, 684, 1956.
21. **O'Brien, R. D.,** The design of organophosphate and carbamate inhibitors of cholinesterase, in *Drug Design*, Vol. 2, Ariens, E. J., Ed., Academic Press, New York, London, 1971.
22. **Oosterbaan, R. A., Jansz, H. S., and Cohen, J. A.,** The chemical structure of the reactive group of esterases, *Biochem. Biophys. Acta*, 20, 402, 1956.
23. **Koelle, G. B.,** Ed., *Handbook of Experimental Pharmacology*, Vol. 15, Cholinesterases and Anticholinesterase Agents, Springer-Verlag, Berlin, 1963.
24. **Belleau, B. and Lacasse, G.,** Aspects of the chemical mechanism of complex formation between acetylcholinesterase and acetylcholine-related compounds, *J. Med. Chem.*, 7, 768, 1964.
25. **Wilson, I. B., Harrison, M. A., and Ginsburg, S.,** Carbamyl derivatives of acetylcholinesterase, *J. Biol. Chem.*, 236, 1498, 1961.
26. **Davies, J. H., Campbell, W. R., and Kearns, C. W.,** Inhibition of fly head acetylcholinesterase by bis [(*m*-hydroxyphenyl)-trimethylammonium iodide] esters of polymethylenedicarbamic acids, *Biochem. J.*, 117, 221, 1970.
27. **Reiner, E.,** Spontaneous reactivation of phosphorylated and carbamylated cholinesterases, *Bull. W. H. O.*, 44, 109, 1971.
28. **Yu, C. C., Kearns, C. W., Metcalf, R. L., and Davies, J. D.,** Acetylcholinesterase inhibition by bis-carbamates, *Pestic. Biochem. Physiol.*, 1, 241, 1971.
29. **Yu, C. C., Kearns, C. W., and Metcalf, R. L.,** Acetylcholinesterase inhibition by substituted phenyl N-alkyl carbamates, *J. Agric. Food Chem.*, 20, 537, 1972.
30. **Hellenbrand, K. and Krupka, R. M.,** Kinetic studies on the mechanism of insect acetylcholinesterase, *Biochemistry*, 9, 4665, 1970.
31. **Bender, M. L. and Stoops, J. K.,** Titration of the active sites of acetylcholinesterase, *J. Am. Chem. Soc.*, 87, 1622, 1965.
32. **Kitz, R. J., Broswell, L. M., and Ginsburg, S.,** On the question: Is acetylcholinesterase an allosteric protein? *Mol. Pharmacol.*, 6, 108, 1970.
33. **O'Brien, R. D.,** Binding sites of cholinesterases, *Biochem. J.*, 113, 713, 1969.
34. **Hetnarski, B. and O'Brien, R. D.,** The role of charge-transfer complex formation in the inhibition of acetylcholinesterase by aromatic carbamates, *Pestic. Biochem. Physiol.*, 2, 132, 1972.
35. **Metcalf, R. L. and March, R. B.,** Studies on the mode of action of parathion and its derivatives and their toxicity to insects, *J. Econ. Entomol.*, 42, 721, 1949.
36. **Stedman, E.,** Studies on the relationship between chemical constitution and physiological action. Part I. Position isomerism in relation to miotic activity of some synthetic urethanes, *Biochem. J.*, 20, 719, 1926.
37. **Aeschlimann, J. A. and Reinert, M.,** The pharmacological action of some analogues of physostigmine, *J. Pharmacol. Exp. Ther.*, 43, 413, 1931.
38. **Metcalf, R. L. and Fukuto, T. R.,** Some effects of molecular structure upon anticholinesterase and insecticidal activity of substituted phenyl N-methylcarbamates, *J. Agric. Food Chem.*, 15, 1022, 1967.
39. **Knowles, C. O. and Arthur, B. W.,** Biological activity of N-methylcarbamates and dimethyl phosphorothionate esters of various phenols and monoterpenoids, *J. Econ. Entomol.*, 60, 1417, 1967.
40. **Macko, J. and Gaher, S.,** Aliphatic esters of carbamic acid, *Chem. Zvesti*, 13, 436, 1959.
41. **Hadaway, A. B., Barlow, F., Grose, J. E. H., Turner, C. R., and Flower, L. S.,** Evaluation of compounds for insecticidal activity on adult mosquitoes. 3. Properties of some N-methylthiolcarbamates, *Bull. W. H. O.*, 42, 377, 1970.
42. **Kilsheimer, J. R., Kaufman, H. A., Foster, H. M., Driscoll, P. R., Glick, L. A., and Napier, R. P.,** Benzo(*b*)thienyl carbamate insecticides, *J. Agric. Food Chem.*, 17, 91, 1969.
43. **Ferguson, G. R. and Alexander, C. C.,** Heterocyclic carbamates having systemic insecticidal action, *J. Agric. Food Chem.*, 1, 888, 1953.
44. **Gubler, K., Margot, A., and Gysin, H.,** Alicyclic and heterocyclic carbamates, *J. Sci. Food Agric. Suppl.*, p. 13, 1968.
45. **Metcalf, R. L., Fukuto, T. R., and Winton, M. Y.,** Insecticidal carbamates: comparison of the activities of N-methyl- and N,N-dimethylcarbamates of various phenols, *J. Econ. Entomol.*, 55, 345, 1962.
46. **Metcalf, R. L., Fukuto, T. R., Frederickson, M., and Peak, L.,** Insecticidal activity of alkylthiophenyl N-methylcarbamates, *J. Agric. Food Chem.*, 13, 473, 1965.
47. **Sacher, R. M. and Olin, J. F.,** Insecticidal and the anticholinesterase activity of *m*-acylamidophenyl and *m*-thioureidophenyl N-methylcarbamates, *J. Agric. Food Chem.*, 20, 354, 1972.

48. **Durden, J. A., Jr. and Weiden, M. H. J.,** Insecticidal 2-(methylcarbamoyloxyphenyl)-1,3-dioxolanes, -oxathiolanes, and -dithiolanes, *J. Agric. Food Chem.*, 17, 94, 1969.
49. **Nikles, E. F.,** N-Methyl and N,N-dimethylcarbamates of hydroxybenzaldehyde acetals and mercaptals, *J. Agric. Food Chem.*, 17, 939, 1969.
50. **Brooker, P. J., Gates, P. S., Gillon, J., Newbold, G. T., and Saggers, D. T.,** Insecticidal activity of certain benzodioxolyl carbamates, *Pestic. Sci.*, 3, 735, 1972.
51. **Kazano, H., Asakawa, M., Tanaka, T., and Fukunaga, K.,** Carbamate insecticides. I. Insecticidal activities of substituted phenyl N-methyl- and N,N-dimethylcarbamates to several species of insects, *Nippon Oyo Dobutsu Konchu Gakkai Shi,* 12, 202, 1968.
52. **Wang, C. L.,** N,N-Dimethylcarbamates as insecticides. I. The synthesis of phenyl and naphthyl N,N-dimethylcarbamates and their chloro analogues, *Yao Hsueh Hsueh Pao,* 10, 237, 1963.
53. **Bachmann, F. and Legge, J. B.,** Insecticidal properties of a new group of carbamates, *J. Sci. Food Agric. Suppl.*, p. 39, 1968.
54. **Sacher, R. M., Alt, G. H., and Darlington, W. A.,** Insecticidal and anticholinesterase activity of benzotriazolyl methyl- and dimethylcarbamates, *J. Agric. Food Chem.*, 21, 132, 1973.
55. **Baranyovits, F. L. C. and Ghosh, R.,** Pirimicarb (PP 062): a new selective insecticide, *Chem. Ind.*, 30, 1018, 1969.
56. **Casida, J. E., Augustinsson, K. B., and Johnson, G.,** Stability, toxicity, and reaction mechanism with esterases of certain carbamate insecticides, *J. Econ. Entomol.*, 53, 205, 1960.
57. **O'Brien, R. D., Hilton, B. D., and Gilmour, L.,** The reaction of carbamates with cholinesterase, *Mol. Pharmacol.*, 2, 593, 1966.
58. **Dittert, L. W. and Higuchi, T.,** Rates of hydrolysis of carbamate and carbonate esters in alkaline solution, *J. Pharm. Sci.*, 52, 852, 1963.
59. **Mel'nikov, N. N., Shvetsova-Shilovskaya, D., Maksimova, Z. I., Scharova, L. P., and Shapovalova, G. K.,** Aryl N-alkylcarbamate insecticides, *Khim. Promst.*, p. 679, 1961.
60. **Thayer, H. I., Hartle, R. J., and Mallis, A.,** N-Alkyl carbamates as insecticides and pyrethrins synergists, *J. Agric. Food Chem.*, 13, 43, 1965.
61. **Fahmy, M. A. H., Metcalf, R. L., Fukuto, T. R., and Hennessy, D. J.,** Effects of deuteration, fluorination, and other structural modifications of the carbamyl moiety upon the acetylcholinesterase and insecticidal activities of phenyl N-methylcarbamates, *J. Agric. Food Chem.*, 14, 79, 1966.
62. **Robertson, W. A. H., Fraser, J., and Clinck, G. C.** (Bootes Pure Drug Co.), British Patent 982,235, February 3, 1965.
63. **Fraser, J., Clinch, P. G., and Reay, R. C.,** N-Acylation of N-methylcarbamate insecticides and its effect on biological activity, *J. Sci. Food Agric.*, 16, 615, 1965.
64. **Fahmy, M. A. H. and Fukuto, T. R.,** Insecticidal activity of N-acylcarbamates on susceptible and carbaryl-resistant strains of the Egyptian cotton leafworm, *J. Econ. Entomol.*, 63, 1783, 1970.
65. **Fraser, J., Harrison, I. R., and Wakerley, S. B.,** Synthesis and insecticidal activity of N-acyl-N-methyl carbamates, *J. Sci. Food Agric. Suppl.*, p. 8, 1968.
66. **Miskus, R. P., Look, M., Andrews, T. L., and Lyon, R. L.,** Biological activity as an effect of structural changes in aryl N-methylcarbamates, *J. Agric. Food Chem.*, 16, 605, 1968.
67. **Boulton, J. J. K., Boyce, C. B. C., Jewess, P. J., and Jones, R. F.,** Comparative properties of N-acetyl derivatives of oxime N-methylcarbamates and aryl N-methylcarbamates as insecticides and acetylcholinesterase inhibitors, *Pestic. Sci.*, 2, 10, 1971.
68. **Wolfenbarger, D. A.,** Toxicity of methylcarbamates and their acylated and chloroacetylated analogs against the bollworm and the tobacco budworm, *J. Econ. Entomol.*, 166, 272, 1973.
69. **Fraser, J., Greenwood, D., Harrison, I. R., and Wells, W. H.,** The search for a veterinary insecticide. II. Carbamates active against sheep blowfly, *J. Sci. Food Agric.*, 18, 372, 1967.
70. **Barlow, F. and Hadaway, A. B.,** Chemical structure and physical properties of candidate insecticides in relation to residual toxicity to adult mosquitoes, *Pestic. Sci.*, 1, 117, 1970.
71. **Lewis, D. K.,** In vitro acetylcholinesterase activity of certain N-methylcarbamate insecticides compared with their N-acyl derivatives, *Nature*, 213, 205, 1967.
72. **Reay, R. C. and Lewis, D. K.,** Evidence for differences in mode of action between two related carbamate insecticides, *J. Sci. Food Agric.*, 17, 17, 1966.
73. **Miskus, R. P., Andrews, T. L., and Look, M.,** Metabolic pathways affecting toxicity of N-acetyl Zectran, *J. Agric. Food Chem.*, 17, 842, 1969.
74. **Schaffer, C. H. and Wilder, W. H.,** New carbamate insecticide, RE 11775 [*m-sec*-butylphenyl N-methyl-N-(phenylthio)carbamate], with promising activity against organophosphorus-resistant mosquitoes, *J. Econ. Entomol.*, 63, 480, 1970.
75. **Kohn, G. K.,** Target-specific Pesticides an Industrial Case History, paper presented at the 166th American Chemical Society Meeting, Chicago, August, 1973.
76. **Black, A. L., Chiu, L., Fahmy, M. A. H., and Fukuto, T. R.,** Selective toxicity of N-sulfenylated derivatives of insecticidal methylcarbamate esters, *J. Agric. Food Chem.*, 21, 747, 1973.
77. **Fahmy, M. A. H., Fukuto, T. R., Myers, R. O., and March, R. B.,** The selective toxicity of new N-phosphorothioylcarbamate esters, *J. Agric. Food Chem.*, 18, 793, 1970.

78. **Chiu, Y. C., Fahmy, M. A. H., and Fukuto, T. R.,** Aryl N-hydroxy- and N-methoxy- N-methylcarbamates as potent reversible inhibitors of acetylcholinesterase, *Pestic. Biochem. Physiol.*, 3, 1, 1973.
79. **Metcalf, R. L. and Fukuto, T. R.,** Effects of chemical structure on intoxication and detoxication of phenyl N-methylcarbamates in insects, *J. Agric. Food Chem.*, 13, 220, 1965.
80. **Iverson, F. and Main, A. R.,** Effect of charge on the carbamoylation and binding constants of eel acetylcholinesterase in reaction with neostigmine and related carbamates, *Biochemistry*, 8, 1889, 1969.
81. **Fukuto, T. R., Metcalf, R. L., and Winton, M. Y.,** Carbamate insecticides: insecticidal properties of some optically active substituted phenyl N-methylcarbamates, *J. Econ. Entomol.*, 57, 10, 1964.
82. **Fukuto, T. R.,** Relationships between the structure of organophosphorus compounds and their activity as acetylcholinesterase inhibitors, *Bull. W.H.O.*, 44, 31, 1971.
83. **Metcalf, R. L., Fukuto, T. R., and Winton, M. Y.,** Insecticidal carbamates: position isomerism in relation to activity of substituted phenyl N-methylcarbamates, *J. Econ. Entomol.*, 55, 889, 1962.
84. **Kohn, G. K., Ospenson, J. N., and Moore, J. E.,** Some structural relationships of a group of simple alkyl phenyl N-methylcarbamates to anticholinesterase activity, *J. Agric. Food Chem.*, 13, 232, 1965.
85. **Wilson, I. B.,** Acetylcholinesterase. XII. Further studies of binding forces, *J. Biol. Chem.*, 197, 215, 1952.
86. **Hastings, F. L., Main, A. R., and Iverson, F.,** Carbamylation and affinity constants of some carbamate inhibitors of acetylcholinesterase and their relation to analogous substrate constants, *J. Agric. Food Chem.*, 18, 497, 1970.
87. **Meltzer, J. and Welle, H. B. A.,** Insecticidal activity of substituted phenyl N-methylcarbamates, *Entomol Exp. Appl.*, 12, 169, 1969.
88. **Kazano, H., Kurosu, Y., Asakawa, M., and Fukunaga, K.,** Uses of carbamate insecticides. 4. Insecticidal activities and characteristics of butyl- and amyl-substituted phenyl carbamates in controlling *Laodelphax striatellus* and *Nephotettix cincticeps, Nippon Oyo Dobutsu Konchu Gakhai Shi*, 14, 173, 1970.
89. **Kurosu, Y.,** Insecticidal activity of carbamate insecticides against leafhoppers and planthoppers, *Jap. Pestic. Inf.*, 10, 60, 1972.
90. **Metcalf, R. L., Fukuto, T. F., and Winton, M. Y.,** Alkoxyphenyl N-methylcarbamates as insecticides, *J. Econ. Entomol.*, 53, 828, 1960.
91. **Alexander, E. K. and Cope, A. C.,** OSRD Report No. 3877, Library of Congress, Washington, D. C., 1944.
92. **Schrader, G.,** New insecticides of low toxicity on the basis of phosphoric acid, *Angew. Chem.*, 73, 331, 1961.
93. **Mahfouz, A. M. M., Metcalf, R. L., and Fukuto, T. R.,** Influence of the sulfur atom on the anticholinesterase and insecticidal properties of thioether N-methylcarbamates, *J. Agric. Food Chem.* 17, 917, 1969.
94. **Metcalf, R. L., Fuertes-Polo, C., and Fukuto, T. R.,** Carbamate insecticides: multisubstituted chloro- and methylphenyl N-methylcarbamates, *J. Econ. Entomol.*, 56, 862, 1963.
95. **Kaeding, W. W., Shulgin, A. T., and Kenaga, E. E.,** Alkyl- and amino-substituted phenyl N-methylcarbamate insecticides, *J. Agric. Food Chem.*, 13, 215, 1965.
96. **Shulgin, A. T.,** (Dow Chemical Co.), U.S. Patent 3,060,225, October 23, 1962.
97. **Shulgin, A. T.,** (Dow Chemical Co.), U.S. Patent 3,084,098, April 2, 1963.
98. **Kaeding, W. W.,** (Dow Chemical Co.), U.S. Patent 2,984,682, May 16, 1961.
99. **Kaeding, W. W. and Kenaga, E. E.,** Schiff base phenyl N-methylcarbamates, *J. Agric. Food Chem.* 13, 378, 1965.
100. **Kaeding, W. W.,** (Dow Chemical Co.), U.S. Patent 3,009,909, November 21, 1961.
101. **Kaeding, W. W. and Kenaga, E. E.,** Azobenzene carbamates as insecticides, *J. Agric. Food Chem.*, 15, 512, 1967.
102. **Knowles, C. O.,** Metabolism of two acaricidal chemicals, N′-(4-chloro-*o*-tolyl)-N,N-dimethylformamidine (chlorphenamidine) and *m*-[(dimethylamino)methylene]amino phenyl methylcarbamate hydrochloride (formetanate), *J. Agric. Food Chem.*, 18, 1038, 1970.
103. **Hetnarski, B. and O'Brien, R. D.,** Preparation and properties of phenyl carbamates with phosphoryl and carbamyl substituents, *J. Agric. Food Chem.*, 20, 543, 1972.
104. **Lemin, A. J., Boyack, G. A., and MacDonald, R. M.,** Alkyl- and halophenyl N-methylcarbamates, *J. Agric. Food Chem.*, 13, 214, 1965.
105. **Gilbert, E. E., Peterson, J. O., and Walker, G. L.,** Insecticidal evaluation of isomeric methylthio isopropylphenyl N-methylcarbamates, *J. Agric. Food Chem.*, 16, 787, 1968.
106. **Fukuto, T. R., Metcalf, R. L., Winton, M. Y., and Roberts, P. A.,** The synergism of substituted phenyl N-methylcarbamates by piperonyl butoxide, *J. Econ. Entomol.*, 55, 341, 1962.
107. **Lambrecht, J. A.,** (Union Carbide Corp.), U.S. Patent 2,903,478, September 8, 1959.
108. **Metcalf, R. L., Fukuto, T. R., Collins, C., Borck, K., Abd-El-Aziz, S., Munoz, R., and Cassil, C. C.,** Metabolism of 2,2-dimethyl-2,3-dihydrobenzofuranyl-7 N-methylcarbamate (Furadan) in plants, insects, and mammals, *J. Agric. Food Chem.*, 16, 300, 1968.
109. **Yu, C. C., Park, K. S., and Metcalf, R. L.,** Correlation of toxicity and acetylcholinesterase (AChE) inhibition in 2-alkyl-substituted 1,3-benzodioxolyl-4 N-methylcarbamates and related compounds, *Pestic. Biochem. Physiol.*, 4, 178, 1974.
110. **Metcalf, R. L., Fukuto, T. R., Wilkinson, C. F., Fahmy, M. H., Abd-El-Aziz, S., and Metcalf E. R.,** Mode of action of carbamate synergists, *J. Agric. Food Chem.*, 14, 555, 1966.
111. **Nikles, E., Dittrich, V., and Pinter, L.,** (Ciba, Ltd.), Belgian Patent 670,630, April 7, 1966.
112. **Weil, E. D. and Schlichting, H. L.,** (Hooker Chemical Co.), Canadian Patent 742,706, September 13, 1966.

113. **Kilsheimer, J. R. and Manning, D. T.**, (Union Carbide Corp.), South African Patent Application 62/3821, 1962.
114. **Payne, L. K., Jr., Stansbury, H. A., Jr., and Weiden, M. H. J.**, The synthesis and insecticidal properties of some cholinergic trisubstituted acetaldehyde O-(methylcarbamoyl)oximes, *J. Agric. Food Chem.*, 14, 356, 1966.
115. **Weiden, M. H. J.**, Insecticidal carbamoyloximes, *J. Sci. Food Agric. Suppl.*, p. 19, 1968.
116. **Fridinger, T. L., Mutsch, E. L., Bushong, J. W., and Matteson, J. W.**, Synthesis and biological activity of α-mercaptol O-(carbamoyl)oximes, *J. Agric. Food Chem.*, 19, 422, 1971.
117. **Payne, L. K., Jr., Stollings, H. W., and Strow, C. B., Jr.**, Synthesis, biological activity, and properties of the *syn* and *anti exo*-3-chloro-*endo*-6-cyano-2-norbornanone O-(methylcarbamoyl)oximes. Absolute structure of Tranid miticide, *J. Agric. Food Chem.*, 15, 883, 1967.
118. **Felton, J. C.**, Insecticidal activity of some oxime carbamates, *J. Sci. Food Agric. Suppl.*, p. 32, 1968.
119. **Addor, R. W.**, (American Cyanamid Co.), British Patent 920,757, March 13, 1963.
120. **Addor, R. W.**, Synthesis of dithiolane oxime carbamate insecticides, *J. Agric. Food Chem.*, 13, 207, 1965.
121. **D'Silva, T. D. J. and Durden, J. A., Jr.**, Insecticidal Thiazolidin-4-one-2-carbamoyloximes, paper presented at the 166th American Chemical Society Meeting, Chicago, August, 1973.
122. **Fukuto, T. R., Metcalf, R. L., Jones, R. L., and Myers, R. O.**, Structure, reactivity, and biological activity of O-(diethyl phosphoryl)oximes and O-(methylcarbamoyl)oximes of substituted acetophenones and α-substituted benzaldehydes, *J. Agric. Food Chem.*, 17, 923, 1969.
123. **Jones, R. L., Fukuto, T. R., and Metcalf, R. L.**, Structure, reactivity, and biological activity of O-(methylcarbamoyl)–oximes of substituted benzaldehydes, *J. Econ. Entomol.*, 65, 28, 1972.
124. **Rosenfield, D. D. and Kilsheimer, J. R.**, Insecticidal activity of α,α,α-trifluoroacetophenone oxime carbamates and thiophosphates, *J. Agric. Food Chem.*, 22, 926, 1974.
125. **Moorefield, H. H.**, Synergism of the carbamate insecticides, *Contrib. Boyce Thompson Inst.*, 19, 501, 1958.
126. **Weiden, M. H. J. and Moorefield, H. H.**, Synergism and species specificity of carbamate insecticides, *J. Agric. Food Chem.*, 13, 200, 1965.
127. **Hammett, L.**, *Physical Organic Chemistry*, McGraw-Hill Book Co., New York, 1940.
128. **Jaffe, H. H.**, Re-examination of the Hammett's equation, *Chem. Rev.*, 53, 191, 1953.
129. **Metcalf, R. L., Fukuto, T. R., and Frederickson, M.**, *para*-Substituted *meta*-xylenyl diethyl phosphates and N-methylcarbamates as anticholinesterases and insecticides, *J. Agric. Food Chem.*, 12, 231, 1964.
130. **Furdik, M., Drabek, J., Locigova, I., and Ondrejka, J.**, Effect of substituents and their position in the benzene nucleus on the insecticide activity of the phenyl-substituted O,O-dimethyl O-phenyl thiophosphates, O,O-diethyl O-phenyl thiophosphates, and N-methyl phenyl carbamates, *Chem. Zvesti*, 20, 650, 1966.
131. **Hansch, C. and Deutsch, E. W.**, The use of substituent constants in the study of structure–activity relations in cholinesterase inhibitors, *Biochem. Biophys. Acta*, 126, 117, 1966.
132. **Metcalf, R. L. and Fukuto, T. R.**, *m*-Sulfurpentafluorophenyl diethyl phosphate and *m*-sulfurpentafluorophenyl N-methylcarbamate as insecticides and anticholinesterases, *J. Econ. Entomol.*, 55, 340, 1962.
133. **Gonzalez, E., Municio, A. M., and Ribera, A.**, Anticholinesterases. Influence of electronic and steric factors on the biological activity of aryl N-methylcarbamates, *An. R. Soc. Esp. Fis. Quin. Ser. B*, 62, 975, 1966.

Chapter 5

TOXICOLOGY

INTRODUCTION

Although toxicology is broadly defined as the basic science of poisons, a strict definition of the term relative to insecticides would vary drastically, depending upon one's own interest and relationship to this area. The most fundamental concept of insecticide toxicology might be summed up by merely considering "what they kill and how much does it take." This simplistic approach is taken literally by very few, but there are those who must consider it quite seriously.

For example, one responsible for the preliminary screening of candidate insecticides must be concerned at that point primarily with the product's toxicity to target pest species and whether its effective concentration is low enough to be commercially competitive. On the other hand, a physician treating a patient for acute insecticide poisoning would naturally question the compound's inherent lethality to man and the quantity to which his patient has been exposed.

There are many other facets of insecticide toxicology, however, and their complexity creates perhaps the most interesting and challenging area in the entire discipline of toxicology. Usually, a commercially successful insecticide is a potential contaminant of man at low levels throughout its many years of use. Unlike drugs and many other potent chemicals which are contacted by a small percentage of the total population, insecticides, as residues in food, air, etc., may become contaminants of most people, young and old, and on a completely involuntary basis. Consequently, the acute toxicity of insecticides to man, though important, is secondary to and much easier to determine than the more subtle harmful effects which might occur subsequent to contact with large nonlethal doses, or upon continued long-term exposure to minute quantities of the toxicant. Thus, insecticide toxicology includes an almost endless number of evaluations relative to the effects these chemicals may have on man. Among the more important are genetic, reproductive, teratogenic, carcinogenic, and behavioral effects; the diagnosis of these requires very special expertise on the part of the investigators and involves studies ranging from the subcellular to the whole-body level.

The diverse and complex nature of insecticide toxicology becomes even more evident when the essential ingredients of insecticide effects on nontarget organisms other than man, and on the total quality of the environment, are properly placed under this general category. In addition to the concerns directly applicable to man, specific attention must also be given to such things as the balance of nature and the very survival of certain species as influenced by insecticide usage.

Carbamate insecticide toxicology differs little from that of other major groups of classical insecticides. The acute toxicity of carbamate compounds to insects and mammals covers about the same range as that exhibited by members of the chlorinated hydrocarbon and organophosphorus insecticides. Generally, carbamates are not as persistent as chlorinated toxicants, and thus do not pose as serious a threat to the environment. They are, however, much like organophosphorus insecticides in that some are virtually nonpersistent, whereas others may persist in or on treated substrates, especially soil, for several weeks or even months. They are also similar in that both are acetylcholinesterase inhibitors and produce symptoms of poisoning that are not easily distinguishable. The two groups of insecticides do differ in the stability of the enzyme–inhibitor complex formed, the organophosphorus complex formed being much more stable (see Chapter 3).

Many aspects of carbamate insecticide toxicology (mode of action, metabolism, environmental stability, etc.) are integral parts of other chapters in this book and will not be repeated here. Rather, emphasis is placed on the toxicity of carbamate insecticides to selected pest species and on their actual or potential harmful effects to nonpest species, including man.

TOXICITY TO INSECTS

Because many factors, both physiological and biochemical, influence the amount of insecticide required to kill a particular insect,[1] it is virtually impossible to predict the degree of toxicity which a particular compound might exhibit toward one species on the basis of its action toward another.[2,3] Quite naturally, there are reams of data available concerning the effects of hundreds of carbamate materials on an even greater number of insects. However, it is necessary to consider only a very limited amount of information to demonstrate the general nature of carbamate toxicity to insects. Additional examples are given in other chapters, especially Chapters 4 and 10.

The housefly is commonly used for laboratory investigations and is usually included among those insect species against which potential insecticides are evaluated. Results of one such investigation, using some of the better known methylcarbamates, are shown in Table 5.1. Although carbaryl is recognized as a broad-spectrum insecticide, it is ineffective against even susceptible strains of houseflies.[4] Other carbamates are only mildly toxic to certain insect species and highly toxic to others, but the large difference exhibited by carbaryl when administered to houseflies in comparison to other pest species is seldom observed with carbamates.

For many insect pests, the variation in toxicity of a series of methylcarbamate insecticides to any particular species will be similar to that shown in Table 5.1 for flies, except for the data with carbaryl. Some compounds, such as aldicarb and carbofuran, are extremely toxic and require as little as 0.1 μg/housefly to kill 50% (LD_{50}) of the test organisms. It is not uncommon that other insecticidal carbamates require a 5- to 10-fold dosage increase to achieve the same degree of mortality, e.g., Matacil and Baygon.

There are, however, instances where a given insect species is almost equally susceptible to a number of carbamates exhibiting a range of toxicity to other organisms. The adult western corn rootworm was shown to respond similarly to several carbamate insecticides topically applied at similar dosage rates[5] (Table 5.1). LD_{50} values of the compounds extended only from 2 to 10 μg/g, although the chemical configuration of this group of carbamates and their range of toxicity to other animals are as varied as any which may have been selected for study.

TABLE 5.1

Toxicity of Topically Applied Methylcarbamate Insecticides to Certain Adult Insects

LD_{50}, μg/g			
Housefly[4]		Western Corn Rootworm[5]	
Carbaryl	>500	Baygon	10
Matacil	80	Carbaryl	6
Baygon	25	Carbofuran	5
Carbofuran	7	Bux	5
Aldicarb	6	Methomyl	2

The most notable species-specificity demonstrated by the data in Table 5.1 occurs, of course, with carbaryl. Not only was the LD_{50} of carbaryl to houseflies approximately one hundred times greater than to the adult corn rootworm, it was just as lethal to the latter as Baygon and carbofuran. Against adult *Lygus hesperus,* carbaryl was more toxic than Matacil, and only slightly less toxic than carbofuran.[6] However, when the two pests listed in Table 5.1 were treated with carbofuran, both died at the same dosage level.

Other studies have shown that carbaryl has perhaps the widest spectrum of insect toxicity of any of the carbamate insecticides, with LD_{50} values ranging from more than 1,000 μg/g to as little as 0.1 μg/g.[7,8] Though most carbamates display a similar, but less dramatic, spectrum of activity, there are some "general trends" among species. For example, most aryl methylcarbamates are only moderately toxic to houseflies, quite effective against the Mexican bean beetle and the bean aphid, but essentially nontoxic to southern armyworm and two-spotted spider mite.[9] Despite a few of these trends, however, it is still extremely difficult to predict a carbamate's potential insect toxicity.

It is obvious from the data already presented that carbamate insecticides possess contact toxicity characteristics. In this regard they are no different from most chlorinated hydrocarbon and organophosphorus insecticides. Similarly, representatives of these three groups of chemicals are toxic to insects when ingested. In fact, most of the materials are more active when administered orally or by injection than when applied topically.[1,10,11] Since the topical toxicity of insecticides correlates with increased lipophilicity of the compounds,[11,12] it may be surmised that the greater toxicity via the oral and injected routes occurs because the necessity of cuticle penetration is eliminated. There are exceptions to this generality. For example, 3-isopropylphenyl N-methylcarbamate (UC 10854) was less toxic to the cotton leafworm as a stomach poison (LD_{50} =

0.8 μg/g) than as a contact poison (LD_{50} = 0.1 μg/g).[10] On the other hand, carbaryl was ten times more toxic when ingested (LD_{50} = 0.4 μg/g) by the same insect species.

Although the combination of contact and stomach poisoning broadens the scope of possible toxic action of the carbamates to insects, stomach poisoning probably plays a minor role in the efficacy of most of these materials. If applied as contact poisons, their rapid action provides little opportunity for susceptible exposed insects to continue feeding in a normal manner. Moreover, many of the carbamates dissipate and/or degrade so rapidly that ingestion of toxic levels by insects escaping exposure by contact would be unlikely.

Contrary to the point that most carbamates act primarily via contact action, those which exhibit plant-systemic activity most certainly are true stomach poisons. Compounds such as aldicarb and carbofuran, though extremely potent contact toxicants, are also effective when consumed by plant-feeding insects. These insecticides are unique in that they are taken from the soil by plants in a form and quantity that are lethal to insects feeding thereon. Complex and critical physical, chemical, and biochemical interactions between the pesticide and plant obviously must take place in order that certain compounds are active systemically while others of similar structure are completely inactive. Much remains to be learned about these interactions, but it is suspected, and in most cases known, that the production of relatively stable toxic metabolites in the soil and/or plant is an essential ingredient for plant-systemic activity. This facet is discussed at some length in Chapter 7.

Another aspect of insecticide toxicology which is important is the variation in susceptibility that exists among different life stages. Many insects are pests to crops and animals only in the larval stage, and control measures aimed at the adults are ineffective. In such cases, the insecticide must be toxic to the larvae or eggs, if adequate control is to be achieved. Unfortunately, the larval stages, particularly later instars, of many insects are often much more tolerant to methylcarbamates and other insecticides than are the adults. Baygon, for example, is extremely lethal to certain adult insects,[13] but is a relatively weak larvicide.[14] This does not negate its use as a control agent, but it does limit the manner in which it may be used to attain sufficient control.

The comparative toxicity of aldicarb to adult and larval stages of the housefly[15] is typical of that observed with many other methylcarbamates, and with comparable stages of other insect species.[16] When applied topically to 5-day-old housefly adults of a susceptible strain, aldicarb gave an LD_{50} of 0.31 μg/insect. The LD_{50} to third-instar larvae of the same strain was not approached even at treatments of 100 μg/larva. No mortality occurred in the immature insects until the third day following treatment, whereas the adults died within a few hours of exposure, as is usual with methylcarbamate insecticides. Similar slow action and high-dosage requirements were observed with larvae of the pecan weevil and of the pink bollworm.

The same study[15] considered the possibility that sublethal doses of aldicarb to housefly larvae would prevent normal development to the adult stage. Data summarized in Table 5.2 show that this indeed was the case. As little as 1.5 μg/larva reduced development to the adult stage significantly, and no adults at all emerged when the larvae were treated at doses above 12 μg/insect. The compound affected pupation only at very high treatment rates, demonstrating that there was a latent toxicity not expressed until after the

TABLE 5.2

Effect of a Single Application of Aldicarb on Survival and Further Development of Third-Instar Housefly Larvae[15]

Dose, μg/larva	Mortality of larva after 3 days, %	Pupation, %	Larva to adults, %	Adults surviving for 5 days, %
0	0	100	99	100
1.5	0	100	25	100
12.0	0	100	2	100
100.0	40	60	0	–

larvae had pupated. It appeared that the delayed toxic action did not extend into the adult stage. Other studies have shown that adult houseflies tend to become more susceptible to carbamate poisoning as they age.[16,17] Much greater changes in response, however, have been reported for growing immature forms of holometabolous insects; for example, the topical LD_{50} to third-instar European corn borers was found to be 3 μg carbaryl/insect, whereas the LD_{50} to fifth-instar larvae was 200 to 300 μg carbaryl/insect.[18] Additional factors, such as sex and diet, also influence carbamate susceptibility, making toxicity forecasts even more precarious.

Other methyl- and dimethylcarbamates are known to adversely affect the reproductive potential, food intake, and survival rate of the housefly.[19] This likely occurs in many insect species and could represent a degree of control that is not usually attributed to a compound when only isolated life stages are included in efficacy evaluations. A related response elicited by certain carbamates is an antifeeding reaction which causes eventual starvation or dessication (see Chapter 3). Certainly, there are more behavior patterns disrupted by sublethal carbamate exposure which might also contribute to the ultimate control of insect pests.

TOXICITY TO MAMMALS

Factors Affecting Toxicity

Acute oral LD_{50} values are the most commonly used indices of carbamate insecticide toxicity to mammals. Although it may be argued that results of dermal exposure are more representative of the hazards encountered during commercial use of a compound, acute oral toxicity data are not without considerable merit. Since such data are routinely available for all pesticides, one can quickly relate the inherent oral toxicity of one compound, particularly if it is new, to others with which one is familiar. Dermal toxicity values do not serve the same purpose, because most carbamates are relatively nontoxic by this route, and the toxicity levels are often expressed as being greater than a designated dose. This is demonstrated in Table 5.3, where a few compounds are given as examples.[20] Additional acute-toxicity data are presented in Chapter 1.

The differential in toxicity (oral vs. dermal) of Baygon, Zectran, and carbaryl is typical of most methylcarbamates, but there is almost always a meaningful difference between the oral toxicities of different carbamates. Again, this permits one to make a mental comparison of pesticide toxicity and to take necessary precautions or recommendations relative to handling a chemical. From a practical standpoint, one is well advised to use extreme caution when dealing with a toxicant whose acute oral LD_{50} to rats is 25 mg/kg or below, extra caution if the value is above 25 mg/kg, but less than 100 mg/kg, and due caution if the LD_{50} value exceeds 100 mg/kg. No carbamate, irrespective of its acute oral or dermal LD_{50} value, should be considered innocuous.

It is rare that the dermal toxicity of a carbamate insecticide approaches that of the oral route. Such cases do exist, however, as exemplified by aldicarb (Table 5.3). Even more unusual is Isolan, which is more toxic to rats when applied dermally than when administered orally. Isolan, and especially aldicarb, are very potent toxicants by both routes, and it is clear that their high dermal toxicity enhances their potential hazard to mammals. There is no carbamate insecticide which is extremely toxic dermally and relatively nontoxic orally. Thus, establishing general safety precautions for a carbamate compound on the basis of its acute oral toxicity should not result in a false sense of security.

In addition to oral and dermal applications,

TABLE 5.3

Acute Toxicity of Some Carbamate Insecticides to Rats[20]

Insecticide and sex of animal	LD_{50}, mg/kg	
	Oral	Dermal
Aldicarb		
Male	0.8	3
Female	0.6	1
Isolan		
Male	23	5.6
Female	13	6.2
Baygon		
Male	83	>2,400
Female	86	>2,400
Zectran		
Male	37	>1,500
Female	25	>1,500
Carbaryl		
Male	850	>4,000
Female	500	>4,000

researchers often administer pesticides to mammals by intraperitoneal, subcutaneous, intravenous, and even by intracerebral injections.[21] Carbamates and other insecticides vary in their toxicity with the different modes of delivery, but the relative order of activity tends to be the same, regardless of the route of exposure. This does not mean that the degree of change in toxicity is constant for all compounds. Carbaryl, for example, has an oral LD_{50} of 300 to 400 mg/kg in mice,[22-24] but when injected intraperitoneally, the LD_{50} is approximately 24 mg/kg; the corresponding values for dimetilan are 63 and 12 mg/kg following oral and intraperitoneal treatments respectively.[25,26] It appears that the greater variation in acute toxicity of carbamates to mammals is exhibited when the insecticides are administered orally than by any other route.[27]

Method of exposure is but one of many factors which may influence the mammalian toxicity of a carbamate insecticide. However, all of the factors involved are common to each of the major classes of conventional insecticides, and to most toxicants in general.[28] Accordingly, only brief mention concerning these factors need be made at this point.

With respect to differences in susceptibility between species, the carbamate insecticides do not demonstrate too wide a spectrum of activity as a general rule. Certainly the differences between mammalian species are not nearly as great as those found among insects, as discussed in the preceding section. Of the two most common experimental animals, mice are usually more sensitive to carbamates than are rats. The following data demonstrate this generality:

Acute oral LD_{50}, mg/kg

	Rats	Mice
1-Naphthyl N-methylcarbamate (carbaryl)[24]	515	438
2-Ethylthiomethylphenyl N-methylcarbamate (Croneton)[29]	338	71
3-Methyl-5-isopropylphenyl N-methylcarbamate[30]	60	40

These values were taken from studies comparing the toxicity of the carbamates to rats and mice, and should reflect relative toxicities fairly accurately. Because of the variations in LD_{50} values obtained in different laboratories, under different conditions, etc., it is difficult to establish real differences in toxicity by selecting data for each species from different investigations.

As might be expected, some carbamates do not appear to be any more lethal to mice than to rats or to other animals. For example, dimetilan is equitoxic to mice, rats, and guinea pigs, with an acute oral LD_{50} of approximately 63 mg/kg for all species.[26] A similar situation is found with carbofuran, where the acute oral LD_{50} to rats, dogs, and chickens is about 11 mg/kg.[31]

The cat is one species which is often more susceptible to toxicants than are other mammals, and this seems to hold true for the methylcarbamate insecticides, at least as exemplified by carbaryl. In a study in which this carbamate was tested against cats, mice, and rats, the respective acute oral LD_{50} values were 150, 363, and 721 mg/kg.[22] In addition to showing the relative toxicities of carbaryl to these three species, the data also illustrate, when compared to the rat and mice data shown above, the degree of variation commonly obtained from different laboratories using the same carbamate and test organisms. Mention is made of this again in order to stress the importance of choosing comparative toxicity data carefully before too much emphasis is placed on the role any factor plays in determining toxicity.

It was pointed out in the previous section on insects that the sex of the animal may influence susceptibility to carbamate insecticides. For most insect species, males appear to be less tolerant than females. In rats, the opposite is generally true, although the magnitude of the difference is rather slight (Table 5.3). Insufficient data are available to determine if this is true for the majority of mammalian species. It certainly does not hold for all cases, as attested by the fact that female swine are more tolerant to carbaryl than are the males.[32] Although based almost solely on rat toxicity studies, the order of activity of most carbamates supports the observed trend toward greater susceptibility in females, as well as a general conclusion that the sex of mammals is of little, if any, practical significance insofar as acute toxicity of carbamates is concerned.

That the vehicle used in administering pesticides to animals influences toxicity is well recognized.[28,33] In dermal applications, the solvent or carrier effect on toxicity is related primarily to increased or decreased rates of skin absorption.[34] A similar absorption relationship may exist in the gut. At least, it is known that carbaryl is more toxic when given orally to mice in oil than as a suspension in water; the respective LD_{50} values were 157 and 278 mg/kg.[35] Although enhanced absorption may result from the presence of the oil, it is possible that the oil protects the carbamate from enzymic and spontaneous degradation in the gut, which otherwise would lessen its toxicity.

Because of low protein levels in the food of some peoples of the world, there has been some concern about the effects of diet on the toxicity of carbamate insecticides to man. Experiments in rats suggest that the amount of protein in the diet has no significant effect on the acute toxicity of carbamates, except when it is drastically different from that required for normal growth. With carbaryl, significant increases in toxicity were noted only in those animals fed diets containing 10% or less of the normal level of protein.[36,37] Increasing the protein intake above the usual range did not change the animals' susceptibility to carbaryl.[38] Lowering the level of protein content in the diet also increased the acute toxicity of Banol to rats, whereas animals fed a casein-supplemented diet were less susceptible than the control animals.[39,40] Similar to the carbaryl studies, these results were evident only when the protein content of the diet was vastly different from normal.

As evidenced by the discussion presented in Chapter 10, repeated exposure of insects to carbamate insecticides through several generations can, and often does, result in resistance of the organism to these toxicants. Such is not the case with mammals. Loss of sensitivity to dimetilan, within the same generation, was not demonstrated in rats when this dimethylcarbamate was given orally at doses approaching the LD_{50} value on a daily basis for 4 weeks.[26] Furthermore, resistance did not occur in a strain of mice after 14 generations of selection with LD_{50} doses of carbaryl.[41] Since the mechanisms of metabolic degradation of carbamates in mammals are far more efficient than in nonresistant insects, it is not surprising that resistance in mammals was not exhibited, even if some increase in metabolic activity was elicited by selection with these insecticides.

The signs and symptoms of carbamate and organophosphorus insecticide poisoning in mammals are essentially identical. The nature of these signals can be appreciated by examining the results reported when swine were fed large doses of carbaryl.[32,42] Single oral treatments of 1,000 or 2,000 mg/kg produced salivation, vomiting, difficult breathing, central-nervous-system depression, muscle tremors, and cyanosis. When 13-week-old pigs were fed a diet containing carbaryl, 150 to 300 mg/kg per day, for up to 12 weeks, intoxication was evidenced by progressive myasthenia, incoordination, ataxia, and clonic muscular contractions. Continued exposure of the animals to carbaryl resulted in paraplegia and prostration.

There is a very critical difference in the mode of action of organophosphorus and carbamate insecticides which is manifested by a more rapid recovery from poisoning by carbamates. This is due to the reversible nature of carbamate-acetylcholinesterase inhibition, whereas organophosphorus inhibition is more permanent (see Chapter 3). The ability of the blocked nerve to "recover" allows for poisoning symptoms to appear far in advance of an accumulation of a lethal quantity of carbamate. For example, consider the data in Table 5.4, where acute LD_{50} and ED_{50} values are compared following injection of rats with the methylcarbamate insecticide Baygon or the organophosphorus compound methyl paraoxon.[43] In this case, ED_{50} is that dose evoking the slightest tremor in 50% of the treated animals. This symptom was found to be a more reliable criterion of low-level exposure to anticholinesterase agents than the appearance of any other classical cholinergic symptom. It is obvious from the data (Table 5.4) that tremors are induced by doses of Baygon which do not approach the acute LD_{50} value, and that such a large differential is

TABLE 5.4

LD_{50} and ED_{50} Values for Baygon and Methyl Paraoxon after Injection into Rats[43]

Route	Baygon	Methyl paraoxon
Intravenous		
LD_{50}, mg/kg	10.60	0.46
ED_{50}, mg/kg[a]	0.33	0.08
Ratio, LD_{50}/ED_{50}	32.12	5.75
Intramuscular		
LD_{50}, mg/kg	53.00	1.69
ED_{50}, mg/kg	1.66	0.40
Ratio, LD_{50}/ED_{50}	31.92	4.22

[a]ED_{50} = dose evoking slightest tremor in 50% of the injected animals.

not evident with methyl paraoxon. Other carbamate and organophosphorus insecticides yield similar results (see Table 3.9), but one is cautioned not to assume that all members of these chemical groups would behave in exactly the same manner. The practical importance of early-warning signs with carbamates, as opposed to organophosphorus compounds, should not be disregarded, however, in selecting a product for insect control.

Effect on Man

In man, the signs and symptoms of carbamate intoxication follow those described in other mammals, as discussed in detail in Chapter 3 for a male volunteer ingesting 1.5 mg Baygon/kg body weight.[27] It is important to note that the symptoms, although severe, began to lessen progressively 1 hr after ingestion; by 2 hr the subject was feeling well, his cholinesterase levels had returned to normal, and he was able to consume a meal without discomfort. The ability of man to tolerate high doses of Baygon was clearly demonstrated by this experiment, but another study showed that symptoms could be evoked at far lower levels than was the case in rats (Table 5.4). A single carbamate dose of 0.36 mg/kg to another volunteer produced mild symptoms, but the subject returned to normal within 3 hr. In attempted suicide cases, atropine has been used successfully to save the lives of patients who ingested large quantities of Baygon.[44]

Controlled human studies with carbaryl have been conducted, but not at dosages sufficiently high to cause significant adverse effects. One such investigation showed that daily administration of carbaryl at 0.13 mg/kg for 6 weeks produced no signs attributable to the insecticide.[45] However, there was a slight decrease in the ability of the proximal convoluted tubule to reabsorb amino acids. Notwithstanding its relative safety, carbaryl nonetheless may be fatal. A 39-year-old man died 6 hr after ingestion of about 500 ml of "Sevin 80" solution.[46] Death occurred in spite of immediate hospitalization, prompt gastric lavage, and administration of the antidote atropine every half hour. However, it must be noted that the patient's condition deteriorated rapidly after he was given another antidote, 2-PAM. The possible contribution of the latter "antidote" to the fate of the patient will become evident in discussions of antidotes presented later in this chapter. Although atropine cannot overcome the effects of all carbamate poisoning, its use is certainly recommended when carbaryl is involved, and it has proved to be a life-saving measure in reversing severe intoxication of humans by Baygon,[44] Zectran,[47] and other methylcarbamate insecticides.

Exposure of man to carbamate insecticides via penetration through the skin is far more common than the ingestion of large quantities as just discussed. This is a particularly important route of exposure for pesticide applicators, farm workers, and certain villagers of developing nations where pesticides are commonly used in dwellings as part of the malaria eradication program. Limited information is available on the kinetics of skin absorption of pesticides in general, and relatively little of this relates specifically to carbamates. Some of the more definitive data reported on the subject are presented in Tables 5.5 and 5.6.

Radioactive carbaryl was absorbed from the ventral forearm much faster and in much greater total quantities than some of the commonly used organophosphorus and chlorinated hydrocarbon insecticides (Table 5.5).[48] It is impossible to determine whether the almost complete absorption of carbaryl, 73.9% of the applied dose after 5 days, is typical of all methylcarbamates. However, the similarity of the data among the other compounds, ranging only between 8 and 10% of the dose, surely suggests that methylcarbamates could be unique insofar as skin penetration is concerned. Studies of permeability as a function of anatomic region (Table 5.6) substantiate the high absorption rate observed with carbaryl as compared to parathion.[49] The data also show that

TABLE 5.5

Absorption of ^{14}C-labeled Carbaryl and Certain Non-carbamate Insecticides Applied to the Ventral Forearm of Man

	Amount of absorption	
Insecticide	Maximum rate (% of dose/hr)	% of total dose after 5 days
Carbaryl	3.00	73.9 ± 21.0
Parathion	0.22	9.7 ± 5.7
Lindane	0.25	9.3 ± 3.7
Malathion	0.41	8.2 ± 2.6
Dieldrin	0.14	7.7 ± 2.9
Aldrin	0.09	7.6 ± 2.9

Adapted from Reference 48, with permission.

the penetration of pesticides through the skin is very much influenced by the specific anatomic region exposed. Data were less complete with carbaryl than with parathion, but it is likely that the order of absorption rates in the different areas would be similar for both materials.

The consequences of skin contamination of spraymen and villagers have been the topic of several excellent investigations by the World Health Organization Insecticide Testing Unit. Extensive trials where candidate carbamates were applied to dwellings for vector control showed that carbaryl caused no noticeable adverse effects among spray crews or village inhabitants, whereas UC 10854 tests had to be suspended very shortly after initiation because of the appearance of gross symptoms in the spray operators[50] (refer to Chapter 3 for more details). Baygon has also been the subject of considerable testing, and occasional symptoms have appeared.[51-53] As an example, field trials in Iran involved applying Baygon as a 2.5% suspension to walls of dwellings.[54] Workmen who sprayed the material 8 hr/day for 17 days and inhabitants were evaluated for carbamate intoxication. Some of the spraymen complained of headache, giddiness, and nausea, these being the most common symptoms. Other symptoms included muscular weakness, abdominal cramps, diarrhea, and severe sweating. Among 927 inhabitants questioned (about 14% of the population in the treated areas), 45 complained of headaches, and sometimes of other mild symptoms, all of short duration. Baygon was considered safe as a residual spray in the home, assuming that proper precautions were taken during application.

Carbamate insecticides – especially carbaryl, because of its widespread use – have been implicated as causal agents of eye disease[55,56] and of suppression of various immunologic responses in animals.[57,58] To date, the data relating to carbamates and human disease are too incomplete to judge the importance of the few reported relationships. It is an area deserving further study, but one which does not appear to merit alarm of any kind.

TABLE 5.6

Absorption of Topical ^{14}C-Carbaryl and Parathion by Man According to Site of Application[49]

	Percent of applied dose	
Insecticide and anatomic region	0–8 hr	0–5 days
Carbaryl		
Forearm	5.1	73.9 ± 21.0
Jaw angle	15.7	69.9 ± 19.6
Parathion		
Forearm	0.5	8.6 ± 5.6
Jaw angle	1.2	33.9 ± 12.9
Palm	0.4	11.8 ± 3.9
Hand dorsum	2.0	21.0 ± 8.1
Axilla	6.7	64.0 ± 32.5
Scrotum	20.9	101.6 ± 18.9

Biochemical and Pharmacological Effects

The biochemical, pharmacological, and pathological effects on warm-blooded animals which are attributed to carbamate insecticides are numerous. Often it is difficult to assess the significance of these changes, especially if they are slight, but it is essential that they be noted and considered in any discussion of carbamate safety to man. A general idea of the types of effects on various animal systems reported for carbamates may be gained by perusing the data presented in Table 5.7. These data refer only to carbaryl, but they are indicative of results of similar studies with other carbamate insecticides. Much of the information is taken from the works of Russian scientists, whose studies are commonly conducted on a chronic

TABLE 5.7

Summary of Certain Biochemical and Pharmacological Effects Produced in Warm-blooded Animals by Carbaryl

1. Cholinesterase depression in blood, brain, liver, etc.
2. Other enzyme activity altered: alkaline phosphatase, alanine and aspartate aminotransferase, arginase, aldolase, glucose-6-phosphate dehydrase, NAD-hydrolase, phosphofructokinase, transaminase, transketolase.
3. Increased levels of nicotinamide coenzymes, RNA, DNA, and glycogen in blood and tissues; tissue respiratory processes altered; reduced phospholipid synthesis and metabolism.
4. Increased erythrocytes, leukocytes, hemoglobin, and blood serotonin; decreased eosinophils, lymphocytes, and altered K^+ content in erythrocytes and plasma.
5. Necrotic changes in liver and in metabolizing capacity; hexobarbital sleep time increased by acute doses, decreased by chronic exposure.
6. Other effects include increased thyroid activity, increased adrenal gland weight, induced hypocalcemia and hypernatremia, and functional depression of the cerebral cortex.

basis, with exposure of the animals to the insecticide usually spanning a period ranging from several weeks up to 1 year or more. Results from investigations performed in the United States are seldom reported in the scientific literature.

Of course, cholinesterase (ChE) depression (Table 5.7, No. 1) is common to all insecticidal carbamates.[59,60] Data are often presented relative to the inhibition of ChE in the blood, but the enzyme is also inhibited in other tissues of the body. When assayed 6 hr after treatment of rats with carbaryl at 100 mg/kg, it was found that erythrocyte ChE was reduced by 22%, plasma ChE by 35%, and liver ChE by 82%.[61]

Clinically, the blood ChE level is an important factor in determining exposure of animals to carbamates. The reversible nature of the carbamylated enzyme has presented many problems in assessing human poisoning by carbamates, however, since complete regeneration of the enzyme may occur in a blood sample shortly after it is taken from the body. In addition, it has been demonstrated that a sample should be diluted 300-fold or more and assayed within 4 hr if the results are to be valid.[62,63] Although comparisons of ChE levels in individuals before and after possible contact are the most reliable means of determining exposure to an anticholinesterase agent, an inhibition of 20 to 30% as compared to the population average is usually considered conclusive.[64]

Many other enzyme systems are altered in various degrees by methylcarbamates (Table 5.7, No. 2). Of these, alkaline phosphatase appears to be most sensitive,[65-70] with activity in liver, kidney, and adrenals generally reduced by chronic exposure to carbaryl, although its activity is increased in serum.[68] The other enzymes listed in Table 5.7, No. 2, are usually less active in carbaryl-treated animals,[66-72] and levels of RNA, DNA, and glucogen tend to be lower (Table 5.7, 3).[67,69,70] Higher levels of nicotinamide coenzymes and enhanced NAD-hydrolase activity have been observed in the blood, liver, and brain of carbaryl-treated rats.[73] Tissue oxidative processes are usually stimulated,[74-76] whereas phospholipid synthesis and metabolism in liver, brain, and heart are inhibited as a result of carbamate treatment.[77]

Blood components are influenced by carbaryl in a variety of ways (Table 5.7, No. 4). In cattle, a dose of 100 mg/kg raised the erythrocyte count by 17% after 24 hr, and it also increased leukocytes, hemoglobin, and serum calcium, but it lowered eosinophils, lymphocytes, blood sugar, and inorganic phosphorus.[78]

Blood serotonin levels are significantly increased by doses of carbaryl which do not inhibit cholinesterase. Whereas a dose of 60 mg/kg orally administered to rats was required to decrease cholinesterase activity, a single dose of 10 mg/kg increased the serotonin level by 21%.[79] Daily feedings at the same rate doubled the serotonin titer after 2 months, but by 6 months the blood serotonin level was about the same or somewhat below that of the control animals.[80] Other changes in the blood of animals dosed with carbaryl include alteration of the potassium content. Following an oral dose of 200 mg/kg, the potassium levels increased in the erythrocytes as well as in the myocardium and skeletal muscles.[81] It was suggested that electrolyte concentration in the blood might be used clinically to test for carbamate poisoning, but it does not appear as promising as measuring serotonin levels.

The effects of carbaryl on liver function and morphological condition are not as pronounced as those of DDT, but they are recognizable (Table 5.7, No 5). Proliferative processes were observed in the livers of rats and rabbits treated daily with 7.6 mg/kg for up to 11 months, and as little as 0.38 mg/kg was said to decrease metabolic detoxication.[82] At the higher dose, carbaryl prolonged the sleep time in animals treated with hexobarbital as compared to animals which had not been treated with carbaryl. Since hexobarbital is degraded by the liver microsomal system, it would appear that the system is impaired by carbaryl. Conflicting evidence is available, however, and in a form that is highly unusual. That is, divergent results were reported by the same authors[83,84] in the same year (1972), using the same compounds (including carbaryl) and the same animal species (mouse), but in different journal articles, without acknowledgment of the conflicting study. The authors' abstract of one of the reports[83] states that parathion, paraoxon, disulfoton, and carbaryl produced a *decrease* in hexobarbital sleeping time and an *increase* in the rate of metabolism of hexobarbital and aniline. Their other abstract[84] states that the same compounds *increased* hexobarbital sleeping time and *decreased* hexobarbital metabolism. Examination of the data presented in the two papers shows that carbaryl does indeed influence the hypnotic effect of hexobarbital. In animals given carbaryl orally at half the LD_{50} 1 hr before intraperitoneal injection of hexobarbital at 100 mg/kg, sleep time was increased from 30 min for control animals to almost 80 min for those treated with carbaryl. When given 24 hr before the barbiturate, sleep time was extended for only 8 min. However, mice which were treated daily for 3 to 10 days with carbaryl prior to hexobarbital administration actually slept for shorter periods than control animals. A similar trend was evident with in vitro hexobarbital metabolism studies. Thus, it appears that the influence carbaryl has on liver function is initially of an inhibitory nature, but upon repeated exposure it may be slightly stimulatory. The same conclusion could be reached regarding the organophosphorus insecticides included in the above studies.

Other effects of carbaryl on animal systems, in addition to those already mentioned, are changes in cortical rhythm, suggesting depression of cerebral-cortex function,[85] increased weight of the adrenals,[79] hypocalcemia and hypernatremia,[81] and reduced thyroid gland activity[86,87] (Table 5.7, No. 6). Of these, the effect of carbaryl on the thyroid appears to be of greatest potential importance in that it has been reported to enhance the development of endemic goiter.

Most literature reports concerning the reproductive and teratogenic effects of carbamate insecticides deal with carbaryl, and they usually show that there is some degree of activity. Many papers from Russia state that carbaryl alters the reproductive system rather drastically and may induce abnormalities ranging from slight alteration in the estrous cycle to total sterility.[88-92] Based upon spermatogenesis and oogenesis evaluations, the minimum active daily dose to rats and dogs was shown to be 0.7 mg/kg.[93] Embryotoxicity, stillborn rate, increased death rate among offspring, and delayed development are other outcomes attributed to carbaryl.[94,95] Attesting to the significance of these effects are results of field studies which show that certain mammals do not reproduce normally in carbaryl-treated areas.[96-98] Not all of the adverse reports for carbaryl relative to reproduction and teratology come from Russian scientists, however. Reduction of litter size, malformation of fetuses, and reduced survival have been reported by scientists of the Food and Drug Administration in the United States.[99-101] Contrary to these results, other studies with carbaryl[102,103] and carbofuran[104] show the methylcarbamates to be virtually free of any adverse reproductive effects or teratogenic activity when tested against several mammalian species.

Like most chemicals and chemical groups, the carbamates contain certain members which are reportedly carcinogenic. The noninsecticidal ethyl carbamate (urethane) produced skin carcinogenesis in mice, but only in the presence of croton oil as a promoting agent; other very similar analogs were ineffective, with the exception of ethyl N-hydroxycarbamate, which induced about the same number of tumors as an equivalent dose of ethyl carbamate.[105,106] A noninsecticidal analog of carbaryl, the N-nitroso derivative, is a very potent mutagen[107,108] in bacterial systems that have proven very sensitive to known carcinogens.[109] N-Nitroso carbaryl is formed by reacting the parent insecticide with nitrous acid under conditions which were likened to those possible in the stomach.[107] However, in vivo formation of N-nitroso carbaryl has not been demonstrated experimentally.

Carbaryl *per se* and the dithiocarbamate fungicides maneb, ziram, and zineb have been claimed to possess a marked ability to induce malignant tumors in rats.[110] This was based on a 22-month study in which large doses of the pesticides were administered to the animals orally or in pellets of paraffin placed under the skin. For example, carbaryl was given at 30 mg/kg orally and in 20-mg quantities subcutaneously. With each of the pesticides, the doses were sufficient to cause death in more than 80% of the animals by the end of the experiment. Less than 5% of the control animals died, and only one control rat out of 46 surviving after 22 months developed tumors. Of the 22 carbaryl-treated animals surviving, 6 were said to develop tumors. Almost identical results were reported for the other compounds. Because of the high doses employed and the resulting high incidence of death, this experiment would hardly be the type to offer sound evidence that these pesticides pose a threat to humans who might ingest minute quantities as residues in the diet.

Antidotes

A cursory examination of published reports concerning the effect of classical "antidotes" on carbamate insecticide toxicity to animals is apt to leave one more confused than convinced. Results often appear to be contradictory, even when the same animal, carbamate, and antidote are evaluated by different laboratories. Changing either of these parameters may result in an effect vastly different from those anticipated on the basis of prior work. These seemingly unpredictable results immediately raise the question whether there are any real antidotes for carbamate insecticide poisoning. This question may be partially answered, and the apparent discrepancies partially explained, by closer examination of the available data.

Considering first the apparent discrepancies, one finds that these stem primarily from the interpretation of the results rather than from the raw data themselves. One author may classify a compound as an effective antidote if it slightly reduces the toxicity of the carbamate or lessens the severity of the symptoms resulting from a nonlethal dose. Another may consider this as insignificant antidotal action and classify a compound "effective" only if it is therapeutically capable of preventing death of the exposed animal. Others, of course, would take a stand somewhere in the middle between these two extremes. The point is that there is no established level of protection which designates the degree of antidotal action of a compound, and therefore the degree is largely a matter of personal judgment. This by no means explains all of the apparent discrepancies, for – as we shall see later – some of the differences are very real and result from the compound-specificity exhibited by most of the antidotes. It is this specificity that provides the basis upon which the aforementioned question may be answered. That is, the toxic action of certain carbamate insecticides may be reduced by certain antidotes, but the degree of protection provided the animal is dependent upon the specific toxicant and antidote involved. Support for this statement is presented by the data in Table 5.8,[111-113] which typify the findings of numerous investigators working in the area of antidotal action.

The data designated as Series A in Table 5.8 show that atropine sulfate reduces the toxicity of aldicarb, Baygon, and carbaryl. Granted that the degree of protection afforded by this antidote varies with different carbamates and with different routes of administration, the compound consistently gives some protection. Moreover, it is almost always more effective than other compounds commonly used as antidotes for anticholinesterase toxicants.

Evidence of atropine sulfate's clinical value is well demonstrated in a report concerning the treatment of a 1-year old dog which had been poisoned with the methylcarbamate Mesurol.[114] When examined by a veterinarian, the animal showed muscle tremors, profuse salivation, marked pupillary constriction, and gross irregularity of the pulse. Atropine sulfate, 2.4 mg in 4 ml of solution, was administered via slow intravenous injection until pupillary dilation and pulse returned to normal. No further treatment was necessary; the animal responded completely and was released after 4 hr. The therapeutic value of atropine sulfate under experimental conditions is further indicated by the data in Table 5.9.[115] With each of the methylcarbamates tested, and for both rats and mice, atropine sulfate lessened the severity of the symptoms, which ranged from fasciculations and salivation (++) to the former plus prostration (+++), and usually death (++++). Certainly, these results demonstrate that atropine sulfate can prove the difference between life and death in carbamate-poisoned animals.

In 1968 it was reported that tetraethylammonium chloride (TEAC) was a more potent antidote for carbamate insecticide poisoning in mice than was atropine sulfate.[116] However, subsequent studies using male rats and nine different monomethylcarbamates revealed that in no case was TEAC more effective than atropine sulfate as an antidote.[112] Some of the results of this 1971 study are summarized in Table 5.8, Series A and B. Other data presented by the author[112] showed that a combination of TEAC and atropine sulfate produced no better antidotal action than atropine sulfate alone. It was concluded that the use of TEAC instead of atropine sulfate was not justified. The same conclusion was reached by other investigators, who compared the antidotal action of tetraethylammonium bromide (TEAB) with atropine sulfate on rats and mice poisoned with various methylcarbamate insecticides (Table 5.9).[115]

There is no doubt that the most confusing issue in the entire carbamate-insecticide–antidotal-action area resides with the pyridinium oximes. As amply demonstrated by references cited in a previous report,[111] it is possible to show that the toxicity of certain carbamate esters is reduced by parti-

TABLE 5.8

Action of "Antidotes" on Methylcarbamate Insecticide Activity

Carbamate, test animals, and route	Antidote, dose (mg/kg), and route	LD_{50} of carbamate	
		Without antidote	With antidote
Series A			
Aldicarb; rats; sc[111]	Atropine; 17.4; sc	4	18
Baygon; rats; or[112]	Atropine; 50; ip	191	543
Carbaryl; rats; ip[111]	Atropine; 17.4; sc	397	2,572
Carbaryl; rats; or[112]	Atropine; 50; ip	338	1,215
Series B[112]			
Baygon; rats; or	TEAC; 20; ip	191	248
Carbaryl; rats; or	TEAC; 20; ip	338	891
Zectran; rats; or	TEAC; 20; ip	73	415
Series C[111]			
Aldicarb; rats; sc	Obidoxime; 90; sc	4	8
Landrin; rats; ip	Obidoxime; 90; sc	705	760
Aldicarb; rats; sc	P2S; 50; sc	4	5
Landrin; rats; ip	P2S; 50; sc	705	1,243
Series D			
Carbaryl; mice; or[113]	2-PAM; 50; sc	224	128
Carbaryl; mice; or[113]	Toxogonin; 3; sc	224	59
Carbaryl; rats; ip[111]	Obidoxime; 90; sc	397	101
Carbaryl; rats; ip[111]	P2S; 50; sc	397	173

sc = subcutaneous; ip = intraperitoneal; or = oral.
TEAC = tetraethylammonium chloride
P2S = pyridine-2-aldoxime methanesulfonate
2-PAM = pyridine-2-aldoxime methiodide

TABLE 5.9

Therapeutic Value of Atropine Sulfate and TEAB for Methylcarbamate Insecticide Poisoning in Mice and Rats[115]

Insecticide[a]	Animals	Dose, mg/kg	Severity of symptoms/therapy[b]		
			None	Atropine	TEAB
Zectran	Mice	40	++++ (4)	+++ (1)	+++ (1)
Zectran	Rats	25	++++ (1)	++ (0)	+++ (0)
Baygon	Mice	100	++++ (4)	+++ (1)	++++ (3)
Carbofuran	Mice	68	++++ (4)	+++ (0)	+++ (0)
Aldicarb	Rats	1	++++ (0)	++ (0)	+++ (0)

[a]Insecticides were administered as single doses, and antidotes were given intraperitoneally (10 mg/kg for atropine and 20 mg/kg for TEAB, tetraethylammonium bromide) after 5 to 10 min.
[b]Relative severity of symptoms, and, in parentheses, the number of animals out of 4 treated which died during a 6-hr period after dosing.

cular oximes, whereas carbamate toxicity may not be affected, or may even be enhanced, by other oximes. These inconsistencies, plus a report in 1961[117] that the protection provided by atropine sulfate against carbaryl poisoning in dogs was lessened by 2-PAM (pyridine-2-aldoxime methiodide), have generally excluded oximes as antidotes for carbamate insecticide intoxication. Whether such a blanket indictment is justified remains unclear. Most studies tend to suggest that oximes are rather weak antidotes for carbamate poisoning or that they have no effect at all on carbamate toxicity. This premise is based on results of the type shown in Table 5.8, Series C. An exception to the general trend appears to be restricted to carbaryl. With this insecticide, oximes potentiate the toxicity to animals (Table 5.8, Series D) and often negate the antidotal action of atropine sulfate.[111] Even here, however, conflicting evidence may be found. A Russian scientist reported in 1969 that the antidotal action of 2-PAM and DAM (diacetyl monoxime) against carbaryl poisoning in mice was insignificant, or maybe even slightly therapeutic.[118,119] On the other hand, TBM-4 (1,1-trimethylenebis[4-formylpyridinium bromide]) tended to increase the toxicity of carbaryl to mice.[118] Notwithstanding the results of these latter studies, it is obviously undesirable to use the oximes as antidotes for treatment of humans who may have been poisoned by a carbamate insecticide. First, the treatment would not be as effective as atropine sulfate, and second, if the toxicant was carbaryl, treatment with an oxime might well be fatal to the patient, or at the very least provide no protection against the toxic effects of this carbamate.

In considering the subject of antidotes for anticholinergic carbamates, it is intriguing to note that the carbamates themselves are often demonstrated to be effective antidotes against organophosphorus toxicants. Their antidotal action is attributed to the reversibility of the carbamylated acetylcholinesterase which serves to protect the enzyme from the more permanent phosphorylation.[120] Unfortunately, relief is dependent upon administering the carbamate prior to the organophosphorus compound, since no protection is provided if they are given in reverse order.[120,121] Although it is of no practical significance to the present discussion, the proposed mode of action of the carbamate antidotes makes one wonder if carbamylcholine, a compound which yields a highly unstable carbamylated acetylcholinesterase, would similarly act as an antidote for the mono- and dimethylcarbamates. While poorly competing with carbamylcholine for acetylcholinesterase, the free carbamate insecticides might be detoxified in sufficient quantities to reduce intoxication.

Behavorial Toxicology

Of the numerous parameters investigated relative to the effects of pesticides on animals, behaviorial toxicology is possibly one of the most important and least studied. There is little doubt that certain modifications in behavior could be more deleterious to the survival of a species than many of the more classicial effects commonly attributed to these toxicants. It takes little imagination to envision the consequences should a pesticide severely alter the escape and/or defense mechanisms of an animal, deter the act of mating, or foster the improper care of offspring. Such responses, however, are rarely evaluated, even in those few studies which consider pesticides as a potential detriment to animal behavior.

More often, investigators concern themselves with factors which may very well be important, but whose actual significance is difficult to assess. The ability to learn is one type of behavior frequently evaluated, and carbamate insecticides, like many cholinesterase inhibitors, can interfere with the learning process.[122] Whether these effects are sufficient to jeopardize the existence or general well-being of a species is not known, but the mere fact that carbamate chemicals can decrease the learning ability of animals is certainly a cause for concern. There is a real need for behavioral studies to be expanded to include more thorough examination of the ultimate consequences of altered behavior brought about by insecticides.

Very few of the carbamate insecticides have been subjected to critical behavioral toxicological evaluations. However, the data available are sufficient to suggest that these toxicants may alter animal behavior in a variety of ways. Baygon, for example, seriously interferes with the ability of rats to negotiate a labyrinth and also reduces their spontaneous vertical-orientation ability.[123] In this case, the behavioral effects were noted as an increase in the time required for the animals to traverse a labyrinth and reach food which served as the incentive. Rats receiving Baygon in the diet, 3.3 mg/kg/day, immediately displayed difficulty in finding the food, and after 7 days of treatment required almost twice as much time to run the labyrinth as did the control rats. Such a marked reduction in learning ability at a relatively low exposure rate could obviously be of great significance, especially when one considers that some of the animals under treatment were completely unsuccessful in locating the food source.

Similar debilitating effects on the learning process in rats have been attributed to other methylcarbamates, using avoidance of electric shock as the criterion. Carbaryl in the diet at 100, 500, and 2,500 ppm for 35 days caused a progressive decrease in the animals' ability to avoid electric punishment.[124] In another study, a single subcutaneous injection of carbaryl, 10 mg/kg, given to rats also lowered the avoidance rate to foot shock when evaluated 2 hr after treatment.[125] During a 20-min session of grid shocks, control animals which had been deprived of water for 2 days continually reduced their shock rate by avoiding a water-lever press that activated the grid. Carbaryl-treated animals, on the other hand, pressed the water lever significantly more often. Visual observations of the rats indicated that there was an increased tolerance to electrical stimuli by those animals exposed to carbaryl.

The ability of rats to learn to avoid electrical shock by pressing a lever was severely curtailed in rats fed a protein-deficient diet containing 1,000 ppm Banol (Table 5.10).[126] Basically, the disability was a function of decreased protein, but the effect was greatly potentiated by the presence of the carbamate. The number of animals avoiding shock increased only 11% over a 10-week training period when fed a 5% casein diet, but increased 24% when fed a diet containing 18% casein. The addition of Banol to the high-protein food resulted in no impairment of learning. However, the multiple stresses imposed by simultaneous exposure to low protein and Banol caused a 30% decline in the number of animals avoiding shock over the 10-week training period. Thus, carbamates not only have the potential for decreasing learning ability, but also may lessen the memory capabilities of exposed species.

As just noted, diet can influence the effect carbamate insecticides may have on animal behavior. Undoubtedly, other pesticides, drugs, and chemicals, acting singly or in combination, may induce similar responses. As an example, one study evaluated the effects of carbaryl, caffeine, and atropine sulfate on the activity of rats in a wheel cage.[127] In short, carbaryl and atropine sulfate decreased the number of wheel revolutions per unit time, and caffeine increased the movement. The effects of the latter compound were antagonized by carbaryl, which was reversed by atropine sulfate. The findings were not particularly surprising or of apparent practical significance. However, the author stated repeatedly that the data "should be regarded as an important symptom in determining the clinical psychotoxicological toxicity of a compound." Although such importance is debatable, the study serves to illustrate the obvious; that is, mixtures of chemicals may influence animal behavior in a manner completely different from that observed when each compound is administered alone, and antagonism as well as potentiation may result.

TABLE 5.10

Retarded Behavioral Performance in Rats Fed Banol, 1,000 ppm, in a Protein Adequate (18%) and Inadequate (5%) Diet

	Percent pressing lever to avoid shock/diet			
	Without Banol		With Banol	
Weeks of training	Casein 18%	Casein 5%	Casein 18%	Casein 5%
1	68	64	63	51
4	90	64	91	25
8	90	65	97	16
10	92	77	99	16

Adapted from Reference 126, with permission.

TOXICITY TO NONTARGET ORGANISMS

The perfect pesticide would be one that selectively kills only those organisms which are considered pests. Such compounds are indeed the goal of scientists and laymen alike, but one must face reality. There is extreme commonality among all living systems, and to adversely affect pest species exclusive of all others is currently a seemingly impossible task. Many biochemical mechanisms vital for the survival of an insect pest, for example, are equally essential for the survival of nonpest insect species as well as of other animal life, including man.

Carbamate insecticides are nerve poisons (see Chapter 3), and as such are potentially damaging to all animals whose nervous system depends upon a proper balance of acetylcholinesterase for normal nerve–impulse transmission. Moreover, carbamates are the same as most other toxicants in that they may also interfere with vital mechanisms in addition to that generally accepted as the primary site of action. Thus, it is possible that certain species, because of penetration barriers, less sensitive nervous systems, etc., could cope with the anticholinergic action of the carbamates, but would succumb to other types of action exhibited by these chemicals.

Bleak as prospects may appear on the surface, there is good reason to be encouraged about the relative safety of carbamate insecticides to nontarget, or nonpest, organisms. Unfortunately, the encouragement is not the result of success in developing compounds which are pest-specific; rather, it is a function of the limited chemical stability of the carbamates in concert with the natural variation which exists among and between animal species. One would have to conclude, however, that it is the carbamates' nonpersistent nature which is primarily responsible for their favorable position over chlorinated hydrocarbons regarding toxicity to nontarget organisms.[128] Their acute toxicities to most animals, pest and nonpest, are too great to suggest that inherent differences are sufficient to protect most nontarget species from exposure to levels of carbamates which prove lethal to the target species.

In some cases, it is the nontarget organisms which are most sensitive to insecticides. This is vividly demonstrated by the data given in Table 5.11, which showed *Daphnia* to be much more susceptible to carbamate insecticides than mosquito larvae.[129] A similar toxicity differential was also noted for insecticides other than the carbamates. With DDT, the LC_{50} to *Daphnia* was 0.004 ppm, and to the mosquito larvae it was 0.070 ppm. There is little doubt that the control of mosquito larvae with any of these compounds would be detrimental to *Daphnia* populations. Those less persistent materials, however, would allow *Daphnia* to reestablish themselves in the treated locale much earlier than would the persistent chemicals. Regrettably, the same would hold true for the pest species.

TABLE 5.11

Toxicity of Certain Methylcarbamate Insecticides to *Daphnia* and Mosquito Larvae[129]

	LC_{50}, ppm	
Insecticide	*Daphnia*	Mosquito larvae
Carbaryl	0.011	1.000
Carbofuran	0.020	0.054
Baygon	0.029	0.354
Aldicarb	0.035	0.160
Mobam	0.170	0.580

The toxicity of carbaryl to honeybees is perhaps the best-known example of carbamates adversely affecting a nontarget species. Considering that in some major crops honeybees are responsible for more than 75% of the pollination,[130] it is clear that any compound highly toxic to these insects would definitely be undesirable. The sensitivity of honeybees to carbaryl has been documented by many investigators and from their data it appears that the average LD_{50} value approximates 1.5 μg/insect.[130–135] Bees are not protected from the toxic action of carbaryl by either atropine sulfate or 2-PAM,[136] but are made dramatically more susceptible to the carbamate by the synergist, piperonyl butoxide.[137] Although other methylcarbamate insecticides have not been associated particularly with honeybee deaths, many are more toxic to the insect than carbaryl is. Methomyl, Matacil, and Baygon each have an LD_{50} of less than 1.0 μg/insect,[138] but because they are not as widely used as carbaryl, no severe problems have as yet surfaced. If used on the same crops as carbaryl, it is likely that they too would result in immediate mortality of foraging bees and of emerging bees feeding on contaminated pollen for several days following treatment.[139] Carbaryl also exhibits high toxicity to leaf-cutting bees and to alkali bees.[140,141] In the leaf-cutting bees, the extreme sensitivity was attributed to a very slow degradation rate of the carbamate by the insects.[142] Other desirable insect species, predators and parasites, may be adversely affected by carbamate insecticides,[143-146] but the consequences are not as dramatic as with the pollinating bees, and the populations, if depressed, usually return to the pretreatment level in a few weeks.

Nontarget soil organisms are particularly vulnerable to carbamate insecticides applied directly to the soil for control of insect pests therein, or for control of plant-feeding pests by systemic action. Carbofuran is an effective soil insecticide and plant-systemic material whose action on earthworm populations has been studied in some detail. By the oral route, LD_{50} to earthworms was reported as 1.3 mg/kg,[147] whereas the LD_{50} value by injection was 2.4 mg/kg.[148] In the soil, the LC_{50} was approximately 12.5 mg/kg. Under field conditions, carbofuran, as well as Bux and carbaryl, reduced the number and biomass of earthworms when applied to pasture soil (Table 5.12),[149] but carbofuran, considering its higher treatment rate, was probably no more detrimental to the earthworms on a weight/weight basis than the other two carbamates. Carbaryl applied to soil at 2 lb active/acre reduced both biomass and number of arthropods by 95%, although after 7 weeks total biomass had returned to normal.[145]

Aquatic animals vary tremendously in their sensitivity to methylcarbamate insecticides,[128,150-157] but most indications are that carbamates, as a class, pose no serious threat to this segment of the environment. For example, among 18 different species of fish, the LC_{50} for carbaryl ranged from 0.8 to 20 ppm.[155,156] Zectran tested against 12 species gave LC_{50} values from 1.8 to 19 ppm.[156] Many organochlorine insecticides, including DDT, and certain organophosphorus compounds are known to be 100 to 1,000 times more toxic to the same species.[128,156] Acute oral LD_{50} values to bullfrogs are about 600 mg/kg for Baygon and Zectran, and more than 4,000 mg/kg for carbaryl.[157] Perhaps the most sensitive water creatures are molluscs and crustaceans, especially when carbaryl is involved.[153-155] Among clams, mussels, crabs, and shrimp, it is the latter which appear to be most vulnerable, with LC_{50} values for carbaryl against several species only about ten times higher than against *Daphnia* (Table 5.11). However, there is increasing evidence that the real toxic agent may be 1-naphthol and other hydrolytic metabolites rather than carbaryl *per se*.

Limited data are available on the effects of carbamates on terrestrial nontarget species. However, there have been no reports of major wildlife kills as a result of using carbamate insecticides, and the few controlled studies which have been conducted show the compounds to be surprisingly safe. In one study, carbaryl was applied at the rate of 1.25 g/acre for control of the gypsy moth, and a very elaborate monitoring system was devised to evaluate the effect of the insecticide on wildlife.[158] Many small animals were trapped, and 49 bird species were monitored. The findings revealed no adverse effects on mammals, birds, toads, salamanders, or snakes as a result of the carbaryl treatment. Similar results have been reported from a 3-year study in which Zectran was used for control of the spruce budworm.[159]

Because of the difficulty in trapping sufficient numbers of wildlife for toxicity determinations, there are very few data that reflect comparative LD_{50} values on wild species. One elaborate project conducted with Zectran on 16 captivity-acclimated wildlife species showed that 40% of the acute oral LD_{50} dose administered daily for 30 days had no apparent effect on any of the animals.[160] The acute oral LD_{50} values for some of the animals

TABLE 5.12

Effect of Various Methylcarbamate Insecticides on Number and Biomass of Earthworms[149]

Treatment (lb AI/a)	Mean number of earthworms (% of controls)	Grams of biomass (% of controls)
Carbofuran, 4 lb W.P.	3.1 (17%)	160 (39%)
Bux, 1 lb granular	7.9 (44%)	242 (59%)
Carbaryl, 2 lb W.P.	7.2 (40%)	128 (32%)
Untreated control	17.9 (100%)	405 (100%)

used in the study are shown in Table 5.13. Although the number of animals was too low for statistical treatment, the data indicate that birds are generally more sensitive to Zectran than mammals, which are more sensitive than amphibians. However, with the exception of Zectran and carbofuran, birds appear fairly tolerant to most carbamate insecticides (Table 5.14).

TABLE 5.13

Acute Oral Toxicity of Zectran to Various Vertebrate Species

Animal	Number tested	Approximate LD_{50}, mg/kg
Lesser sandhill crane	5	1.0–4.5
Canada goose	12	2.6
Mallard duck	12	4.2
Coturnix quail	12	3.2
Ring-necked pheasant	12	4.5
Chukar partridge	12	5.2
Domestic pigeon	12	6.5
Albino rat	12	14.1
Domestic goat	2	15–30
Mule deer	6	20–30
House sparrow	16	50.0
Bullfrog	14	283–800

Adapted from Reference 160, with permission.

TABLE 5.14

Acute Oral Toxicity of Various Methylcarbamate Insecticides to Birds[157]

	LD_{50} (mg/kg)			
Insecticide	Mallard	Pheasant	Pigeon	House sparrow
Baygon	12	20	60	13
Carbaryl	>2,000	>2,000	2,000	–
Carbofuran	0.4	4.2	–	–
Landrin	17	35	170	55
Matacil	23	42	–	–
Methomyl	16	15	–	–
Mobam	1,130	230	270	58
Zectran	3	4.5	6.5	50

REFERENCES

1. **Winteringham, F. P. W.**, Mechanisms of selective insecticidal action, *Annu. Rev. Entomol.*, 14, 409, 1969.
2. **Asakawa, M.**, Carbamate insecticides, *Noyaku Seisan Gijutsu*, 24, 1, 1971.
3. **Riviere, J. L.**, The carbamate insecticides, *Ann. Zool. Ecol. Anim.*, 2, 479, 1970.
4. **Metcalf, R. L., Osman, M. F., and Fukuto, T. R.**, Metabolism of ^{14}C-labeled carbamate insecticides to $^{14}CO_2$ in the house fly, *J. Econ. Entomol.*, 60, 445, 1967.
5. **Ball, H. J.**, Topical toxicity of insecticides to *Diabrotica virgifera* adults, *J. Econ. Entomol.*, 62, 1172, 1969.

6. **Leigh, T. F. and Jackson, C. E.**, Topical toxicity of several chlorinated hydrocarbon, organophosphorous and carbamate insecticides to *Lygus hesperus*, *J. Econ. Entomol.*, 61, 328, 1968.
7. **Brattsten, L. B. and Metcalf, R. L.**, The synergistic ratio of carbaryl with piperonyl butoxide as an indicator of the distribution of multifunction oxidases in the Insecta, *J. Econ. Entomol.*, 63, 101, 1970.
8. **Brattsten, L. B. and Metcalf, R. L.**, Synergism of carbaryl toxicity in natural insect populations, *J. Econ. Entomol.*, 66, 1347, 1973.
9. **Weiden, M. H. J.**, Toxicity of carbamates to insects, *Bull. W.H.O.*, 44, 203, 1971.
10. **Salama, A. E., Eldefrawi, M. E., Toppozada, A., and Hassan, S.**, Toxicological studies on the Egyptian cotton leafworm, *Prodenia litura.* V. Contact versus stomach toxicity of insecticides to larvae, *J. Econ. Entomol.*, 59, 21, 1966.
11. **Weiden, M. H. J.**, Insecticidal carbamoyloximes, *J. Sci. Food Agric. Suppl.*, p. 19, 1968.
12. **Hadaway, A. B.**, Some factors affecting the distribution and rate of action of insecticides, *Bull. W.H.O.*, 44, 221, 1971.
13. **Burnett, G. F.**, The susceptibility of tsetse flies to topical applications of insecticides. V. Young adults of *Glossina morsitans* Westw. and some substituted N-methylcarbamates, *Bull. Entomol. Res.*, 54, 747, 1963.
14. **Georghiou, G. P.**, Considerations on the relationship of larval and adult tolerance to insecticides in mosquitoes, *Proc. Annu. Conf. Calif. Mosq. Control*, 38, 55, 1970.
15. **Bruce, D. W.**, Response of Immature Insects to the Carbamate Insecticide Temik, M. S. thesis, Texas A & M University, 1968.
16. **El-Aziz, S. A., Metcalf, R. L., and Fukuto, T. R.**, Physiological factors influencing the toxicity of carbamate insecticides to insects, *J. Econ. Entomol.*, 62, 318, 1969.
17. **Green, L. R. and Dorough, H. W.**, Housefly age as a factor in their response to certain carbamates, *J. Econ. Entomol.*, 61, 88, 1968.
18. **Kuhr, R. J. and Davis, A. C.**, Toxicity and metabolism of carbaryl in the European corn borer, *Pestic. Biochem. Physiol.*, 5, 330, 1975.
19. **Georghiou, G. P.**, Effects of carbamates on house fly fecundity, longevity and food intake, *J. Econ. Entomol.*, 58, 58, 1965.
20. **Gaines, T. B.**, Acute toxicity of pesticides, *Toxicol. Appl. Pharmacol.*, 14, 515, 1969.
21. **Natoff, I. L.**, Influence of the route of exposure on the acute toxicity of cholinesterase inhibitors, *Arh. Hig. Rada Toksikol.*, 21, 347, 1970.
22. **Yakim, V. S.**, Data for substantiating the maximum permissible concentration of Sevin in the air, *Gig. Sanit.*, 32, 29, 1967.
23. **Bukin, A. L. and Filatov, G. V.**, Sevin toxicity for mammals and birds, *Veterinariya*, 42, 93, 1965.
24. **Rybakova, M. N.**, Toxic effect of Sevin on animals, *Gig. Sanit.*, 31, 42, 1966.
25. **Baron, R. L., Casterline, J. L., Jr., and Fitzhugh, D. G.**, Specificity of carbamate-induced esterase inhibition in mice, *Toxicol. Appl. Pharmacol.*, 6, 402, 1964.
26. **Stenger, E. G.**, Toxicology of dimetilan (2-dimethylcarbamoyl-3-methyl-5-pyrazolyl dimethyl carbamate), *Med. Exp.*, 6, 331, 1962.
27. **Vandekar, M., Plestina, R., and Wilhelm, K.**, Toxicity of carbamates for mammals, *Bull. W.H.O.*, 44, 241, 1971.
28. **Brown, V. K. H. and Muir, C. M. C.**, Some factors affecting the acute toxicity of pesticides to mammals when absorbed through the skin and eyes, *Int. Pest Control*, 13, 16, 1971.
29. **Nye, D. E., Hurst, H. E., and Dorough, H. W.**, The metabolism of Croneton in rats, *J. Agric. Food Chem.*, in press.
30. **Miyaa, N., Ishiguro, S., Kono, I., and Yasuda, N.**, Toxicity of carbamate compounds. I. Subchronic toxicity of 3-methyl-5-isopropylphenyl N-methylcarbamate, *Kagoshima Daigaku Nogakubu Gakujutsu Hokoku*, 22, 131, 1972.
31. **Tobin, J. S.**, Carbofuran: a new carbamate insecticide, *J. Occup. Med.*, 12, 16, 1970.
32. **Smalley, H. E., O'Hara, P. J., Bridges, C. H., and Radeleff, R. D.**, Effects of chronic carbaryl administration on the neuromuscular system of swine, *Toxicol. Appl. Pharmacol.*, 14, 409, 1969.
33. **Maes, R.**, Chemical epidemiological aspects of pesticides, *Arch. Belg. Med. Soc. Hyg. Med. Trav. Med. Leg.*, 29, 244, 1971.
34. **Trochimowicz, H. J. and Rinehart, W. E.**, The dermal absorption of four solid formulations of carbaryl in rabbits, *Toxicol. Appl. Pharmacol.*, 12, 304, 1968.
35. **Goncharova, N. I.**, Toxicity and effect of Sevin on guinea pigs, *Uch. Zap. Vitebsk. Vet. Inst.*, 20, 214, 1968.
36. **Boyd, E. M. and Boulanger, M. A.**, Augmented susceptibility to carbaryl toxicity in albino rats fed purified casein diets, *J. Agric. Food Chem.*, 16, 834, 1968.
37. **Boyd, E. M. and Krijnen, C. J.**, Influence of protein intake on the acute oral toxicity of carbaryl, *J. Clin. Pharmacol. New Drugs*, 9, 292, 1969.
38. **Boyd, E. M.**, Dietary protein and pesticide toxicity in male weanling rats, *Bull. W.H.O.*, 40, 801, 1969.
39. **Casterline, J. L., Jr. and Williams, C. H.**, The effect of pesticide administration on serum and tissue esterases of rats fed diets of varying casein, calcium and magnesium content, *Toxicol. Appl. Pharmacol.*, 15, 532, 1969.
40. **Casterline, J. L., Jr. and Williams, C. H.**, Effect of pesticide administration upon esterase activities in serum and tissues of rats fed variable casein diets, *Toxicol. Appl. Pharmacol.*, 14, 266, 1969.
41. **Guthrie, F. E., Monroe, R. J., and Abernathy, C. O.**, Response of the laboratory mouse to selection for resistance to insecticides, *Toxicol. Appl. Pharmacol.*, 18, 92, 1971.

42. **Smalley, H. E., Jr.,** Diagnosis and treatment of carbaryl poisoning in swine, *J. Am. Vet. Med. Assoc.*, 156, 339, 1970.
43. **Vandekar, M., Reiner, E., Svetlicic, B., and Fajdetic, T.,** Value of ED_{50} testing in assessing hazards of acute poisoning by carbamates and organophosphates. *Br. J. Ind. Med.*, 22, 317, 1965.
44. **Romirska, T. and Winiarska, A.,** Toxicology of carbamates as exemplified by intoxication with Baygon, *Pol. Tyg. Lek.*, 27, 1448, 1972.
45. **Wills, J. H.,** Effects of oral doses of carbaryl on man, *Clin. Toxicol.*, 1, 265, 1968.
46. **Farago, A.,** Suicidal, fatal Sevin (1-naphthyl N-methylcarbamate) poisoning, *Arch. Toxikol.*, 24, 309, 1969.
47. **Richardson, E. M. and Batteese, R. I.,** An incident of Zectran poisoning, *J. Maine Med. Assoc.*, 64, 158, 1973.
48. **Feldmann, R. J. and Maibach, H. I.,** Pesticide percutaneous penetration in man, *J. Invest. Dermatol.*, 54, 435, 1970.
49. **Maibach, H. I., Feldmann, R. J., Milby, T. H., and Serat, W. F.,** Regional variation in percutaneous penetration in man, *Arch. Environ. Health*, 23, 208, 1971.
50. **Vandekar, M.,** Observations on the toxicity of carbaryl, Folithion and 3-isopropylphenyl N-methylcarbamate in a village-scale trial in Southern Nigeria, *Bull. W. H. O.*, 33, 107, 1965.
51. **Vandekar, M. and Svetlicic, B.,** The toxicity of three anticholinesterase insecticides in a village-scale trial and comparison of methods used for determining cholinesterase activity, *Arh. Hig. Rada Toksikol.*, 17, 135, 1966.
52. **Vandekar, M., Hedayat, S., Plestina, R., and Ahmady, G.,** A study of the safety of *o*-isopropoxyphenyl methylcarbamate in an operational field trial in Iran, *Bull. W. H. O.*, 38, 609, 1968.
53. **Wright, J. W., Fritz, R. F., Hocking, K. S., Babione, R., Gratz, N. G., Pal, R., Stiles, A. R., and Vandekar, M.,** *ortho*-Isopropoxyphenyl methylcarbamate (OMS-33) as a residual spray for control of anopheline mosquitoes *Bull. W. H. O.*, 40, 67, 1969.
54. **Motabar, M.,** Observations on the effect on operators and inhabitants following first round spraying with OMS-33 in Jareh, Iran, *Acta Med. Iran.*, 14, 81, 1971.
,55. **Ohishi, S.,** High incidence of blindness of cattle in Ehime prefecture: discussion of causal relationship to pesticide poisoning, *Nippon Saigai Igakukai Kaishi*, 20, 191, 1972.
56. **Ohto, K.,** Diagnosis and treatment of eye disease caused by organophosphorus pesticides, *Ganka (Ophthalmology)*, 14, 17, 1972.
57. **Olefir, A. I.,** Effects of acute carbamate pesticide poisoning on immunologic responses, *Vrach. Delo*, 4, 138, 1973.
58. **Shtenberg, A. I., Ashmenskas, Y. I., and Kusevitskiy, I. A.,** Immunobiological reactivity changes under the influence of some pesticides belonging to the groups of carbamate and dithiocarbamate compounds, *Vopr. Pitan.*, 31, 58, 1972.
59. **Williams, C. H. and Casterline, J. L., Jr.,** Comparison of two methods for the measurement of erythrocyte cholinesterase inhibition after carbamate administration to rats, *Food Cosmet. Toxicol.*, 7, 149, 1969.
60. **Guilbault, G. C., Sadar, M. H., Kuan, S. S., and Casey, D.,** Effect of pesticides on liver cholinesterases from rabbit, pigeon, chicken, sheep and pig, *Anal. Chim. Acta*, 51, 83, 1970.
61. **Khomich, N. V.,** Toxicology and pharmacology of some naphthyl and xylenol carbamates, *Gig. Toksikol. Novykh Pestits. Klin. Otravlenii Sb.*, 1, 358, 1962.
62. **Lin, T. H.,** Stability of Carbamate-inhibited Cholinesterases, M. S. thesis, Texas A & M University, 1969.
63. **Wilhelm, K. and Reiner, E.,** Effect of sample storage on human blood cholinesterase activity after inhibition by carbamates, *Bull. W. H. O.*, 48, 235, 1973.
64. **Cornish, H. H.,** Problems posed by observations of serum enzyme changes in toxicology, *Crit. Rev. Toxicol.*, 1, 1, 1971.
65. **Sasinovich, L. M. and Voronina, L. Y.,** Enzyme activity as a diagnostic test in liver injuries from some pesticides, *Fermenty Nar. Khoz. Med.*, 1, 213, 1971.
66. **Kagan, I. S., Sasinovich, L. M., and Voronina, L. Y.,** The chronic action of some pesticides on the functional state of the liver, *Gig. Sanit.*, 35, 36, 1970.
67. **Makouskaya, E. I. and Rappoport, M. B.,** Significance of histochemical studies in a toxicological evaluation of insecticides, *Vopr. Gig. Tr. Profpatol. Ezan. Sb.*, 88, 1963.
68. **Onikiyenko, F. A.,** The activity of certain blood and liver enzymes in animals subjected to acute and chronic Sevin (carbaryl) poisoning, *Farmakol. Toksikol.*, 35, 237, 1972.
69. **Anina, I. A.,** Effect of repeated administration of small doses of Sevin and TMTD (thiram) on nucleic acid metabolism in rat liver and spleen, *Gig. Tr.*, 1, 94, 1969.
70. **Anina, I. A.,** The influence of pesticide derivatives of carbamic acid on nucleic acid metabolism in rat liver and spleen, *Byull. Eksp. Biol. Med.*, 66, 48, 1968.
71. **Kuzminskaya, U. A. and Yakushko, V. E.,** Effect of DDT and Sevin on the activity of some liver enzymes, *Gig. Sanit.*, 35, 97, 1970.
72. **Kuzminskaya, U. A. and Pavlova, I. I.,** Effect of DDT and Sevin on the pyridine nucleotide level in animal tissues, *Gig. Primen. Toksikol. Pestits. Klin. Otravlenii*, 8, 110, 1970.
73. **Pavlova, I. I.,** Effect of DDT and Sevin on the level of nicotinamide coenzymes and the activity of blood NAD-hydrolases, *Farmakol. Toksikol.*, 32, 738, 1969.
74. **Proklina-Kaminskaya, T. L.,** The state of tissue oxidative processes in poisonings with pesticides – derivatives of carbamic, thio- and dithiocarbamic acids, *Vrach. Delo*, 6, 101, 1969.

75. **Proklina-Kaminskaya, T. L.**, Effect of carbamates on tissue respiration and the activity of cytochrome oxidase and succinate dehydrogenase in rat liver, *Farmakol. Toksikol.*, 31, 153, 1968.
76. **Proklina-Kaminskaya, T. L.**, The effect of the prolonged action of carbamate pesticides on oxidative processes in animals, *Gig. Tr. Resp. Mezhved. Sb.*, 1, 91, 1969.
77. **Kuzminskaya, U. A.**, Effect of DDT and Sevin on the content and metabolism of phospholipids in warm-blooded tissues, *Farmakol. Toksikol.*, 31, 609, 1968.
78. **Goncharova, N. I.**, Morphological changes in animal blood in Sevin poisoning, *Veterinariya*, 2, 82, 1968.
79. **Mogilevchik, Z. K., Rusyaev, A. P., Vyatchannikov, K. A., and Buslovich, S. Y.**, Importance of nonspecific manifestations of pesticide poisoning in setting norms for their residual quantities, *Voprosy Gigieny Toksikologii Pestitsidov Truda Nauchno Sessii Akademii Meditsinskikh Nauk SSSR, 1967*, 1970, 49.
80. **Butygin, V. A. and Vyatchannikov, K. A.**, The effect of prolonged administration of small doses of Sevin on the serotonin content in the blood, brain, tissue and enterochromaffinic cells of the duodenum in white rats, *Zdravookhr. Beloruss.*, 15, 44, 1969.
81. **Voitenko, G. A., Karpenko, V. N., and Vladimirskaya, A. I.**, Changes of some indices of the electrolyte metabolism as a result of poisoning with carbamate pesticides, *Vrach. Delo*, 11, 112, 1969.
82. **Kagan, I. S., Rodionov, G. A., Voronina, L. I., Velichko, L. S., Kulagin, O. M., and Peremitina, A. D.**, The effect of Sevin on the functional state and structure of the liver, *Farmakol. Toksikol.*, 33, 219, 1970.
83. **Stevens, J. T., Stitzel, R. E., and McPhillips, J. J.**, Effects of anticholinesterase insecticides on hepatic microsomal metabolism, *J. Pharmacol. Exp. Ther.*, 181, 576, 1972.
84. **Stevens, J. T., Stitzel, R. E., and McPhillips, J. J.**, The effects of subacute administration of anticholinesterase insecticides on hepatic microsomal metabolism, *Life Sci. Part II*, 11, 423, 1972.
85. **Belonozhko, G. A. and Kuchak, I. U.**, Electrocortical reactions in rats to the administration of Sevin, *Gig. Sanit.*, 34, 111, 1969.
86. **Shtenberg, A. I. and Khovaeva, L. A.**, The effect of Sevin on the thyroid gland of rats receiving an artificial diet with varying content of iodine, *Gig. Sanit.*, 35, 119, 1970.
87. **Kusevitskii, I. A., Kirlich, A. E., and Khovaeva, L. A.**, The action of maneb and Sevin on the thyroid gland, *Veterinariya*, 47, 73, 1970.
88. **Mandzhgaladze, R. N. and Vashakidze, V. I.**, Action of some chemicals on offspring and sexual characteristics of rats, *Soobshch., Akad. Nauk Gruz. SSR*, 65, 485, 1972.
89. **Rybakova, M. N.**, Comparative toxic effects of Sevin and DDT on food crops, *Vopr. Pitan.*, 26, 9, 1967.
90. **Rybakova, M. N.**, The effect of certain pesticides on the hypophysis and its gonadotropic function, *Gig. Sanit.*, 33, 27, 1968.
91. **Shtenberg, A. I. and Rybakova, M. N.**, Effect of carbaryl on the neuroendocrine system of rats, *Food Cosmet. Toxicol.*, 6, 461, 1968.
92. **Vashakidze, V. I.**, Effect of Sevin intoxication on sexual function of experimental animals, *Soobshch. Akad. Nauk Gruz. SSR*, 39, 471, 1965.
93. **Shtenberg, A. I. and Orlova, N. V.**, Mode of toxicity of Sevin, a carbamate pesticide, *Vestn., Akad. Med. Nauk SSSR*, 25, 72, 1970.
94. **Shtenberg, A. I. and Ozhovan, M. V.**, Effect of low Sevin doses on the reproductive function of animals in a number of generations, *Vopr. Pitan.*, 30, 42, 1971.
95. **Shtenberg, A. I., Orlova, N. V., and Torchinskiy, A. M.**, The action of pesticides of different chemical structure on the gonads and embryogenesis of experimental animals, *Gig. Sanit.*, 38, 16, 1973.
96. **Shilova, S. A., Smirhov, A. A., Ognev, V. A., and Voronova, L. D.**, The effect of Sevin on the sexual functions of warm-blooded animals, *Med. Parazitol. Parazit. Bolezni*, 35, 604, 1968.
97. **Smirnov, A. A., Lobachev, V. S., and Denisova, A. V.**, Effect of Sevin on the reproductive ability of the great gerbil after Sevin treatment of plants among which it settles, *Biol. Nauki*, 14, 29, 1971.
98. **Krylova, T. V. and Denisova, A. V.**, Effect of carbaryl on the Mongolian tree creeper, *Biol. Nauki*, 10, 25, 1973.
99. **Smalley, H. E., Curtis, J. M., and Earl, F. L.**, Teratogenic action of carbaryl in beagle dogs, *Toxicol. Appl. Pharmacol.*, 13, 392, 1968.
100. **Robens, J. F.**, Teratologic studies of carbaryl, diazinon, norea, disulfiram and thiram in small laboratory animals, *Toxicol. Appl. Pharmacol.*, 15, 152, 1969.
101. **Collins, T. F., Hansen, W. H., and Keeler, H. V.**, Effects of carbaryl on reproduction of the rat and gerbil, *Toxicol. Appl. Pharmacol.*, 19, 202, 1971.
102. **Dougherty, W. J., Golberg, L., and Coulston, F.**, The effect of carbaryl on reproduction in the monkey *(Macacca mulatta)*, *Toxicol. Appl. Pharmacol.*, 19, 365, 1971.
103. **Weil, C. S., Woodside, M. D., Carpenter, C. P., and Smyth, H. F., Jr.**, Current status of tests of carbaryl for reproductive and teratogenic effect, *Toxicol. Appl. Pharmacol.*, 21, 390, 1972.
104. **McCarthy, J. F., Faucher, O. E., Kennedy, G. L., Keplinger, M. L., and Calandra, J. C.**, Reproduction and teratology studies with the insecticide carbofuran, *Toxicol. Appl. Pharmacol.*, 19, 370, 1971.
105. **Pound, A. W.**, Initiation of skin tumor formation in mice by N-hydroxycarbamates, *Pathology*, 1, 27, 1969.
106. **Pound, A. W.**, The initiation of skin tumors in mice by homologs and N-substituted derivatives of ethyl carbamate, *Aust. J. Exp. Biol. Med. Sci.*, 45, 507, 1967.

107. **Elespuru, R. K. and Lijinsky, W.,** The formation of carcinogenic nitroso compounds from nitrite and some types of agricultural chemicals, *Food Cosmet. Toxicol.,* 11, 807, 1973.
108. **Dorough, H. W.,** *Biological Activity of Pesticide Conjugates,* (American Chemical Society Symposium Series, 1975), American Chemical Society, Washington, D.C., in press.
109. **Ames, B. N., Lee, F. D., and Durston, W. E.,** An improved bacterial test system for the detection and classification of mutagens and carcinogens, *Proc. Natl. Acad. Sci. U.S.A.,* 70, 782, 1973.
110. **Andrianova, M. M. and Alekseev, I. V.,** The carcinogenic properties of the pesticides Sevin, maneb, ziram and zineb, *Vopr. Pitan.,* 29, 71, 1970.
111. **Natoff, I. L. and Reiff, B.,** Effect of oximes on the acute toxicity of anticholinesterase carbamates, *Toxicol. Appl. Pharmacol.,* 25, 569, 1973.
112. **Kimmerle, G.,** Comparison of the antidotal action of tetraethylammonium chloride and atropine in acute poisoning of carbamate insecticides in rats, *Arch. Toxikol.,* 27, 311, 1971.
113. **Akamatsu, K. and Kohgo, K.,** Antidotal effect of bis(4-hydroxyiminomethylpyridiniomethyl)ether dichloride (Toxogonin) on acute toxicity caused by several insecticides, particularly in comparison with 2-pyridine aldoxime methiodide (PAM), *Showa Yakka Daigaku Kiyo,* 4, 15, 1966.
114. **Udall, N. D.,** The toxicity of the molluscicides metaldehyde and methiocarb to dogs, *Vet. Rec.,* 93, 420, 1973.
115. **Verschoyle, R. D. and Barnes, J. M.,** Therapeutic effect of atropine and tetraethylammonium bromide against anticholinesterase insecticides in mice and rats, *Bull. W.H.O.,* 41, 306, 1969.
116. **Andrews, T. L. and Miskus, R. P.,** Tetraethylammonium chloride as an antidote for insecticides in mice, *Science,* 159, 1367, 1968.
117. **Carpenter, C. P., Weil, C. S., Palm, P. E., Woodside, M. W., Nair, J. H., and Smyth, H. F.,** Mammalian toxicity of 1-naphthyl N-methylcarbamate (Sevin insecticide), *J. Agric. Food Chem.,* 9, 30, 1961.
118. **Aleksashina, Z. A.,** Use of cholinesterase reactivators during the poisoning of laboratory animals with Sevin, *Zdravookhr. Beloruss.,* 15, 74, 1969.
119. **Aleksashina, Z. A.,** The therapeutic properties of cholinolytics in combination with cholinesterase reactivators in Sevin poisoning of animals, *Zdravookhr. Beloruss.,* 15, 513, 1969.
120. **Barry, W. K. and Davies, D. R.,** Use of carbamates and atropine in the protection of animals against poisoning by 1,2,2,-trimethylpropyl methylphophonofluoridate, *Biochem. Pharmacol.,* 19, 927, 1970.
121. **Dorough, H. W.,** Effect of Temik on methyl parathion toxicity to mice, *Tex. Agric. Exp. Stn. Prog. Rep.,* No. 2771, 6, 1970.
122. **Giachetti, A., Chippi, O., and Guerra, M.,** The effect of parathion on the response conditioning of avoidance in albino rats, *Ric. Sci.,* 36, 1078, 1966.
123. **Desi, I., Szelenyi, I., and Sos, J.,** The influence of magnesium orotate and orotic acid on learning processes of normal and neurotoxically injured rats, *Arzneim. Forsch.,* 21, 780, 1971.
124. **Tachibana, H., Hashimoto, K., Ishibashi, T., Ando, T., and Kitagawa, M.,** Studies of the learned behavior of rats as an index for toxicity, *Jap. Ind. Hyg. Soc. Proc.,* 46, 408, 1973.
125. **Sideroff, S. I. and Santolucito, J. A.,** Behavioral and physiological effects of the cholinesterase inhibitor carbaryl, *Physiol. Behav.,* 9, 459, 1972.
126. **Casterline, J. L., Jr., Brodie, R. E., and Sobotka, T. J.,** Effect of Banol and parathion on operant learning behavior of rats fed adequate and inadequate casein diets, *Bull. Environ. Contam. Toxicol.,* 6, 297, 1971.
127. **Singh, J. M.,** Decreased performance behavior with carbaryl – an indication of clinical toxicity, *Clin. Toxicol.,* 6, 97, 1973.
128. **Pimentel, D.,** Ecological Effects of Pesticides on Nontarget Species, Report to the Executive Office of the President, Office of Science and Technology, U.S. Government Printing Office, Stock No. 4106-0029, 1971.
129. **Sangha, G.,** Environmental Effects of Carbamate Insecticides as Assayed in the "Model Ecosystem" – a Comparison with DDT, Ph.D. dissertation, University of Illinois, 1971.
130. **Wafa, A. K. and Ibrahim, S. H.,** The honey bee as an important insect for pollination, *Bull. Fac. Agric. Cairo Univ.,* 162, 1, 1957.
131. **Anderson, L. D. and Atkins, E. L.,** Toxicity of pesticides to honey bees in laboratory and field tests in southern California, 1955–1956, *J. Econ. Entomol.,* 51, 103, 1958.
132. **Johansen, C. A.,** Toxicity of certain insecticides to honeybees in the laboratory, *J. Econ. Entomol.,* 54, 1008, 1961.
133. **Shaw, F. R. and Fischang, W. J.,** The effects of certain newer pesticides on honey bees, *Am. Bee J.,* 102, 312, 1962.
134. **Ibrahim, S. H., Madkour, A. M., and Selim, H. A.,** Contact toxicity of some new insecticides and acaricides to honeybees, *Agric. Res. Rev.,* 45, 141, 1967.
135. **Stevenson, J. H.,** Laboratory and Field Assessment of Pesticide Poisoning of Honeybees *(Apis mellifera),* Proceedings, 5th British Insecticide and Fungicide Conference, Vol. 2, 1969, 378.
136. **Barker, R. J.,** Cholinesterase reactivators tested as antidotes for use on poisoned honey bees, *J. Econ. Entomol.,* 63, 1831, 1970.
137. **Georghiou, G. P. and Atkins, E. L., Jr.,** Temperature coefficient of toxicity of certain N-methylcarbamates against honeybees, and the effect of the synergist piperonyl butoxide, *J. Apic. Res.,* 3, 31, 1964.
138. **Clinch, P. G. and Ross, J. G. M.,** Laboratory assessment of the speed of action on honey bees of orally dosed insecticides, *N.Z. J. Agric. Res.,* 13, 717, 1970.

139. **Strang, G. E., Nowakowski, J., and Morse, R. A.,** Effect of carbaryl on honeybees, *J. Econ. Entomol.,* 61, 1103, 1968.
140. **Johansen, C., Jaycox, E. R., and Hutt, R.,** The effect of pesticides on the alfalfa leafcutting bee, *Wash. Agric. Exp. Stn. Circ.,* No. 418, 1963.
141. **Johansen, C. and Eves, J.,** Toxicity of insecticides to the alkali bee and the alfalfa leaf-cutting bee, *Wash. Agric. Exp. Stn. Circ.,* No. 475, 1967.
142. **Guirguis, G. N. and Brindley, W. A.,** Carbaryl penetration into and metabolism by alfalfa leafcutting bees, *Megachile pacifica, J. Agric. Food Chem.,* 23, 274, 1975.
143. **Dinkins, R. L., Brazzel, J. R., and Wilson, C. A.,** Effect of early season insecticide applications on major predaceous arthropods in cotton fields under an integrated control program, *J. Econ. Entomol.,* 64, 480, 1971.
144. **Dondale, C. D.,** Effects of carbofuran on arthropod populations and crop yield in hayfields, *Can. Entomol.,* 104, 1433, 1972.
145. **Barrett, G. W.,** The effects of an acute insecticide stress on a semienclosed grassland ecosystem, *Ecology,* 49, 1019, 1968.
146. **Doane, C. C. and Shaefer, P. W.,** Aerial application of insecticides for control of the gypsy moth: with studies of effects on nontarget insects and birds, *Conn. Agric. Exp. Stn. New Haven Bull.,* No. 724, 5, 1971.
147. **Stenersen, J., Gilman, A., and Vardanis, A.,** Carbofuran: its toxicity to and metabolism by earthworm *(Lumbricus terrestris), J. Agric. Food Chem.,* 21, 166, 1973.
148. **Gilman, A. P. and Vardanis, A.,** Carbofuran: comparative toxicity and metabolism in the worms *Lumbricus terrestris* L. and *Eisenia foetida* S., *J. Agric. Food Chem.,* 22, 625, 1974.
149. **Thompson, A. R.,** Effects of nine insecticides on the numbers and biomass of earthworms in pasture, *Bull. Environ. Contam. Toxicol.,* 5, 577, 1970.
150. **Cornman, I.,** Inhibition of sea-urchin egg cleavage by a series of substituted carbamates, *J. Natl. Cancer Inst.,* 10, 1123, 1950.
151. **Mulla, S.,** Toxicity of new organic insecticides to mosquito fish and some other aquatic organisms, *Mosq. News,* 26, 87, 1966.
152. **Hendrick, R. D., Everett, T. R., and Caffey, H. R.,** Effects of some insecticides on the survival, reproduction, and growth of the Louisiana red crawfish, *J. Econ. Entomol.,* 59, 188, 1966.
153. **Butler, J. A., Millemann, R. E., and Stewart, N. E.,** Effects of the insecticide Sevin on survival and growth of the cockle clam *Clinocardium nuttalli, J. Fish. Res. Board Can.,* 25, 1621, 1968.
154. **Grosch, D. S.,** Reproduction tests: the toxicity for *Artemia* of derivatives from non-persistent pesticides, *Biol. Bull.,* 145, 340, 1973.
155. **Stewart, N. E., Milleran, R. E., and Breese, W. P.,** Acute toxicity of the insecticide Sevin and its hydrolytic product 1-naphthol to some marine organisms, *Trans. Am. Fish. Soc.,* 96, 25, 1967.
156. **Macek, K. J. and MacAllister, W. A.,** Insecticide susceptibility of some common fish family representatives, *Trans. Am. Fish. Soc.,* 99, 20, 1970.
157. **Tucker, R. K. and Crabtree, D. G.,** *Handbook of Toxicity of Pesticides to Wildlife,* Research Publication No. 84, U.S. Fish and Wildlife Service, Sport Fisheries and Wildlife Bureau, Washington, D.C., 1970.
158. **Conner, P. F.,** A study of small mammals, birds and other wildlife in an area sprayed with Sevin, *N.Y. Fish Game J.,* 7, 26, 1960.
159. **Pillmore, R. D., Flickinger, E. L., and Richmond, M. L.,** Forest spraying of Zectran and its safety to wildlife, *J. For.,* 69, 721, 1971.
160. **Tucker, R. K. and Crabtree, D. G.,** Toxicity of Zectran insecticide to several wildlife species, *J. Econ. Entomol.,* 62, 1307, 1969.

Chapter 6

METABOLIC REACTIONS

INTRODUCTION

The success of a compound as a commercial insecticide is largely dependent on its ability to resist or succumb to chemical and biochemical transformations. Ideally, the insecticide must be sufficiently stable to effect pest control, and yet its degradation profile should be such that no toxic residues persist to create health and/or environmental hazards. This is, of course, impossible to balance completely, and one can only attempt to establish the optimal ratio between efficacy and safety. The latter can be more reliably assured, however, through the use of selective toxins, i.e., compounds which are particularly lethal to one class, order, or, preferably, species of animal.

Because target selectivity and stability are often determined by metabolic reactions operating on the applied pesticide, knowledge of potential degradation pathways is essential for the design of safe, effective insecticides. In fact, prediction of the biological fate of a candidate insect poison is a very real part of the development process. Mapping out probable metabolic pathways does not necessarily imply that all potential products will exist. However, failure to anticipate the type of biotransformations likely to occur increases the chances that metabolites may go undetected or, at the very least, may not be properly identified. Thus, prediction, isolation, identification, and toxicological evaluation of insecticide metabolites begin with an understanding of the biochemical reactions mediating their formation.

Carbamate insecticides, like most xenobiotics, are susceptible to in-vivo biochemical alterations through a multitude of enzyme-catalyzed reactions. However, the principal modes of detoxication are hydrolysis, oxidation, and conjugation. Variations in susceptibility of different carbamate structures to these biochemical pathways allow for compounds of long persistence on one hand and for compounds demonstrating rapid and complete degradation in a very short time on the other. These differences contribute greatly to the commercial versatility of carbamate insecticide chemicals.

A detailed examination of the metabolites formed in plants and animals appears in the next chapter for many experimental and commercial carbamate insecticides. The purpose of this chapter is to acquaint the reader with what is known about the enzymes and mechanisms governing the biotransformations. Most information has been drawn from in-vitro studies with tissue, cellular, and subcellular preparations where carbamates were used as substrates. Such "test-tube" reactions allow the use of excessive levels of insecticide to encourage biosynthesis of large quantities of metabolites. Furthermore, in-vitro experiments favor the isolation of oxidative and hydrolytic products which are not always recoverable from living organisms because of the very efficient conjugation reactions occurring in vivo.

HYDROLYSIS

The immediate response of any plant or animal challenged by a lipid-soluble foreign substance is to convert the intruder into a more polar material for excretion or storage away from sensitive tissues.[1] One of the most effective means used to accomplish this feat is hydrolysis. In many cases, particularly with pesticide esters, the result is a "breaking off" of the toxophoric portion of the molecule, leaving nontoxic products or products that can easily be deactivated by conjugation (see below). Alternatively, the free hydrolytic metabolites, if not excreted or conjugated and eliminated, may form a part of the organism's metabolic pool, with eventual incorporation into natural products. Usually, in-vivo hydrolysis reactions are biochemical in nature, i.e., they are catalyzed by enzymes, although some chemical hydrolysis may also occur.

Since carbamate insecticides are esters, they are prone to cleavage by esterases, giving products which are identical to those formed via chemical hydrolysis, namely, parent phenol, oxime, or enol, plus methyl- or dimethylcarbamic acid. The latter materials are biologically short-lived and instantaneously decompose into carbon dioxide and methyl- or dimethylamine:

$$(R)ArOC(=O)NHCH_3 \xrightarrow[\text{esterase}]{\text{carbamate}} (R)ArOH + CO_2 + NH_2CH_3$$

Moreover, oxidative metabolites which still possess an intact carbamate ester linkage are also subject to hydrolysis by the same or possibly different enzymes. In many cases it is impossible to ascertain whether oxidation takes place prior to or after hydrolysis.

Differences in enzymic hydrolysis rates depend on the structure of the carbamate insecticide and on the species contacting the chemical. Usually, the carbamate ester bond is relatively stable in plants and insects; mammals, on the other hand, are more apt to cleave this linkage. Indeed, such differences are partly responsible for the selective toxicity of carbamates (see Chapter 5). However, there are exceptions; for example, carbaryl is readily hydrolyzed in rat, sheep, guinea pig, and dog, but the same insecticide is fairly resistant to hydrolysis when inside monkey or pig.[2] Furthermore, carbaryl hydrolysis represents a minor metabolic pathway in American cockroaches, cabbage loopers, boll weevils, bollworms, cotton leafworms, houseflies, and rice weevils, but it is a major pathway in the German cockroach.

The significance of chemical structure can be appreciated by the fact that in rats only one fourth of an injected dose of carbaryl is hydrolyzed, compared with one third for Baygon, and as much as three fourths for Isolan and Zectran.[2] With compounds such as aldicarb, Meobal, and carbofuran, hydrolysis of oxidative metabolites appears to be more extensive than hydrolysis of parent toxicants. Many other compounds will be examined more thoroughly in the next chapter, but it should be evident that hydrolysis must be considered when investigating the fate of any carbamate insecticide in any living being.

What are the nature of the esterases and the mechanism responsible for carbamate hydrolysis? Unfortunately, we know very little about the biochemistry of this process. It has been shown that cholinesterases and aliphatic esterases (aliesterases) residing in rabbit blood hydrolyze N,N-dimethylcarbamyl fluoride,[3] though carbaryl was untouched when incubated with rabbit serum fractions containing arylesterases, aliesterases, or cholinesterases.[4] On the other hand, a fraction associated with plasma albumin was able to effect ester cleavage of both carbaryl and 4-nitrophenyl N-methylcarbamate. More specifically, attack by albumin was directed at the carbamate ester bond, and not at the amide bond.

Later experiments with purified human serum confirmed the importance of albumin in the hydrolysis of several carbamate structures by plasma.[5,6] Some substrate specificity was apparent because human serum albumin did not catalyze hydrolysis of various carbamyl cholines or N,N-dimethylcarbamyl fluoride,[7] and carbaryl and UC 10854 were much more susceptible to hydrolysis than Baygon.[6] However, a good correlation ($r = 0.81$) between rate of hydrolysis of carbaryl and of UC 10854 suggested that the same enzyme(s) catalyzed degradation of both carbamates. In addition, the activity of albumin was shown to vary from animal to animal.[7] For example, among 15 species of birds and mammals, plasma from sheep, rabbit, and swine was noticeably more effective at hydrolyzing carbaryl.[4] When compared with other body tissues on a fresh-weight basis, rabbit blood was more than twice as active as any other organ. Finally, incubation of other enzymes, such as egg albumin, trypsin, chymotrypsin, pepsin, lipase, papain, or human serum butylcholinesterase, with 4-nitrophenyl N-methylcarbamate did not result in any carbamate hydrolysis.[5]

Almost all of the above conclusions are based on results generated in the late 1950's. It is amazing that, since then, only a few experiments have been performed relative to defining the biochemistry associated with carbamate hydrolysis in mammals. The same is essentially true for insects. Some reports in 1960[8,9] spoke of a "non-specific esterase," possibly an aromatic esterase,[10] which was primarily responsible for carbamate hydrolysis in insects. At that time, of course, hydrolysis was thought to be the predominant detoxication mechanism in arthropods, with differences in susceptibility to carbamate poisoning a result of variations in carbamate esterase titers. Even resistance to carbamate poisoning was blamed on a change in the balance of esterases among resistant and susceptible strains[11] (see Chapter 10 for more details).

More recent studies involving enzymic hydrolysis of carbamates have been conducted with homogenates of American cockroaches,[12] mouse brain tissues,[13] and human brain tissues.[14] A variety of electrophoretically separated esterases from each source were incubated with carbaryl or Baygon under various conditions. Several arylesterases and at least one aliesterase from the roach were able to catalyze the degradation of carbaryl. Hydrolysis zymograms from mouse and human brain tissues revealed that a number of enzymes exhibited carbamate esterase activity. However, the patterns were different for each substrate and for each animal.

Thus, with a few exceptions, research on carbamate hydrolysis mechanisms essentially stopped in the early 1960's after the exciting discovery that N,N-dimethylcarbamates were oxidatively metabolized by rat liver preparations,[15,16] and, that N-methylcarbamate insecticides, such as carbaryl and Baygon, formed a series of oxidation products in rat liver microsomes as well as in living mammals, insects, and plants.[17,18] Since that time, a concerted effort has been made toward establishing the biochemical mechanisms regulating oxidative metabolism at the expense of further elucidation of the enzyme systems catalyzing carbamate hydrolysis.

OXIDATION

Since about 1950, more than 100 review articles and books describing various aspects of enzymic oxidation have been published. Substrates studied range from natural endogenous chemicals such as steroids or fatty acids to a wide range of exogenous intruders, including drugs and pesticides. The enzyme most intensively investigated has been microsomal mixed-function oxidase, a unique, relatively broad-spectrum enzyme capable of catalyzing aromatic or aliphatic hydroxylation, N-, O-, or S-dealkylation, deamination, desaturation, epoxidation, and N- or S-oxidation. The primary purpose of these biological conversions seems to be enhancement of the polarity of the substrate and/or provision of an active functional group for conjugation.

General Mechanism

Research in this area began in 1938 with the centrifugal isolation and characterization of a high-speed subcellular sediment from perfused mammalian liver homogenates.[19] To distinguish these particles from other cell components, the term "microsome" was proposed. When viewed through the electron microscope several years later, microsomes were found to consist mostly of fragments or vesicles of endoplasmic reticulum (ER).[20] Furthermore, the hepatic microsomes could be subfractioned into rough-surfaced vesicles (ER plus attached ribosomes) and smooth vesicles (naked ER); it is the latter which are most often associated with the deactivation and elimination of xenobiotics from animals.[21] However, most investigators do not separate the ER fractions, but use microsomes isolated by the general procedure outlined in Figure 6.1.

The connection between microsomes and the biochemical alteration of pharmacologically active chemicals became evident in 1954, when it was learned that demethylation and deamination of certain drugs could be accomplished in vitro by incubating microsomes with substrate in the presence of oxygen and reduced nicotinamide adenine dinucleotide phosphate (NADPH).[22] At the same time, oxidative desulfuration of organophosphorus insecticides was also demonstrated to occur in the microsome fraction of liver cells fortified with NADPH and oxygen.[23] By 1958, it was clear that a large group of oxidative reactions operating on biologically active foreign compounds was catalyzed by a microsomal-NADPH-O_2 system.[24]

Meanwhile, other scientists were characterizing the chemical constituents of the multicomponent

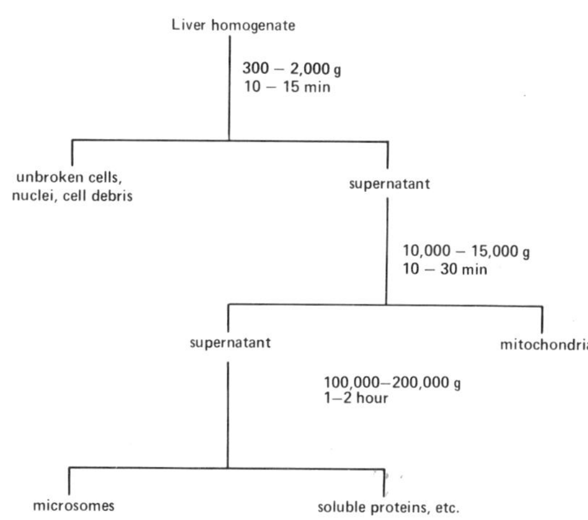

FIGURE 6.1. Centrifugal isolation of hepatic microsomes.

oxidase. Their experiments revealed that the system was comprised of two hemoproteins, namely cytochrome b_5[25,26] and, in particular, cytochrome P-450;[27-29] at least two flavoproteins, mainly cytochrome b_5 reductase[30] and cytochrome c reductase;[31] and the reduced pyridine nucleotides NADPH and NADH (nicotinamide adenine dinucleotide). However, all of the pieces have not been found, and those that are available can be arranged in a variety of patterns to account for most observations. One hypothetical sequence is pictured in Figure 6.2.[32,33] Starting at the top of the cycle, the substrate to be oxidized is bound to cytochrome P-450,[34-36] which in turn picks up molecular oxygen for activation.[37-39] This is accomplished by acceptance of one electron from the NADPH-flavoprotein respiratory chain and donation of a second electron from the interconnecting NADH-flavoprotein-cytochrome b_5 chain.[32] Next, one atom of activated oxygen is reduced to water while the other oxidizes the substrate, releasing P-450 for another cycle. The latter reactions categorize cytochrome P-450 as a typical monooxygenase[40] or mixed-function oxidase (mfo),[41] described by the sequence below, where SH is the substrate:

$$SH + O_2 + NADPH + H^+ \xrightarrow{mfo} SOH + H_2O + NADP^+$$

The presence of cytochrome P-450 in a subcellular preparation is ascertained by a UV-difference spectrum between reduced and carbon monoxide-treated reduced microsomes. The CO combines with reduced cytochrome P-450 to give a light-reversible complex with a single peak at 450 nm.[29,37,38] Hence, in-vitro oxidations catalyzed by P-450 can be inhibited by CO and reversed by light.

The proposed pathway in Figure 6.2 is only theoretical, and much of the fine detail remains to be learned. For example, the precise mechanism by which P-450 activates molecular oxygen to a high-energy electrophile containing six electrons is not understood, though several speculations exist. The role of NADH-cytochrome b_5 reductase, if indeed there is one, needs much more clarification. The location and morphology of the site(s) where substrate and P-450 interact requires characterization to account for the apparent broad specificity of the oxidase. It appears certain that there are at least two binding sites on the cytochrome, one near the heme iron and another on the iron itself. This is based on the fact that two distinct spectral traces appear when oxidized microsomes are combined with different substrates, i.e., Type I spectrum with a peak at 385 nm and trough at 420 nm for drugs, pesticides, steroid hormones, and some synergists, and Type II with a maximum at 430 nm and a minimum near 400 nm for amines, isocyanides, and certain steroids and alcohols.[34,36,42,43] A third kind of interaction occurs with synergists which inhibit drug or insecticide oxidations[44,44a] (synergists are described in Chapter 10).

Another possibility is that different substrates are oxidized by different or altered cytochrome P-450's; or one cytochrome may serve a number of more specific supporting oxidases; or a multitude of sites may dot a single P-450.[45-50] Finally, there is support for one general point of interaction where all substrates bind.[51] None of the above prophecies, however, explain why the amount of cytochrome P-450 in the endoplasmic recticulum of liver cells does not always correlate with the amount of oxidation of foreign compounds.[52,53]

The discussion thus far has been limited to liver cells, where most detoxication occurs in mammals.

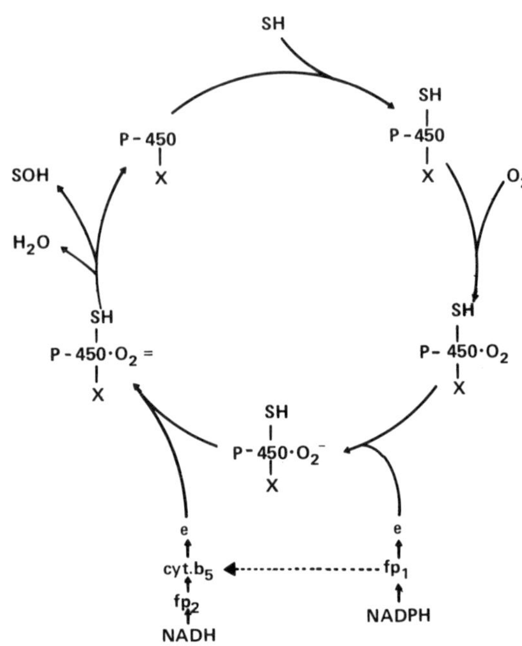

FIGURE 6.2. Mechanism of substrate (SH) oxidation in liver microsomes. (Adapted from Reference 33, with permission.)

Other mammalian tissues and other animal species, particularly insects, are known to possess an active cytochrome P-450 mfo system which catalyzes the oxidation of drugs, insecticides, and synergists.[33,43,54-59] Comparisons between insects and mammals indicate that the biochemistry and physiology of oxidation is basically very similar, but some notable differences have been uncovered. First of all, the tissue location for most detoxications in insects changes from species to species, although fat body and gut are often principal sites.[58-60] Despite the fact that both animals possess similar endoplasmic reticulum, the homogenization and centrifugation techniques needed for sedimentation of microsomal oxidases from most insect tissues deviate substantially from the general scheme shown in Figure 6.1.[61-67] At the enzyme level, hepatic and housefly cytochrome P-450 respond differently to inhibition by CO and reversal by light,[68,69] and they vary in their sensitivity to methylendioxyphenyl synergists.[33] Many other specific differences and similarities between mammalian and arthropod microsomes have recently been exhaustively listed and compared.[33,43,57,58]

In spite of these differences, all animals appear to utilize a very similar ER-attached mfo system for oxidation of exogenous xenobiotics. Although plants are capable of performing the same pesticide oxidations as those occurring in animals,[70] there is a paucity of information regarding the mechanism and enzymes responsible. A few in-vitro studies with plant tissues and subcellular isolates have shed some light on the subject. For example, substrate hydroxylation is catalyzed by a spinach leaf oxidase[71] and by a microsome fraction from developing castor beans,[72] both of which require NADH as a cofactor. In the case of castor beans, oxidation definitely involves activation of molecular oxygen, but, interestingly, the reaction is not inhibited by carbon monoxide. Hydroxylation via the "NIH shift" occurs in pea seedling microsomes in the presence of FH_4 and NADPH.[73,74]

The mode of epoxidation in plant tissues is most puzzling. The first study reported that epoxidation of the organochlorine insecticide aldrin in pea roots was catalyzed primarily by a soluble enzyme which did not require NADPH, NADP, or NAD, and which was not inhibited by synergists.[75] On the other hand, a second study showed that bean root fractions responded like typical animal counterparts; epoxidation was associated primarily with a synergist-sensitive microsomal mfo enzyme requiring O_2 and NADPH.[76] A later attempt to resolve these inconsistencies by investigating a variety of plants confirmed the findings of the second study, implying that the mechanism and cellular location of epoxidation in plants was not unlike that present in animals.[77]

Plant microsomes from leaf tissues of various plants have also been found capable of oxidative dealkylation. Thus, N-demethylation of substituted 3-(phenyl)-1-methylureas took place principally in smooth microsomes fortified with NADPH and oxygen.[78,79] Activity was inhibited by CO, indicating the presence of cytochrome P-450, but mfo action was not blocked by synergists.

Carbamate Insecticides

The growing interest in microsomal oxidations during the 1950's coincided with the development of carbamate insecticides. Therefore, one of the first carbamate metabolism studies involved an incubation of rat liver microsomes plus NADPH and oxygen with 4-nitrophenyl N,N-dimethylcarbamate.[15,16] Strong evidence was obtained favoring the formation of an N-hydroxymethyl analog as a metabolite. The same reaction apparently detoxified the dimethylcarbamate insecticides dimetilan, Isolan, and Pyramat, and, to a lesser degree, the methylcarbamates Zectran and carbaryl. Inhibition of oxidation was demonstrated by adding insecticide synergists, such as piperonyl butoxide and sesamex, or drug extenders, such as SKF 525-A and MGK 264. Thus, the importance of carbamate N-methyl hydroxylation was established, as pictured for the hypothetical N-methylcarbamate insecticide shown in Figure 6.3.

A few years later, in 1963, the first report of extensive nonhydrolytic metabolism of carbaryl and Baygon appeared in the literature.[17] Since similar products were recovered from living animals and from rat liver microsomes fortified with NADPH and oxygen, the metabolites were stated to result from oxidation of the carbamates. Later, ring-hydroxylated products of carbaryl and the N-hydroxymethyl derivatives of carbaryl and Baygon were identified.[18] The same oxidative pathways appeared when carbaryl was exposed to liver mixed-function oxidases from guinea pigs,[80] mice,[81,82] and rabbits.[82] Hence, ring hydroxylation became the second known oxidative reaction utilized by animals to metabolize carbamate insecticides (see Figure 6.3).

By that time, at least 20 oxidative products had been isolated following incubation of carbaryl with rat liver microsomes.[82] In addition to monohydroxylated products, one metabolite was identified as 5,6-dihydro-5,6-dihydroxy carbaryl. Despite the presence of two hydroxy groups, however, the derivative was believed to have been formed by mixed-function oxidase catalysis via an epoxide intermediate. Such a pathway is recognized for naphthalene oxidation in liver microsomes, as shown in Figure 6.4.[83] Note that the intermediate epoxide is broken down in the presence of microsomal epoxide hydrase to the *trans* diol, or by nonenzymic means to 1-naphthol. It is possible that this reaction may in fact be responsible for formation of all ring-hydroxylated metabolites of carbamates.

In 1966, many of the other oxidations affecting carbamate stability within an animal body were characterized by the use of a rat hepatic mfo system and a series of insecticidal carbamates.[84] The importance of N-methyl hydroxylation was confirmed by the appearance of N-hydroxymethyl analogs of carbaryl, Baygon, Banol, Matacil, and Zectran. Carbaryl ring hydroxylation was also demonstrated again, and the lack of direct oxidation or hydroxylation on the carbamate N atom was established. A new discovery was oxidative N-dealkylation of a dimethylamino ring substituent occurring with Zectran, Matacil, and dimetilan. An isolated intermediate in this biotransformation was the formamido analog, as shown below:

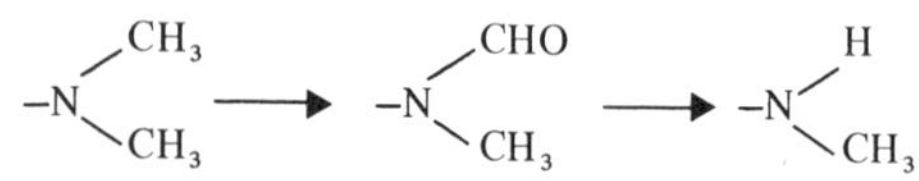

An analogous O-dealkylation took place when Baygon was incubated with the liver microsomes, whereby the propyl group was oxidatively removed. Finally, sulfoxidation of Mesurol proved

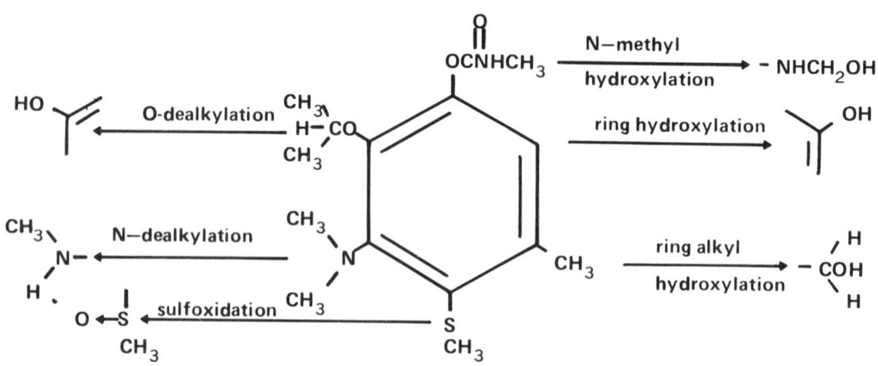

FIGURE 6.3. Sites of oxidation on hypothetical methylcarbamate insecticide.

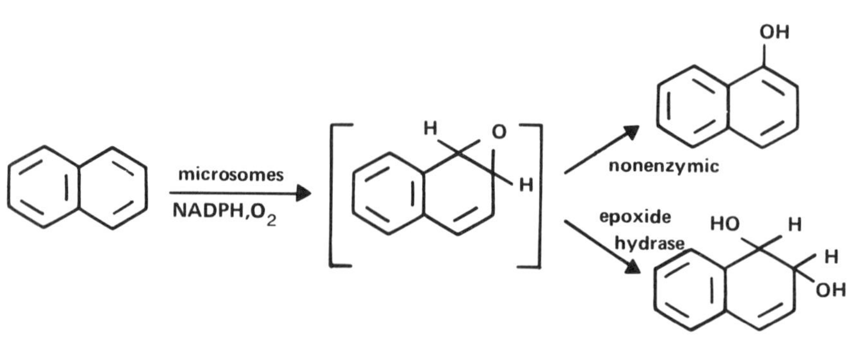

FIGURE 6.4. Metabolism of naphthalene in mammals. (Adapted from Reference 83, with permission.)

to be another reaction catalyzed by the mixed-function oxidase (Figure 6.3). This pathway, as will be seen in the next chapter, is particularly critical in the metabolic fate of the oxime carbamate aldicarb. Interestingly, aldicarb sulfoxide has been found when either microsomes or soluble enzymes were incubated with parent compound in the company of oxygen and NADPH.[85]

Elucidation of the major oxidative pathways summarized in Figure 6.3 was completed in 1967 with the suggestion that ring methyl groups of cresyl N-methylcarbamates were hydroxylated by mouse liver microsomes coupled with O_2 and either NADPH or NADH.[86] Of particular note was the fact that further oxidation to the –COOH analogs did not follow. The importance of this oxidative detoxication route has since been realized for Meobal in rat liver[87] and for Landrin in mouse liver.[88] However, in both instances extended oxidation of the $-CH_2OH$ groups to –COOH occurred, though housefly microsomal oxidation stopped at the alcohol stage.[88] Other ring alkyl groups, such as the isopropyl moiety of UC 10854, also fall victim to hydroxylation in hepatic microsomes.[89]

Turning to insects, the first attempts[81,82] at microsomal metabolism of carbamates were disappointing, especially because of the fact that subcellular preparations from German roaches and houseflies were known to hydroxylate DDT[90] and naphthalene respectively.[91] However, when it was discovered that endogenous mfo inhibitors contaminated housefly microsomes, much more active enzyme preparations were obtained following dialysis[81] or use of fly abdomens instead of whole flies as the microsome source.[92,93] In the presence of oxygen and NADPH, which proved to be the best pyridine cofactor, all of the oxidations shown in Figure 6.3, together with epoxidation, acted on a wide variety of N-methylcarbamates introduced into the insect microsomal in-vitro system.[92]

At about the same time (1967), mfo degradation of Banol was reported to occur in a microsomal fraction from isolated cockroach fat bodies.[94] The system displayed typical dependence on oxygen and NADPH, and action was inhibited by the synergists piperonyl butoxide and sesamex. Because of these studies (and others), avoidance of endogenous inhibitors and the use of selected insect tissues have allowed the separation of very active insect microsomes.

Therefore, by the late 1960's it was apparent that oxidative metabolism of carbamate insecticides in mammals and insects was catalyzed by a NADPH-dependent mixed-function oxidase, requiring molecular oxygen and showing inhibition in the presence of methylenedioxyphenyl synergists. But at that point, evidence for participation of cytochrome P-450 was lacking. This led to the speculation that another oxidase, namely tyrosinase, was responsible for carbamate insecticide detoxication in insects.[95,96] Besides some circumstantial evidence, a convincing argument was put forth based on the findings that housefly soluble tyrosinase could hydroxylate certain phenols and phenyl N-methylcarbamate, and that the tyrosinase was inhibited by methylenedioxyphenyl synergists. Unaccounted for was the fact that the soluble enzyme did not oxidize insecticides such as carbaryl, Baygon, or carbofuran.[69]

The tyrosinase theory was disproved almost as soon as it was proposed, however, because of three almost simultaneous experiments. First, cytochrome P-450 was found in housefly microsomes.[97] Second, when flies treated with carbaryl were exposed to an atmosphere of 40% CO:40% N_2: 20% O_2, the toxicity of carbaryl was greatly enhanced (presumably due to CO inhibition of carbaryl detoxication).[98] And third, fly microsomes containing P-450 hydroxylated carbaryl and phenyl N-methylcarbamate in the presence of NADPH and O_2, but not in the added presence of CO.[69] In addition, CO inhibition of carbamate metabolism was partially reversed by light.

Further support for P-450 involvement soon followed through experiments showing a close relationship between cytochrome P-450 and oxidative metabolism of carbaryl in blowfly fat body[62] and cabbage looper fat body.[99] Moreover, levels of housefly abdomen microsomal P-450 were correlated with Baygon toxicity,[100] and inhibition of carbamate oxidation by synergists was shown to be due to interactions with the cytochrome.[100,101] Lastly, differences in the nature of cytochrome P-450 from different strains of flies appeared to be responsible for carbamate resistance.[101-103]

One other oxidative pathway not mentioned thus far is apparently catalyzed by the microsomal mfo of both insects and mammals. A recent study has revealed that substrates such as Baygon,

Butacarb, or 3-*t*-butylphenyl N-methylcarbamate are oxidatively cleaved to yield the same products normally associated with carbamate ester bond hydrolysis.[104] There is a definite requirement for NADPH and oxygen, though "hydrolysis" is not inhibited by piperonyl butoxide, suggesting that another oxidative enzyme may be responsible. Finally, a soluble enzyme from rat liver, namely formamidase, has been shown to assist in the N-demethylenation of the $-NCHN<^{CH_3}_{CH_3}$ group on formetanate, though typical mfo oxidation also takes place.[105,106]

Undoubtedly, there are other enzymes contributing to carbamate insecticide oxidation in animals, but the microsomal cytochrome P-450 mfo system(s) is clearly responsible for production of most oxidative metabolites. Whether the same is true for plants remains to be established, but studies with other substrates tend to support a similar mechanism. The types and identity of these oxidative products will be fully explored in the next chapter.

CONJUGATION

Conjugation, by evolutionary design, is a metabolic process whereby endogenous as well as exogenous chemicals are converted to water-soluble components in order to facilitate their excretion or storage. In animals elimination may be via urine and/or feces, whereas in plants conjugates tend to be stored as terminal metabolites in various tissues. Although generally referred to as detoxication when speaking of pesticide metabolism, conjugation more often than not simply enhances excretion of metabolites previously formed from hydrolysis, oxidation, or another mechanism. Because of this, conjugates are labeled "secondary metabolites," and their toxicological significance is often ignored. However, there are sufficient instances where either the parent pesticide or toxic primary metabolite is deactivated by conjugation, and for this reason an understanding of the biochemical mechanism controlling pesticide conjugation is needed.

Until very recently, our concept of the types of conjugates associated with pesticide metabolism was based largely on general biochemical principles and on quasi-identification via "specific" enzymic cleavage. The former tell us that foreign substances containing functional groups, such as hydroxyl, carboxyl, sulfhydryl, or amino, react with natural body components to form glucuronides, glycosides, sulfates, mercapturic acids, amino acid conjugates, and methylated conjugates. Which of these reactions is involved following a particular pesticide application is usually determined after incubation of isolated water-soluble metabolites with an enzyme preparation, for example, β-glucosidase or aryl sulfatase. If a pesticide-derived fragment is released, then the conjugate is assumed to be, in this case, a glucoside or sulfate. Unfortunately, most of the enzyme preparations used in these situations have not been pure and usually consist of a mixture of hydrolases,[2] so that the true identity of the conjugate is not really known.

The obvious way to answer questions concerning pesticide conjugate structure is to isolate and chemically characterize the metabolic products. Although this may appear to be an easy task, the water-soluble nature and chemical constitution of conjugates make them all very much akin to each other and to a host of endogenous materials found in living systems. Thus, the problems involved in their isolation, purification, and positive identification are magnified tremendously as compared to the more lipid-soluble primary metabolites. In addition to this, difficulties in synthesis of potential conjugates have severely limited characterization. Nevertheless, in the last few years intact pesticide conjugates, particularly those derived from methylcarbamate insecticides, have been more completely identified, and a rudimentary knowledge of the associated biochemical mechanisms has evolved.

One other type of secondary metabolite deserves some attention. These are the so-called "bound" or "unextractable" residues which appear in almost every metabolism study. Little is known about their structure, although most are probably natural products or closely related substitutes, and nothing is known about the mechanism of their formation. Many pesticide scientists agree that bound residues are toxicologically insignificant, but recent concern by government regulatory agencies about possible toxicological significance may cause more steps to be taken toward positive identification.

The first legitimate attempts at chemical characterization of water-soluble metabolites formed from carbamate insecticides began in the mid-1960's. Based mostly on other studies with related compounds and on enzymic hydrolysis procedures, end products of carbaryl metabolism in several mammals were reported to include glucuronides and sulfates;[80] water-soluble carbamate products in bean plants were suggested to consist mostly of various glycosides;[107] and Baygon conjugates in houseflies apparently included glucosides, glucuronides, sulfates, and phosphates.[108] Many more recent results with other compounds and other species indicate that oxidative and hydrolytic metabolites of methylcarbamate insecticides are usually conjugated as β-glucuronides in mammals and as β-glucosides or other glycosides in plants and insects. Sulfate conjugates are also biosynthesized in mammals, whereas phosphates may be more common in insects; plants may not form either. All of the organisms studied appear to be capable of amino acid conjugation, though direct evidence for this reaction is minimal. An acount of the mechanisms controlling these conjugations follows.

Glucuronide Formation

The biosynthesis of β-glucuronides occurs in the particulate fracture of many cells throughout the animal body.[109-111] Glucuronic acid is passed from the cofactor, uridine disphospho-α-D-glucuronic acid (UDPGA), to a suitable acceptor via catalysis by glucuronyl transferase. Thus, phenols, alcohols, and acids form O-β-glucuronides (Figure 6.5, top), and aromatic amines yield N-β-glucuronides. Although this mechanism has been implicated on a few occasions in conjugation of exogenous phenols in insects,[112-114] glucuronidation is usually associated with mammals, where it serves as one of the primary conjugation mechanisms.[109-111,115,116]

The assumption follows that carbamate insecticide metabolites are conjugated in mammals as β-glucuronides via the same biosynthetic pathway. Proof for this postulation exists as a result of a few

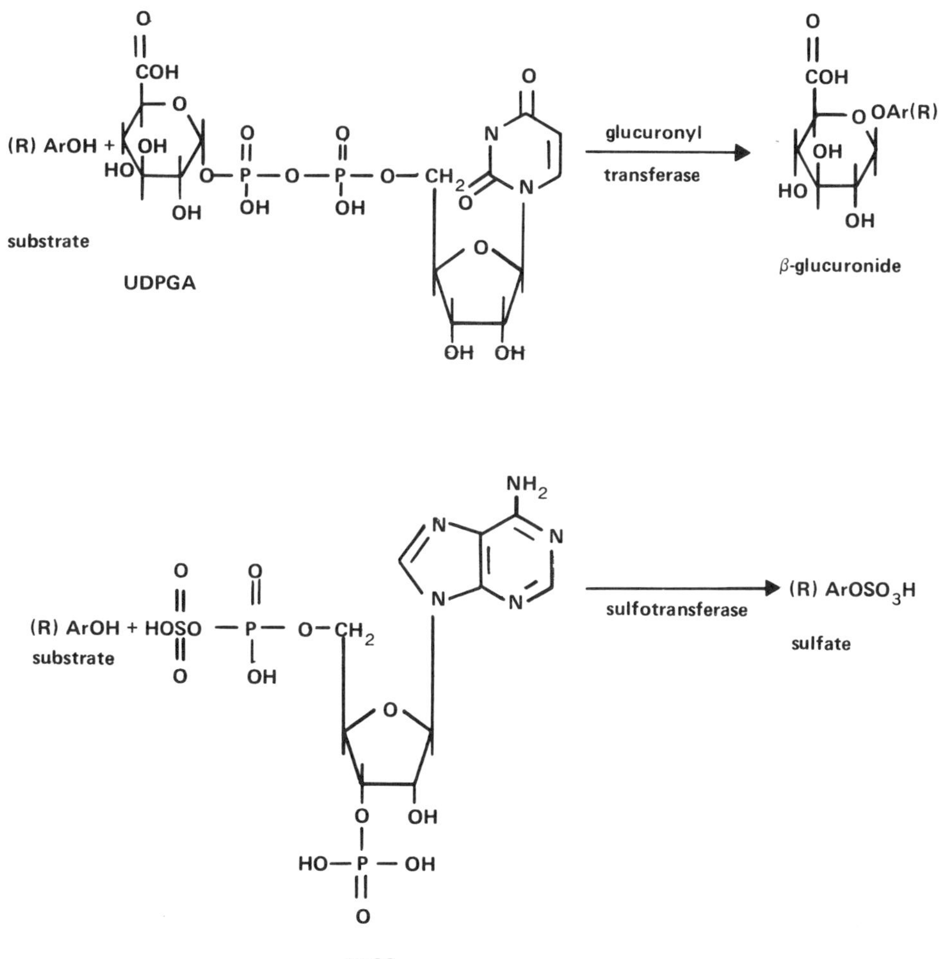

FIGURE 6.5. Mechanism of glucuronide (top) and sulfate (bottom) conjugation of phenols and alcohols.

in-vitro studies. When carbaryl was introduced into rat or guinea pig liver microsome preparations fortified with UDPGA, the methylcarbamate was hydrolyzed, and the released 1-naphthol was conjugated as a glucuronide.[80] Direct addition of hydroxylated carbaryl metabolites to a rat liver microsome or soluble fraction containing UDPGA gave considerable conjugation of 1-naphthol and 4-hydroxy carbaryl, intermediate glucuronidation of 5-hydroxy carbaryl and 5,6-dihydro-5,6-dihydroxy 1-naphthol, but very little conjugation of N-hydroxymethyl carbaryl or 5,6-dihydro-5,6-dihydroxy carbaryl.[82]

With the exception of the hydroxymethyl analog, the same substrate specificity has been reported for conjugation of carbaryl metabolites by another rat liver microsome plus soluble enzyme system, as shown in Table 6.1.[117,118] Further characterization of the glucuronyl transferase, using 1-naphthol as substrate, indicated maximum activity at 37°C in tris-HCl buffer at pH 7.0. Cofactors other than UDPGA failed to yield conjugates, and the only metal ion that enhanced glucuronic acid transfer was magnesium (almost all other metals inhibited the transferase). Interestingly, insecticide synergists, which are usually thought of as inhibitors of oxidative reactions, also inhibited conjugation, although the levels needed were greater than $10^{-3} M$. Finally, varying the level of UDPGA in enzyme preparations fortified with NADPH implied that continued hydrolysis and oxidation of carbaryl were dependent on continuous removal of the primary metabolites by conjugation. This supports an earlier finding which showed a similar in-vivo decrease in carbaryl metabolism when rats were treated with monamine oxidase inhibitors, compounds known to interfere with glucuronidation in mammals.[119,120]

TABLE 6.1

Specificity of Glucuronide Conjugation of Carbaryl Metabolites by Rat Liver in vitro[117,118]

Substrate	Percent conjugation
Carbaryl	0
1-Naphthol	92
N-Hydroxymethyl carbaryl	89
5-Hydroxy carbaryl	44
6-Hydroxy carbaryl	38
5,6-Dihydro-5,6-dihydroxy carbaryl	0

Thus, although far from overwhelming, the available information favors typical glucuronide formation as a mechanism for conjugation of primary carbamate insecticide metabolites in mammals. Of importance is the fact that direct conjugation of carbaryl or Banol was not found to occur in these rat in-vitro systems, even though O- and N-glucuronides of carbaryl,[80,121,122] Banol,[123] and formetanate[124] have been suggested as metabolites in various vertebrate species, including the rat. On the other hand, there is also evidence showing that N-glucuronidation does not require a transferase, or at least not the same transferase utilized in phenol glucuronidation.[125]

Glucoside Formation

The counterpart for mammalian glucuronide conjugation in plants[126] and insects[115,116] is glucosidation. The mechanism, cofactors, and products are similar, with uridine diphospho-α-D-glucose (UDPG) substituting for UDPGA, and glucosyl transferase replacing glucuronyl transferase to yield glucosides instead of glucuronides by the same mechanism as that shown at the top of Figure 6.5. Although biosynthesis of glucosides from exogenous phenols has been observed in a number of insect species,[127-133] in-vitro glucosylation by insect enzymes has only recently been indicated.[118,134] Active preparations were extracted from several tissues of tobacco hornworm larvae and from whole housefly adults. For example, when hornworm midgut glucosyl transferases were fortified with UDPG, rapid conversion of 1-naphthol to 1-naphthyl glucoside occurred. Maximum conjugation took place at 37°C in 0.05 *M* tris-HCl buffer, pH 8.5, in the presence of magnesium ion (other metals inhibited the reaction). Cofactors other than UDPG, including UDPGA, did not support conjugation of the substrate.

When different hornworm tissues were compared for conjugating ability, the following order of activity was found: fat body > midgut > blood > hindgut and foregut. The glucosyl transferase was associated with soluble enzymes in hornworms, but in houseflies maximum enzyme activity was removed with particulate fractions. These results indicate that insect glucosyl transferase activity may vary considerably from tissue to tissue and from species to species.

Sulfate and Phosphate Formation

Whenever a substance is conjugated as a glucoside in arthropods or as a glucuronide in other animals, chances are good that the molecule will also be found as a sulfate and occasionally, particularly in insects, as a phosphate. Though sulfate and phosphate synthesis are believed to be less important conjugation pathways, there are cases where certain substrates are recovered from treated animals almost exclusively as acid salts. For example, 4-nitrophenol is conjugated almost entirely with sulfuric acid in Southern armyworms, with no glucoside evident and with indications of only minor amounts of phosphate.[135] The sole conjugate of isopropyl- or ethyl-N-(4-hydroxyphenyl) carbamate excreted from rats is the sulfate.[1,136] On the other hand, phosphate conjugates of 1-naphthol are the primary detoxication products in flies dosed topically with 1-naphthol.[132]

The mechanism of sulfoconjugation apparently varies somewhat in different animal species, particularly with respect to the cofactor carrying the transferable sulfate and the nature or specificity of the accompanying soluble sulfotransferase.[137] For most mammals, the active donor is adenosine-3′-phosphate-5′-phosphosulfate (ADPS), and there are at least four known transferases, depending on whether the acceptor is a phenol, alcohol, amine, or steroid.[137,138] An example of the reaction with a phenol or alcohol as substrate is given at the bottom of Figure 6.5. Such a system has been shown to conjugate 1-naphthol together with 4- and 5-hydroxy carbaryl during incubation with rat[82] or chicken[122] liver soluble fractions fortified with adenosine triphosphate (ATP), potassium sulfate, and magnesium chloride. Interestingly, in living chickens, 1-naphthyl sulfate is excreted in the urine as a combination of potassium, sodium, and ammonium salts.[122]

The same mechanism is presumed to conjugate phenols in insects,[115,116,127,131,132,135] though only one in-vitro study serves as confirmation.[139] Working with Southern armyworm tissues, it was established that maximum sulfation of 4-nitrophenol occurred with whole homogenates of larval guts; less activity was present in Malpighian tubules or fat body, and no conjugation was evident in haemolymph.[139] Further characterization of the sulfotransferase in gut tissues defined it as a soluble enzyme operating best at pH 7.5 in 0.1 *M* tris-HCl buffer at 38°C. The cofactors ATP and K_2SO_4 presumably formed adenosine-3′-phosphate-5′phosphosulfate *in situ* for subsequent conjugation of 4-nitrophenol and a variety of plant and animal steroids. Transferase activity required the presence of magnesium, and glutathione further enhanced sulfation. Finally, the degree of conjugation varied with larval life stage[140] and species of insect.[141]

Despite the presence of more phosphate than sulfate in the animal body, phenols are not often excreted as phosphates.[142] In fact, the occurrence of aryl phosphates in living systems is not at all common.[132] Nevertheless, recent evidence points toward phosphate formation as a major detoxication pathway for many phenolic compounds in certain insect species.[130–132,143] The only in-vitro indication for such an occurrence involves the armyworm system discussed above, where phosphorylation was demonstrated when inorganic phosphate replaced sulfate in the reaction medium.[139]

Mercapturic Acid Formation

Animals employ mercapturic acid conjugation to enhance urinary excretion of alkyl and benzyl halides, halonitrobenzenes, and allylic compounds.[144] Thus, biosynthesis involves reaction of substrate with glutathione to form an intermediate glutathione conjugate. The latter is broken down into the corresponding cysteine analog, which is subsequently acetylated to give the mercapturic acid.

For carbaryl, this type of degradation has been observed in the living rat and in mouse liver preparations, although the mercapturic acid *per se* was not isolated.[145,146] Presumably, conjugation begins by oxidation of carbaryl across the 5,6-position, and the epoxide then reacts with glutathione (GS), as shown in Figure 6.6. The glutathione conjugate is excreted in the bile or is subsequently converted to the mercapturic acid. However, on isolation and treatment with acid, the mercapturic acid and glutathione conjugate recovered from animals appear as S-(5-hydroxy-1-naphthyl)cysteine (Figure 6.6).

The enzymes catalyzing the formation of glycosides and other conjugates of carbamate metabolites in plants are not known. Presumably, UDPG or other UDP sugars together with a sugar transferase constitute the major components, but in-vitro preparations establishing such pathways have not been isolated.

Taken together, the information on conjugation reaction mechanisms associated with methylcarbamate detoxication and excretion is meager. Part of the reason lies in the difficulty surrounding isolation and identification of products, but most is a result of lack of interest in this particular facet of pesticide metabolism. However, where studies have been completed, some unexpected and possibly toxicologically significant findings were uncovered. Particularly important are the implications that species differences in conjugation rates and routes may add to the selective toxicity of carbamate insecticides.

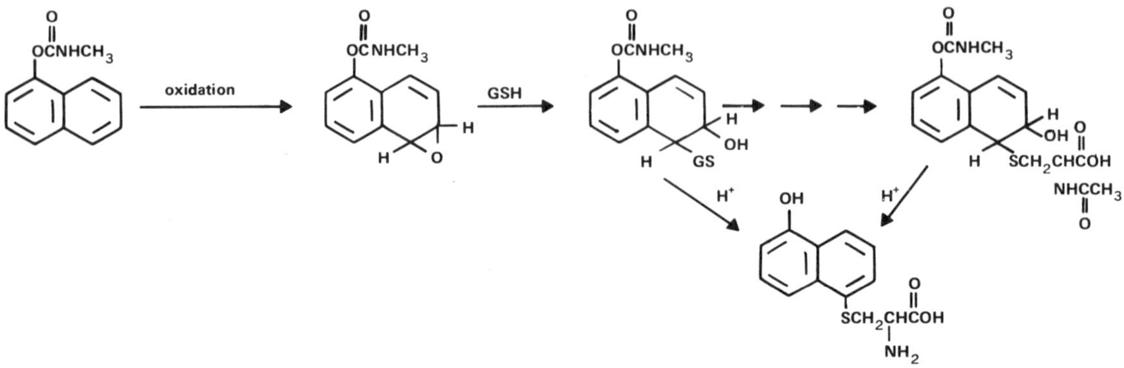

FIGURE 6.6. Formation of thiol conjugates and their hydrolysis by acid.[145,146]

REFERENCES

1. **Williams, R. T.,** *Detoxication Mechanisms,* 2nd ed., Chapman and Hall, London, 1959.
2. **Kuhr, R. J.,** The formation and importance of carbamate insecticide metabolites as terminal residues, *Pure Appl. Chem. Suppl.,* p. 199, 1971.
3. **Augustinsson, K.-B. and Casida, J. E.,** Enzymic hydrolysis of N:N-dimethylcarbamoyl fluoride, *Biochem. Pharmacol.,* 3, 60, 1959.
4. **Casida, J. E. and Augustinsson, K.-B.,** Reaction of plasma albumin with 1-naphthyl N-methylcarbamate and certain other esters, *Biochim. Biophys. Acta,* 36, 411, 1959.
5. **Casida, J. E., Augustinsson, K.-B., and Jonsson, G.,** Stability, toxicity, and reaction mechanism with esterases of certain carbamate insecticides, *J. Econ. Entomol.,* 53, 205, 1960.
6. **Reiner, E. and Skrinjaric-Spoljar, M.,** Hydrolysis of some monomethylcarbamates in human sera, *Chroat. Chem. Acta,* 40, 87, 1968.
7. **Augustinsson, K.-B., Fredriksson, T., Sundwall, A., and Jonsson, G.,** Biochemical and pharmacological studies on N-methylated carbamoylcholines, *Biochem. Pharmacol.,* 3, 68, 1959.
8. **Metcalf, R. L., Fukuto, T. R., and Winton, M. Y.,** Alkoxyphenyl N-methylcarbamates as insecticides, *J. Econ. Entomol.,* 53, 828, 1960.
9. **Gordon, H. T. and Eldefrawi, M. E.,** Analog-synergism of several carbamate insecticides, *J. Econ. Entomol.,* 53, 1004, 1960.
10. **Metcalf, R. L., Maxon, M., Fukuto, T. R., and March, R. B.,** Aromatic esterase in insects, *Ann. Entomol. Soc. Am.,* 49, 274, 1956.
11. **Plapp, F. W., Jr., Chapman, G. A., and Bigley, W. S.,** A mechanism of resistance to Isolan in the housefly, *J. Econ. Entomol.,* 57, 692, 1964.
12. **Matsumura, F. and Sakai, K.,** Degradation of insecticides by esterases of the American cockroach, *J. Econ. Entomol.,* 61, 598, 1968.

13. **Sakai, K. and Matsumura, F.,** Esterases of mouse brain active in hydrolyzing organophosphate and carbamate insecticides, *J. Agric. Food Chem.*, 16, 803, 1968.
14. **Sakai, K. and Matsumura, F.,** Degradation of certain organophosphate and carbamate insecticides by human brain esterases, *Toxicol. Appl. Pharmacol.*, 19, 660, 1971.
15. **Hodgson, E. and Casida, J. E.,** Biological oxidation of N,N-dialkyl carbamates, *Biochim. Biophys. Acta*, 42, 184, 1960.
16. **Hodgson, E. and Casida, J. E.,** Metabolism of N:N-dialkyl carbamates and related compounds by rat liver, *Biochem. Pharmacol.*, 8, 179, 1961.
17. **Dorough, H. W., Leeling, N. C., and Casida, J. E.,** Nonhydrolytic pathway in metabolism of N-methylcarbamate insecticides, *Science*, 140, 170, 1963.
18. **Dorough, H. W. and Casida, J. E.,** Nature of certain carbamate metabolites of the insecticide Sevin, *J. Agric. Food Chem.*, 12, 294, 1964.
19. **Claude, A.,** Microsomes, endoplasmic reticulum and interactions of cytoplasmic membranes, in *Microsomes and Drug Oxidations*, Gillette, J. R. et al., Eds., Academic Press, New York, London, 1969.
20. **Palade, E. S. and Siekevitz, P.,** Liver microsomes. An integrated morphological and biochemical study, *J. Biophys. Biochem. Cytol.*, 2, 171, 1956.
21. **Gram, T. E. and Fouts, J. R.,** Studies on the intramicrosomal distribution of hepatic enzymes which catalyze the metabolism of drugs and other foreign compounds, in *Enzymatic Oxidation of Toxicants*, Hodgson, E., Ed., North Carolina State University Press, Raleigh, 1968.
22. **Axelrod, J.,** Enzymatic demethylation of sympathomimetic amines, *Fed. Proc.*, 13, 332, 1954.
23. **Davison, A. N.,** Conversion of Schradan and parathion by an enzyme system of rat liver, *Nature*, 174, 1056, 1954.
24. **Brodie, B. B., Gillette, J. R., and LaDu, B. N.,** Enzymic metabolism of drugs and other foreign compounds, *Annu. Rev. Biochem.*, 27, 427, 1958.
25. **Strittmatter, C. F. and Ball, E. G.,** A hemochromogen component of liver microsomes, *Proc. Natl. Acad. Sci. U.S.A.*, 38, 19, 1952.
26. **Strittmatter, P. and Velick, S. F.,** The isolation and properties of microsomal cytochrome, *J. Biol. Chem.*, 221, 253, 1956.
27. **Klingenburg, M.,** Pigments of rat liver microsomes, *Arch. Biochem. Biophys.*, 75, 376, 1958.
28. **Garfinkel, D.,** Studies on pig liver microsomes. I. Enzymic and pigment composition of different microsomal fractions, *Arch. Biochem. Biophys.*, 77, 493, 1958.
29. **Omura, T. and Sato, R.,** The carbon monoxide-binding pigment of liver microsomes. I., *J. Biol. Chem.*, 239, 2370, 1964; II., *J. Biol. Chem.*, 239, 2379, 1964.
30. **Strittmatter, P.,** The interaction of nucleotides with microsomal cytochrome reductase, *J. Biol. Chem.*, 233, 748, 1958.
31. **Masters, B. S. S., Kamin, H., Gibson, Q. H., and Williams, C. H., Jr.,** Studies on the mechanism of microsomal triphosphopyridine nucleotide-cytochrome c reductase, *J. Biol. Chem.*, 240, 921, 1965.
32. **Estabrook, R. W. and Cohen, B.,** Organization of the microsomal electron transport system, in *Microsomes and Drug Oxidations*, Gillette, J. R. et al., Eds., Academic Press, New York, London, 1969.
33. **Casida, J. E.,** Mixed-function oxidase involvement in the biochemistry of insecticide synergists, *J. Agric. Food Chem.*, 18, 753, 1970.
34. **Imai, Y. and Sato, R.,** Substrate interaction with hydroxylase system in liver microsomes, *Biochem. Biophys. Res. Commun.*, 22, 620, 1966.
35. **Imai, Y. and Sato, R.,** Substrate interactions with P-450 in drug hydroxylation by liver microsomes, *J. Biochem.*, 62, 239, 1967.
36. **Remmer, H., Schenkman, J. B., Estabrook, R. W., Sasame, H., Gillette, J., Narasimhulu, S., Cooper, D. Y., and Rosenthal, O.,** Drug interaction with hepatic microsomal cytochrome, *Mol. Pharmacol.*, 2, 187, 1966.
37. **Estabrook, R. W., Cooper, D. Y., and Rosenthal, O.,** The light-reversible carbon monoxide inhibition of the steroid C21-hydroxylase system of the adrenal cortex, *Biochem. Z.*, 338, 741, 1963.
38. **Cooper, D. Y., Levin, S., Narasimhulu, S., Rosenthal, O., and Estabrook, R. W.,** Photochemical action spectrum of the terminal oxidase of mixed-oxidase systems, *Science*, 147, 400, 1965.
39. **Omura, T., Sato, R., Cooper, D. Y., Rosenthal, O., and Estabrook, R. W.,** Function of cytochrome P-450 of microsomes, *Fed. Proc.*, 24, 1181, 1965.
40. **Hayaishi, O.,** Enzymic hydroxlyation, *Annu. Rev. Biochem.*, 38, 21, 1969.
41. **Mason, H. S.,** Mechanism of oxygen metabolism, *Adv. Enzymol.*, 19, 79, 1957.
42. **Schenkman, J. B., Remmer, H., and Estabrook, R. W.,** Spectral studies of drug interaction with hepatic microsomal cytochrome, *Mol. Pharmacol.*, 3, 113, 1967.
43. **Hodgson, E. and Kulkarni, A. P.,** Interactions of pesticides with cytochrome P-450, in *Mechanism of Pesticide Action*, Kohn, G. K., Ed., American Chemical Society, Washington, D.C., 1974.
44. **Hodgson, E., Philpot, R. M., Baker, R. C., and Mailman, R. B.,** Effect of synergists on drug metabolism, *Drug Metab. Dispos.*, 1, 391, 1973.

44a. **Hodgson, E. and Philpot, R. M.,** Interaction of methelenedioxyphenyl (1,3-benzodioxole) compounds with enzymes and their effects on mammals, *Drug Metab. Rev.*, 3, 231, 1974.

45. **Mannering, G. J., Sladek, N. E., Parli, C. J., and Shoeman, D. W.,** Formation of a new P-450 hemoprotein after treatment of rats with polycyclic hydrocarbons, in *Microsomes and Drug Oxidations,* Gillette, J. R. et al., Eds., Academic Press, New York, London, 1969.
46. **Hildebrandt, A. G. and Estabrook, R. W.,** Spectrophotometric studies of cytochrome P-450 of liver microsomes after induction with phenobarbital and 3-methylcholanthrene, in *Microsomes and Drug Oxidations,* Gillette, J. R. et al., Eds., Academic Press, New York, London, 1969.
,47. **Kuntzman, R., Levin, W., Schilling, G., and Alvares, A.,** The effects of 3-methylcholanthrene and phenobarbital on liver microsomal hemoproteins and on the hydroxylation of benzpyrene, in *Microsomes and Drug Oxidations,* Gillette, J. R. et al., Eds., Academic Press, New York, London, 1969.
48. **Philpot, R. M. and Hodgson, E.,** The effect of piperonyl butoxide concentration on the formation of cytochrome P-450 difference spectra in hepatic microsomes from mice, *Mol. Pharmacol.,* 8, 204, 1972.
49. **Kuwatsuka, S.,** Biochemical aspects of methylenedioxyphenyl compounds in relation to the synergistic action, in *Biochemical Toxicology of Insecticides,* O'Brien, R. D. and Yamamoto, I., Eds., Academic Press, New York, London, 1970.
50. **Remmer, H.,** Enzyme induction phenomenon: effects in vertebrate livers, in *Insecticide Resistance, Synergism, Enzyme Induction,* Tahori, A. S., Ed., Gordon and Breach, New York, 1971.
51. **Kamin, H. and Masters, B. S. S.,** Electron transport in microsomes, in *Enzymatic Oxidation of Toxicants,* Hodgson, E., Ed., North Carolina State University Press, Raleigh, 1968.
52. **Schenkman, J. B., Frey, I., Remmer, H., and Estabrook, R. W.,** Sex differences in drug metabolism by rat liver microsomes, *Mol. Pharmacol.,* 3, 516, 1967.
53. **Remmer, H., Estabrook, R. W., Schenkman, J., and Greim, H.,** Induction of microsomal liver enzymes, in *Enzymatic Oxidations of Toxicants,* Hodgson, E., Ed., North Carolina State University Press, Raleigh, 1968.
54. **Dewaide, J. H.,** *Metabolism of Xenobiotics,* Drukkerij Leijn, Nijmegen, Netherlands, 1971.
55. **Terriere, L. C.,** The oxidation of pesticides: the comparative approach, in *Enzymatic Oxidation of Toxicants,* Hodgson, E., Ed., North Carolina State University Press, Raleigh, 1968.
56. **Hook, G. E. R., Jordan, T. W., and Smith, J. N.,** Factors affecting insect microsomal oxidations, in *Enzymatic Oxidation of Toxicants,* Hodgson, E., Ed., North Carolina State University Press, Raleigh, 1968.
57. **Hodgson, E. and Plapp, F. W., Jr.,** Biochemical characteristics of insect microsomes, *J. Agric. Food Chem.,* 18, 1048, 1970.
58. **Wilkinson, C. F. and Brattsten, L. B.,** Microsomal drug-metabolizing enzymes in insects, *Drug Metab. Rev.,* 1, 153, 1972.
59. **Kuhr, R. J.,** Mechanisms of insecticide metabolism in insects: tissue, cellular, and subcellular studies, in *Mechanisms of Pesticide Action,* Kohn, G. K., Ed., American Chemical Society, Washington, D.C., 1974.
60. **Kuhr, R. J.,** Metabolism of carbamate insecticide chemicals in plants and insects, *J. Agric. Food Chem.,* 18, 1023, 1970.
61. **Schonbrod, R. D. and Terriere, L. C.,** Improvements in the methods of preparation and storage of housefly microsomes, *J. Econ. Entomol.,* 59, 1411, 1966.
62. **Price, G. M. and Kuhr, R. J.,** The metabolism of the insecticide carbaryl (1-naphthyl N-methylcarbamate) by fat body of the blowfly larva, *Calliphora erythrocephala, Biochem. J.,* 112, 133, 1969.
63. **Cassidy, J. D., Smith, E., and Hodgson, E.,** An ultrastructural analysis of microsomal preparations from *Musca domestica* and *Prodenia eridania, J. Insect Physiol.,* 15, 1573, 1969.
64. **Brindley, W. A. and Dahm, P. A.,** Microscopic examination of parathion-activating microsomes from American cockroach fat body, *J. Econ. Entomol.,* 63, 31, 1970.
65. **Benke, G. M. and Wilkinson, C. F.,** Microsomal oxidation in the house cricket, *Acheta domesticus* (L.), *Pestic. Biochem. Physiol.,* 1, 19, 1971.
66. **Benke, G. M., Wilkinson, C. F., and Telford, J. N.,** Microsomal oxidases in a cockroach, *Gromphadorhina postenosa, J. Econ. Entomol.,* 65, 1221, 1972.
67. **Gilbert, M. D. and Wilkinson, C. F.,** Microsomal oxidases in the honey bee, *Aphis mellifera* (L.), *Pestic. Biochem. Physiol.,* 4, 56, 1974.
68. **Lewis, S. E., Wilkinson, C. F., and Ray, J. W.,** The relationship between microsomal epoxidation and lipid peroxidation in houseflies and pig liver and the inhibitory effect of derivatives of 1,3-benzodioxole (methylene-dioxybenzene), *Biochem. Pharmacol.,* 16, 1195, 1967.
69. **Kuhr, R. J.,** Possible role of tyrosinase and cytochrome P-450 in the metabolism of 1-naphthyl methylcarbamate (carbaryl) and phenyl methylcarbamate by houseflies, *J. Agric. Food Chem.,* 17, 112, 1969.
70. **Casida, J. E. and Lykken, L.,** Metabolism of organic pesticide chemicals in higher plants, *Annu. Rev. Plant Physiol.,* 20, 607, 1969.
71. **Vaughan, P. F. T. and Butt, V. S.,** The suppression of catechol oxidase activity during the enzymic hydroxylation of *p*-coumaric acid by spinach leaf phenolase, *Biochem. J.,* 111, 32P, 1969.
72. **Galliard, T. and Stumpf, P. K.,** Fat metabolism in higher plants, *J. Biol. Chem.,* 241, 5806, 1966.
73. **Russell, D. W. and Conn, E. E.,** The cinnamic acid 4-hydroxylase of pea seedlings, *Arch. Biochem. Biophys.,* 122, 256, 1967.

74. **Russell, D. W., Conn, E. E., Sutter, A., and Griesebach, H.,** Hydroxylation-induced migration and retention of tritium on conversion of [4-^{3}H] cinnamic acid to 4-hydroxycinnamic acid by an enzyme from pea seedlings, *Biochim. Biophys. Acta,* 170, 210, 1968.
75. **Lichtenstein, E. P. and Corbett, J. R.,** Enzymatic conversion of aldrin to dieldrin with subcellular components of pea plants, *J. Agric. Food Chem.,* 17, 589, 1969.
76. **Yu, S. J., Kiigemagi, U., and Terriere, L. C.,** Oxidative metabolism of aldrin and isodrin by bean root fractions, *J. Agric. Food Chem.,* 19, 5, 1971.
77. **Mehendale, H. M., Skrentny, R. F., and Dorough, H. W.,** Oxidative metabolism of aldrin by subcellular root fractions of several plant species, *J. Agric. Food Chem.,* 20, 398, 1972.
78. **Frear, D. S., Swanson, H. R., and Tanaka, F. S.,** N-Demethylation of substituted 3-(phenyl)-1-methylureas: isolation and characterization of a microsomal mixed-function oxidase from cotton, *Phytochemistry,* 8, 2157, 1969.
79. **Tanaka, F. S., Swanson, H. R., and Frear, D. S.,** Mechanism of oxidative N-demethylation by cotton microsomes, *Phytochemistry,* 11, 2709, 1972.
80. **Knaak, J. B., Tallant, M. J., Bartley, W. J., and Sullivan, L. J.,** The metabolism of carbaryl in the rat, guinea pig, and man, *J. Agric. Food Chem.,* 13, 537, 1965.
81. **Matthews, H. B. and Hodgson, E.,** Naturally occurring inhibitors of microsomal oxidations from the house fly, *J. Econ. Entomol.,* 59, 1286, 1966.
82. **Leeling, N. C. and Casida, J. E.,** Metabolites of carbaryl (1-naphthyl methylcarbamate) in mammals and enzymatic systems for their formation, *J. Agric. Food Chem.,* 14, 281, 1966.
83. **Jerina, D. M., Daly, J. W., Witkop, B., Zaltzman-Nirenberg, P., and Udenfriend, S.,** The role of arene oxide-oxepin systems in the metabolism of aromatic substrates. III. Formation of 1,2-naphthalene oxide from naphthalene by liver microsomes, *J. Am. Chem. Soc.,* 90, 6525, 1968.
84. **Oonnithan, E. S. and Casida, J. E.,** Metabolites of methyl- and dimethylcarbamate insecticide chemicals as formed by rat liver microsomes, *Bull. Environ. Contam. Toxicol.,* 1, 59, 1966.
85. **Andrawes, N. R., Dorough, H. W., and Lindquist, D. A.,** Degradation and elimination of Temik in rats, *J. Econ. Entomol.,* 60, 979, 1967.
86. **Hook, G. E. R. and Smith, J. N.,** Oxidation of methyl groups by grass grubs and vertebrate liver enzymes, *Biochem. J.,* 102, 504, 1967.
87. **Miyamoto, J., Yamamoto, K., and Matsumoto, T.,** Metabolism of 3,4-dimethylphenyl N-methylcarbamate in white rats, *Agric. Biol. Chem.,* 33, 1060, 1969.
88. **Slade, M. and Casida, J. E.,** Metabolic fate of 3,4,5- and 2,3,5-trimethylphenyl methylcarbamates, the major constituents in Landrin insecticide, *J. Agric. Food Chem.,* 18, 467, 1970.
89. **Oonnithan, E. S. and Casida, J. E.,** Oxidation of methyl- and dimethylcarbamate insecticide chemicals by microsomal enzymes and anticholinesterase activity of metabolites, *J. Agric. Food Chem.,* 16, 28, 1968.
90. **Agosin, M., Michaeli, M. D., Miskna, R., Nagasawa, S., and Hoskins, W. H.,** A new DDT-metabolizing enzyme in the German cockroach, *J. Econ. Entomol.,* 54, 340, 1961.
91. **Arias, R. D. and Terriere, L. C.,** The hydroxylation of naphthalene-1-C^{14} by house fly microsomes, *J. Econ. Entomol.,* 55, 925, 1962.
92. **Tsukamoto, M. and Casida, J. E.,** Metabolism of methylcarbamate insecticides by the $NADPH_2$-requiring enzyme system from houseflies, *Nature,* 213, 49, 1967.
93. **Tsukamoto, M. and Casida, J. E.,** Albumin enhancement of oxidative metabolism of methylcarbamate insecticide chemicals by the house fly microsome-$NADPH_2$ system, *J. Econ. Entomol.,* 60, 617, 1967.
94. **Gemrich, E. G.,** Enzymatic degradation of 2-chloro-4,5-dimethylphenyl N-methylcarbamate by the fat bodies of *Blaberus giganteus, J. Agric. Food Chem.,* 15, 617, 1967.
95. **Metcalf, R. L., Fukuto, T. R., Wilkinson, C., Fahmy, M. H., Abd El-Aziz, S., and Metcalf, E. R.,** Mode of action of carbamate synergists, *J. Agric. Food Chem.,* 14, 555, 1966.
96. **Metcalf, R. L.,** Mode of action of insecticide synergists, *Annu. Rev. Entomol.,* 12, 229, 1967.
97. **Ray, J. W.,** The epoxidation of aldrin by housefly microsomes and its inhibition by carbon monoxide, *Biochem. Pharmacol.,* 16, 99, 1967.
98. **Lewis, S. E.,** Effect of carbon monoxide on metabolism of insecticides in vivo, *Nature,* 215, 1408, 1967.
99. **Kuhr, R. J.,** Comparative metabolism of carbaryl by resistant and susceptible strains of the cabbage looper, *J. Econ. Entomol.,* 64, 1373, 1971.
100. **Perry, A. S. and Buckner, A. J.,** Studies on microsomal cytochrome P-450 in resistant and susceptible houseflies, *Life Sci. Part II,* 9, 335, 1970.
101. **Matthews, H. B. and Casida, J. E.,** Properties of housefly microsomal cytochromes in relation to sex, strain, substrate specificity, and apparent inhibition and induction by synergist and insecticide chemicals, *Life Sci. Part I,* 9, 989, 1970.
102. **Philpot, R. M. and Hodgson, E.,** Differences in the cytochrome P-450 from resistant and susceptible house flies, *Chem. Biol. Interact.,* 4, 399, 1971.
103. **Perry, A. S., Dale, W. E., and Buckner, A. J.,** Induction and repression of microsomal mixed-function oxidases and cytochrome P-450 in resistant and susceptible houseflies, *Pestic. Biochem. Physiol.,* 1, 131, 1971.
104. **Douch, P. G. C., Smith, J. N., and Turner, J. C.,** NADPH-dependent cleavage of carbamates, *Life Sci. Part II,* 10, 1327, 1971.

105. **Ahmad, S. and Knowles, C. O.**, Degradation of formetanate acaricide by rat liver preparations, *J. Econ. Entomol.*, 63, 1690, 1970.
106. **Ahmad, S. and Knowles, C. O.**, Formamidase involvement in the metabolism of chlorophenamidine and formetanate acaricides, *J. Econ. Entomol.*, 64, 792, 1971.
107. **Kuhr, R. J. and Casida, J. E.**, Persistent glycosides of metabolites of methylcarbamate insecticide chemicals formed by hydroxylation in bean plants, *J. Agric. Food Chem.*, 15, 814, 1967.
108. **Shrivastava, S. P., Tsukamoto, M., and Casida, J. E.**, Oxidative metabolism of C^{14}-labeled Baygon by living houseflies and by house fly enzyme preparations, *J. Econ. Entomol.*, 62, 483, 1969.
109. **Dutton, G. J.**, *Glucuronic Acid, Free and Combined*, Academic Press, New York, London, 1966.
110. **Parke, V. D.**, *The Biochemistry of Foreign Compounds*, Pergamon Press, London, 1968.
111. **Fishman, W. H., Ed.**, *Metabolic Conjugation and Metabolic Hydrolysis*, Vol. 1, Academic Press, New York, 1970.
112. **Terriere, L. C., Boose, R. B., and Roubal, W. T.**, The metabolism of naphthalene and 1-naphthol by houseflies and rats, *Biochem. J.*, 79, 620, 1961.
113. **Hassan, A., Zayed, S. M. A. D., and Abdel-Hamid, F. M.**, Metabolism of organophosphorus insecticides. V. Mechanism of detoxication of dipterex in *Prodenia litura* F., *Biochem. Pharmacol.*, 14, 1577, 1965.
114. **Inayami, K.**, On the tryptophan metabolism of the silkworm, *Nippon Nogei Kagaku Kaishi*, 29, 918, 1955.
115. **Smith, J. N.**, Comparative biochemistry of detoxication, in *Comparative Biochemistry*, Vol. 6, Flokin, M. and Mason, H. S., Eds., Academic Press, New York, 1964.
116. **Smith, J. N.**, The comparative metabolism of xenobiotics, *Adv. Comp. Physiol. Biochem.*, 3, 173, 1968.
117. **Mehendale, H. M. and Dorough, H. W.**, Glucuronidation mechanisms in the rat and their significance in the metabolism of insecticides, *Pestic. Biochem. Physiol.*, 1, 307, 1971.
118. **Mehendale, H. M. and Dorough, H. W.**, Conjugative metabolism and action of carbamate insecticides, in *Insecticides*, Tahori, A. S., Ed., Gordon and Breach, New York, London, Paris, 1971.
119. **Culver, D. J., Lin, T., and Dorough, H. W.**, The effect of monamine oxidase inhibitors on carbaryl metabolism, *J. Econ. Entomol.*, 63, 1369, 1970.
120. **Dorough, H. W., Mehendale, H. M., and Lin, T.**, Modification of carbaryl metabolites in rats with monamine oxidase inhibitors, *J. Econ. Entomol.*, 65, 958, 1972.
121. **Knaak, J. B., Tallant, M. J., Kozbelt, S. J., and Sullivan, L. J.**, The metabolism of carbaryl in man, monkey, pig and sheep, *J. Agric. Food Chem.*, 16, 465, 1968.
122. **Paulson, G. D., Zaylskie, R. G., Zher, M. V., Portnoy, C. E., and Feil, V. J.**, Metabolites of carbaryl (1-naphthyl methylcarbamate) in chicken urine, *J. Agric. Food Chem.*, 18, 110, 1970.
123. **Baron, R. L. and Doherty, J. D.**, Metabolism and excretion of an insecticide (6-chloro-3,4-dimethylphenyl-N-methylcarbamate) in the rat, *J. Agric. Food Chem.*, 15, 830, 1967.
124. **SenGupta, A. K. and Knowles, C. O.**, Fate of formetanate-^{14}C acaricide in the rat, *J. Econ. Entomol.*, 63, 10, 1970.
125. **Shuster, L.**, Metabolism of drugs and toxic substances, *Annu. Rev. Biochem.*, 33, 571, 1964.
126. **Yamaha, T. and Cardini, C. E.**, The biosynthesis of plant glucosides. 1. Monoglucosides, *Arch. Biochem. Biophys.*, 86, 127, 1960.
127. **Smith, J. N.**, Ethereal sulphate and glucoside conjugations in insects, *Biochem. J.*, 60, 436, 1955.
128. **Dutton, G. J.**, The mechanism of O-aminophenyl glucoside formation in *Periplaneta americana*, *Comp. Biochem. Physiol.*, 7, 39, 1962.
129. **Trivelloni, J. C.**, Formation of β-glucosides in the locust *(Shistocerca cancellata)*, *Enzymologia*, 26, 329, 1963.
130. **Smith, J. N. and Turbet, H. B.**, Conjugations of 1-naphthol and some other phenols in houseflies and locusts, *Biochem. J.*, 92, 127, 1964.
131. **Darby, F. J., Heenan, M. P., and Smith, J. N.**, The absence of glucuronide conjugates from 1-naphthol dosed flies and grass grubs; detection of 1-naphthylphosphate, *Life Sci.*, 5, 1499, 1966.
132. **Binning, A., Darby, F. J., Heenan, M. P., and Smith, J. N.**, The conjugation of phenols with phosphate in grass grubs and flies, *Biochem. J.*, 103, 42, 1967.
133. **Boose, R. B. and Terriere, L. C.**, Quantitative aspects of the detoxication of naphthalene by resistant and susceptible house flies, *J. Econ. Entomol.*, 60, 580, 1967.
134. **Mehendale, H. M. and Dorough, H. W.**, In-vitro glucosylation of 1-naphthol by insects, *J. Insect Physiol.*, 18, 981, 1972.
135. **Yang, R. S. and Wilkinson, C. F.**, Conjugation of *p*-nitrophenol with sulfate in larvae of the Southern armyworm *(Prodenia eridania)*, *Pestic. Biochem. Physiol.*, 1, 327, 1971.
136. **Holder, G. M. and Ryan, A. J.**, Metabolism of propham (isopropyl-N-methylcarbamate) in the rat, *Nature* (London), 220, 77, 1968.
137. **Dodgson, K. S. and Rose, F. A.**, Sulfoconjugation and sulfohydrolysis, in *Metabolic Conjugation and Metabolic Hydrolysis*, Vol. 1, Fishman, W. H., Ed., Academic Press, New York, 1970.
138. **Bernstein, S. and Solomon, S., Eds.**, *Chemical and Biological Aspects of Steroid Conjugation*, Springer-Verlag, New York, Berlin, 1970.
139. **Yang, R. S. H. and Wilkinson, C. F.**, Enzymic sulphation of *p*-nitrophenol and steroids by larval gut tissues of the Southern armyworm (*Prodenia eridania* Cramer), *Biochem. J.*, 130, 487, 1972.

140. **Yang, R. S. H., Pelliccia, J. G., and Wilkinson, C. F.,** Age-dependent arylsulphatase and sulphotransferase activities in the Southern armyworm: a possible insect endocrine regulatory mechanism?, *Biochem. J.,* 136, 817, 1973.
141. **Yang, R. S. H. and Wilkinson, C. F.,** Sulfotransferases and phosphotransferases in insects, *Comp. Biochem. Physiol.,* 46, 717, 1973.
142. **Boyland, E., Kinder, C. H., and Manson, D.,** The biochemistry of aromatic amines, *Biochem. J.,* 78, 175, 1961.
143. **Heenan, M. P. and Smith, J. N.,** Conjugation of 1-naphthol and *p*-nitrophenol in flies: formation of glucoside-6-phosphates, *Life Sci.,* 6, 1753, 1967.
144. **Boyland, E. and Chasseaud, L. F.,** The role of glutathione and glutathione-S-transferases in mercapturic acid biosynthesis, *Adv. Enzym. Relat. Areas Mol. Biol.,* 32, 173, 1969.
145. **Ryan, A. J.,** The metabolism of pesticidal carbamates, *Crit. Rev. Toxicol.,* 1(1), 33, 1971.
146. **Bend, J. R., Holder, G. M., Protos, E., and Ryan, A. J.,** Water-soluble metabolites of carbaryl (1-naphthyl N-methylcarbamate) in mouse liver preparations and in the rat, *Aust. J. Biol. Sci.,* 24, 535, 1971.

Chapter 7

METABOLISM

INTRODUCTION

In less than ten years, the biochemistry of carbamate insecticides has developed from infancy into a mature and sophisticated science in its own right. It is difficult to imagine that in the published minutes of a meeting held by the IUPAC Commission on Terminal Residues as recently as 1966 it was stated that "A review of *Chemical Abstracts* citations for the past 10 years has revealed but a single citation each for the metabolism of carbamates in plants and mammals."[1] This statement referred to carbamates other than carbaryl, the other compound cited being Zectran. Even with carbaryl, the Commission listed less than a dozen reports involving plant and mammalian studies, and about half of these were residue studies using a colorimetric method of analysis. This situation did not remain very long, for in 1966 reports on carbamate metabolism began to appear regularly in the scientific literature. The pace increased rapidly, and now there are hundreds of published papers on carbamate metabolism. The subject is one of the most often reviewed areas of pesticide science and is a standard topic in general pesticide metabolism discussions.[2-10] Moreover, metabolism of carbamate insecticides is included in specific reviews dealing with carbamate pesticides of all types,[11,12] and many reviews consider the insecticidal carbamates exclusively.[13-22]

Metabolism studies of carbamate insecticides are conducted for a variety of reasons, as they are for any other group of toxic compounds. First, it is important to recognize that the carbamate insecticides are designed to be toxic to pest species and that this toxicity is unfortunately, not restricted to the target organism. This being the case, it is necessary to determine the manner in which these insecticides may be used so that their desirable insecticidal characteristics are expressed without inflicting hazards upon man and the environment.

The fact that metabolic processes can convert a toxic compound into another which still may be toxic, transform it to a product having greater toxicity, or degrade the compound to virtually innocuous products, clearly negates any idea that the toxicological significance of a compound can be evaluated solely by determining the toxicity and residual nature of the applied compound. It is imperative that the chemical transformations which toxicants undergo be defined and that the toxicity of individual transformation products be determined and included in any evaluation relative to the safe use of the applied compounds. Thus, a major reason for studying metabolism of carbamate insecticides is to ensure their safe use, and even those studies not designed specifically for this purpose are useful in this regard. For example, many of the in-vitro investigations conducted with both animal and plant systems often give the first indication of the types of metabolites which might be formed under practical conditions. In addition, these studies may signify, either directly or indirectly, compounds that have a high potential for being toxicologically significant.

Metabolites which retain the carbamate ester moiety are generally considered of more potential importance than those resulting from hydrolysis of the carbamate linkage. This is based on the observation that hydrolysis destroys the anticholinesterase activity of the carbamates; consequently, the hydrolysis products are expected to be less toxic than metabolites structurally similar to the initial compound. In-vitro studies also assist in the development of techniques which allow for the detection and quantitation of metabolites. In this way, the same metabolites can be included in the in-vivo evaluation of the metabolism of the toxic agent under consideration.

One should bear in mind, however, that in-vitro studies, by nature and design, are limited in their metabolic capability and do not represent, especially quantitatively, the total metabolic pathway of a compound in living organisms. This is particularly true when using microsomal enzymes fortified with NADPH, where the system is geared to yield predominantly oxidative metabolites. Results from enzyme preparations might indicate that toxicologically significant metabolites occur at high levels in vivo when, in fact, the living organisms may metabolize the compound predominantly by an alternate route, such as hydrolysis, or may degrade a metabolite so rapidly that the in-vitro pathway would be of no significance. Therefore, while extremely useful biochemical tools, in-vitro studies should not be used to support claims of safety or hazards of candidate pesticide materials. The final evaluation should be based on detailed examinations of the fate of the toxicants in living organisms.

Another very important reason for studying metabolism of insecticides is to determine the role of chemical transformation in the efficacy and selectivity of the materials. Although the carbamate insecticides are generally metabolized by the same major pathways in vertebrates, insects, plants, and microorganisms the rate of metabolic degradation is species-dependent. This rate difference can very definitely affect the toxicity of a compound, and it can be very desirable when nontarget species, including man, are capable of rapidly detoxifying the insecticide. On the other hand, rapid metabolism by the target organism would be undesirable, as is the case when pest species develop an enhanced capability of detoxifying an insecticide and thereby become resistant to its action, and possibly to the action of other insecticides as well (see Chapter 10). Although there has been little success in designing selective insecticides based on different metabolic rates of different organisms, it is a meritorious objective, and one that should not be ignored.

Although safety and selectivity of carbamate insecticides are important reasons for studying their metabolism, this explains only partially the vast accumulation of literature on this subject during the past ten years. One must admit that the success of the carbamates as commercial insecticides has been less than overwhelming when compared to the total number of compounds evaluated. Yet this dearth is not the result of lack of investigations into the biochemistry of the carbamate insecticides. Therefore, it must be assumed that this group of chemicals has provided a challenging and intriguing area of research for scientists who have a basic interest in the biological fate of xenobiotics.

In the following discussion, the metabolism of carbamate insecticides is considered on an individual compound basis, providing a ready and we hope, complete reference to specific compounds or to compounds with similar chemical characteristics. The insecticides are identified by the name most commonly referred to in the literature, whether a trade name, common name, or other designation. Although alternate designations are also given, this is not intended as a complete listing, but rather reflects those names which came to the authors' attention during a review of the literature on carbamate metabolism. Where possible, the chemical identity of the metabolites is given, and the percentage distribution of the biotransformation products is tabulated, so that the magnitude of each material can be realized.

DIMETHYLCARBAMATES (Dimetilan, Isolan, and Pyramat)

Dimethylcarbamate insecticides developed thus far are of little practical utility, and therefore little information exists on the metabolism of these insecticides in either plants or animals. It is significant, however, that our present knowledge of the non-hydrolytic pathway of carbamate insecticide metabolism was first suggested by studies with the dimethylcarbamates. Several insecticidal compounds were among 45 N,N-dimethylcarbamates which were metabolized by rat liver enzymes to form a product that could be extracted into organic solvent and that decomposed in strong acid to yield formaldehyde.[23,24] This very simple but effective experimental approach provided strong evidence that N-dealkylation of the dimethylcarbamates proceeded through an N-methyl-N-hydroxymethyl intermediate and, more important, demonstrated that some of these intermediates were very effective inhibitors of cholinesterase.

The enzyme system responsible for oxidative metabolism was localized in the microsomes, and the production of the formaldehyde-yielding metabolite was enhanced by addition of NADPH. As was later proven, these early studies indicated that N-hydroxymethyl metabolites of the dimethylcarbamates were formed in vivo by both insects and mammals. It was also demonstrated that the monomethylcarbamate carbaryl underwent the same type of metabolism and that the process of N-methyl hydroxylation was inhibited by several insecticide synergists. These early findings indicated that hydrolysis possibly was not the only metabolic route for the carbamate insecticides, as had been previously assumed, and that other toxicologically significant metabolites of carbamates might exist.

Structures of some dimethylcarbamates whose

metabolism has been investigated are pictured in Figure 7.1. In-vitro studies with rat liver microsomes showed that ^{14}C-carbonyl-labeled Isolan (1-isopropyl-3-methyl-5-pyrazolyl N,N-dimethylcarbamate) was almost completely metabolized.[25] Only 3% of the added radioactivity was recovered as Isolan; 24% was recovered as ether-soluble metabolites consisting of one major and five minor products, none of which were identified; 40% was converted to water-soluble metabolites; and 33% of the added radiocarbon was lost. Each of the ether-extractable metabolites was less active than Isolan as a cholinesterase inhibitor.

Dimetilan-carbonyl-^{14}C (2-dimethylcarbamoyl-3-methyl-5-pyrazolyl N,N-dimethylcarbamate) was not metabolized by the rat liver enzymes as thoroughly as was Isolan. Of the added radioactivity, 36% was recovered as dimetilan, 17% as ether-extractable metabolites, and 40% as water-soluble materials; there was a 10% loss of radiocarbon during extraction, but none during incubation, suggesting that ^{14}C-carbon dioxide was not liberated. Eight metabolites were detected in the ether-extractable fraction, but the only one identified was 2-methylcarbamoyl-3-methyl-5-pyrazolyl N,N-dimethylcarbamate (Figure 7.2, I), which occurred in relatively small quantities. Although the other metabolites were not identified, it was established that neither 2-dimethylcarbamoyl-3-methyl-5-pyrazolyl N-methylcarbamate nor 3-methyl-5-pyrazolyl N,N-dimethylcarbamate was formed by microsomes. The identified metabolite and the major ether-extractable material were equal to dimetilan as anticholinesterase agents; the other materials were poorer cholinesterase inhibitors.

There was less hydrolysis of dimetilan than of Isolan when the ^{14}C-carbonyl-labeled carbamates were injected into rats.[26] With dimetilan, 49% of the dose was exhaled as ^{14}C-carbon dioxide

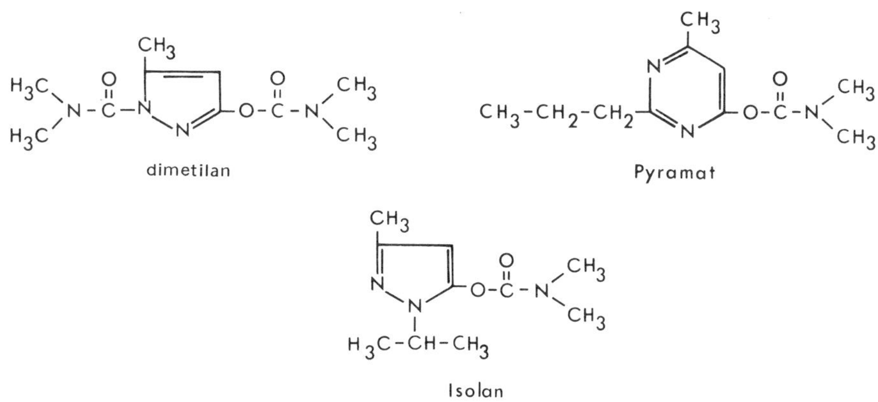

FIGURE 7.1. Dimethylcarbamate insecticides whose metabolism has been investigated to some degree.

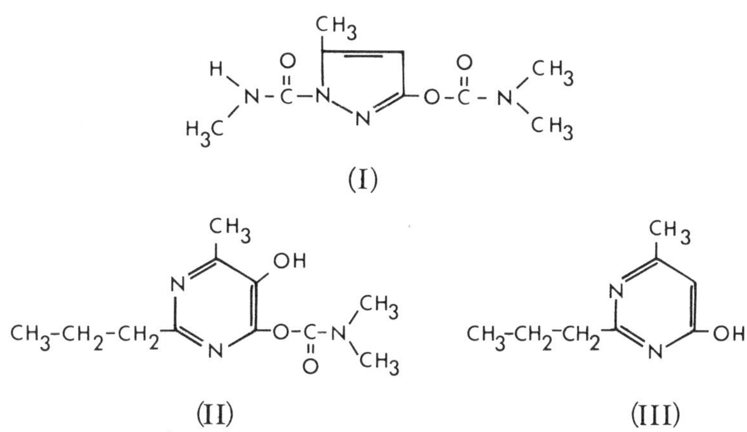

FIGURE 7.2. Dimetilan metabolite (I) formed by rat liver enzymes, and Pyramat metabolites (II, III) isolated from the urine of rats.

(Figure 7.3). The amount of urinary ^{14}C metabolites from dimetilan was approximately twice that observed with Isolan, making the total elimination of applied radiocarbon essentially the same, 96 to 99%, for both compounds.

Six days after injection of ^{14}C-carbonyl-labeled dimetilan or Isolan into bean plants, approximately 25% of the applied radiocarbon had been converted to water-soluble metabolites.[27] Treatment with β-glucosidase cleaved 67% of the water-soluble Isolan metabolites, but only 26% of the dimetilan materials. Six aglycones were recovered from each of the water-soluble preparations by enzyme treatment, but none were identified.

Pyramat (4-methyl-2-propyl-6-oxypyrimidyl N,N-dimethylcarbamate) was metabolized by mice, rats, guinea pigs, and rabbits via ring hydroxylation to form 4-methyl-2-propyl-5-hydroxy-6-oxypyrimidyl N,N-dimethylcarbamate (Figure 7.2, II), and by hydrolysis of the carbamate ester to form 4-methyl-2-propyl-6-hydroxy pyrimidine (Figure 7.2, III).[28] The decarbamylated analog of metabolite II was also produced by the animals.

The metabolism of dimetilan-carbonyl-^{14}C by insects yielded nine radioactive metabolites in German cockroaches, six in American cockroaches, and four in houseflies.[29] Although the identification of the metabolites was not sufficient to allow formulation of a metabolic pathway, N-methyl hydroxylation appeared to be a major mechanism in the degradation of the insecticide by all species. Two metabolites tentatively identified were 2-dimethylcarbamoyl-3-methyl-5-pyrazolyl N-hydroxymethyl-N-methylcarbamate and 2-methylcarbamoyl-3-methyl-5-pyrazolyl N,N-dimethylcarbamate. Evidence was obtained that three of the other metabolites were also N-hydroxymethyl derivatives.

MONOMETHYLCARBAMATES
CARBARYL
(Sevin, UC 7744, Denapon, Karbosep, Carpolin, NAC)

Carbaryl was synthesized in 1953,[30] and introduced as a commercial insecticide in 1958.[31] Its success as an insecticide has been truly phenomenal, and for many years it constituted almost the sole component of the class of carbamate commercial insecticidal compounds. In the United States alone, carbaryl is registered for use in controlling more than 160 different insect pests on 85 food and fiber crops.[32] It is also effective in controlling pests on a wide range of ornamental plants, pets, poultry, and in buildings. On a world-wide basis, the use of carbaryl is extended to more than 100 different crops. In 1965 it was reported that approximately 40% of the millions of pounds used around the world was applied to cotton.[31] The other two major uses of carbaryl are for the control of pests on soybeans and on corn.

As one might expect, the metabolism of carbaryl has been the subject of most investigations dealing with carbamate insecticides. This does not mean, however, that its metabolic fate is better known than that of some newer carbamate materials. In fact, the metabolism of carbaryl is extremely complex, and there is still much to be learned about the biochemistry of this compound.[18]

As indicated on previous pages, it was first assumed that hydrolysis of the carbamate ester

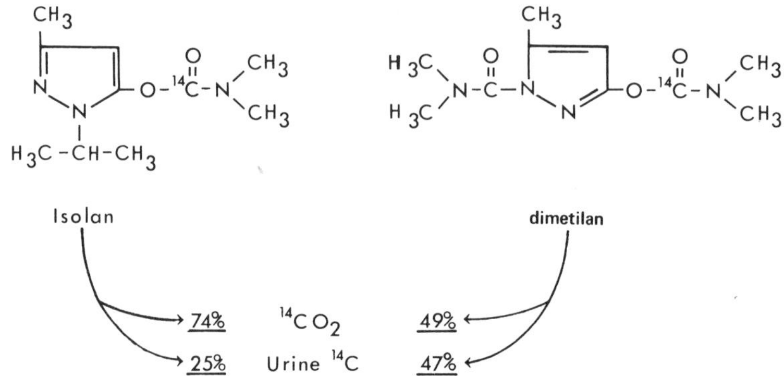

FIGURE 7.3. ^{14}C-carbon dioxide exhaled and urinary ^{14}C (percent of dose after 48 hr) from rats treated with ^{14}C-carbonyl-labeled Isolan or dimetilan.[26]

linkage was the predominant metabolic pathway.[33] As a result of this line of thinking, a great deal of emphasis was placed on determining the important role hydrolysis played in the mechanism of action and efficacy of a carbamate.[34-37] The metabolic pathway of carbaryl as envisioned prior to the mid 1960's was a simple process of hydrolysis, yielding 1-naphthol, carbon dioxide, and methylamine (Figure 7.4). In view of the data available at that time, methods for the detection of carbaryl residues were based on the colorimetric quantitation of 1-naphthol before and after alkaline treatment of the substrate being analyzed.[38-40]

Although it was recognized that methyl hydroxylation, ring hydroxylation, and conjugation before or after hydrolysis of the carbamate moiety were known pathways for certain biologically active carbamates,[33,41] the first hint that such mechanisms were involved in the metabolism of carbaryl was not reported until 1960.[23,24] Carbaryl was shown to be metabolized by a rat liver microsome system, requiring NADPH and oxygen, to a formaldehyde-yielding derivative without loss of anticholinesterase activity. The data implied that carbaryl was oxidatively metabolized to the N-hydroxymethyl derivative (Figure 7.5), a pathway later confirmed with carbaryl and other monomethyl and dimethylcarbamates.

That a non-hydrolytic pathway of carbaryl metabolism did indeed exist was further demonstrated by using separate preparations of carbaryl radiolabeled with carbon-14 at three different sites on the molecule: at the 1-position of the naphthyl ring,[42] on the carbonyl carbon, and on the N-methyl carbon.[43] Each radiolabeled sample of carbaryl was subjected to identical in-vivo and/or in-vitro experimental conditions, and the presence or absence of each labeled carbon on a particular metabolite was determined. Such multilabeling is now routinely used when the metabolic pathways of carbamate and other insecticides are studied. These techniques demonstrated that a variety of carbaryl metabolites were produced in mammals, plants, and insects, and that many of the metabolites were formed from each radiocarbaryl preparation, thereby demonstrating that hydrolysis of the carbamate acid ester did not precede their formation.[44,45] Non-hydrolytic metabolites tentatively identified were 4-hydroxy-1-naphthyl N-methylcarbamate (Figure 7.6, I), 5-hydroxy-1-naphthyl N-methylcarbamate (Figure 7.6, II), 1-naphthyl N-hydroxymethylcarbamate (Figure 7.5) and 3,4-dihydro-3,4-dihydroxy-1-naphthyl N-methylcarbamate. Subsequent studies showed that the latter metabolite was more likely 5,6-dihydro-5,6-dihydroxy-1-naphthyl N-methylcarbamate (Figure 7.6, III),[46] and in 1972, chemical synthesis of this metabolite confirmed its identification.[47]

Water-soluble metabolites of carbaryl have been shown to consist largely of conjugates of the hydroxylated carbamate metabolites shown in Figures 7.5 and 7.6. In mammals they appear to be present primarily as glucuronide and sulfate conjugates.[46,48,49] At one time it was proposed that direct coupling of carbaryl with glucuronic acid yielded 1-naphthyl N-methylcarbamate N-glucuronide and 1-naphthyl methylimidocarbonate O-glucuronide.[48] However, the latter metabolite has been identified as the glucuronide of the 5,6-dihydrodihydroxy derivative of carbaryl (Figure 7.6, III).[50] The isolation and characterization of carbaryl N-glucuronide has never been achieved. In plants the same hydroxylated carbaryl metabolites probably conjugate primarily as glycosides,[13-15,27,51,52] and in insects evidence favors formation of a combination of sulfate, phosphate, and sugar conjugates.[14]

Though there is ample documentation that carbaryl is metabolized by oxidative mechanisms, it must not be assumed that hydrolysis is unimportant in the degradation of this carbamate. Most animals readily hydrolyze a portion of the administered carbaryl,[18-22] although hydrolysis does not seem to be a significant pathway in the metabolism of carbaryl in monkeys or pigs.[53] Partial hydrolysis of carbaryl was reported to occur in cotton plants,[54] but it is not a major

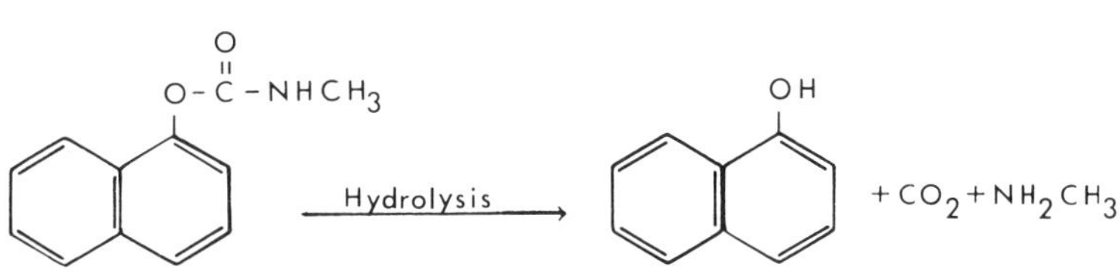

FIGURE 7.4. Hydrolytic pathway for carbaryl.

pathway of metabolism in plants; in fact, the carbamate ester linkage appears to be quite stable.[27] In insects, the degree of hydrolysis of the carbamate ester is dependent upon the species involved.[14] However, oxidative metabolism is usually the predominant pathway of metabolism.

In addition to the non-hydrolytic metabolites of carbaryl mentioned in the discussion above, several others have been identified from various organisms. These metabolites, and the organisms in which they are formed, are included in the following presentation of specific studies of carbaryl metabolism.

Metabolism of Carbaryl in Vertebrates

Degradation studies of carbaryl using subcellular liver preparations have played a major role in determining the metabolic pathway of this carbamate in animals. Several investigations revealed that carbaryl was metabolized by hepatic enzymes primarily to 1-naphthyl N-hydroxymethylcarbamate, 4-hydroxy-1-naphthyl N-methylcarbamate, and 5-hydroxy-1-naphthyl N-methylcarbamate, and to conjugates of these hydroxylated materials. Maximum rates were achieved with the microsome plus soluble fractions of rat liver in the presence of NADPH.

The nature of the conjugate metabolites was also uncovered in early studies using rat liver enzymes when it was shown that both hydrolysis products and hydroxylated carbamate metabolites were probably conjugated as glucuronides and sulfates.[46] Glucuronide conjugation appeared to be of greatest significance, except with 1-naphthyl N-hydroxymethylcarbamate and 5,6-dihydro-5,6-dihydroxy-1-naphthyl N-methylcarbamate. More recent studies have shown that the N-hydroxymethyl derivative of carbaryl was almost 90% conjugated with glucuronic acid after incubation with rat liver microsomes fortified with UDPGA.[55] However, this system would not conjugate the 5,6-dihydro-5,6-dihydroxy metabolite. Thus the earlier report[46] suggesting that this metabolite would not form a glucuronide under in-vitro conditions may have played a part in the later erroneous identification of 1-naphthyl methylimidocarbonate O-glucuronide as a urinary metabolite[48] rather than the glucuronide of 5,6-dihydro-5,6-dihydroxy-1-naphthyl N-methylcarbamate.[50] It is not clear why the in-vitro glucuronidation of this carbaryl metabolite was unsuccessful, since the 5-hydroxy, 6-hydroxy, and N-hydroxymethyl derivatives of carbaryl, as well as 1-naphthol, were all conjugated by liver microsomes containing UDPGA.[55]

In addition to glucuronide and sulfate conjugates, evidence exists for the occurrence of amino acid conjugates of carbaryl metabolites.[56] However, further proof is needed before positive structural assignments are made for these water-soluble metabolites of carbaryl.

In studies comparing the degradation of carbaryl by human and rat liver preparations, some of the results imply that metabolism is almost the same in the two species,[57,58] whereas other reports show that the human liver metabolizes very small quantities of carbaryl and that 1-naphthol is the major product formed.[59] A more promising method for estimating species variation in the metabolism of carbaryl and other insecticides has been developed, utilizing liver tissue or profused whole liver.[60-62] With these organ-maintenance techniques, variations encoun-

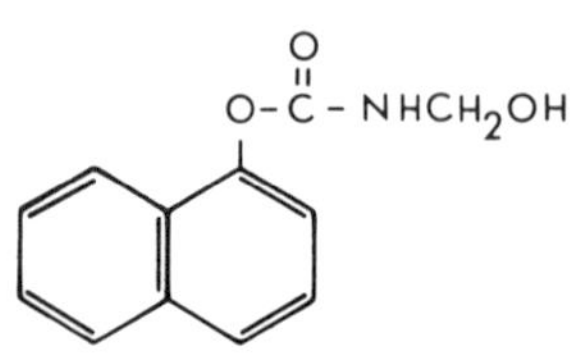

FIGURE 7.5. Structure of 1-naphthyl N-hydroxymethylcarbamate; the first proposed non-hydrolytic metabolite of carbaryl.

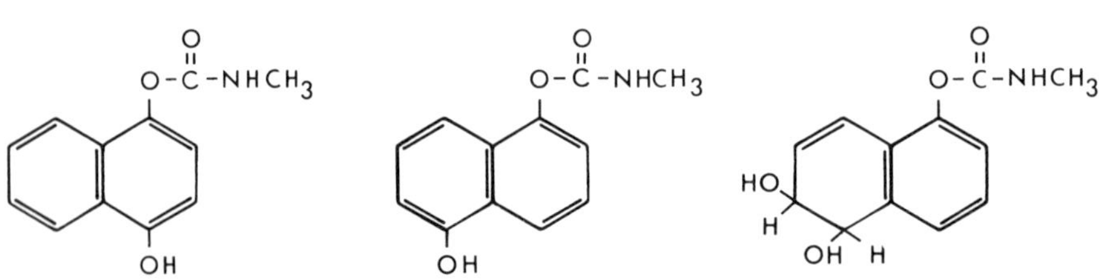

(I) (II) (III)

FIGURE 7.6. Commonly found oxidative metabolites of carbaryl.

tered in the metabolism of carbaryl by different livers of living animals can be more closely reproduced in an in-vitro setting. Indeed, the technique has recently uncovered some variations in metabolism of carbaryl by different isolated human tissues.[63]

The use of mammalian cells in culture has also been employed to determine the metabolism of carbaryl by different species and by different animal organs.[64-68] However, the agreement of results obtained by the cell culture method has not been especially good in comparison with results obtained by in-vivo studies. For example, human embryonic lung cells metabolized carbaryl primarily by hydrolysis to 1-naphthol.[68] Also, cubes of rat liver formed a compound that has tentatively been identified as the N-O-conjugate of N-hydroxycarbaryl.[66] This metabolite (1-naphthyl N-hydroxy-N-methylcarbamate) has also been reported by the same researcher[66] as being formed by tobacco cells, and it was pointed out that N-hydroxy metabolites are frequently carcinogenic and/or mutagenic.[67] However, the premature suggestion that 1-naphthyl N-hydroxy-N-methylcarbamate may pose a serious toxicological hazard to man, simply on the basis of its proposed similarity to 1-naphthylamine,[66] is strictly a hypothesis. Authentic 1-naphthyl N-hydroxy-N-methylcarbamate has been used as a metabolite standard in both in-vivo and in-vitro studies. It was not formed by liver microsomes,[45,69] and it was not present as the glucuronide in the urine of rats, guinea pigs, or man.[48]

Several in-vitro systems have shown that animals are entirely capable of hydrolyzing carbaryl. For example, rabbit plasma very effectively hydrolyzed carbaryl and was more than twice as active in this regard as were comparable fresh weights of a series of tissues, including liver, kidney, brain, and erythrocytes.[35] The most active plasma was that from swine, rabbits, and sheep; human plasma was only slightly less active; and plasma from mice and horses was the least active. Human sera hydrolyzed carbaryl and 3-isopropoxyphenyl N-methylcarbamate, but they did not hydrolyze Baygon (2-isopropoxyphenyl N-methylcarbamate).[70,71] Hydrolysis of the carbamates was not affected by heat treatment of the whole serum or of an isolated active fraction of the sera.

Carbaryl was hydrolyzed by two of the 14 human brain esterases separated by thin-layer agar gel electrophoresis techniques.[72] Three esterases isolated from mouse brain, which differed in chromatographic behavior from the two active human esterases, also hydrolyzed carbaryl.[73] One of the human esterases operating on carbaryl was shown to degrade malathion (diethyl mercaptosuccinate S-ester with O,O-dimethylphosphorodithioate) at the carboxyl ester linkage. Two other human brain carboxyl esterases did not hydrolyze carbaryl.

Penetration rates of carbaryl and other insecticides through the isolated midgut of insects and mice showed that no pattern existed which could explain the selective action of the insecticide between insects and mammals.[74,75] Movement of carbaryl through the colon was more rapid than through the small intestine or rectum.[76] With respect to age, absorption through the small intestine was the same when the intestine was isolated from mice ranging in age from 12 days to 12 weeks. In the case of rats, it has been demonstrated that carbaryl is probably metabolized to some degree before absorption from the intestine. The predominant metabolite found in the incubation mixtures was 1-naphthyl glucuronide.[77-79]

To determine the gastric absorption of 1-naphthyl-^{14}C carbaryl, the pylorus of rats was ligated, and an intragastric dose of the insecticide was administered.[80] After 22 min 53% of the dose was absorbed from the stomach, and after 67 min 82% was absorbed. Similar treatment of rats with the same compound and collection of the portal blood showed that 66 min after application 70% of the dose was absorbed from the stomach, with 67% present in the blood. Therefore, 97% of the total absorbed radioactivity was recovered in the portal blood. Analysis of benzene extracts of this blood yielded 89% of the radiolabeled material in the form of the administered compound. In normal female rats, approximately 85% of an oral dose of ratioactive carbaryl was absorbed from the intestinal tract after 24 hr. This was reduced to 50% when the pylorus was ligated, demonstrating that much of the absorption occurred in the intestine.[81]

Turning now to in-vivo degradation, it is evident that the metabolic fate of carbaryl in different living animals has generally shown little species variation. An examination of the fate of various radioactive preparations of carbaryl provides an excellent measure of the rate of metabolism and excretion as well as some insight into the nature of the metabolites formed by the animals. Since 1-naphthyl-^{14}C-labeled carbaryl is not degraded in the animal system to appreciable amounts of ^{14}C-carbon dioxide, results of studies using this radiolabeled preparation of carbaryl best reflect the excretion rate from different animals (Table 7.1).

Most animals excrete between 65 and 75% of an administered dose of 1-naphthyl-^{14}C carbaryl in the urine within 24 hr after treatment. This, of course, represents both hydrolytic and non-hydrolytic metabolites of carbaryl. The dog is unique in eliminating only 30% of the dose in the urine and excreting 15% in the feces; usually, only 5 to 10% of an administered dose is found in the feces of animals. Maximum excretion of carbaryl was noted in chickens, where 94% was eliminated in the urine after 24 hr, and the remainder of the dose was removed via the feces.

Assuming that all of the radioactive metabolites in the urine of animals treated with carbonyl-^{14}C carbaryl are carbamate-type metabolites, the data from different animal studies show that, of the 1-naphthyl-^{14}C carbaryl equivalents in the urine, approximately half is composed of hydrolytic products and half of carbamate metabolites (Table 7.1). In chickens, however, much more of the ^{14}C-residues in the urine were composed of hydrolytic products than was generally observed in other animals. This increased rate of carbaryl hydrolysis by chickens obviously enhanced the rate of elimination of radiocarbon.

TABLE 7.1

Elimination of ^{14}C-Carbaryl Equivalents from Animals

Position of ^{14}C label	Animal	% of orally administered dose after 24 hr: $^{14}CO_2$	Urine	Feces	Reference
1-Naphthyl-1-^{14}C	Rat	0	70	8	48
1-Naphthyl-1-^{14}C	Rat	–	68	–	50
1-Naphthyl-1-^{14}C	Rat	0.2	74	9	26
N-Methyl-^{14}C	Rat	5	63	4	48
N-Methyl-^{14}C	Rat	–	70	–	50
N-Methyl-^{14}C	Rat[a]	12	54[b]	4	26
N-Methyl-^{14}C	Rat[a]	24	53[b]	–	49
Carbonyl-^{14}C	Rat	30	45	8	48
Carbonyl-^{14}C	Rat	26	72[b]	4	26
Carbonyl-^{14}C	Rat[a]	30	34	–	49
1-Naphthyl-1-^{14}C	Dog	–	30	15	86
N-Methyl-^{14}C	Dog	–	10	5	86
1-Naphthyl-1-^{14}C	Pig	–	68	2	53
N-Methyl-^{14}C	Pig	–	56	1	53
1-Naphthyl-1-^{14}C	Sheep	–	70	3	53
N-Methyl-^{14}C	Sheep	–	52	5	53
Carbonyl-^{14}C	Goat	–	45	–	45
1-Naphthyl-1-^{14}C	Cow	–	68	11	88
Carbonyl-^{14}C	Cow	–	30	–	90
1-Naphthyl-1-^{14}C	Chicken	0	94	7	90a
Carbonyl-^{14}C	Chicken	49	31	6	90a
Nonradioactive	Man	–	38	–	90b
Nonradioactive	Man	–	28	–	53

[a]Injected.
[b]After 48 hr.

Quantitation of ^{14}C-carbon dioxide from animals treated with carbonyl-^{14}C carbaryl is also an excellent method of estimating the rate and degree of hydrolysis of carbaryl in animals. Usually, the amount of carbon dioxide exhaled very closely approximates the difference in the radioactive residues in the urine of animals treated with 1-naphthyl-^{14}C carbaryl and with carbonyl-^{14}C carbaryl. Though it has been shown that, following hydrolysis of ^{14}C-carbonyl-labeled carbaryl and carbofuran, a small quantity of ^{14}C-carbon dioxide was incorporated into natural milk constituents,[82,83] it is obvious that most of the ^{14}C-carbon dioxide produced in the animal is exhaled.

Studies using N-methyl-^{14}C-labeled carbaryl have shown that radioactivity is expired in the exhaled air. However, the amount of material collected varies considerably more than that observed in different studies with carbonyl-^{14}C carbaryl; this may mean that not all of the trapped ^{14}C-material is actually carbon dioxide.

The variation may be due to the efficiency of the carbon dioxide traps used in different studies to collect volatile components other than carbon dioxide. This is a point that should be clarified, since production of "^{14}C-carbon dioxide" from N-methyl-^{14}C carbamates is often used to denote the degree of N-demethylation of these insecticides. There is strong evidence that ^{14}C-methylamine hydrochloride injected into rats is expired as ^{14}C-carbon dioxide, and not as methylamine.[26] Thus, one could not tell whether the conversion of the radioactive carbon to ^{14}C-carbon dioxide occurred before or after hydrolysis of the carbamate ester, even if ^{14}C-carbon dioxide is liberated from N-methyl-^{14}C carbamates.

Results from in-vitro studies comparing ^{14}C-carbonyl- and N-methyl-^{14}C carbaryl tend to indicate that the production of ^{14}C-carbon dioxide from N-methyl-^{14}C carbaryl in the living animal probably occurs subsequent to hydrolysis of the ester linkage.[69] Both radiolabeled preparations are extensively metabolized and form the same oxidative-type metabolites, the N-hydroxymethyl derivative being the major metabolic product. The fact that there was no loss of radiocarbon during incubation with N-methyl-^{14}C carbaryl implies that 1-naphthyl N-hydroxymethyl-^{14}C is not metabolized to ^{14}C-carbon dioxide subsequent to its formation in this in-vitro system.

The nature of the metabolites in rat urine has not been studied in detail. However, the results of the available data (Table 7.1) show that from 85 to 90% of the urinary ^{14}C-metabolites are water-soluble. Based on the behavior of the products on DEAE-cellulose columns, the metabolites consist of sulfate and glucuronide conjugates of 1-naphthol (Figure 7.7, IV), 4- and 5-hydroxy-1-naphthyl N-methylcarbamate (Figure 7.7, V), and the glucuronide of 5,6-dihydro-5,6-dihydroxy-1-naphthyl N-methylcarbamate (Figure 7.7, VI).[48,50] Because the metabolites as resolved by DEAE-cellulose chromatography are not composed entirely of the designated components,[57,84] it is not possible to determine the absolute quantity of each material.

The metabolite identified as the glucuronide of 5,6-dihydro-5,6-dihydroxy carbaryl[48] has been subjected to further analyses and was shown to contain only 40% of the designated material.[50] From this study it was concluded that up to 10% of a carbaryl dose was eliminated in the urine of rats in the form of this glucuronide conjugate.

The failure of the water-soluble metabolites to be converted completely to ether-extractable materials by enzyme and weak-acid treatment has led to the speculation that some of the metabolites might be the N-glucuronide of carbaryl (Figure 7.8, VIII)[48] and various thioether amino acid conjugates of the type shown in Figure 7.8 (IX).[56] The existence of an N-glucuronide of carbaryl has not been confirmed, but, in fact, evidence has been presented that suggests another structure for this metabolite. The original identification of the N-glucuronide of carbaryl was based largely on the fact that a metabolite tentatively identified as 1-naphthyl methylimidocarbonate O-glucuronide (Figure 7.9, X) could be converted to a similar material, thought to be the N-glucuronide conjugate (Figure 7.9, XI).[48] It was later demonstrated, however, that the tentatively identified 1-naphthyl methylimidocarbonate O-glucuronide was actually the glucuronide of 5,6-dihydro-5,6-dihydroxy-1-naphthyl N-methylcarbamate (Figure 7.9, VI).[50] Although not stated by the authors,[50] this would negate the formation of carbaryl-N-glucuronide and would suggest the formation of free 5,6-dihydro-5,6-dihydroxy-1-naphthyl N-methlycarbamate, as shown in Figure 7.9 (III).

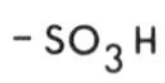

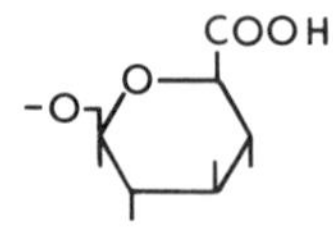

X Y

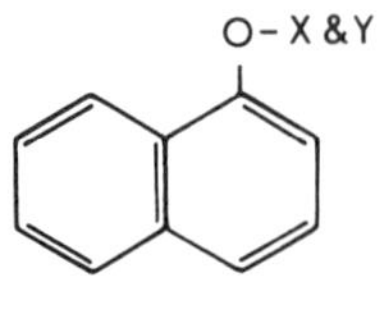

(IV)

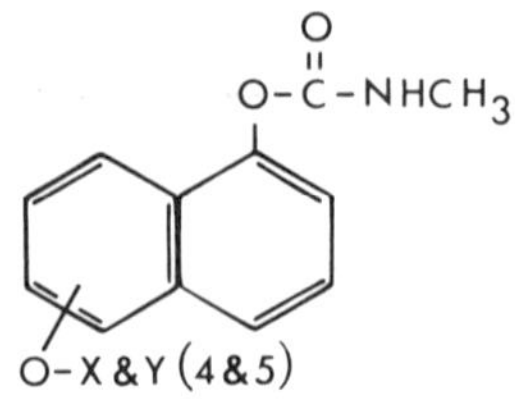

(V)

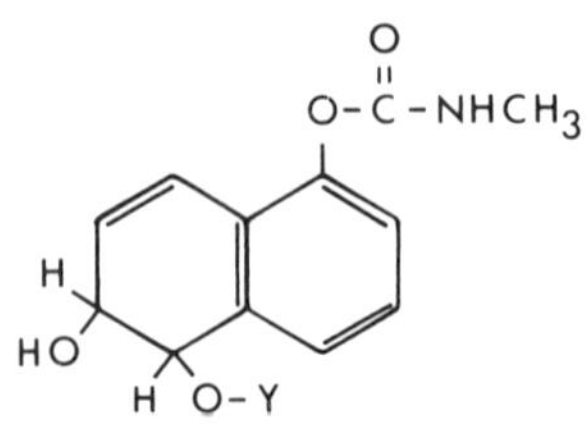

(VI)

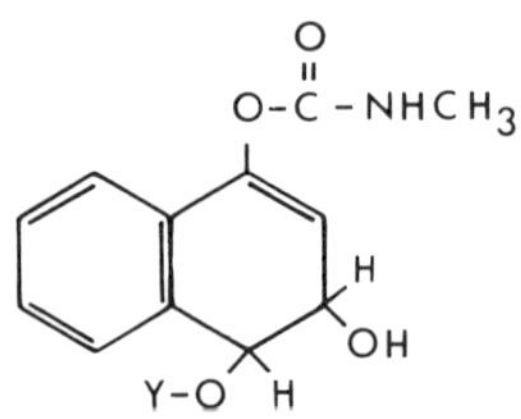

(VII)

FIGURE 7.7. Carbaryl metabolites found in the urine of different animals. The majority of materials are usually in the conjugated form, as shown, but they also exist at low levels in the free form.

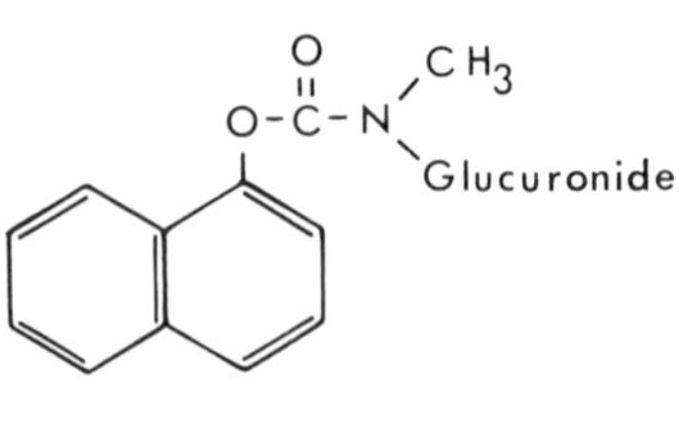

(VIII)

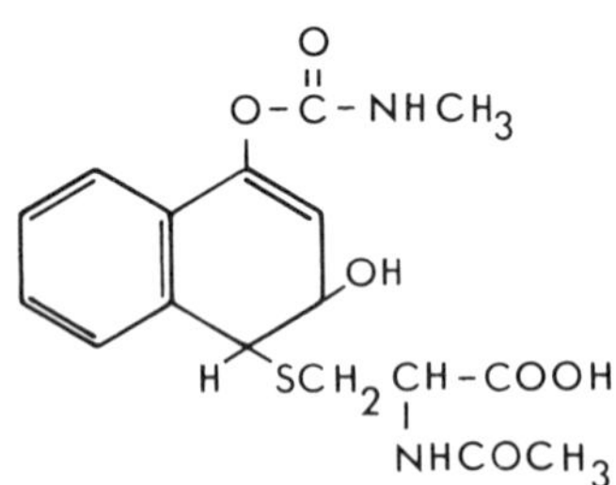

(IX)

FIGURE 7.8. Proposed but unconfirmed metabolites of carbaryl found in the urine of rats.

The latter material did elute from the DEAE-cellulose column in the neutral fraction[48] rather than from the acid fraction, as would have been the case if it had been in the form of a glucuronide.

The possibility that glutathione conjugates and mercapturic acids are formed from certain ring-hydroxylated metabolites of carbaryl is certainly viable. However, research conducted,[11,56] which suggested the existence of this type of carbaryl metabolite, is not convincing. First, much of the evidence is based on the fact that enzyme treatment did not cleave the water-soluble metabolites, or at least render them ether-soluble. Without proof that the hydrolases were active in cleaving known substrates under the conditions employed, this type of evidence is incomplete. A second bit of corroboration presented to show that the metabolites were thioether amino acid conjugates was that the water-soluble metabolites were not converted to ether-soluble materials by acid hydrolysis. The experimental conditions employed involved heating the solutions for 3 hr after adjusting them to an acidity of 3*N* with hydrochloric acid. Under these conditions, 5,6-dihydro-5,6-dihydroxy- and 3,4-dihydro-3,4-dihydroxy-1-

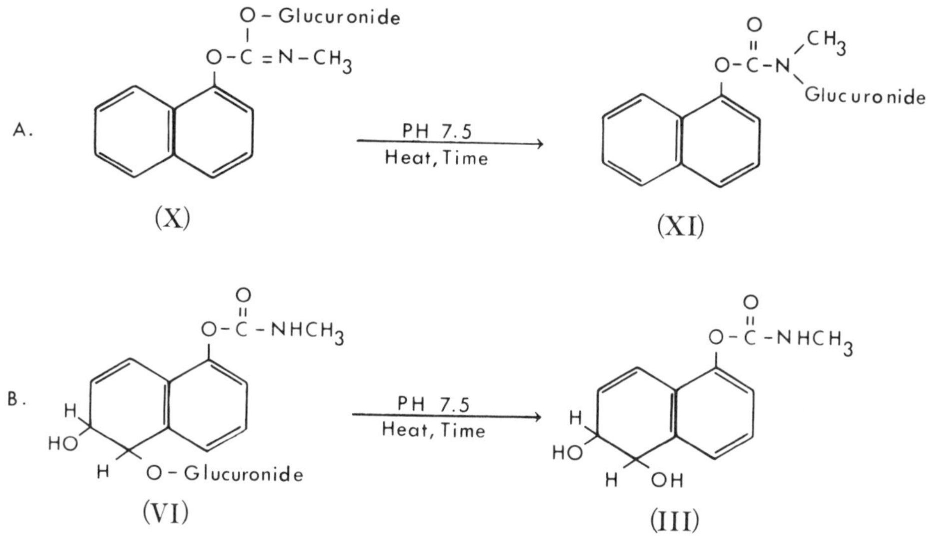

FIGURE 7.9. A. Urinary metabolite of carbaryl (X) and method of formation of carbaryl N-glucuronide as proposed in 1965.[48] B. Probable actual reaction, based on identification of metabolite X as metabolite VI in 1972.[50]

naphthyl N-methylcarbamate and their hydrolytic products would, if present, be almost completely degraded to materials that could not be extracted from the aqueous phase with ether.[85] It is quite likely that some of these dihydroxy materials were present in the original water fractions, because ether extraction removes only a portion of the 5,6-dihydrodihydroxy derivatives, and even less of the 3,4-dihydrodihydroxy metabolites. Since 5,6-dihydro-5,6-dihydroxy-1-naphthyl N-methylcarbamate has been shown to exist in rat urine in both the free and conjugated form,[50] and since both forms were shown to be unextractable prior to treatment of the urine with acid, it is evident that at least some of the metabolites identified as thioether amino acid conjugates[11,56] actually consisted of degradation products of the dihydrodihydroxy metabolites of carbaryl.

Generally, the nature of the metabolites in the urine of sheep, guinea pig, monkey, pig, and man is similar to that in rat,[48,53] although the monkey and the pig hydrolyze very little of the administered carbaryl. The dog, however, is totally unlike the rat insofar as the identity of the metabolites eliminated in the urine is concerned.[86] None of the major metabolites found in rat urine were present in the urine of dogs. Unfortunately, characterization of the metabolites in the dog urine was not accomplished.

Metabolites in the urine of poultry consisted primarily of sulfate and of glucuronide conjugates of 1-naphthol and hydroxylated naphthalene.[87] The major difference between the products in urine from chickens and from rats was that the chicken urine contained no 5,6-dihydro-5,6-dihydroxy carbaryl metabolites. However, considerable amounts of 5,6-dihydroxy derivatives, both carbamate and hydrolysis products, were detected following acetylation of the isolated metabolites. Because of the instability of 5,6-dihydro-5,6-dihydroxy-1-naphthyl N-methylcarbamate under acid conditions, it is very likely that the 5,6-dihydroxy carbaryl metabolites were present, at least in part, in the urine as the dihydrodihydroxy analogs.

More than 90% of the radioactive residues in the urine of a cow treated orally with 1-naphthyl-^{14}C carbaryl was water-soluble.[88] The radiocarbon extractable into chloroform consisted of up to 18 components, but approximately 60% was divided equally between 5,6-dihydro-5,6-dihydroxy-1-naphthyl N-methylcarbamate and its hydrolytic product. The nature of the water-soluble metabolites was not determined. In a subsequent study, where 1-naphthyl-^{14}C carbaryl was fed to cows in the diet for up to 14 days, the metabolites excreted in the urine were examined in more detail.[85] At equilibrium, which occurred after 4 days of feeding, 10 to 15% of the urinary

^{14}C-metabolites was extracted into organic solvent. About 45% of these materials consisted of equal amounts of 5,6-dihydro-5,6-dihydroxy-1-naphthyl N-methylcarbamate and its hydrolysis product, whereas 3,4-dihydro-3,4-dihydroxy-1-naphthyl N-methylcarbamate and its hydrolysis product accounted, respectively, for 5% and 15% of the radiocarbon in the urine. The identification of these latter products was confirmed by mass-spectral analysis and by the fact that weak-acid treatment yielded a mixture of 3- and 4-hydroxy carbaryl from the 3,4-dihydro-3,4-dihydroxy-1-naphthyl N-methylcarbamate.

Incubation of the water-soluble metabolites with sulfatase converted 15% of the radiocarbon into ether-extractable material, which was identified as 1-naphthol. Subsequent treatment with glusulase yielded an additional 45% that could be extracted with ether. This was composed of a series of metabolites similar in thin-layer chromatographic behavior to those found in the original extract of the urine. The predominate metabolite, 40% of the glusulase-released radioactivity, was identified as 1-naphthol, whereas 5,6-dihydro-5,6-dihydroxy-1-naphthyl N-methylcarbamate and its hydrolytic product each accounted for approximately 25%. Some of the metabolites remaining in the water after enzyme treatment may have been the 3,4-dihydro-3,4-dihydroxy carbaryl and its hydrolytic product, since these materials are not easily extracted from water with organic solvent, as determined later. They may be recovered quantitatively from an aqueous phase after saturation with sodium chloride and extraction with a 1:1 mixture of chloroform and acetonitrile. A closer examination of the nature of the metabolites in the urine of rats and other animals treated with carbaryl could reveal the presence of the 3,4-dihydro-3,4-dihydroxy metabolites of carbaryl.

Radioactive residues in the milk of a cow fed 100 ppm 1-naphthyl-^{14}C carbaryl in the diet for 14 days reached a plateau after about 3 days, at which time the residues in the milk were approximately 0.3 ppm ^{14}C-carbaryl equivalents.[85] Approximately 35% of the residues consisted of 5,6-dihydro-5,6-dihydroxy-1-naphthyl N-methylcarbamate, and 3% was present as its hydrolytic product. The 5,6-dihydrodihydroxy derivative of carbaryl was also the major metabolite in the milk of cows treated with a single oral dose of the carbamate.[88,90] In the continuous-feeding study, 5% of the residues in the milk was composed of the 3,4-dihydro-3,4-dihydroxy analog of carbaryl, and carbaryl, *per se,* accounted for 6% of the residues. Small amounts of 5-hydroxy-1-naphthyl N-methylcarbamate, 1% or less, were also detected in the milk.

The remainder of the radioactivity in the milk consisted of conjugated metabolites. 1-Naphthyl sulfate constituted 26% of the ^{14}C-residues, and 23% of the residues was in the form of 1-methoxy-5-(methylcarbamoyloxy)-2-naphthyl sulfate and/or glucuronide (Figure 7.10, XII). Both sulfatase and β-glucuronidase treatment yielded 5-methoxy-6-hydroxy-1-naphthyl N-methylcarbamate, but the sulfatase was twice as effective in this regard. A small amount (2%) of the ^{14}C in the milk was identified as the conjugate of the hydrolytic product of this metabolite (Figure 7.10, XIII) Such products have not been reported for other organisms, although they may be present, since it is extremely difficult to separate 5-methoxy-6-hydroxy-1-naphthyl N-methylcarbamate from 4- and 6-hydroxy carbaryl by thin-layer chromatographic techniques. A colorimetric method of detecting carbaryl residues in milk of cows fed 450 ppm in the diet for 14 days failed to show the presence of carbaryl, 1-naphthol, or conjugates of 1-naphthol.[91]

Metabolites of carbaryl in the tissues of cows

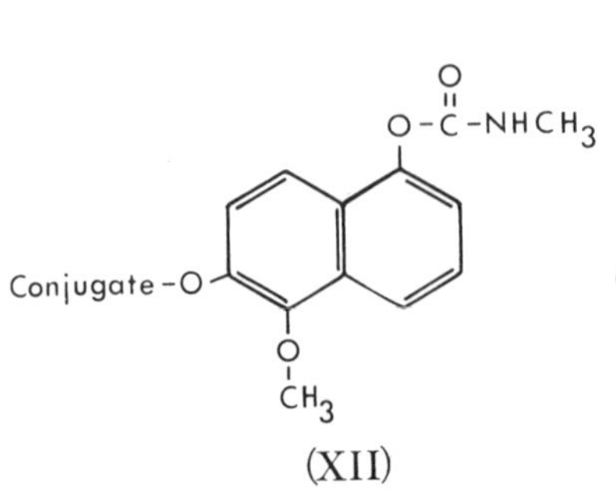

(XII)

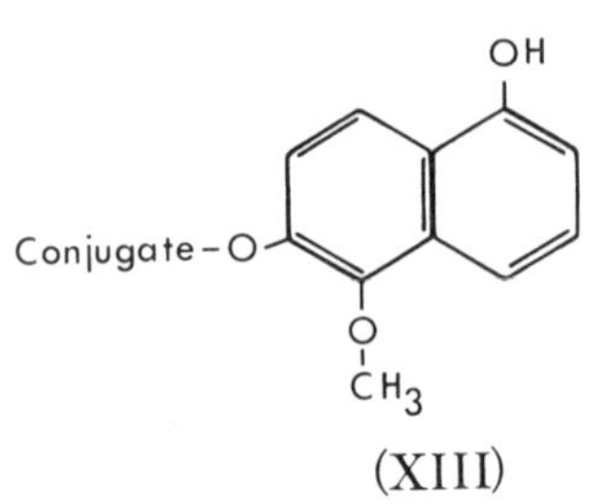

(XIII)

FIGURE 7.10. Metabolites of carbaryl isolated from the milk of cows.

ingesting 1-naphthyl-^{14}C carbaryl for 14 days included carbaryl and 5,6-dihydro-5,6-dihydroxy-1-naphthyl N-methylcarbamate and its hydrolytic product.[85] These materials were present in relatively small amounts; the majority of the radiocarbon in most tissues consisted of 1-naphthyl sulfate and of unidentified metabolites that were water-soluble in nature or could not be extracted with organic solvent.

Feeding 1-naphthyl-^{14}C carbaryl to laying hens in the diet for 14 days at 70 ppm resulted in maximum residues of approximately 0.4 ppm ^{14}C-carbaryl equivalents in the combined egg yolk and white.[92] The largest portion of the residues (60%) was found to be 1-naphthol in the free and conjugated forms. Carbaryl and 1-naphthyl N-hydroxymethylcarbamate each accounted for about 4% of the total. Another major portion (14%) was found to be an unknown material chromatographing on thin-layer plates in a manner very similar to that of 5,6-dihydro-5,6-dihydroxy-1-naphthyl N-methylcarbamate; however, additional chromatographic analyses showed that the unknown metabolite and the authentic standard and of 5,6-hydrodihydroxy carbaryl could be resolved, and the metabolite has not been identified.

The effect of other chemicals on the metabolism of carbaryl in animals has been the topic of several reports. For example, a series of monoamine oxidase inhibitors were shown to interfere with the rate of carbaryl methobolism by rats, resulting in a marked decrease in the rate of elimination of the administered carbamate.[93] Both in-vivo and in-vitro studies indicate that these drugs act primarily by inhibiting the conjugative mechanisms of metabolism, although oxidative metabolism rates are also reduced.[93,94]

An opposite effect was noted when rats were treated with methylmercury hydroxide and/or chlordane for two days, starved for 1 day, and then treated with ^{14}C-carbaryl.[84] Under these conditions, an increased initial rate of excretion of ^{14}C-carbaryl equivalents was observed. However, total elimination via the urine was essentially the same in all animals 3 days after treatment. The nature of the metabolites in the 0–6-hr urine was similar in animals treated with chlordane, with methylmercury hydroxide, and in the controls, which received no pretreatment at all. Liver microsomes from animals treated with chlordane degraded carbaryl more effectively than microsomes from control animals. The increased metabolism was reflected in a higher concentration of oxidative metabolites as well as in an increase in conjugate metabolites. Microsomes from rats treated with methylmercury hydroxide were less effective in metabolizing carbaryl than the microsomal enzymes from livers of rats that received no treatment.

Administering carbofuran (0.5 mg/kg) simultaneously with 1-naphthyl-^{14}C carbaryl (50 mg/kg) to rats resulted in a decrease in the rate of elimination of carbaryl from the animals.[95] However, the reduction in excretion rate was minor, and there was no difference in the nature of the metabolites in the urine.

Metabolism of Carbaryl in Insects

In 1961, prior to the realization that carbamate insecticides could be metabolized by non-hydrolytic mechanisms, it was proposed that the critical step in carbaryl metabolism by insects was its hydrolysis to 1-naphthol and methylamine.[96] This was based primarily on the fact that the milkweed bug and the German cockroach metabolized 1-naphthyl-^{14}C carbaryl to the same series of metabolites as when treated with 1-naphthol-^{14}C. In each case the metabolites were of a polar nature, and those remaining at the origin were very likely conjugates of oxidative metabolites.

In 1964 it was discovered that carbaryl and Baygon were metabolized by American cockroaches and by houseflies through non-hydrolytic pathways.[45] Using carbaryl labeled at three different sites on the molecule with carbon-14, investigations proved that the carbamate was converted by the insects primarily to water-soluble metabolites containing the intact carbamate moiety. Non-conjugated metabolites were tentatively identified as 5,6-dihydro-5,6-dihydroxy-1-naphthyl N-methylcarbamate, 1-naphthyl N-hydroxymethylcarbamate, and the 4- and 5-hydroxy analogs of carbaryl. It was assumed that the water-soluble metabolites were conjugates of these hydroxylated derivatives of carbaryl.

The same general metabolic pathway of carbaryl was found in other in-vivo studies with houseflies, boll weevils, bollworms, rice weevils, cotton leaf worms, silk worms, cabbage loopers, and European corn borers.[14,45,97-102] In each of these insect species carbaryl is converted to hydroxylated carbamate metabolites, which in turn are transformed largely to water-soluble materials. Metabolites identified as hydrolysis products seldom exceed 10% of the applied dose, and in some cases hydrolysis of the carbamate ester is virtually nonexistent.[14,102] One possible exception to this non-hydrolytic pathway of carbaryl metabolism by insects was reported for the German cockroach.[103] Using colorimetric means of analysis, 1-naphthol in the free and conjugated form was recovered from the treated insects. Because the analytical method did not allow for the quantitation of total carbaryl equivalents, it was not possible to determine the ratio of hydrolytic to non-hydrolytic products.

Microsomal preparations from insects convert carbaryl almost exclusively to oxidative products.[104-107] Investigations of the comparative metabolism of carbaryl by different insect larval tissues revealed that the most efficient metabolic sites were fat body and gut.[14,101,102,108] Although the fat body generally predominated, the gut from one strain of corn borers was found to contain a higher capacity for carbaryl metabolism than its fat body.[102] Interestingly, the reverse was true for a different strain of the same insect species. Such subtle variations in metabolism of carbamate and other insecticides by different strains and species of insects have been noted in several studies.[109] Qualitatively, however, the metabolites formed in many of these in-vitro preparations are similar to the unconjugated products found in living insects.

Differences in carbaryl metabolism have also been found for various ages or developmental stages of insects. Tissues from larvae of blowflies,[108] cabbage loopers,[14,101] and European corn borers[102] all increased in carbaryl metabolic activity as the larvae proceeded from early instars through late instars. Then, as feeding ceased and the immature insects prepared for pupation, metabolic activity dropped drastically. This pattern has also been found with other insects and other insecticides.[109] Metabolic variations with age of adult insects, sex, and diet exist, but they have not been clearly defined.[14,109]

Though hydrolysis of carbaryl is a minor metabolic route in most insects, it has been shown to take place in isolated midguts of certain insect species.[75] In addition, various American cockroach esterases, isolated by agar gel electrophoresis, catalyze hydrolysis of carbaryl.[110]

In studies designed to compare the metabolism of carbaryl by resistant and susceptible strains of insects, it has been shown that there is less absorption, greater metabolism, and more rapid excretion of the insecticide from the resistant or more tolerant insects.[14,96,101,103,111-113] For some insects, the greater metabolic activity in resistant species appears to result from higher levels of NADPH-dependent oxidative enzymes in the resistant insects as compared to the susceptible ones.[101,113] The ability of the resistant insects to oxidatively metabolize carbaryl and subsequently convert these metabolites to water-soluble conjugates is definitely an important factor in the mechanism of resistance. This will be discussed in more detail in Chapter 10.

In addition to the carbamate metabolites already mentioned as being formed by insects, there are several others that deserve some mention. For example, the silk worm is reported to metabolize carbaryl to several metabolites, one of which is considered the primary route of carbaryl metabolism in this insect.[114] Good evidence was presented, suggesting that the metabolite contained the intact carbamate moiety and that it contained a hydroxyl group at some position. The researchers identified the metabolite as 2-hydroxy-1-naphthyl N-methylcarbamate, although the evidence presented would not rule out the possibility of the metabolite being the more commonly occurring 4- or 5-hydroxy-1-naphthyl N-methylcarbamate. Cattle ticks apparently differ considerably from their insect relatives in that they hydrolyze large amounts of carbaryl. In one study, extracts of the ticks treated with 1-naphthyl-^{14}C carbaryl contained 55% 1-naphthol and 45% 1,5-dihydroxynaphthalene after acid hydrolysis.[115] Some 1-naphthyl N-hydroxymethylcarbamate also was evident. Other investigators[116] showed that hydrolysis was important in the metabolism of carbaryl by cattle ticks, but that oxidative metabolites were also present. The major metabolite isolated from the treated ticks was tentatively identified as 5,6,7,8-tetrahydrotetrahydroxy-1-naphthyl N-methylcarbamate. It was proposed that this metabolite originated from the 5,6-dihydrodihydroxy carbaryl analog, which was also present in the ticks in small amounts. There have been no reports confirming the identification of the proposed metabolite.

1-Naphthol Metabolism in Animals

As evidenced by the above discussion,

1-naphthol is a major hydrolytic product of carbaryl in some animal systems. It has been used in a variety of studies, which have contributed significantly to our knowledge of ring-hydroxylation as it pertains to the metabolism of xenobiotics in general. The compound is relatively nontoxic to mammals,[18] but it has been observed that 1-naphthol is more toxic than carbaryl to some marine organisms.[117]

The metabolism of 1-naphthol by rats,[118-123] by rabbits,[119,124,125] and by chickens[126] has been determined. Results obtained in one of the more recent studies[123] are typical of the findings. When 1-naphthol-^{14}C was administered orally to rats, 90% of the dose was eliminated in the urine within 24 hr. Of the excreted radiocarbon, 81% was present as 1-naphthyl glucuronide, 16% as 1-naphthyl sulfate, and 1.6% as free 1-naphthol. A small amount, 0.1%, behaved on TLC in the same manner as 1-naphthyl glucoside, and the remainder of the radiocarbon consisted of three other minor unknown metabolites. Houseflies fed 1-naphthol in the diet excreted the compound primarily as 1-naphthyl glucoside and 1-naphthol sulfate.[127] In addition to these metabolites, phosphate conjugates of 1-naphthol have been isolated from the housefly, blowfly, and New Zealand grass grub.[128]

From these studies it is obvious that hydroxylation of the naphthyl ring is at best only a minor pathway in the metabolism of 1-naphthol by animals. On the other hand, this type of metabolism has been reported for the unsubstituted naphthalene in insects and animals, but comparative experiments indicated that it occurred to a lesser degree with 1-naphthol.[129,130] The metabolism of 1-naphthol by soil microorganisms is discussed in the section on carbaryl metabolism in soil.

Metabolism of Carbaryl by Plants

The metabolism of carbaryl has been examined with a variety of plant species, and the results are surprisingly similar. A pathway for carbaryl breakdown in plants that is now considered quite typical, regardless of the species involved, is illustrated in Figure 7.11. Carbaryl is degraded primarily by oxidation of the N-methyl group and of the naphthalene ring to form several hydroxylated metabolites, which are immediately conjugated as glycosides. Metabolites formed by all plants studied thus far include 4-hydroxy-1-naphthyl N-methylcarbamate, 5-hydroxy-1-naphthyl N-methylcarbamate, and 1-naphthyl N-hydroxymethylcarbamate. The 5,6-dihydro-5,6-dihydroxy-1-naphthyl N-methylcarbamate is usually isolated, unless strong acid is used to cleave the conjugates. In one instance, 7-hydroxy-1-naphthyl N-methylcarbamate was found as a metabolite. Amounts of these metabolites of carbaryl vary according to the time after treatment and, to some degree, with the species of plants treated. The glycoside conjugates appear to be further metabolized to very stable materials that are not extractable from the plant.

The rate of degradation of carbaryl by bean plants and the formation of water-soluble and unextractable products are shown in Figure 7.12. Carbaryl is readily degraded within the plant, and there is immediate formation of water-soluble metabolites which reach maximum concentration about 10 days after treatment.[131] The quantity of water-soluble metabolites then declines at the expense of an increase in unextractable material. Even after 90 days, approximately 90% of the injected radiocarbon was still present in the mature bean plant, with almost all of the 1-naphthyl-^{14}C carbaryl equivalents existing as metabolites which could not be extracted from the plant tissue. The very persistent nature of ^{14}C-residues in bean plants treated with carbaryl labeled at various sites on the molecule has been demonstrated by several investigators.[27,44,45,52] The same metabolic pathway of carbaryl occurs in

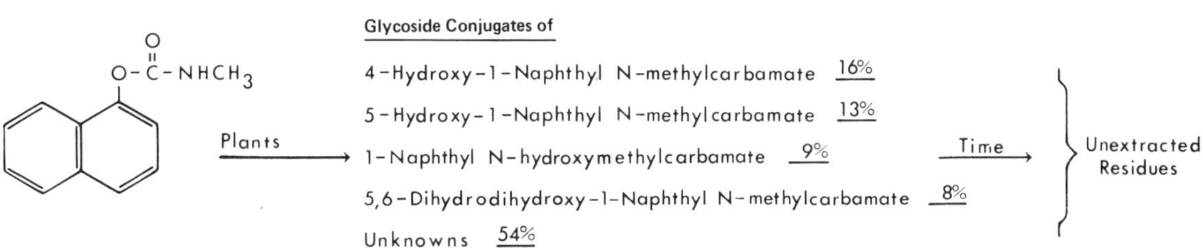

FIGURE 7.11. Distribution of carbaryl residues in plants (percent of total water-soluble metabolites) 6 days after treatment by stem injection. Unknowns consist of water-soluble materials not cleaved by β-glucosidase.[27]

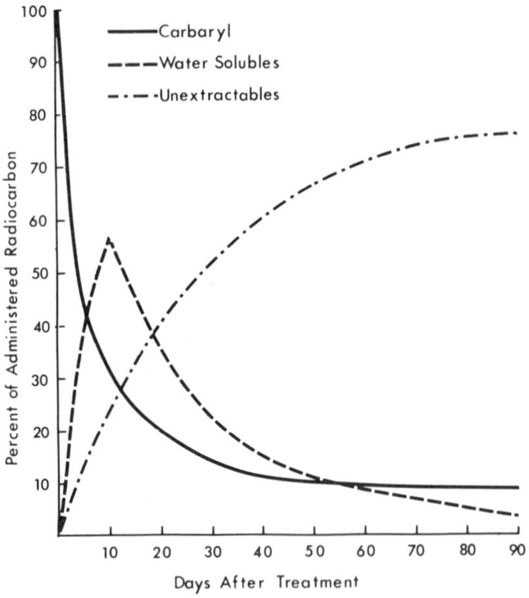

FIGURE 7.12. Fate of 1-naphthyl-^{14}C carbaryl following injection into stems of young bean plants.[131]

peas, peppers, and corn.[52] When metabolites were isolated in sufficient quantities for mass and ultraviolet spectroscopic analysis, it was demonstrated that N-hydroxymethyl carbaryl, 4-hydroxy carbaryl, and 5-hydroxy carbaryl recovered from acid-treated plant extracts were identical to the synthetic compounds.[52] The glucosides of 4- and 5-hydroxy carbaryl have been chemically synthesized[132] and give characteristic mass spectra, which should prove useful in the identification of intact plant conjugates. Interestingly, it was possible to obtain a weak molecular ion of the two intact glucosides.

Cotton plants injected with ^{14}C-carbonyl carbaryl metabolized the carbamate in the same manner as that given above.[44] More than 90% of the injected radiocarbon was still present in the plants 28 days after treatment. When carbaryl was introduced into cotton plants through the roots, about half of the total ^{14}C-residues from a 1-naphthyl-^{14}C carbaryl treatment was present as hydrolytic products.[54] Since the radioactive carbaryl was in an aqueous solution for several days while being absorbed by the cotton plants, much of the hydrolysis may have occurred in the treatment solution.

It is important to consider that most of these investigations of carbaryl metabolism by plants involved the injection of the insecticide into the stem or feeding the compound through the roots in an attempt to obtain maximum metabolites for purposes of identification. However, carbaryl is not used commercially in such a manner, but is usually applied by foliar treatment. Under these conditions, the residual life of carbaryl on bean leaf surfaces was shown to be relatively short.[51] Approximately 50% of the ^{14}C-carbonyl carbaryl placed on the leaf surfaces was lost after 68 hr, and all of the remaining radioactivity was identified as the original compound. Foliar treatment of wheat, alfalfa, peanuts, potatoes, and tomatoes with 1-naphthol-^{14}C carbaryl resulted in up to 85% loss of the applied material from some of the plants after 14 days.[133] The metabolism of the absorbed carbaryl was essentially the same as that described for bean plants. It was demonstrated that the N-hydroxy-N-methyl derivative of carbaryl was not present in any of the plants receiving foliar treatments of carbaryl. When water-soluble metabolites of carbaryl formed by bean plants were adminstered orally to rats, the radioactive residues were rapidly eliminated.[131] After 24 hr; 72% of the administered dose had appeared in the urine and 4% in the feces; after 96 hr, 86% had been removed in the urine and 8% in the feces.

Carbaryl Metabolism in Soil

Carbaryl metabolism by microorganisms isolated from soil may be predominantly hydrolytic or predominantly oxidative; seldom is there a mixture of hydrolysis products and carbamate metabolites as found in animals. Few studies have been conducted using soil bacteria, but certain isolates, when incubated with carbaryl, appear to hydrolyze the carbamate and completely degrade the naphthyl ring.

A strain of *Achromobacter*[134] and an unidentified bacterium[135] degraded carbaryl to 1-naphthol and to other products, which appeared to be identical to those *Pseudomonas* sp. produced from naphthalene.[136,137] The *Achromobacter* utilized carbaryl as the sole source of carbon and metabolized it to products identified as catechol, pyruvate, and hydroquinone. The fact that there was no lag phase associated with the growth of the organisms with these substrates or with 1-naphthol was presented as additional evidence for their presence as intermediates in the degradation of carbaryl. Hydrolytic metabolism of carbaryl by other bacteria, *Pseudomonas phaseolicola*[138] and *Psuedomonas melophthora,*[139,140] also has been reported.

None of the studies with bacteria presented good quantitative data on the magnitude of the reported metabolites. Therefore, it is impossible to estimate whether 1-naphthol occurred in large enough quantities to suggest that its formation was the first step in carbaryl degradation by the microorganisms. However, in a project using soil fungus *Fusarium solani* it was demonstrated that carbaryl was degraded by 24% in 5 days without any accumulation of 1-naphthol.[141] Since the same strain degraded only 20% of 1-naphthol-^{14}C after 12 days, it would seem that conversion to 1-naphthol is not the initial degradation step, as suggested by the authors;[141] it is more likely that the formation of some unstable or readily metabolized carbamate metabolite served as the first step. The identity of such a metabolite is considered later in this discussion. Cell extracts of *Fusarium solani* were much more effective in degrading 1-naphthol than were the cell cultures.[142,143]

All other soil fungi studied thus far metabolize carbaryl by non-hydrolytic mechanisms and form the same series of metabolites that is formed in plants. A series of soil fungi (*Rhizopus* sp., *Mucor* sp., *Aspergillus terreus, Fusarium oxysporium, Gliocladium roseum, Penicillium* sp. and others) metabolized carbaryl to 1-naphthyl N-hydroxymethylcarbamate, 4-hydroxy-1-naphthyl N-methylcarbamate, and 5-hydroxy-1-naphthyl N-methylcarbamate.[142,144,145] *Pencillium* sp., *Mucor* sp., and *Rhizopus* sp. formed more of the ring-hydroxylated metabolites, whereas the other isolates produced more of the N-hydroxymethyl derivative. When 1-naphthyl N-hydroxymethylcarbamate was added to isolates of *Aspergillus terreus,* the compound was degraded to 1-naphthol and a product identified as 1-naphthyl carbamate.[146] The latter metabolite was suspected of being formed by soil fungi from carbaryl, but upon extensive two-dimensional chromatography, the metabolite and authentic standard were resolved.[147] Although this metabolite was not characterized, the 4- and 5-hydroxy and the N-hydroxymethyl derivatives of carbaryl were identified, as was a small amount of 5,6-dihydro-5,6-dihydroxy-1-naphthyl N-methylcarbamate.

Carbaryl, when added directly to soils, appears to follow the same metabolic pattern observed in isolates of the fungus *Fusarium solani,* which may also be the same in bacterial isolates. Two independent soil studies, one using ^{14}C-carbonyl carbaryl[148] and the other 1-naphthyl-^{14}C carbaryl,[147] illustrate that carbaryl is degraded primarily to unextracted ^{14}C-materials (Table 7.2). After incubation for approximately one month, the only material in the apolar soil extract was carbaryl. This was the case with the 1-naphthyl-^{14}C carbaryl at sampling intervals ranging from 4 to 120 days. The water-soluble metabolites never exceeded 2% of the applied naphthyl-^{14}C-labeled treatment. Based on the single sampling at 32 days, this same pattern was also found for the ^{14}C-carbonyl-labeled material (Table 7.2). These data suggest that hydroxylated metabolites, if formed, are not conjugated as in animals and plants, or if they are, they cannot be extracted with acetone or methanol. Of radioactive water-soluble plant metabolites added to soil, 75% were recovered with acetone and water when extracted immediately. It was found that 40% of the radiocarbon added to the soil as 1-naphthyl-^{14}C carbaryl water-soluble plant metabolites was lost after seven days.[147] Further incubation for up to 120 days did not result in additional dissipation of the residues, although

TABLE 7.2

Fate of Radiocarbon in Clay Loam Soil Treated with Carbonyl-^{14}C and with 1-Naphthyl-^{14}C Carbaryl, and Incubated for 32 and 28 Days Respectively

	% of applied ^{14}C	
^{14}C-Materials	Carbonyl-^{14}C[148]	1-Naphthyl-^{14}C[147]
Carbaryl	6.4	15
Polar metabolites	3.2	1.2
Unextracted	17	23
Loss	73[a]	61

[a]35% recovered as ^{14}C-carbon dioxide.

extractability declined with time. Only 15% of the ^{14}C-residues in the soil after 120 days was extractable with acetone and water. It cannot be stated with certainty, therefore, that the unextracted residues in soils treated with carbaryl (Table 7.2) were not, at least in part, composed of conjugated non-hydrolytic metabolites of carbaryl.

Of particular significance in the two soil studies was the similarity in the percentage loss of radiocarbon (Table 7.2). With the ^{14}C-carbonyl-labeled carbaryl-treated soils, much of the loss, 35% of the applied dose, was attributed to ^{14}C-carbon dioxide.[148] Another experiment showed that 18% of the applied 1-naphthyl-^{14}C carbaryl was liberated as ^{14}C-carbon dioxide after only 3 days (Table 7.3). 1-Naphthol-^{14}C, incubated in a different portion of the same soil, yielded 3% ^{14}C-carbon dioxide. This rate of formation of ^{14}C-carbon dioxide from 1-naphthol-1-^{14}C was similar to that obtained with 1-naphthol-1,4,5,8-^{14}C when incubated with soil.[148]

The organo-soluble radiocarbon from 1-naphthyl-^{14}C carbaryl soils (Table 7.3) was composed entirely of carbaryl, whereas 1-naphthol was the predominant component of the 1-naphthol-^{14}C-derived organo-solubles. Several other metabolites were present in small quantities. Only trace amounts of water-soluble metabolites were formed from carbaryl, but they comprised 17% of the 1-naphthol treatment. Both compounds yielded high quantities of unextractable ^{14}C-residues, there being twice as much from 1-naphthol as from carbaryl. Radiocarbon not accounted for was of similar magnitude with both compounds and suggested the formation of a volatile metabolite other than ^{14}C-carbon dioxide. Such loss was equivalent to 38% of the applied ^{14}C-carbonyl carbaryl when incubated in clay loam soil for 32 days.[148] It should be pointed out that the degree of carbaryl degradation both in the ^{14}C-carbonyl carbaryl study[148] and in the 1-naphthyl-^{14}C carbaryl study[147] was greatly influenced by soil type and by previous exposure of the soils to insecticides, particularly to carbaryl. However, in no case does the degradative pathway appear to differ qualitatively from the results reported herein.

The fact that ^{14}C-carbon dioxide is liberated from 1-naphthyl-^{14}C carbaryl at a more rapid rate than from 1-naphthol-^{14}C supports the existence of an intermediate more susceptible to ring fission than 1-naphthol. Based on the degradative pathway of naphthalene by *Pseudomonas* sp.,[136,137] ring fission and ultimate evolution of carbon dioxide from carbon 1 would result if 1,2-dihydroxynaphthalene was formed. In the case of carbaryl, this would require ring hydroxylation at the 2-position, followed by hydrolysis of the carbamate ester (Figure 7.13). The remainder of the pathway would be the same as for naphthalene and would explain the detection of pyruvate and catechol when carbaryl was incubated with soil microorganisms.[134] 2-Hydroxy-1-naphthyl N-methylcarbamate, if formed in soils, is rapidly degraded and does not occur at detectable levels. Although it is possible that 1-naphthol is converted to 1,2-naphthalenediol, the fate of the radiocarbon from 1-naphthol-^{14}C, compared to that from 1-naphthyl-^{14}C carbaryl (Table 7.3), makes it appear unlikely that hydrolysis of carbaryl to 1-naphthol could be the major pathway for degradation of carbaryl in soils.

TABLE 7.3

Comparative Metabolism of 1-Naphthyl-^{14}C Carbaryl and 1-Naphthol-^{14}C by the Same Soils during 3 Days Incubation[147]

	% of applied ^{14}C	
^{14}C-Materials	Carbaryl	1-Naphthol
Carbon dioxide	18	3.1
Organo-solubles	46	24
Water-solubles	0.4	17
Unextracted	25	48
Loss	11	7

TSUMACIDE
(MTMC, Metacrate, C-3)

Tsumacide (3-methyl-^{14}C-phenyl N-methylcarbamate), when applied orally, was excreted at different rates by male and by female rats.[149,150] In male rats, 97% of the dose was eliminated in the urine and 0.1% in the feces after 24 hr. During the same time, females excreted 70% in the urine and about 4% in the feces. No radioactive carbon dioxide was expired from either the male or the female rodents.

Approximately 70% of the radioactivity in the urine was ether-extractable, with 90% of this composed of 3-carboxyphenyl N-methylcarbamate (Figure 7.14, I). Other carbamate metabolites

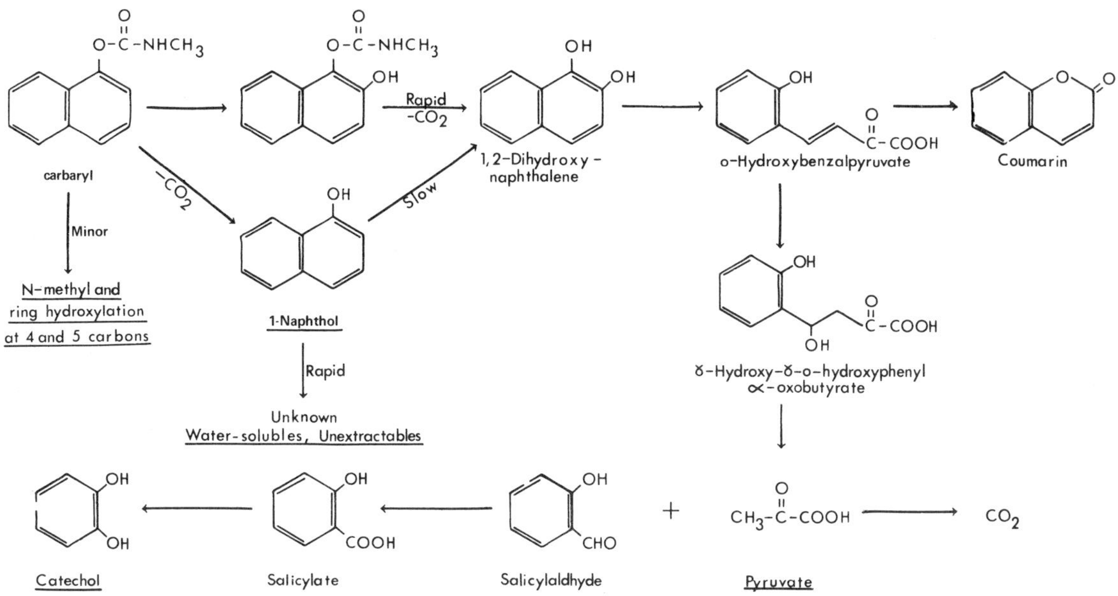

FIGURE 7.13. Hypothetical pathway of carbaryl in soil, based on release of ^{14}C-carbon dioxide from 1-naphthol-^{14}C, carbonyl-^{14}C carbaryl, and 1-naphthyl-^{14}C carbaryl,[147,148] and on known pathway of naphthalene degredation by soil microorganisms.[137] Products underlined have been identified as soil metabolites of carbaryl.

included 3-hydroxymethylphenyl N-methylcarbamate (Figure 7.14, II) and 4-hydroxy-3-methylphenyl N-methylcarbamate (Figure 7.14, III). Small quantities of the hydrolysis product of compounds I and II were also detected as free metabolites. Of the water-soluble radioactivity, 50% was composed of conjugated 4-hydroxy-3-methylphenyl N-methylcarbamate. Only 20% of this was liberated by treatment of the water-solubles with β-glucuronidase, and none by treatment with aryl sulfatase. The remainder was freed only by acid hydrolysis, and the compound was assumed to be a different type of conjugate or a complex with some natural component of the urine. Glucuronides of metabolites II and III were detected in small quantities, as was the sulfate of the 3-methylphenol. Other water-soluble metabo-] lites were not identified. There were no differences in the nature of the urinary metabolites of males and of females. Hydrolysis of the carbamate ester was a minor pathway, less than 5%, in the metabolism of Tsumacide in rats.

Liver microsomes from rabbits, rats, and mice metabolized Tsumacide primarily to the 3-hydroxymethyl (Figure 7.14, II) and 4-hydroxyphenyl (Figure 7.14, III) derivatives. Rabbit liver enzymes yielded three times more of the 3-hydroxymethyl Tsumacide analog than did liver enzymes from rats and mice. The N-hydroxy-

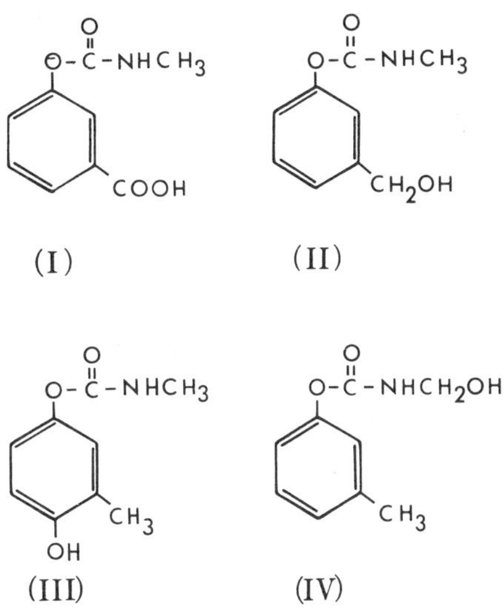

FIGURE 7.14. Carbamate metabolites of Tsumacide formed in rats, houseflies, and/or plants. Metabolite I was not detected in plants, and metabolite IV was not detected in rats.

methyl derivative of Tsumacide (Figure 7.14, IV) was not formed either by the microsomes or by the living animals.

When ring-^{14}C-labeled Tsumacide was applied topically to houseflies, about 70% of the dose was lost after 24 hr. Of the radiocarbon in the flies, 25% was composed of 3-hydroxymethylphenyl N-methylcarbamate (Figure 7.14, II); 92% of this occurring in the free form, and the remainder was as a sugar conjugate. The 3-carboxy derivative (Figure 7.14, I) accounted for 43% of the ^{14}C-residues in the insects, but in this case 58% was as a sugar conjugate. 4-Hydroxy Tsumacide and N-hydroxymethyl Tsumacide appeared at much lower concentrations. Only 0.1% of the applied dose was recovered as the phenolic hydrolysis product of Tsumacide, and it was the only metabolite detected in the houseflies that lacked the carbamate moiety.

There was little metabolism of Tsumacide by bean plants when applied to leaf surfaces. There was, however, very rapid loss of the insecticide from the plants, with only 14% of the applied dose recovered after 1 day, and only 3% after 9 days. No hydrolysis product of Tsumacide was detected at any time after treatment, and only small amounts of carbamate-type metabolites were found. The nature of these latter metabolites suggested that the metabolic fate of Tsumacide in plants was the same as in the housefly, except that oxidation of the ring-methyl group did not proceed beyond the alcohol form. Oxidative metabolites formed by rats, houseflies, or plants were less potent as anticholinesterase agents than the parent compound.

MEOBAL
(S-1046, MPMC)

Meobal (3,4-dimethylphenyl N-methylcarbamate) is used extensively in Japan for the control of various insects attacking rice. The compound is also effective against pests of tea shrubs and vegetables and is used for control of certain types of cattle ticks. The metabolism of the insecticide in rats following oral administration of the 4-methyl-^{14}C-phenyl-labeled material has been thoroughly investigated.[151-153] Urine was the major route of elimination of the carbamate, 92% of the dose being voided 48 hr after administration. Approximately 5% of the radioactivity initially applied was excreted in the feces. After five days, a total of 97% of the treatment had been recovered; only trace amounts of radioactive carbon dioxide were expired.

Direct extraction of the urine with chloroform removed about 3% of the total radiocarbon, which consisted of the intact parent molecule and free 3,4-dimethylphenol. Another product, which had the same chromatographic characteristics as 3-hydroxymethyl-4-methylphenyl N-methylcarbamate (Figure 7.15, I), was present in chloroform extracts of β-glucuronidase hydrolysates from the urine. Of the total urinary metabolites, 40% consisted of the 4-carboxy analog of Meobal (Figure 7.15, II) and its component phenol. Other metabolites that were identified include the N-hydroxymethyl analog of 4-carboxy Meobal (Figure 7.15, III), glucuronide conjugates of 4-hydroxymethyl Meobal (Figure 7.16, IV), and glucuronide conjugates of its phenol (Figure 7.16, V).

The free phenol of Meobal, 3,4-dimethylphenol, and its sulfate (Figure 7.16, VI) and glucuronide conjugates were minor components of the urine. Since rats metabolized this phenol, when administered orally, mainly to sulfate and glucuronide conjugates, and not to its oxidized derivative, it was concluded that the metabolism of Meobal in rats followed primarily an oxidative pathway rather than hydrolysis of the carbamate acid ester.

Of the oxidized metabolites of Meobal, the 4-hydroxymethyl analog was an active inhibitor of mammalian cholinesterase and was nearly equal to Meobal in its toxicity to insects and mammals. The 3-hydroxymethyl Meobal (Figure 7.15, I) was of low biological activity, whereas the 4-carboxy analog (Figure 7.15, II) was essentially inactive.

Figure 7.16 shows some of the types of

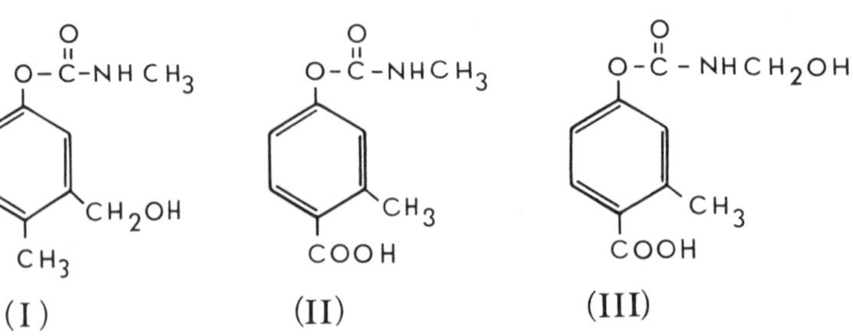

FIGURE 7.15. Oxidative carbamate metabolites of Meobal formed in the rat.

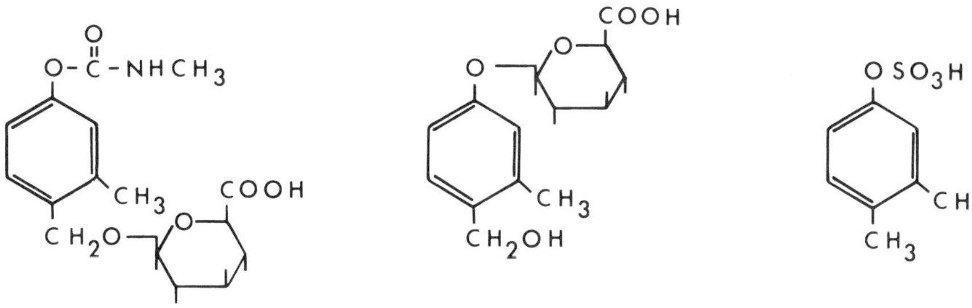

(IV) (V) (VI)

FIGURE 7.16. Examples of conjugated metabolites isolated from the urine of rats treated with Meobal.

conjugate metabolites found in the urine of rats. Since β-glucuronidase treatment of the polar urinary metabolites liberated 95% of the radiocarbon, none of which was 3,4-dimethylphenyl N-methylcarbamate, it was evident that N-conjugation of the parent compound was of little, if any, significance in the metabolism of Meobal.

When 4-methyl-^{14}C Meobal was applied to the leaf surface of bean plants, there was a 70% loss of radioactivity after 1 day, and a 95% loss after 12 days.[149] Meobal appeared as the major component of the residues remaining on and in the bean leaves, but the glucosides of 4-hydroxymethyl and N-hydroxymethyl Meobal were present in small amounts. Unextracted residues in the plant pulp after 12 days were equivalent to 1.6% of the applied treatment. No hydrolysis of the carbamate ester was detected, nor were any carboxy derivatives formed in the plant. Studies of the residual persistence of Meobal on treated rice showed that the compound dissipated rapidly and that mature rice grain contained levels of Meobal considerably lower than the acceptable daily intake for man.[153]

In-vitro studies of the metabolism of Meobal in rats demonstrated that a kidney homogenate degraded only 3% of the compound, whereas a similarily prepared sample of liver tissue metabolized 48% of the material.[151] A lysosome-free supernatant of the liver homogenate was metabolically more active than the whole homogenate, and the microsomal fraction was slightly less effective than the whole homogenate in metabolizing Meobal. Coenzyme-requirement studies for the formation of hydroxymethylphenyl carbamate metabolites from Meobal by the rat liver microsomal fraction showed that a NADPH-generating system (NADP + glucose-6-phosphate + glucose-6-phosphate dehydrogenase) was the most effective. NADPH was only half as efficient, and NADH and NAD were not active as cofactors. The metabolites formed by this in-vitro system included a mixture of 3-hydroxymethyl-4-methylphenyl and 3-methyl-4-hydroxymethylphenyl N-methylcarbamate in a ratio of roughly 3 to 2. In-vitro conversion of these hydroxymethylphenyl carbamates by liver preparations to the corresponding carboxy analogs was generally very low and never exceeded 10%. Although not isolated and identified, there was some indication that the N-methylcarbamate of a phenol containing an aldehyde group was derived from the hydroxymethylphenyl carbamate materials.

3,5-DIMETHYLPHENYL N-METHYLCARBAMATE (XMC, Macbal, Cosban)

A study of the metabolism of ^{14}C-carbonyl-labeled 3,5-dimethylphenyl N-methylcarbamate in five different soil types demonstrated that the main pathway of degradation was ester hydrolysis (Figure 7.17).[148] This was determined in microbiologically active soils containing 80% of the field capacity moisture with an incubation time of 32 days at 25°C.

Total recovery of radiocarbon after incubation ranged from 25% to 100%. The difference in recoveries was probably attributable to inefficient trapping of ^{14}C-carbon dioxide, since recovery generally declined as ^{14}C-carbon dioxide production increased. Because of the position of the ^{14}C-label, it is unlikely that the loss represents the formation of a highly volatile metabolite which escaped the carbon dioxide traps.

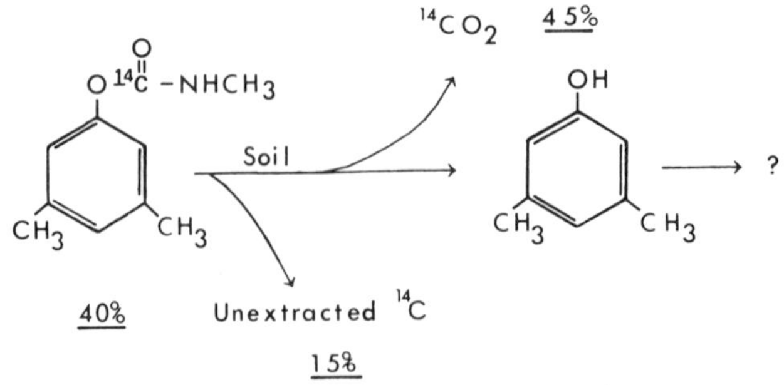

FIGURE 7.17. Metabolism of 3,5-dimethylphenyl N-methylcarbamate in clay loam soils.[148] Percentages refer to the portion of initial radioactivity remaining after 32 days of incubation at 25°C.

Hydrolysis, as indicated by the production of ^{14}C-carbon dioxide, was greatest (40%) in clay loam soil which contained 4% organic matter, and lowest (18%) in sandy loam soil containing 1.5% organic matter. The percentage of organic matter in the soils, however, was not the sole controlling factor in hydrolysis of the carbamate, as evidenced by less production of ^{14}C-carbon dioxide by soils containing more organic matter than that present in clay loam. For example, loam soil hydrolyzed only 35% of the insecticide, even though it contained three times the organic matter of clay and of loam sand soils with a percentage of organic matter similar to that of clay loam.

All of the radiocarbon extracted from the soil was in the form of the administered carbamate, and its concentrations in the different soil types were inversely proportional to the amounts of ^{14}C-carbon dioxide produced. Although these findings clearly show that hydrolysis is the primary means of degradation of 3,5-dimethylphenyl N-methylcarbamate in soil, hydroxylation of the ring or the N-methyl carbon could precede hydrolysis. The use of the carbonyl-labeled compound prevented the evaluation of the chemical fate of the phenolic moiety once hydrolysis had occurred.

Surprisingly, the amounts of unextracted ^{14}C-materials in the soils were not influenced by the properties of the five soil types. Materials of this nature constituted 10% to 25% of the applied treatment, but with the exception that the lowest concentration was in sandy loam soil (1.5% organic matter), there were no correlations between the amounts of unextracted ^{14}C-residues and soil properties, such as percentage of sand, silt, and organic matter, or soil pH.

Some antimicrobial and/or antihydrolytic enzyme activity by 3,5-dimethyl N-methylcarbamate was implied from data obtained when the compound was added to soil in a proportion of 200 ppm. At this concentration, ^{14}C-carbon dioxide production was only half that in soils fortified with 2 ppm of the carbamate. Likewise, the amount of parent compound recovered was greater in soils treated at the higher level. The amounts of unextracted ^{14}C-materials were not effected by concentration leading to the speculation that these residues were not metabolites, but unchanged 3,5-dimethylphenyl N-methylcarbamate bound in some manner to the soil particles. The same two concentrations of carbaryl were metabolized equally in the soils whereas 1-naphthol at the 200-ppm level was broken down to a lesser extent than at the 2-ppm concentration.

BANOL
(Carbanolate, Chlorxylam)

Rats hydrolyzed 52% of a single oral dose of ^{14}C-carbonyl-labeled Banol (2-chloro-4,5-dimethylphenyl N-methylcarbamate) within 48 hr, as indicated by expired ^{14}C-carbon dioxide; another 35% of the dose was eliminated in the urine, and 10% in the feces (Figure 7.18).[154] Very similar results were obtained when the ^{14}C-carbonyl-labeled insecticide was injected into rats.[26] The radioactivity from ring-^{14}C-methyl-labeled Banol, on the other hand, was almost quantitatively excreted in the urine 48 hr after ingestion.[154] With the N-methyl-^{14}C material, 34% of the dose was eliminated via the urine and 4% via the feces.

Because the urinary excretion of radiocarbon was approximately the same with ^{14}C-carbonyl- and N-methyl-^{14}C-labeled Banol, it was concluded

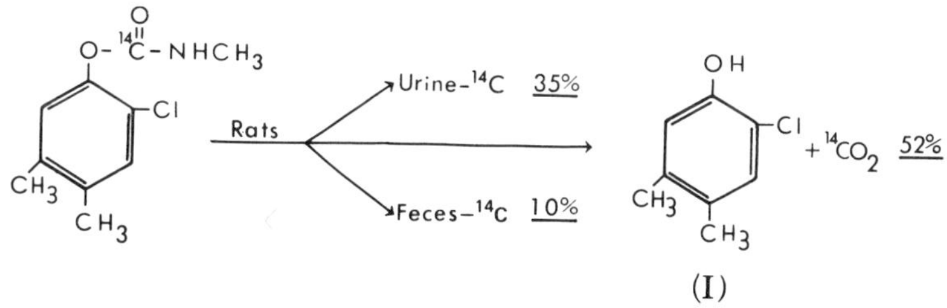

FIGURE 7.18. Elimination of radiocarbon (percent of dose after 48 hr) by rats treated orally with ^{14}C-carbonyl Banol. (Adapted from Reference 154, with permission.)

that approximately 35% of the administered dose was released in the urine as metabolites containing the intact carbamate ester. An equivalent of 18% of the treatment appeared in the urine as a metabolite thought to be the N-glucuronide of Banol (Figure 7.19, II). No direct evidence for the structure was provided, but the metabolite was not hydrolyzed by glucuronidase, yielded formaldehyde upon degradation, was completely alkaline-degradable, and was partially acid-labile. The remainder of the carbamate-type metabolites in the urine were composed of sulfate and of other types of acidic conjugates. The Banol-derived moiety of the conjugates was not characterized.

Rat liver microsomes yielded six carbamate metabolites and several metabolites that did not contain the carbamate grouping.[25] The only product identified was 2-chloro-4,5-dimethylphenyl N-hydroxymethylcarbamate (Figure 7.19, III). A supernatant fraction from rat and human liver homogenates obtained at 15,000 g yielded the N-hydroxymethyl derivative and a series of other carbamate and hydrolysis products.[58] Indirect evidence suggested that oxidation of the ring-methyl groups to the carboxy analog of Banol (Figure 7.19, IV) might be involved in the metabolism of this carbamate. Qualitatively, the metabolism of Banol by the rat and human liver fractions was essentially the same. However, some quantitative differences were noticed.

Human embryonic lung cells metabolized Banol in excess of 90% within 3 days.[65] All of the metabolites were water-soluble and could be converted to the phenol of Banol by treatment with β-glucuronidase and/or by acid treatment. There was no N-glucuronide of Banol formed in this system, as had been proposed for the living rat.[154]

In living houseflies, 80% of topically applied ^{14}C-carbonyl-labeled Banol was absorbed within 24 hr.[155] Of the absorbed dose, 16% was given off

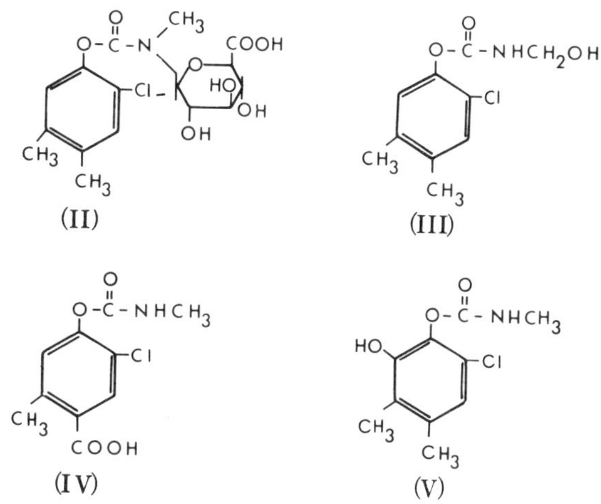

FIGURE 7.19. Structures of some of the proposed metabolites of Banol in animals (II-V) and in plants (III, V).

as ^{14}C-carbon dioxide and 40% was eliminated in the excreta. Pretreatment of the flies with piperonyl butoxide did not alter the penetration rate, but it did reduce the rate of metabolism of Banol, as indicated by a reduction in the amount of ^{14}C-carbon dioxide liberated and by a decrease in the amount of ^{14}C eliminated in the excreta. In-vitro studies have shown that housefly microsomes and fat bodies of cockroaches metabolize Banol to the N-hydroxymethyl derivative (Figure 7.19, III).[105,156]

Six days after ^{14}C-carbonyl-labeled Banol was injected into bean plants, only 60% of the applied radioactivity remained in the plant.[51] Approximately 70% of these materials were soluble in water. Small amounts of organo-soluble metabolites were formed in the plant shortly after treatment, but they never exceeded 2% of the injected dose. β-Glucosidase hydrolysis of the water-soluble metabolites yielded at least eight aglycones, accounting for 90% of the original water-soluble metabolites.[27] More than 80% of the aglycones released by enzyme treatment consisted of a single component, whose identity was not established. The hydroxymethyl derivative of Banol was the only aglycone characterized, and this accounted for only 1% of the total ^{14}C-residues released by enzymic treatment.

A similar pathway of Banol metabolism in bean plants occurred when the material was applied by a root uptake method.[157] The major aglycone released from the water-soluble products was the same with carbonyl-, N-methyl-, or ring-labeled Banol. Mass-spectral analysis indicated that the principal metabolite was a glucoside of a hydroxylated derivative of Banol. It was not definitely established, however, whether hydroxylation had occurred on the ring (Figure 7.19, V) or involved hydroxylation of one of the ring-methyl groups. Because of the large number of water-soluble metabolites containing the carbamate moiety, it is likely that hydroxylation of both the ring and ring-methyl groups occurred in the bean plants.

LANDRIN (SD 8530)

Landrin is composed of 3,4,5-trimethylphenyl N-methylcarbamate (75%) and 2,3,5-trimethylphenyl N-methylcarbamate (18%). Its metabolism in insects and mammals is quite complex with each isomer forming 15 to 20 metabolites.[158] Approximately half of these products have been tentatively identified. In plants, five organo-soluble metabolites are formed from each isomer, but significant levels of residues remain in the plant pulp.

Hydrolysis is an important route in the degradation of Landrin by mice (Figure 7.20), with splitting of the 3,4,5-isomer being twice as great as that of the 2,3,5-isomer. The reason for this is not clear, since the urine contained approximately the same amount of Landrin-equivalents regardless of the isomer administered to the mice. Houseflies hydrolyzed less than 4% of an injected dose of either isomer after 24 hr. Likewise, very little hydrolysis occurred in bean plants treated with the 2,3,5-Landrin, although 22% of the radiocarbon in plants 12 days after treatment with the 3,4,5-isomer was due to cleavage of the carbamate ester.

Hydroxylation of the N-methyl group (Figure 7.21,I) occurred in plants and in enzyme preparations of mouse liver microsomes and housefly

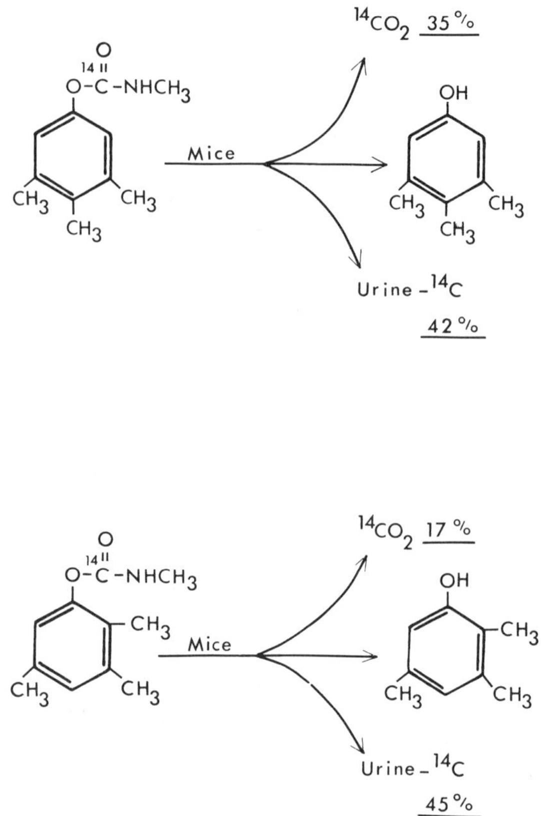

FIGURE 7.20. Evolution of ^{14}C-carbon dioxide and urinary excretion of radiocarbon from mice (percent of dose after 48 hr) treated orally with individual ^{14}C-carbonyl-labeled components of Landrin.[158]

abdomens. The N-hydroxymethyl derivative of the 3,4,5-isomer was also detected as a metabolite in living flies. Various other types of carbamate materials, resulting from ring (Figure 7.21, II) and/or ring-methyl (Figure 7.21, III, IV) hydroxylations, constitute the majority of Landrin metabolites in most of the organisms investigated. Usually, the hydroxylated metabolites were present almost entirely as conjugates. For example, 6 days after treatment of plants, more than 95% of the extractable radiocarbon was in the form of glycosides.

Treatment of the mouse urine and housefly excreta with glusulase yielded various hydroxylated methylcarbamate metabolites. Since glusulase is a crude enzyme mixture containing primarily glucuronidase and sulfatase, the data suggest that both mice and insects formed glucuronides and/or sulfates. This was stated, without question, in a 1971 review,[20] even though it is generally felt that insects form glucosides, not glucuronides[159] (see Chapter 6).

It is the authors' experience that glusulase will liberate "free" metabolites from polar carbamate metabolites formed in plants. Yet these same metabolites can not be cleaved by highly purified β-glucuronidase, but are hydrolyzed by β-glucosidase. Therefore, the designation of a conjugate as a glucuronide based on glusulase treatment should not be accepted as positive proof.

UC 10854
(H-5727, IPFC)

Rats injected with ^{14}C-carbonyl-labeled UC 10854 (3-isopropylphenyl N-methylcarbamate, Figure 7.22, I) expired 53% of the dose as ^{14}C-carbon dioxide during a 48-hr period following treatment.[26] Approximately 45% of the dose was eliminated in the urine and 6% in the feces.

The ^{14}C-carbonyl-labeled carbamate was metabolized by rat liver microsomes to several ether-extractable products.[25,69] The major metabolite (28%) was identified as 3-(1-hydroxy-1-methylethyl)phenyl N-methylcarbamate (Figure 7.22, III). This material equaled the parent compound in anticholinesterase activity. Only one other of the metabolites was identified, which was reported to be 3-isopropylphenyl N-hydroxymethylcarbamate (Figure 7.22, II).

Housefly microsomes metabolized UC 10854 to the hydroxypropyl derivative (Figure 7.22, III), and its formation was enhanced by the addition of bovine serum albumin.[104,105] Human serum was shown to hydrolyze UC 10854, but not as rapidly as it hydrolyzed carbaryl, the Km and V values for carbaryl being about three times higher than those for UC 10854.[70] Nevertheless, the data indicate that the same enzyme(s) that hydrolyzed carbaryl also hydrolyzed UC 10854, although Baygon (2-isopropoxyphenyl N-methylcarbamate) was not touched by the serum.

N-Methyl-^{14}C-labeled UC 10854, when applied topically to houseflies, was degraded and excreted in a similar manner by both resistant and susceptible strains.[155] After 24 hr, 90% of the dose had penetrated the cuticle, and 14% of this was liberated as ^{14}C-carbon dioxide. Approximately 35% of the absorbed radioactivity was eliminated in the excreta. When the phenyl-ring-^{14}C-material was applied, only trace amounts of ^{14}C-carbon dioxide were liberated and only 13% of the absorbed dose was excreted. These data demonstrated that the majority of the radioactive metabolites in the excreta of flies treated with the N-methyl-^{14}C carbamate were derived solely from the carbamoyl moiety of the molecule. In another study, piperonyl butoxide was shown to reduce the absorption of UC 10854 by resistant flies.[160] However, it was concluded that resistance was not

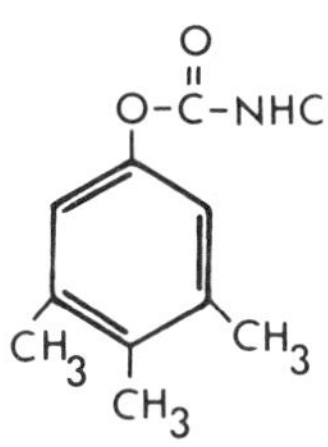

(I)

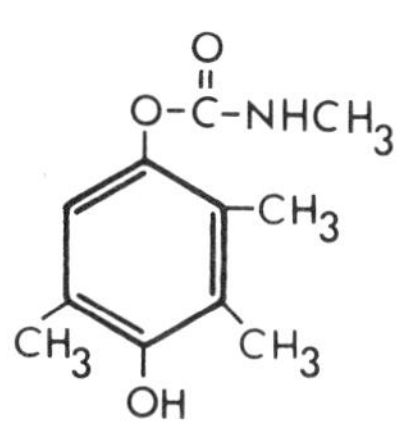

(II)

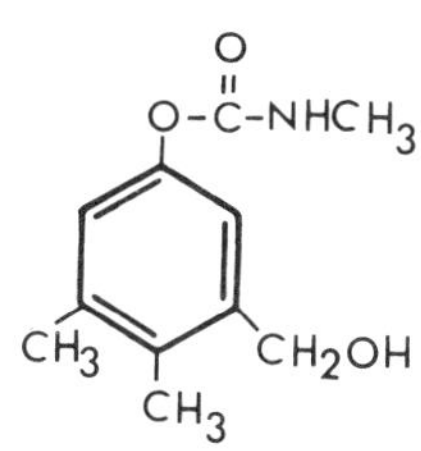

(III)

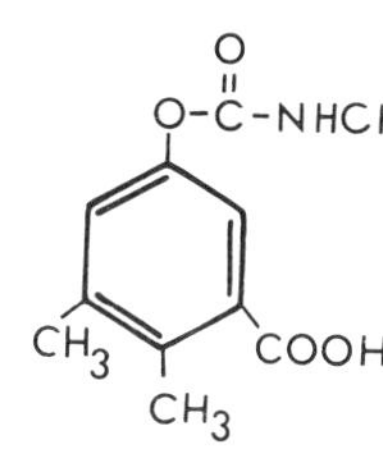

(IV)

FIGURE 7.21. Some examples of carbamate metabolites of Landrin formed by plants and/or animals.

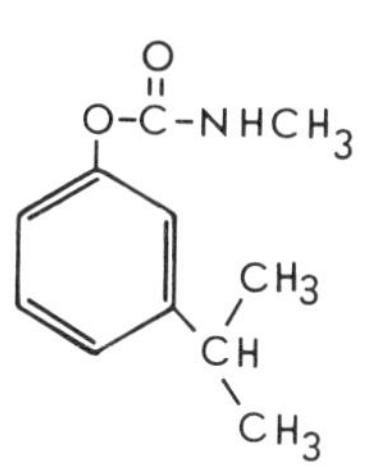

(I)

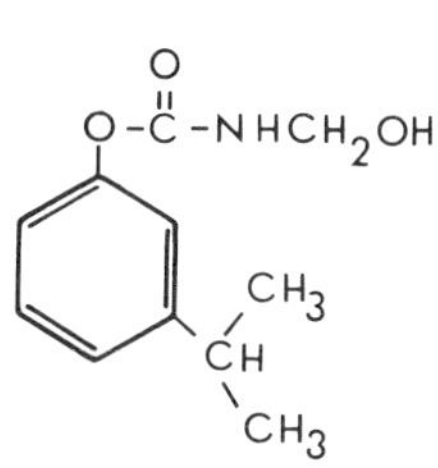

(II)

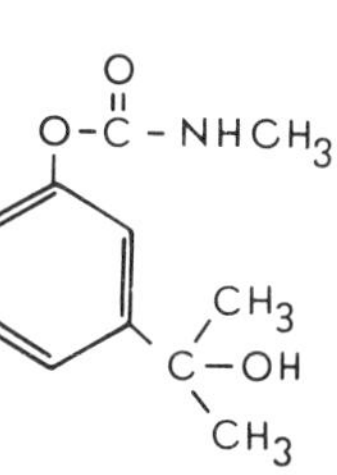

(III)

FIGURE 7.22. UC 10854 (I) and metabolites formed in plants and animals.

caused by a lower absorption rate, but by a detoxication mechanism which destroyed the insecticide at a rate commensurate with its penetration into the insect.

Plants treated with ^{14}C-carbonyl-labeled UC 10854 transformed part of the compound into two organo-soluble materials which accounted for 7% of the applied dose after 12 hr.[51] One of the organo-soluble metabolites was the N-hydroxymethyl derivative (Figure 7.22, II), and the other was the hydroxypropyl derivative (Figure 7.22, III). By 6 days after treatment, 85% of the applied radioactivity had been converted to water-soluble metabolites. β-Glucosidase treatment of the conjugates released at least seven aglycones,[27] the major aglycone being N-hydroxymethyl UC 10854 (Figure 7.22, II). Another freed aglycone was identified as the hydroxypropyl derivative (Figure 7.22, III).

HRS-1422

The carbamate insecticide designated as HRS-1422 (3,5-diisopropylphenyl N-methylcarbamate) has been included in several in-vivo and in-vitro metabolism studies using plants and animals. However, a detailed investigation of the metabolism of HRS-1422 has never been conducted.

In an experiment designed to compare the rates of hydrolysis of a number of carbamate insecticides, it was found that 50% of the ^{14}C-carbonyl-labeled HRS-1422 was hydrolyzed by rats after 48 hr (Figure 7.23).[26] Urinary ^{14}C-metabolites accounted for 37% of the dose, whereas 6% was excreted in the feces. An estimated 4% remained in the body after 48 hr.

Rat liver microsomes fortified with NADPH converted carbonyl-^{14}C-labeled HRS-1422 nine different products, three of which were considered major.[25] The only metabolite identified was 3,5-diisopropylphenyl N-hydroxymethylcarbamate. Several of the minor metabolites whose identity was not established were more potent than the original compound as inhibitors of plasma cholinesterase. HRS-1422 is one of the carbamate insecticides most resistant to degradation by rat liver microsomal enzymes.[25, 69] Homogenates of housefly abdomens were able to metabolize HRS-1422 along various oxidative routes, but no quantitative information on its transformation by the microsomes was reported.[105]

HRS-1422 is a rather persistent carbamate insecticide when applied to plant foliage. Approximately 15 hr was required for 50% of the ^{14}C-carbonyl-labeled compound to dissipate from bean leaf surfaces.[51] Almost all of the radioactivity remaining on the foliage was in the form of the administered compound. When applied to glass surfaces, a more rapid 50% loss of the ^{14}C-carbonyl-labeled insecticide occurred in 5 hr. Since the disappearance from the glass surface was linear for the first 80 to 90%, it is probable that direct volatility of the parent compound rather than conversion to other carbamate components was responsible for the loss. When injected into the stem of bean plants, HRS-1422 was degraded more slowly than seven other N-methylcarbamates included in the study.[51] Of the administered radiocarbon, 35% remained as the parent compound 6 days after injection. An additional 23% of the dose was present as water-soluble metabolites, and 8% could not be extracted from the plant tissue with acetone. When incubated with pH 9.3 buffer, ^{14}C-carbonyl HRS-1422 was degraded by 50% after 7 hr and was the most stable of all carbamates tested under these conditions.

β-Glucosidase treatment of the water-soluble metabolites recovered from bean plants released 11% of the radiocarbon, which was shown to consist of a series of aglycones. More than half of the radioactive aglycones were anticholinesterase agents.[27]

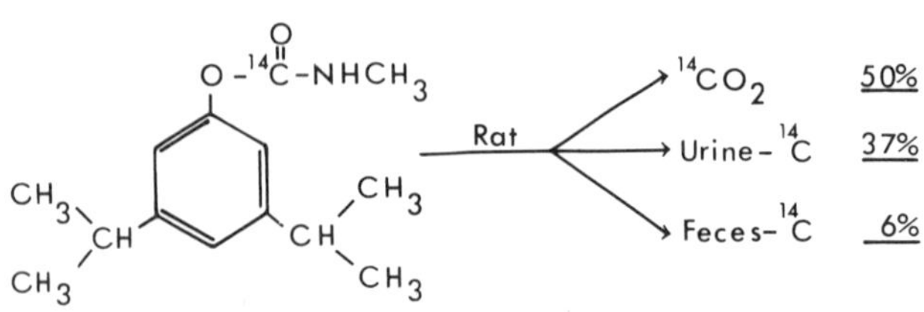

FIGURE 7.23. Disposition of ^{14}C-carbonyl-labeled HRS-1422 (percent of dose after 48 hr) by rats following intraperitoneal treatment.

BUX
(Ortho 5353, Bufencarb, Metalkamate)

Bux, a mixture of 3-(1-methylbutyl)phenyl N-methylcarbamate (Figure 7.24, I, 75%) and 3-(1-ethylpropyl)phenyl N-methylcarbamate (Figure 7.24, II, 25%), was granted a tolerance in corn grain, fodder, and forage of 0.05 ppm in 1969.[161] Surprisingly little information is available on the metabolism of this insecticide. A brief report of some unpublished data submitted in support of the tolerance request to the U.S. Food and Drug Administration[162] stated that no complete study of the metabolism of Bux in plants had been conducted. However, some information pertaining to the biotransformation of Bux in rats showed that the compound was metabolized in a manner similar to other carbamate insecticides.

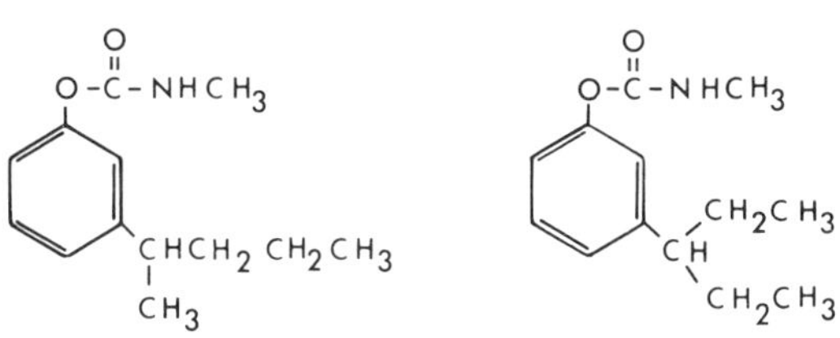

(I) (II)

FIGURE 7.24. Components of Bux insecticide consisting of a 3:1 mixture of I and II.

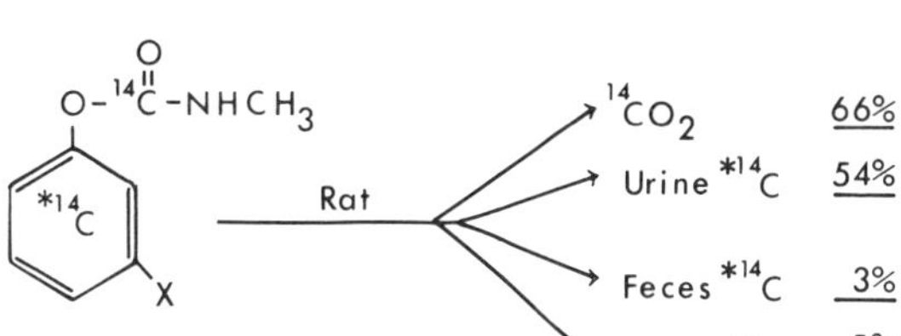

(I)&(II)

FIGURE 7.25. Fate of ^{14}C-carbonyl- and ^{14}C-ring-labeled Bux 48 hr after oral administration to rats. Values are percent of indicated ^{14}C or $^{*14}C$ dose.[162]

Within 48 hr after ^{14}C-carbonyl-labeled Bux was administered to rats, 66% of the dose was exhaled as ^{14}C-carbon dioxide (Figure 7.25). With the ^{14}C-ring-labeled material, 54% of the dose was voided in the urine, 3% in the feces, and approximately 5% remained in the animal tissue. Oxidative metabolites formed in the rats included 3-(1-hydroxy-1-methylbutyl)phenyl N-methylcarbamate (Figure 7.26, III) and 3-(1-methylbutyl)phenyl N-hydroxymethylcarbamate (Figure 7.26, IV). No quantitative data relative to the individual metabolites were reported, although it was stated that conjugates of metabolite III were present in the urine of rats.

A more detailed report of the breakdown of Bux in soils was reported in 1972.[163] Again, these studies were performed using only the ^{14}C-carbonyl-labeled insecticide; thus, no information was obtained on the fate of the phenolic moiety after cleavage of the carbamic acid ester. Hydrolysis was a major route of degradation, as indicated by the release of ^{14}C-carbon dioxide from silt soil fortified with radiolabeled Bux. After one week, 33% of the applied radiocarbon was liberated as ^{14}C-carbon dioxide, and this continued over the 9-week experimental period, finally accounting for 68% of the dose. At the end of the experiment, the parent compound (6% of the dose) and a metabolite (1% of the dose) were the only materials extracted from the soil. The metabolite was identified as 3-(1-hydroxy-1-methylbutyl)-phenyl N-methylcarbamate (Figure 7.26, III). An equivalent of 16% of the applied Bux could not be extracted from the soil with methanol.

The rate of degradation of Bux in soils was related to the presence of microorganisms and to the formulation of the insecticide. ^{14}C-Carbon dioxide evolution in nonsterile soils was twice that in sterile soils. The degradation of Bux in non-

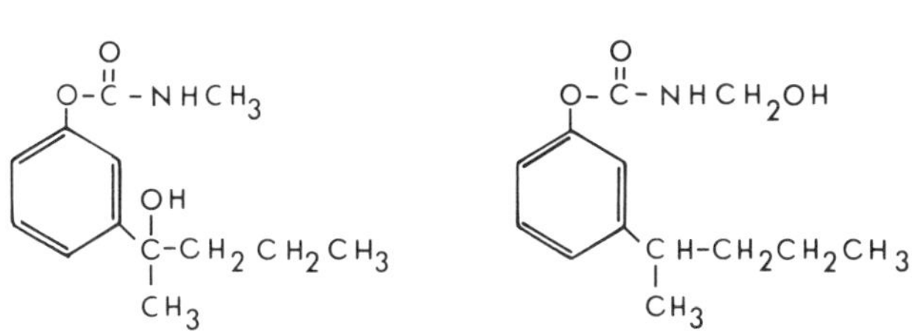

(III) (IV)

FIGURE 7.26. Oxidative metabolic products of Bux. Metabolite III, has been reported in rats and in soil, metabolite IV was found in rats.

sterile soil was slower when the insecticide was applied as granules than when applied as a solution directly to the soil.

RE 11775 AND RE 5365

RE 11775 is the N-benzenesulfenyl derivative of 3-(2-butyl)phenyl N-methylcarbamate (RE 5365). The benzenesulfenyl derivative is less toxic to mammals and a more effective mosquito larvicide than the parent N-methylcarbamate. A comparison study has been conducted on the metabolism of the two carbamate insecticides in rats and in rat liver microsomes, and on their degradation by light.[164]

With a compound like RE 11775, one would expect that the metabolism would be complicated because of the many sites susceptible to attack. These include the sulfur atom, the N-methyl group, the carbamoyl group, the phenyl ring, and each carbon atom of the butyl moiety. Consideration of all the metabolites and their intermediates which could be formed is an excellent exercise in theoretical chemistry. From a practical standpoint, however, the nature and route of formation of metabolites formed in the rat appears simple.[164] There is, for example, no difference in the ultimate fate of RE 11775 and that of RE 5365 (Figure 7.27). With each compound, over 50% of the material is hydrolyzed at the carbamate linkage in a 48-hr period after oral administration. Of a ^{14}C-carbonyl-labeled dose of each compound, 30% was voided in the urine, and almost quantitative excretion by this route was apparent 24 hr after the ^{14}C-butyl-labeled insecticides were given. The compounds, then, are obviously metabolized and rapidly eliminated from the body, a maximum of 30% as carbamate materials and 70% as hydrolytic products.

The results of these studies do not suggest metabolic differences as the reason for the lower mammalian toxicity of the N-benzenesulfenyl derivative (RE 11775). It may be that its enhanced toxicity to insects would be a function of lower rates of hydrolysis, as was the case for N-acetyl Zectran[165] and for N-(2-toluenesulfenyl) carbofuran.[166] However, this point was not considered in the metabolism study of RE 11775.[164]

Possibly, minor changes in the production of carbamate metabolites, such as those shown in Figure 7.28, play a role in the different toxicity of RE 11775 and RE 5365. The metabolites pictured are less than 0.1 times as active as plasma cholinesterase inhibitors than is RE 5365, but another metabolite, the 4-hydroxybutyl analog of RE 5365, is three times more potent. Although this metabolite was not present in the rat urine, or at least only in trace amounts, it very likely was an intermediate in the formation of terminal oxidative and hydrolytic products (Figures 7.28 and

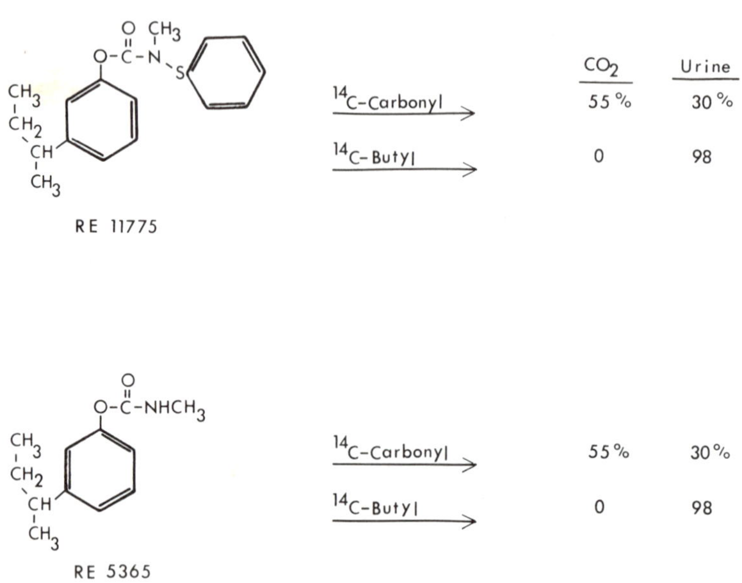

FIGURE 7.27. Fate of ^{14}C-carbonyl- and ^{14}C-butyl-labeled RE 11775 and RE 5365 in orally treated rats. Values are percent of dose after 48 hr.[164]

7.29). It is interesting that RE 11775, although less toxic to mammals, was three times more effective as an inhibitor of plasma cholinesterase than RE 5365. The rapid rate of hydrolysis in the animal likely negates its inherent toxicity.

The rat metabolism study[164] was not designed to determine whether modification and/or cleavage of the N-benzenesulfenyl group of RE 11775 occurred before or after hydrolysis of the carbamic acid ester. If the carbamate ester linkage is broken first, the metabolic route of the compound is probably not responsible for its lower mammalian toxicity. If, on the other hand, RE 11775 is partially metabolized by sulfur oxidation to the sulfoxide and sulfone prior to cleavage of the carbamate ester, it is possible that RE 5365 would never be formed in large quantities in the body. Should the sulfoxide and sulfone derivatives of RE 11775 be nontoxic, the effect of the derivatized compound on animals would be less than that of RE 5365. RE 11775 sulfoxide and sulfone were not isolated and identified as metabolites in the rat, but tentative identification of the sulfoxide was made for a compound formed by liver microsomes. An authentic sample of the sulfoxide was only 1/30 as potent as RE 11775 as an inhibitor of plasma cholinesterase, but it demonstrated high mosquito-larvicidal activity at 0.04 ppm. The sulfone derivative was active at 10 ppm; its anticholinesterase activity was not reported.

RE 5030 AND BUTACARB

In-vitro and/or in-vivo metabolism of 3-*tert*-butylphenyl N-methylcarbamate (RE 5030) and of 3,5-di-*tert*-butylphenyl N-methylcarbamate (Butacarb) have been investigated in insects and mice.[167,168] Structures for both compounds are shown in Figure 7.30. Although ^{3}H-labeled insecticides were used to some extent, the nature of the metabolites was determined primarily on the basis of their reactions to various chromogenic reagents. Inasmuch as this technique has limited utility in the actual identification of metabolites, the data were sufficient to indicate the general metabolic fate of these two insecticides in insects and mammals.

Hydrolysis of the carbamate ester was evident in both insect and mammalian species.[167,168] However, in-vitro studies led to the speculation that most of the "hydrolysis" of these two carbamate insecticides was actually catalyzed by a microsomal NADPH-dependent enzyme system.

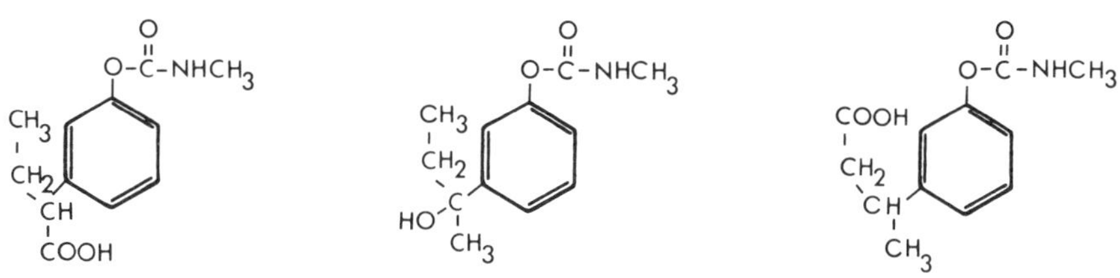

2 %   1 %   5 %

FIGURE 7.28. Carbamate metabolites in the 0–48-hr urine of rats treated with RE 11775 whose levels, conjugated or unconjugated, were equivalent to 1% or more of the administered dose.[164]

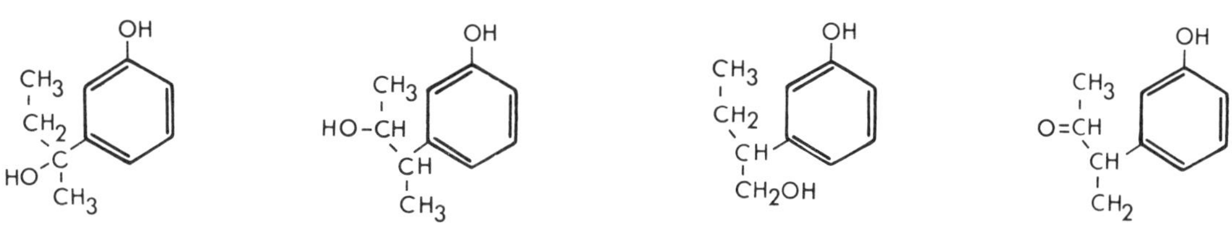

25 %   11 %   7 %   6 %

FIGURE 7.29. Examples of the types of hydrolysis products, free and/or conjugated, and the percent of dose detected in the 0–48-hr urine of rats treated with RE 11775.[164]

Both compounds were also hydroxylated on the N-methyl group and underwent various hydroxylations of the butyl moiety (Figure 7.31). Oxidation of the N-methyl group was less predominant with Butacarb than with RE 5030. Ring-hydroxylation was not evident with either compound.

The degree of oxidative detoxication of the insecticides was correlated with the susceptibility of various insects to the carbamates. Hence, flies were able to tolerate much larger doses of the insecticides than were grass grubs, the latter being less able to oxidize the toxicant. It should be pointed out, however, that the degree of oxidation was based primarily upon the amount of butyl-oxidized products recovered from the organisms or enzyme preparations. Whether the butyl groups were oxidized before or after cleavage of the carbamate ester bond was not established.

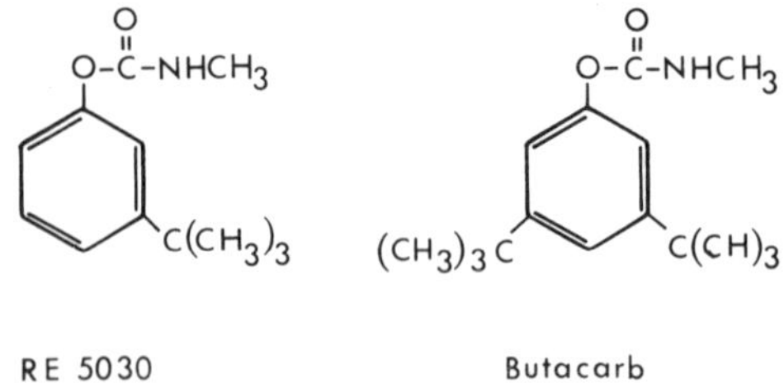

FIGURE 7.30. Structures of RE 5030 and Butacarb.

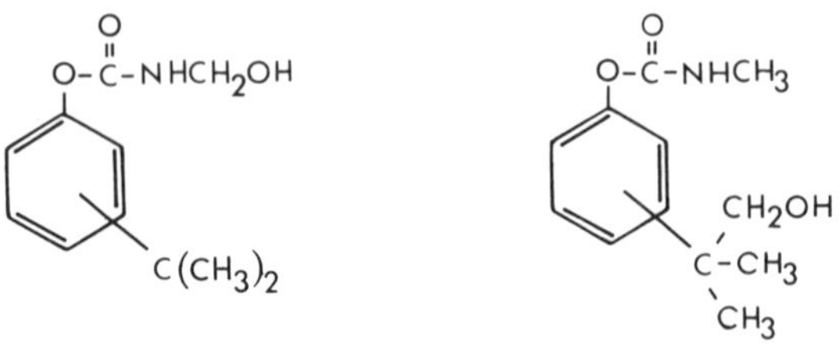

(I) (II)

FIGURE 7.31. Examples of the types of oxidative carbamate metabolites of RE 5030 and Butacarb formed in vivo and/or in vitro by insects and mice.

BAYGON
(Propoxur, Unden, Blattanex, B 39007, Suncide, Arprocarb)

Baygon (2-isopropoxyphenyl N-methylcarbamate) is used extensively for the control of household pests, mosquitoes, and other insects where the use of the material on food crops is not required. Several metabolism studies of Baygon have been conducted with insects, especially houseflies and mosquitoes, but its metabolism in mammals and plants has not received a great deal of attention.

The insecticide was hydrolyzed in rats by about 30% within 48 hr of treatment, based on the quantity of radioactive carbon dioxide liberated after treatment of animals with the ^{14}C-carbonyl-labeled Baygon[26] (Figure 7.32). The majority of the dose, 60%, was eliminated in the urine. Metabolism of the carbamate was essentially the same in animals treated by injection or orally via a stomach tube. Piperonyl butoxide injected along with Baygon appeared to increase the amount of ^{14}C-carbon dioxide liberated from the animals. This effect of piperonyl butoxide on the mammalian metabolism of Baygon has not been further evaluated.

An indication that humans also hydrolyze about 30% of a Baygon dose has been reported.[169] Since the human urine was treated with acid prior to analysis, it was not possible to tell how much of the phenol was present in the free or in the conjugated form. It was evident, however, that excretion was very rapid, with 90% of the phenol detected in urine voided within 10 hr of administration of the insecticide.

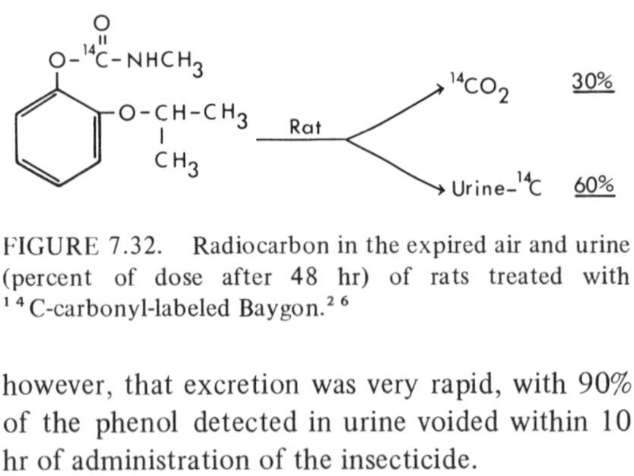

FIGURE 7.32. Radiocarbon in the expired air and urine (percent of dose after 48 hr) of rats treated with ^{14}C-carbonyl-labeled Baygon.[26]

Rat liver microsome studies revealed that Baygon was metabolized primarily to 2-isopropoxyphenyl N-hydroxymethylcarbamate (Figure 7.33, I).[25,69] Other carbamate metabolites identified included 2-hydroxyphenyl N-methylcarbamate (Figure 7.33, II) and 2-isopropoxy-5-hydroxyphenyl N-methylcarbamate (Figure 7.33, III). Baygon was hydrolyzed by a number of mouse and human brain esterases,[72,73] but it was not hydrolyzed by human sera.[70]

Houseflies absorbed 90% of topically applied Baygon within 24 hr.[155] About 30% of the absorbed portion of isopropoxy-^{14}C-Baygon was liberated as ^{14}C-carbon dioxide, and an almost equivalent amount was eliminated in the excreta of resistant flies. On the other hand, susceptible

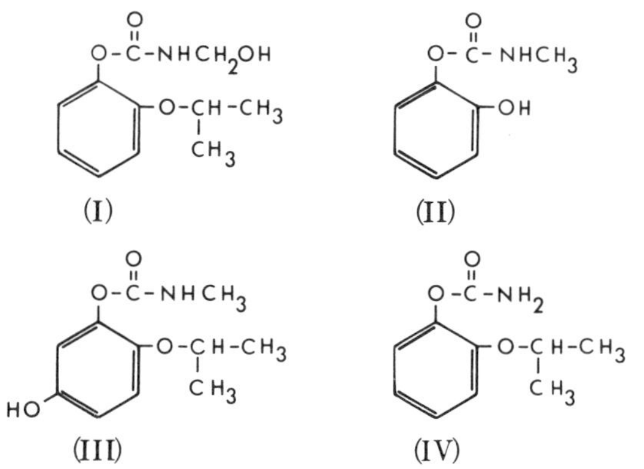

FIGURE 7.33. Oxidative metabolites of Baygon formed in plants and animals. The desmethyl analog (IV) has been reported only in insects.

houseflies eliminated only 9% of the absorbed radiocarbon. Pretreatment of the resistant insects with piperonyl butoxide reduced the ^{14}C-carbon dioxide liberation markedly, but only slightly reduced the amount of radioactivity in the excreta.

A different rate of Baygon metabolism was exhibited by houseflies injected with the carbamate,[170] suggesting that the insect integument plays an important role in the degradation of this carbamate insecticide. In this instance, less than 1% of the dose was liberated as volatile compounds from Baygon radiolabeled in the N-methyl, carbonyl, or isopropoxy positions. The released fragment from the isopropoxy-labeled Baygon was identified as acetone, and its production was identical in the resistant and in the susceptible houseflies.

Hydrolysis of the carbamate ester was not an important metabolic reaction in the detoxication of Baygon by houseflies.[171] The same was true with the yellow fever mosquito adult, but German cockroaches hydrolyzed 21% of a ^{14}C-carbonyl-labeled dose of Baygon. The primary metabolites formed by the housefly were 2-isopropoxy-5-hydroxyphenyl N-methylcarbamate (Figure 7.33, III), 2-hydroxyphenyl N-methylcarbamate (Figure 7.33, II), 2-isopropoxyphenyl N-hydroxymethylcarbamate (Figure 7.33, I), and 2-isopropoxyphenylcarbamate (Figure 7.33, IV). All of the hydroxylated carbamate metabolites existed both in the free and in the conjugated form.

Generally, resistant houseflies metabolized Baygon to a greater degree than did susceptible insects, especially in regard to the formation of conjugate metabolites. The rate of Baygon metabolism by resistant flies was reduced by application of piperonyl butoxide.

The route of degradation of Baygon in mosquitoes was essentially the same as in houseflies.[172] Water-soluble metabolites were cleaved by treatment with β-glucosidase, acid phosphatase, β-glucuronidase, and sulfatase. The β-glucosidase enzyme preparation was the most active, resulting in 50% cleavage of the conjugates. Comparable hydrolysis was obtained with 1*N* hydrochloric acid.

Metabolic O-dealkylation of the isopropoxy group from Baygon varied with the species of insect.[171] For example, production of acetone-^{14}C from isopropoxy-^{14}C Baygon represented 5% of the dose in the navel orange worm adult, 17% in the yellow mosquito adult, and 23% in the German cockroach.

In-vitro metabolism of Baygon by fly abdomen enzymes and by mosquito larval enzymes yielded the same metabolites found in the living insects.[104-106] Biochemical genetic studies[173] indicated that the second chromosome was important in conferring high carbamate resistance and high microsomal oxidase activity to the housefly (see Chapter 10). The fifth chromosome also was shown to be important in the metabolic degradation of Baygon in these insects.

The metabolic fate of Baygon in plants is essentially the same as in insects.[27,51,174] However, there was no indication that the desmethyl analog (Figure 7.33, IV) was formed in the plant, and the hydroxylated metabolites of Baygon were immediately converted to glycoside conjugates. As time after treatment of the plants increased, the percentage of the radioactive residues as water-soluble products declined with a corresponding increase in residues unextractable from the plant pulp. Since 90% of the water-soluble metabolites formed in bean plants was as the glycoside of 2-hydroxyphenyl N-methylcarbamate (Figure 7.33, II),[27] this product probably constituted a significant portion of the terminal residues.

ZECTRAN
(Mexacarbate, Dowco 139)

Zectran (4-dimethylamino-3,5-dimethylphenyl N-methylcarbamate) was metabolized predominantly by hydrolysis when administered orally to mice.[165] By 6 hr after treatment with

^{14}C-carbonyl-labeled Zectran almost 70% of the dose had been expired as ^{14}C-carbon dioxide, and this increased to 75% after 48 hr (Figure 7.34). A total of 17% of the radiocarbon was eliminated in the urine and 7% in the feces. Almost identical results were obtained when the carbonyl-^{14}C Zectran was injected into rats.[26] In this case 77% of the radioisotope was expired as ^{14}C-carbon dioxide, 12% eliminated in the urine, and 2.5% excreted in the feces. The metabolites formed in the animals were not identified.

The nature of radioactive metabolites in the urine of dogs fed a diet containing 20 ppm of ring-3-^{14}C-methyl-labeled Zectran for 7 days has been reported.[175] Urine collected while the animals were on treatment and for 5 days after feeding was combined prior to analysis. The data were expressed as percentages of total radioactivity in the urine; unfortunately, no indication of the total excretion of the administered dose was given. The only free urinary metabolite identified was 4-dimethylamino-3,5-dimethylphenol, which accounted for 2% of the recovered ^{14}C. An additional 6% was extractable into ether, but was not identified. The water-soluble metabolites, comprising 92%, were composed of sulfate and glucuronide conjugates (86%) of the free phenol and of conjugated 2,6-dimethylhydroquinone (6%).

In-vitro studies of the metabolism of Zectran in rat liver and human liver enzyme systems showed that as many as nine metabolites are formed which contain the carbamate moiety.[25,57,58,69] The quantity of individual metabolites varied, but usually 4-dimethylamino-3,5-dimethylphenyl N-hydroxymethylcarbamate (Figure 7.35, IV) and 4-methylamino-3,5-dimethylphenyl N-methylcarbamate (Figure 7.35, II) constituted a majority of the metabolites formed. Other products proposed included 4-methylformamido-3,5-dimethylphenyl N-methylcarbamate (Figure 7.35, I) and 4-amino-3,5-dimethylphenyl N-methylcarbamate (Figure 7.35, III). Human liver enzymes and rat liver microsomes were equally effective in degrading Zectran.[57,58] That the methyl groups of the dimethylamino moiety of Zectran were susceptible to oxidation by rat hepatic enzymes was first shown in 1961.[24] Zectran in the presence of microsomes and NADPH yielded formaldehyde, but its production via the N-hydroxymethylcarbamate route was ruled out, since 4-dimethylamino-3,5-dimethylphenol gave the same results.

Experiments using the blood of dogs and rats showed that plasma proteins had a greater affinity for Zectran than whole blood or serum.[176] In addition, Zectran metabolism was greater in whole blood of humans than in whole blood of the dog or rat. The major ether-extractable metabolite formed in human blood was 4-methylamino-3,5-dimethylphenyl N-methylcarbamate (Figure 7.35, II). Most of the metabolites, however, were of a water-soluble nature and could not be extracted from the aqueous phase with ether.

The fate of Zectran in spruce budworm, tobacco budworm, and housefly larvae did not involve any hydrolysis of the carbamate ester.[177] Following topical treatment, nine metabolites were observed in the spruce budworm and in the tobacco budworm, and ten metabolites were found in the housefly. The major products identi-

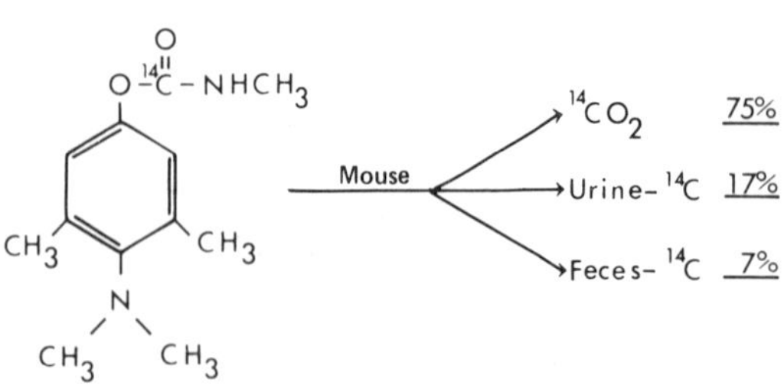

FIGURE 7.34. Elimination of radiocarbon (percent of dose after 48 hr) from mice treated orally with ^{14}C-carbonyl-labeled Zectran.[165]

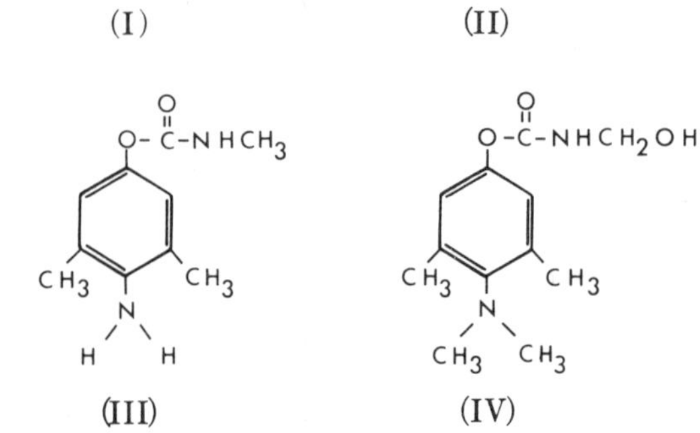

FIGURE 7.35. Some of the proposed oxidative metabolites of Zectran formed by plants and animals.

fied were 4-methylamino-3,5-dimethylphenyl N-methylcarbamate (Figure 7.35, II), 4-amino-3,5-dimethylphenyl N-methylcarbamate (Figure 7.35, III), 4-methylformamido-3,5-dimethylphenyl N-methylcarbamate (Figure 7.35, I), and 4-formamido-3,5-dimethylphenyl N-methylcarbamate. Also, 4-dimethylamino-3,5-dimethylphenyl N-hydroxymethylcarbamate (Figure 7.35, IV) was tentatively proposed as one of the metabolites. Much of the oxidative metabolism of Zectran appeared to occur on the insect cuticle. Housefly microsomes replayed the same metabolic pathway as that found in the spruce budworm and other insects.[104,105]

With the exception of the N-hydroxymethyl derivative (Figure 7.35, IV), plants transformed Zectran to the same products as those found in insects.[51] The majority (80%) of the ^{14}C-carbonyl-labeled metabolites could not be extracted from the plant tissue 6 days after treatment. Of the injected radioactivity, 3% was organo-soluble and 5% was present as water-soluble metabolites. Over the 6-day period, there was a 14% loss of the radioactivity injected into the plants. The conjugates, when treated with β-glucosidase, yielded five aglycones, two of which were inhibitors of cholinesterase.[27]

When ^{14}C-carbonyl-labeled Zectran was applied to the surface of bean plants, there was 50% loss of total radiocarbon after 61 hr. However, half of the parent compound had been destroyed after only 3 hr.[51] Thus, it was apparent that metabolites containing the carbamate moiety were being formed on the surface of the leaves.

Extraction of broccoli plants with methanol 10 days after treatment with ring-^{14}C-labeled Zectran left 18% of the ^{14}C-residues in the pulp.[178] Detailed research into the nature of these pulp residues indicated that the radiocarbon was associated with the plant lignin fraction and consisted of 4-dimethylamino-3,5-dimethylpyrocatechol and 2,6-dimethylhydroquinone. No metabolite containing the intact carbamate ester was identified. Thus, these results differ markedly from those reported for the metabolism of Zectran in bean plants, where hydrolysis was found to be an insignificant pathway.

N-ACETYL ZECTRAN

N-Acetylation of a number of aryl N-methylcarbamates was an ingenious attempt to significantly reduce the mammalian toxicity of these compounds.[179] When 4-dimethylamino-3,5-dimethylphenyl N-methylcarbamate (Zectran) was acetylated, its LD_{50} to mice was changed from 40 mg/kg to more than 1,000 mg/kg.[180] Fortunately, acetylation had little effect on the toxicity of the compound to spruce budworms; the LD_{50} for Zectran was 1.4 μg/g, and for acetyl-Zectran it was 2.4 μg/g.

The selective nature of acetyl-Zectran and its lower toxicity to mammals as compared to Zectran have been investigated relative to the metabolism of the compound in mice and in spruce budworm larvae.[165] In insects, the N–O bond of acetyl-Zectran is cleaved to produce the toxic metabolite Zectran (Figure 7.36). For example, as much as 40% of a topically applied dose of acetyl-Zectran was deacylated by insect larvae. Internal Zectran (Figure 7.36, I) was then detoxified by hydrolysis to yield the parent phenol (Figure 7.36, II). In addition, a small amount of 4-methylamino-3,5-dimethylphenyl N-methylcarbamate (Figure 7.35, II), 4% of the dose, was formed by the insects.

Hydrolysis of the carbamate ester of acetyl-Zectran was much slower in insects than in mice; i.e., collection of ^{14}C-carbon dioxide from animals treated with ^{14}C-carbonyl-labeled material showed that after 2 hr 9% hydrolysis had occurred in the insects, whereas there was 58% hydrolysis by the mice. After 48 hr, the comparable values were 62% and 72%, indicating that splitting the carbamate ester bond is important in both species, but that its rate determines the toxicity of the insecticide to the organism. That hydrolysis of acetyl-Zectran was a major detoxication route by the spruce budworm contradicted the finding that hydrolysis of Zectran in these insects was of only minor importance.[177]

Breaking of the carbamate ester bond did not account for the difference in toxicity of acetyl-Zectran and of Zectran to mice, however.[165] With both compounds, approximately 75% hydrolysis occurred during a 48-hr period after treatment. On the other hand, cleavage of the carbamate ester, at least in part, was a factor in the lower toxicity of N-(2-toluenesulfenyl) carbofuran to mice as compared to carbofuran.[166]

MATACIL
(Aminocarb, B 44646, A 363)

It is necessary to rely heavily on in-vitro studies to gain insight into the metabolic fate of Matacil (4-dimethylamino-3-methylphenyl N-methylcarbamate) in animals. Its metabolism in

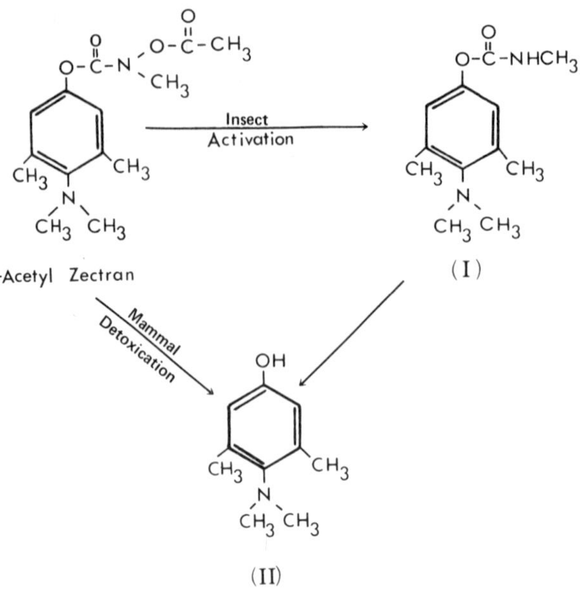

FIGURE 7.36. Major differences in the metabolism of N-acetyl Zectran by mammals and insects which contribute to its selective toxicity.

living plants has been more fully investigated, but even here many of the metabolic products remain unknown.

Hydrolysis predominates in the metabolism of Matacil in animals.[26] Almost 70% of a ^{14}C-carbonyl-labeled dose of Matacil injected into rats was expired as ^{14}C-carbon dioxide after 48 hr; 25% of the administered radiocarbon was eliminated in the urine and 4% was voided in the feces. Although the nature of the metabolites in the excreta was not evaluated, the evidence certainly suggests that carbamate-type metabolites were formed.

In-vitro studies using rat liver microsomes demonstrated that Matacil is metabolized into as many as six products which contain the carbamate moiety.[25,69] Human liver enzymes did not degrade Matacil as thoroughly as did rat liver enzymes, but the same metabolites were formed in each system.[57,58] Metabolites occurring in the largest quantities in both tissues were 4-methylamino-3-methylphenyl N-methylcarbamate (Figure 7.37, II) and 4-dimethylamino-3-methylphenyl N-hydroxymethylcarbamate (Figure 7.37, IV). 4-Amino-3-methylphenyl N-methylcarbamate (Figure 7.37, III) was also identified as a microsomal product of Matacil. The N-hydroxymethyl derivative (Figure 7.37, IV) and the original compound (Figure 7.37, I) were found to have equal anticholinesterase activity, whereas the 4-amino (Figure 7.37, III) and 4-methylamino analogs (Figure 7.37, II) were ten times more potent than the parent compound.[25]

In houseflies, a topically applied dose of ^{14}C-carbonyl-labeled Matacil was not hydrolyzed to any great degree.[155] Although more than 80% of the insecticide was absorbed within 24 hr, only 5% of that absorbed was converted to ^{14}C-carbon dioxide. This amounted to only one fourth of that expired when houseflies were treated with Mesurol or with Banol. The flies dosed with Matacil eliminated 45% of the radioactivity in the excreta after 24 hr.

Although the degradation products formed from Matacil by the houseflies were not identified, in-vitro studies have shown that housefly microsomes yield the same series of metabolites produced by mammalian liver.[104,105] In addition, 4-methylformamido-3-methylphenyl N-methylcarbamate and 4-formamido-3-methylphenyl N-methylcarbamate were tentatively identified as metabolic products of housefly microsomes. These are apparent precursors to the methylamino (Figure 7.37, II) and amino (Figure 7.37, III) analogs of Matacil.

The biotransformation of Matacil in plants is of the same order as that in animal systems. With the exception of the N-hydroxymethyl derivative

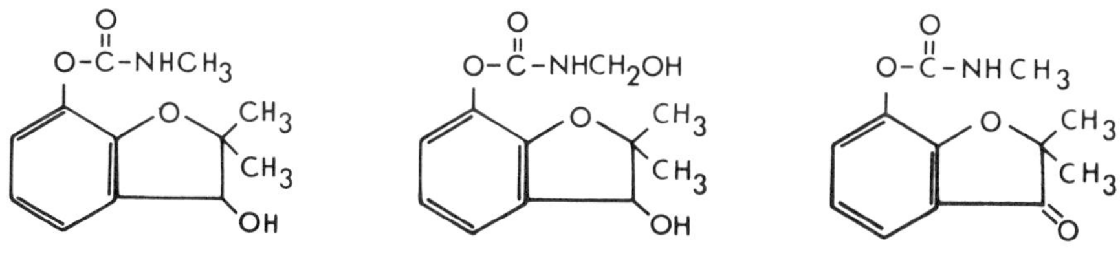

(I) (II) (III)

FIGURE 7.40. Carbamate metabolites of carbofuran and of N-(2-toluenesulfenyl) carbofuran.

carbofuran eliminated 50% of the administered radiocarbon after 24 hr.[184] Less than 2% of the dose was recovered as ^{14}C-carbon dioxide, indicating that N-demethylation was a minor degradative pathway of carbofuran in mice. The major primary metabolites isolated from urine were 3-hydroxy carbofuran (Figure 7.40, I) and a small amount of 3-keto carbofuran (Figure 7.40, III). Conjugated metabolites were mostly derived from hydrolytic products, such as 3-keto phenol and some carbofuran phenol, but the carbamate 3-hydroxy carbofuran (Figure 7.40, I) was also present as a conjugate. No 3-hydroxy-N-hydroxymethyl carbofuran (Figure 7.40, II) appeared either free or conjugated in mice. However, this compound was a major biotransformation product of N-(2-toluenesulfenyl) carbofuran, an insecticide apparently converted immediately to carbofuran in mice.[185] In this instance, 5% of the orally administered compound was eliminated in the urine as nonconjugated 3-hydroxy-N-hydroxymethyl carbofuran, whereas 18% of the dose was excreted as the conjugate. The only other metabolite which existed at higher levels in the urine was the conjugate of 3-hydroxy carbofuran, which constituted almost 30% of the applied radioactivity.

The excretion and metabolism of carbofuran by rats was affected by pretreating the animals with methylmercury hydroxide (10 mg/kg/day) or chlordane (20 mg/kg/day) for 2 days.[84] Hepatic microsomes from animals exposed to the former compound produced less carbofuran ring-^{14}C-labeled metabolites than did the control animals. Chlordane treatments enhanced the microsomal degradation of carbofuran; the greatest increase involved production of 3-hydroxy carbofuran (Figure 7.40, I).

The excretion of carbofuran-^{14}C equivalents from orally treated animals was 30% of the dose after 6 hr in the controls and 70% in animals pretreated with chlordane. Methylmercury hydroxide-treated rats also excreted carbofuran products more rapidly than did control animals, but not as rapidly as the chlordane-treated rats. Final removal of the compound, however, was not affected by treatment of rats with methylmercury hydroxide or with chlordane; after 3 days, all animals had excreted approximately 90% of the administered radioactivity in the urine.

The only significant difference in the nature of the urinary metabolites occurred in rats treated with methylmercury hydroxide. Here, the amount of organo-soluble products was about twice that observed in control or chlordane-treated animals. This possibly means that the methylmercury hydroxide specifically affected the conjugative metabolic mechanisms in the animals, a hypothesis further supported by the ratio of metabolites in the 0–6-hr urine existing as free metabolites, glucoside conjugates, and sulfate conjugates. Based on their concentrations as percent of applied dose, the ratios of these materials in the control animals and in the chlordane-treated animals was 1:4:8, whereas in the methylmercury hydroxide-treated animals the ratio was 1:1:4. The enhanced excretion rate mediated by chlordane treatment probably resulted from an enhanced rate of initial oxidation of the carbofuran molecule, whereas with methylmercury hydroxide initial metabolism was not affected, but the rates of glucuronide and sulfate conjugation of the oxidative metabolites were significantly reduced.

Carbofuran was hydrolyzed much more rapidly in laying hens than has been reported for rats.[186] Studies with ^{14}C-carbonyl- and ^{14}C-ring-labeled carbofuran showed that 72% of the carbamate disappeared after 24 hr and that a total of 66% of the ring-labeled dose was eliminated in the excreta, mostly as hydrolysis products. An estimated 34% of the dose remained in the body after 24 hr and 35% of this was thought to be carbamate-type metabolites. Analysis of the feces showed that 3-hydroxy carbofuran was the predominate material which accounted for 50% of the excreted radiocarbon. Its hydrolysis product was the major non-carbamate residue; other primary metabolites in the excreta included N-hydroxymethyl carbofuran, 3-hydroxy-N-hydroxymethyl carbofuran, and trace amounts of the 3-keto analog (Figure 7.40, III). Water-soluble metabolites constituted 43% of the radiocarbon in the 0–24-hr feces and included conjugated 3-keto phenol as the major acid-cleaved material as well as small quantities of conjugated N-hydroxymethyl carbofuran, 3-hydroxy-N-hydroxymethyl carbofuran and 3-keto carbofuran. A large portion of the water-solubles was not hydrolyzed by acid treatment.

Livers of hens sacrificed 6 hr after treatment with ^{14}C-carbonyl carbofuran contained ^{14}C-residues which were mostly immune to acid hydrolysis. Carbamate metabolites detected at low levels in the livers were 3-hydroxy carbofuran, N-hydroxymethyl carbofuran, and 3-hydroxy-N-hydroxymethyl carbofuran.

Metabolism of carbofuran in lactating cows[187] was shown to duplicate that reported in rats.[183] Within 24 hr after administration of a single oral dose of ^{14}C-carbonyl-labeled carbofuran, 20% was eliminated in the urine and approximately 2% in the milk. With ring-^{14}C carbofuran, 94% of the radiocarbon was collected in the urine and 0.2% in the milk. In both studies, less than 1% was recovered from the feces. These patterns of elimination of ^{14}C-carbonyl- and ^{14}C-ring-labeled carbofuran were generally the same when the compound was administered to lactating cows via a rumen fistula.[188] The percentage of the dose voided in the milk (2.5%) was identical for both radioactive preparations of carbofuran.

Carbamate metabolites free, and/or conjugated, detected in the milk of cows treated orally were 3-hydroxy carbofuran, 3-hydroxy-N-hydroxymethyl carbofuran, 3-keto carbofuran, and small amounts of the applied insecticide. Most of the residues resulting from treatment with ^{14}C-carbonyl-labeled carbofuran consisted of radioactive natural milk components, apparently as a result of incorporation of the liberated ^{14}C-carbon dioxide.[187]

In the milk of fistulated animals,[188] 3-hydroxy carbofuran was the only material identified, and this accounted for only 0.4% of the total ^{14}C-residues. Based on analysis with the ^{14}C-carbonyl-labeled carbofuran study, the remainder of the ^{14}C-residues in the milk was distributed equally between casein and lactose. Milk containing ring-^{14}C carbofuran metabolites was not analyzed although the products were in concentrations equal to those from ^{14}C-carbonyl-labeled carbofuran. It is unlikely that the ring-^{14}C metabolites would have been associated with casein and lactose, as was reported for the carbonyl-^{14}C residues.

Rat liver enzymes degraded 40% of the radioactive carbofuran added to the reaction medium.[183] Of the applied radiocarbon, 20% was oxidized to 3-hydroxy carbofuran and 7% was present as the N-hydroxymethyl derivative; there was 2% conversion to 3-hydroxy-N-hydroxymethyl carbofuran, and 4% of the carbofuran was hydrolyzed.

The major metabolite produced by housefly microsomes was tentatively identified as the 2-hydroxymethyl derivative of 3-keto carbofuran (Figure 7.40, III),[184] the free metabolite accounting for 38% of the added radioactivity. The fact that this product has never been reported to exist in vivo in any plant or animal species makes this one of the few cases where a major metabolite formed from a carbamate insecticide in vitro has not been found at significant levels in the living organism. The other carbamate metabolites isolated from the housefly enzymes were 3-hydroxy-N-hydroxymethyl carbofuran (25%) and trace amounts of 3-keto carbofuran. In addition to these materials, which existed predominantly as the free metabolites, 3-hydroxy carbofuran and another carbamate metabolite, proposed to be a ring-hydroxylated form of 3-keto carbofuran, were also detected as conjugates. About 15% of the added radioactivity was identified as hydrolysis products of carbofuran, primarily as the 3-keto phenol.

Houseflies metabolized carbofuran by oxidation to the 3-hydroxy derivative followed by conjugation of this metabolite as a glucoside.[183] Other metabolites formed, conjugated and/or free, in lesser amounts included 3-hydroxy-N-hydroxymethyl carbofuran, N-hydroxymethyl carbofuran, and 3-keto carbofuran. Carbamate-resistant houseflies absorbed topically applied carbofuran at about the same rate as susceptible insects, but excretion was more rapid in the resistant strains.[155,190] The ^{14}C-components of the body and excreta were of the same general nature in resistant and in susceptible insects, although the percentage present as conjugates was greater in resistant flies.[190] Neither N-methyl-^{14}C-labeled nor ring-^{14}C-labeled carbofuran was metabolized to appreciable quantities of ^{14}C-carbon dioxide.[155,184] These data are in agreement with the metabolic pathway of carbofuran in houseflies as indicated by identification of the metabolites occurring in the body and excreta.

In the salt marsh caterpillar, ring-^{14}C-labeled carbofuran was converted primarily to the glucoside of 3-keto phenol.[184] None of the parent compound was recovered, but there were small amounts of 3-keto carbofuran. It was obvious that 3-hydroxy carbofuran was very rapidly oxidized and hydrolyzed and that the phenolic derivative was immediately conjugated in these insects. Enzymes from mosquitoes metabolized 26% of the added carbofuran.[106] Primary metabolites were 3-hydroxy carbofuran and 3-keto carbofuran, but a number of biotransformations were uncharacterized. Earthworms treated with ring-^{14}C carbofuran excreted only 7% of the dose after 48 hr.[191] Acetonitrile extraction of the worms removed 8% of the radiocarbon; a small portion of this consisted of 3-hydroxy carbofuran, but the majority of the radioactive residues was not identified.

Bean plants injected with ^{14}C-carbonyl- and ^{14}C-ring-labeled carbofuran showed a 20% loss of radiocarbon when analyzed 28 days after treatment as compared to 3-day analyses.[192] During the first 3 days, however, there was 30% loss of the applied radiocarbon from the ^{14}C-carbonyl-labeled material and only 5% loss from the ^{14}C-ring-labeled carbofuran, suggesting initial hydrolysis of the carbamate and transpiration of released ^{14}C-carbon dioxide.

The major residue in the plant after 28 days was 3-hydroxy carbofuran, which accounted for 60% of the total ^{14}C-residues. Only 5% of the metabolite was present *per se;* the remainder was recovered after acid treatment of the water-soluble plant extract. About 6% of the residues was 3-keto carbofuran, and it was there predominantly as the conjugate. Less than 1% of the ^{14}C was conjugated 3-hydroxy-N-hydroxymethyl carbofuran, and 6% was composed of conjugates of carbofuran phenol and the phenolic derivatives of 3-hydroxy and 3-keto carbofuran. The remaining residues were unknown metabolites, both carbamate and hydrolysis products, and some radiocarbon (5%) which could not be extracted from the plant pulp.

The general metabolic fate of carbofuran in isolated cotton and bean leaves was not unlike that found in bean plants.[184] However, the N-hydroxymethyl analogs were not reported, and small quantities of metabolites thought to be ring-hydroxylated forms of 3-keto carbofuran were detected. The same pattern of carbofuran metabolism was observed when the labeled compound was applied to the roots of mugho pines.[193] Most of the radiocarbon was translocated to the needles, and it was noted that the metabolism of carbofuran was somewhat slower than in other plants and in animals.

Tobacco plants, 1 month old, when grown in a nutrient solution containing ring-^{14}C carbofuran, absorbed 80% of the applied radioactivity after 12 days;[194,195] 80% of this was located in the top and 20% in the roots of the plants. Radioactive residues in the latter consisted almost entirely of unextractable metabolites and a small amount of extracted material which was not identified. Products in the plant tops included the unaltered applied compound, small amounts of 3-hydroxy carbofuran, and a series of unidentified metabolites. The major aglycone produced by acid and enzymic hydrolysis was 3-hydroxy carbofuran. Approximately 50% degradation of the carbofuran occurred within 4 days, whether applied through the roots or through the petioles of isolated leaves.

It has been reported that metabolism of carbofuran in barley was inhibited by the herbicide chlorobromuron [3-(4-bromo-3-chlorophenyl)-1-methoxy-1-methylurea], although the herbicide did not affect carbofuran metabolism in corn.[196] Alachlor [2-chloro-2′,6′-diethyl-N-(methoxymethyl)acetanilide)] and butylate (S-ethyl diisobutylthiocarbamate) supposedly had little, if any, affect on metabolism of carbofuran in barley and corn. However, close evaluation of the data reveals that the concentration of individual carbofuran metabolites varied with the plant species and with the application of these three different herbicides. A more palatable conclusion would be that the metabolism of carbofuran in barley and corn was not appreciably affected by either of the herbicides, and instead there was only a tendency towards reduction in the concentration of 3-hydroxy carbofuran, particularly with chlorobromuron. According to reports from the same laboratory, carbofuran also inhibited chlorobromuron metabolism in barley.[197]

Hydrolytic products were the major radioactive constituents of alfalfa plants grown in soils containing labeled carbofuran.[189] Analysis of plants after 30 days revealed that 37% of the residues consisted of 3-hydroxy carbofuran glycoside and that 40% was present as the glycosides of the phenolic derivatives of 3-keto and 3-hydroxy carbofuran, each in equal amounts. Only 5% of the residues was extracted from alfalfa with ether; this consisted primarily of unaltered carbofuran and of the metabolites just mentioned, except that they were in the free form. Bean plants treated in a similar manner for 1 week produced the same products. Since other plant metabolism studies have shown much less hydrolysis of carbofuran, it is apparent that there was a great deal of phenol formation prior to its absorption from the soil by the alfalfa and bean plants.

N-(2-TOLUENESULFENYL) CARBOFURAN

The selective toxicity of N-(2-toluenesulfenyl) carbofuran [2,3-dihydro-2,2-dimethyl-7-benzofuranyl N-methyl-N-(2-toluenesulfenyl)carbamate] prompted a study of its metabolism by houseflies and by mice.[166] Obviously, the compound is a derivative of carbofuran, but N-(2-toluenesulfenyl) carbofuran is $\frac{1}{50}$ as toxic to mice and twice as toxic to the housefly.[185]

Based upon the metabolic patterns of N-(2-toluenesulfenyl) carbofuran in the two animals,[166] it was concluded that selective toxicity probably resulted from different pathways of degradation in insects and mammals. Thus, houseflies rapidly absorbed the insecticide and cleaved the N–S bond to produce carbofuran (Figure 7.41, II). Within 3 hr after topical application, up to 35% of the dose was detected as carbofuran in internal extracts of the flies, but none of the administered compound was found.

That carbofuran, and not N-(2-toluenesulfenyl) carbofuran, was the primary toxicant was supported by the anticholinesterase activity of the two materials. N-(2-Toluenesulfenyl) carbofuran was 76-fold less inhibitory to housefly acetylcholinesterase than carbofuran. In addition, the carbofuran analog was more toxic than carbofuran when applied topically to houseflies; LD_{50}'s are 3.2 and 6.7 μg/g respectively. This suggests that the N-(2-toluenesulfenyl) derivative serves as a slow release mechanism, so that the toxic component, carbofuran, is generated in lethal quantities in the insects at a rate which exceeds its rate of detoxication.

Conversion of N-(2-toluenesulfenyl) carbofuran to carbofuran (Figure 7.41, II) in mice could not be established, because the latter was not detected in the urine or feces. However, other metabolites appeared which may have been derived from carbofuran. Furthermore, the N-(2-toluenesulfinyl) metabolite (Figure 7.41, III) was present at low levels (1% of the dose) in the feces and would be a likely precursor to carbofuran. The metabolism data, however, support the existence of an additional pathway which does not include carbofuran as an intermediate in the metabolic degradation of N-(2-toluenesulfenyl) carbofuran.

According to the authors,[166] the major pathway of metabolism in mice likely involves the

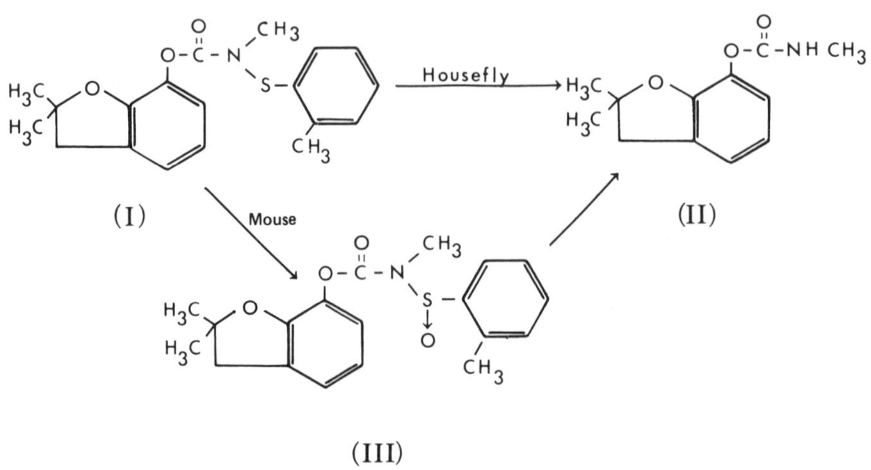

FIGURE 7.41. Proposed pathway of conversion of N-(2-toluenesulfenyl) carbofuran (I) to carbofuran (II) in houseflies and mice.[166]

intermediate, N-(2-toluenesulfenyl) 3-hydroxy carbofuran (Figure 7.42, IV). Although this product was not isolated, the presence of N-(2-toluenesulfenyl) 3-keto carbofuran (Figure 7.42, V) was indicative of the formation of the 3-hydroxy precursor. This may, indeed, be the principal degradation route of N-(2-toluenesulfenyl) carbofuran in mice, but confirmation is difficult, since the terminal residues would be the same whether metabolism proceeded through N-(2-toluenesulfenyl) 3-hydroxy carbofuran or through carbofuran *per se*. The fact that only 0.2% of the 90% or more of the dose voided in the urine was qualitatively unlike the metabolites in the urine of rats treated with carbofuran[183] lends some evidence that carbofuran was formed from N-(2-toluenesulfenyl) carbofuran in mice as it was in insects. Therefore, the rate of conversion of N-(2-toluenesulfenyl) carbofuran to carbofuran and the rate of detoxication of subsequent toxic metabolites would be important considerations in the explanation of the selective toxicity of this insecticide.

Detoxication of 3-hydroxy carbofuran (Figure 7.40, I) and of 3-keto carbofuran (Figure 7.40, III) must be involved in the metabolism of N-(2-toluenesulfenyl) carbofuran in insects and in mammals, since both derivatives are only $\frac{1}{3}$ to $\frac{1}{5}$ less toxic to mice than is carbofuran.[166] No toxicity data were given pertaining to the 3-hydroxy-N-hydroxymethyl analog (Figure 7.40, II), but N-hydroxymethyl carbofuran is approximately $\frac{1}{20}$ as toxic to mice when compared to carbofuran. None of the above hydroxylated compounds are toxic to houseflies, but all are similar to N-(2-toluenesulfenyl) carbofuran with respect to inhibition of housefly head acetylcholinesterase.

Internal extracts of houseflies treated with N-(2-toluenesulfenyl) carbofuran contained 3-hydroxy carbofuran (Figure 7.40, I), accounting for 8% of the dose, and 3-hydroxy-N-hydroxymethyl carbofuran (Figure 7.40, II), representing 1% of the dose, when analyzed 3 hr after treatment. Although not toxic to flies when applied topically, the anticholinesterase activity of these materials would indicate that they could contribute to the action of carbofuran when generated internally. These metabolites were detoxified in the insects by conjugation, but apparently not as rapidly as in mice.

In the urine of mice, the only nonconjugated metabolite of N-(2-toluenesulfenyl) carbofuran of any significance was 3-hydroxy-N-hydroxymethyl carbofuran (Figure 7.40, II). This product accounted for as much as 10% of the administered dose in the free form and for another 20% in the conjugated form. Its total concentration was 10 times greater than the level found in the urine of rats treated with carbofuran.[183]

Perhaps the most important difference in the metabolism of N-(2-toluenesulfenyl) carbofuran in flies and mice in regard to its selective toxicity is the rate of hydrolysis of the carbamate ester.[166] An experiment designed to evaluate this parameter by monitoring the expired ^{14}C-carbon dioxide from aniamls treated with ^{14}C-carbonyl N-(2-toluenesulfenyl) carbofuran showed that after 6 hr 34% of the compound was hydrolyzed by mice and only 6% by insects. Even in mice hydrolysis proceeded very slowly after 6 hr, but with fast-acting compounds like the carbamates the chemical fate of the toxicant shortly after treatment largely determines their effect on the organism.

Carbamate ester hydrolysis rates may also partially explain the difference between toxicity of N-(2-toluenesulfenyl) carbofuran and carbofuran to mice. For example, 3 hr after oral administration of the compounds to mice only 8% of the carbofuran dose was hydrolyzed, whereas 25% of the N-(2-toluenesulfenyl) carbofuran was hydrolyzed. The variation in the hydrolysis of the compounds narrowed as the time after treatment

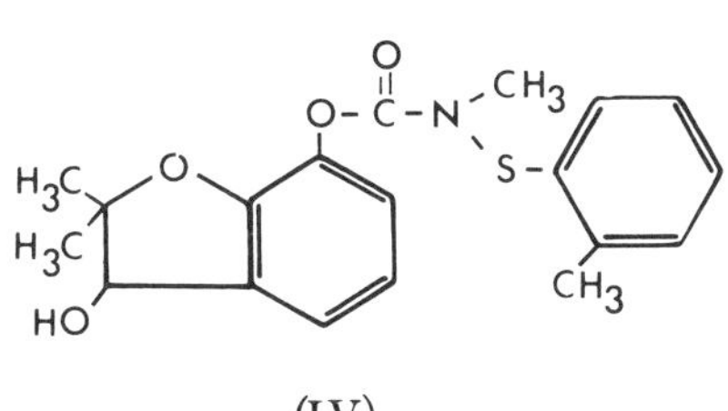

(IV)

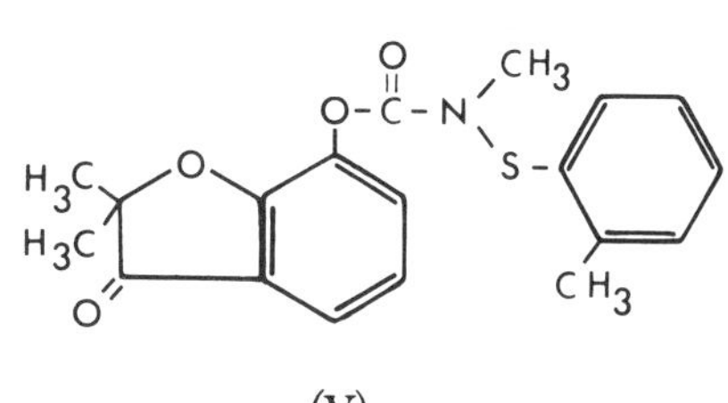

(V)

FIGURE 7.42. Possible intermediate products in the detoxication of N-(2-toluenesulfenyl) carbofuran in mice.

increased. However, an 8-fold difference in the rate of detoxication at such a critical time after exposure certainly should be reflected in reduced toxicity of N-(2-toluenesulfenyl) carbofuran.

In houseflies, carbofuran was hydrolyzed only slightly less than N-(2-toluenesulfenyl) carbofuran; hydrolysis rate, therefore, is not likely to be the explanation for selective action in insects. In-vivo activation of N-(2-toluenesulfenyl) carbofuran by conversion to carbofuran in a relatively slow and continuous manner as the compound is absorbed probably accounts for its enhanced insect toxicity. With carbofuran, the compound is subjected directly to metabolic detoxication once it is applied to the insect.

MOBAM (MCA 600)

In the mid 1960's, Mobam (4-benzothienyl N-methylcarbamate) was considered a prime candidate for commercial use. This potential stimulated several very good studies of its metabolism in mammals as well as more limited studies of its fate in insects and of its degradation by rumen bacteria. The mammalian metabolism studies[198, 199] represent one of the first serious attempts to determine the nature of intact conjugates of the phenolic moiety of a carbamate insecticide.

A summary of the fate of radioactive Mobam in rats, goats, and cows treated orally with the carbamate is shown in Figure 7.43. It is readily apparent that the insecticide is metabolized and rapidly excreted from the body of the mammals. Carbon dioxide evolution from the goat showed that 75% of the carbamate was hydrolyzed within 24 hr. Only 14% of the radiocarbon from carbonyl-labeled Mobam was eliminated in the urine, indicating that this would be the maximum amount of the dose present as carbamate metabolites. This was further substantiated by a similar amount of an N-methyl-^{14}C dose being detected in the urine of the goat. Experiments with ring-labeled Mobam demonstrated that the urine was the major route of metabolite removal in all three mammalian species and that only minor residues were excreted in the feces.

Analysis of the rat urine uncovered the presence of two major metabolites, namely, 4-benzothienylsulfate (65% of dose, Figure 7.44, I) and 4-benzothienylglucuronide (20% of dose, Figure 7.44, III). Only 1 to 2% of the total ^{14}C-residues in the urine was present as the free phenol, 4-hydroxybenzothiophene. Increasing the

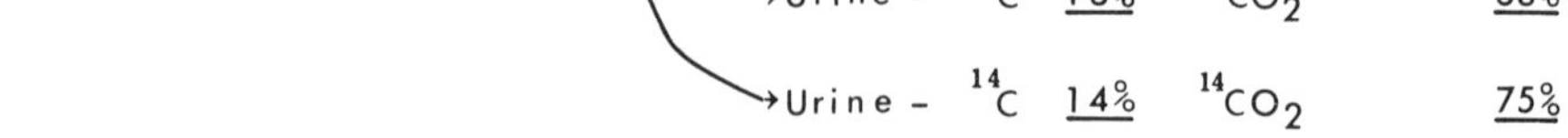

FIGURE 7.43. Fate of radiocarbon in rat, goat, and cow treated orally with Mobam radiolabeled at indicated positions. Values are percent of dose after 24 hr.[198,199]

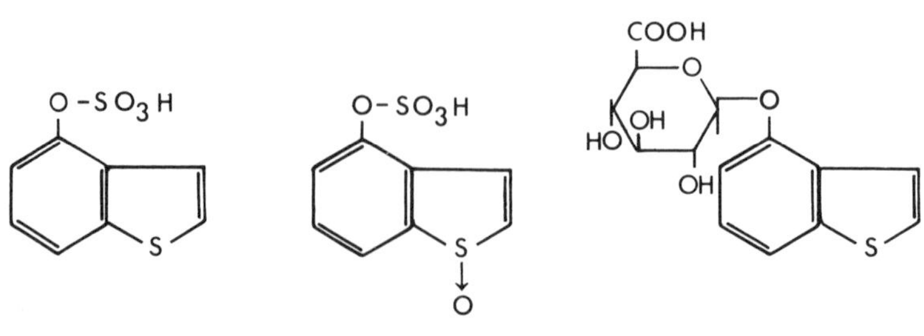

FIGURE 7.44. Conjugate metabolites of Mobam formed by rats.

application rate of Mobam to rats from 2 mg/kg to 13 mg/kg resulted in a decrease in glucuronide formation and an increase in the amount of the dose excreted as 4-benzothienylsulfate. Total elimination of insecticide residues in the urine was reduced by approximately 10% at the higher treatment level.

In lactating goats and cows, approximately 90% of the urinary ^{14}C-residues was identified as 4-benzothienylsulfate (Figure 7.44, I) and 4-benzothienylsulfate-1-oxide (Figure 7.44, II). 4-Benzothienylglucuronide, a major metabolite of Mobam in rats, was not detected as a metabolite in goats or cows. Mobam ^{14}C-equivalents in the milk of the animals consisted almost entirely (>95%) of 4-benzothienylsulfate-1-oxide.

The administration of radioactive Mobam to rats, goats, and cows did not result in the detection and characterization of any metabolite containing the carbamate moiety. However, other unidentified metabolites (10 to 14% of the dose) were present in the urine and could represent ring-hydroxylated and/or N-hydroxymethyl derivatives of Mobam. However, the fact that most of the dose was accounted for as conjugates of Mobam phenol clearly proves that hydrolysis was the major metabolic pathway for Mobam in the mammalian species studied.

Experiments have been conducted which suggest that Mobam undergoes an activation process in the integument of resistant as well as susceptible strains of insects.[200] Preincubation of the carbamate with housefly integument increased the toxicity of Mobam when injected into insects. Neither the activation mechanism nor the ^{14}C-components of the incubations were characterized. Resistant and susceptible insects rapidly metabolized Mobam, and five or six acetone-extractable materials were recovered from the two strains. Susceptible flies formed only two water-soluble metabolites, whereas the resistant strain produced as many as six water-soluble products. Only one metabolite, 4-hydroxybenzothiophene, was identified. The insecticide synergist 2,4,5-trichlorophenyl 2-propanyl ether reduced total metabolism as well as the number of metabolites biosynthesized by both resistant and susceptible houseflies. Pretreatment of the resistant flies with this synergist before injection of Mobam reduced the degradation of the insecticide to a greater extent than when the synergist was injected along with Mobam. On the other hand, the degradation of Mobam in susceptible flies was not altered by pretreatment with synergist.

Radioactive Mobam was degraded by a preparation of mixed rumen bacteria, primarily by hydrolysis of the carbamic acid ester (Table 7.4).[201] The sole metabolite detected after incubation of ring-4,7-^{14}C Mobam was 4-hydroxybenzothiophene. Polar products, material remaining at the TLC origin, appeared only when carbonyl-^{14}C and N-methyl-^{14}C Mobam were incubated with the bacteria. Therefore, these metabolites were derived only from the methylcarbamate moiety of the insecticide molecule, although further characterization was not achieved.

Three bacterial isolates from a rumen sample were also found to metabolize Mobam. However, none of the isolates degraded the insecticide as effectively as did the mixture discussed above. In contrast to these findings, laboratory strains of characterized rumen bacteria failed to degrade Mobam, but it was suggested that these strains might be adapted, over a long period of time, to degrade the toxicant. Such had been the case with isolates of soil bacteria.[202]

TABLE 7.4

Metabolism of Mobam-^{14}C by Mixed Rumen Bacteria During 48-hr Incubation

	% of applied Mobam-^{14}C		
Radioactive components	Carbonyl-^{14}C	N-Methyl-^{14}C	Ring-4,7-^{14}C
Mobam	60	62	65
4-Hydroxybenzothiophene	0	0	34
Polar unknowns	7.2	15	0
^{14}C-Carbon dioxide	31	0.9	0
Loss	2.3	22	1.1

Adapted from Reference 201, with permission.

FORMETANATE
(Schering 36056, Fundal Forte 750, Dicarzol, Carzol SP)

Formetanate (3-[(dimethylamino)-methylene]aminophenyl N-methylcarbamate hydrochloride) is a broad-spectrum acaricide that is especially effective against mites resistant to organophosphorus insecticides.[203] When rats ingested the ring-^{14}C compound, almost 90% of the dose was eliminated from the body within 24 hr (Figure 7.45).[204] Major metabolites in the urine consisted of 3-formamidophenyl N-methylcarbamate (Figure 7.46, I), which was present predominantely in the free form, and 3-hydroxy acetanilide (Figure 7.46, II), which probably existed almost entirely as sulfate and glucuronide conjugates. The hydrolysis product of metabolite I was detected both in the free and in the conjugated form (Figure 7.46, III). N-Conjugation of metabolite I and of formetanate *per se* was proposed, since these products were recovered after treatment of water-soluble metabolites with acid, but not with β-glucuronidase or with aryl sulfatase.

In-vitro studies of the metabolism of formetanate by rat liver homogenates uncovered a pathway which included N-demethylation, deformylation, and cleavage of the N-methyl carbamic acid ester.[205] The resulting metabolites were essentially the same as those isolated from rat urine.[204]

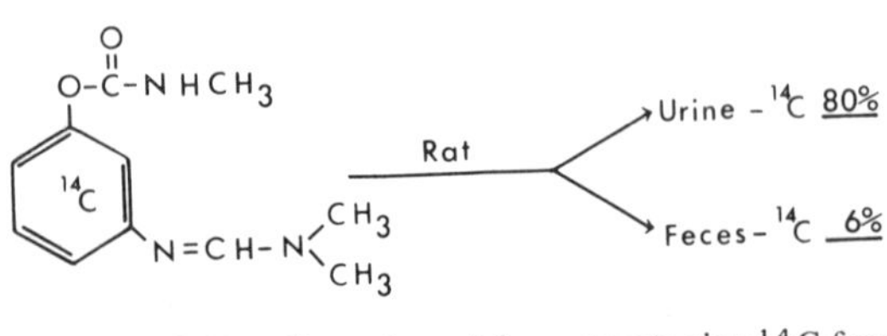

FIGURE 7.45. Excretion of formetanate-ring-^{14}C from rats during the 24-hr period following oral treatment.[204]

Metabolism of the [(dimethylamino)methylene]-amino moiety of formetanate was demonstrated to be catalyzed by a partially purified rat liver formamidase enzyme preparation.[206,207]

The degradation of formetanate in plants and soils was nearly identical to that reported in mammals.[203] However, the unaltered phenol of formetanate and the metabolite 3-aminophenyl N-methylcarbamate were detected only in orange seedlings, whereas 3-hydroxy acetanilide was isolated only from rat urine. In alkaline soil (pH 8.0), formetanate was broken down by more than 50% in less than 2 days.[203] Degradation products isolated from the soil consisted of 3-formamidophenyl N-methylcarbamate (Figure 7.46, I), 3-hydroxy acetanilide (Figure 7.46, III), and 3-aminophenol. Up to 30% of the radioactivity recovered after 16 days was not identified. It was speculated, without elaboration, that the unknown radioactive material was either formetanate or its metabolites complexed with naturally occurring soil components.

UC 34096

Human lung cells, L-13L strain, were unable to metabolize UC 34096 (3-methyl-4-[(dimethylamino)methylene]aminophenyl N-methylcarbamate), but the compound was found to be unstable both in cell culture medium and in water.[65] In culture medium containing calf serum and no lung cells, 80% of the insecticide was converted to 3-methyl-4-formamidophenol (Figure 7.47, I) after 3 days of incubation at 37°C. No other products were detected, and the addition of lung cells did not alter the results.

The same decomposition product of UC 34096 was isolated from water after 40 days of incubation at room temperature. However, two other compounds were also detected, namely, 3-methyl-

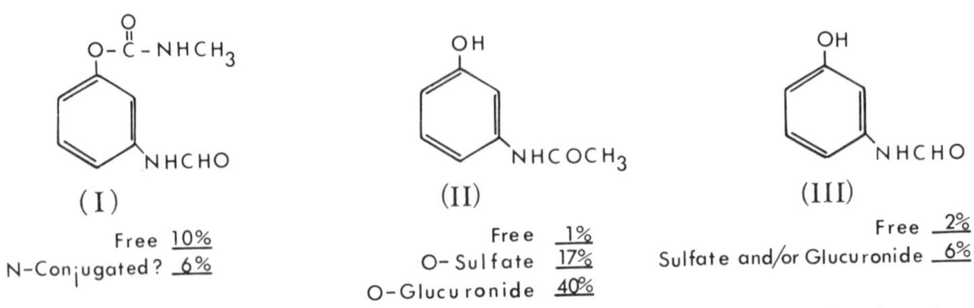

FIGURE 7.46. Major radioactive components (percent of dose after 24 hr) in the urine of rats dosed orally with formetanate.[204]

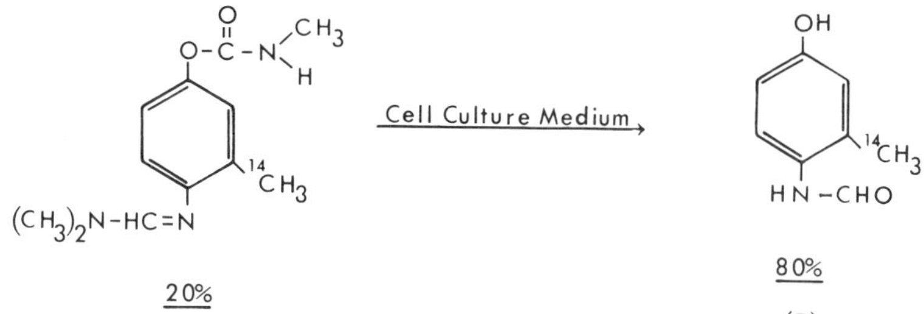

(I)

FIGURE 7.47. Decomposition of UC 34096 in cell culture medium, with or without cells, during incubation for 3 days at 37°C. (Adapted from Reference 65, with permission.)

4-formamidophenyl N-methylcarbamate (Figure 7.48, II) and 3-methyl-4-aminophenyl N-methylcarbamate (Figure 7.48, III). A similar breakdown pattern has been reported for a closely related formamidine pesticide, Galecron [N,N-dimethyl-N′-(2-methyl-4-chlorophenyl)formamidine].[208] Whether calf serum affects the degradation of Galecron as it does UC 34096 has not been reported, but the chances are slight, since Galecron does not contain an ester linkage. It is well documented that human serum quite effectively hydrolyzes carbamic acid esters.[35,70] Therefore, hydrolysis of UC 34096 by calf serum is probably responsible for the appearance of more hydrolysis products in the cell culture medium than in water.

Although the authors[65] did not speculate as to the reason why lung cells failed to metabolize UC 34096, the data suggest that the carbamate and/or metabolite I (Figure 7.47) acted as inhibitors of the enzymes catalyzing the breakdown. Toxicity of either compound to the cells in culture could also be involved, but no mention was made that cell development was impaired when UC 34096 was introduced into the cultures.

The metabolic capability of the cells was demonstrated by using radioactive preparations of Banol and carbaryl at concentrations of 2 and 10 ppm respectively, whereas the concentration of UC 34096 had been 24 ppm. Under these conditions, Banol and carbaryl were metabolized predominantly by hydrolysis and/or conjugation. Thus, it is rather surprising that the same biochemical mechanism did not operate on UC 34096. There has been a precedent for in-vivo metabolism of this type of compound, i.e., formetanate (3-[(dimethylamino)methylene]-aminophenyl N-methylcarbamate), in rats.[203]

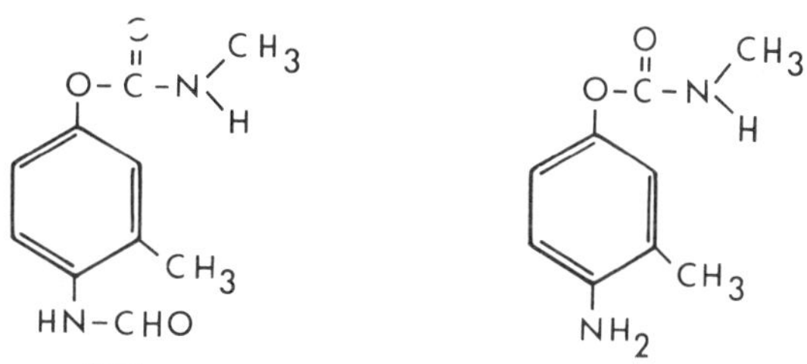

(II) (III)

FIGURE 7.48. Products formed from UC 34096 in addition to that shown in Figure 7.47 when incubated in water at room temperature for 40 days.

ALDICARB
(Temik, UC 21149)

Aldicarb [2-methyl-2-(methylthio)propionaldehyde O-(methylcarbamoyl)oxime] is a very effective systemic insecticide whose metabolism has been studied in a variety of insect and other animal species. Because of the compound's high toxicity to mammals, it is applied commercially mostly in the form of a 10% granule. The majority of plant metabolism studies have been conducted with cotton plants, due to aldicarb's wide use on this crop. Since the compound is applied to the soil, its metabolism and persistence in soils have also been evaluated.

Studies with rat liver enzymes showed that aldicarb was metabolized to its sulfoxide, 2-methyl-2-(methylsulfinyl)propionaldehyde O-(methylcarbamoyl)oxime (Figure 7.49, I) and to water-soluble unknown metabolites.[209] Only a very small amount of the sulfoxide was converted further to the sulfone, 2-methyl-2-(methylsulfonyl)propionaldehyde O-(methylcarbamoyl)-oxime (Figure 7.49, II). However, oxidation of

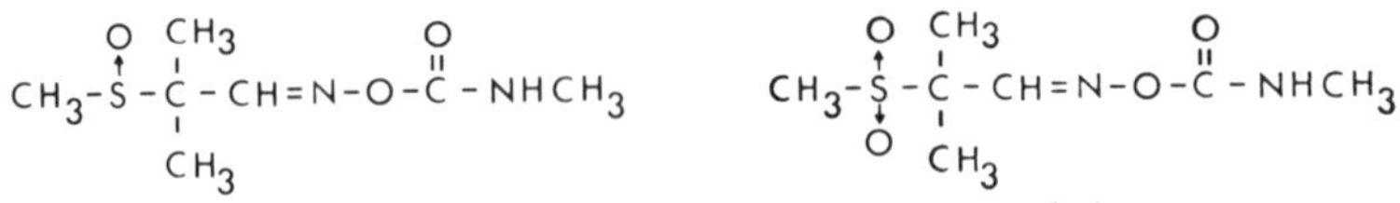

(I) (II)

FIGURE 7.49. Structures of aldicarb sulfoxide (I) and aldicarb sulfone (II).

aldicarb to its sulfoxide is not entirely dependent on liver enzymes.[25] For example, when aldicarb was incubated in buffer alone, it was partially oxidized to the sulfoxide, and when NADPH was added, it was completely converted to the sulfoxide.

Oral treatment of rats with ^{14}C-carbonyl aldicarb resulted in 62% of the dose given off as ^{14}C-carbon dioxide and 30% being eliminated in the urine after 24 hr.[209] With S-methyl-^{14}C material, 80% of the radioactivity was excreted in the urine and none was expired as ^{14}C-carbon dioxide. Approximately 60% of the S-methyl-^{14}C aldicarb equivalents in the urine were organo-extractable, the major metabolite being aldicarb sulfoxide, which constituted 20% of the administered dose. The sulfone accounted for only 1% of the dose, and the remainder of the organo-soluble metabolites was composed primarily of hydrolysis products of aldicarb sulfoxide and nitrile sulfoxide (Figure 7.50, III), with minute amounts of the hydrolysis products of aldicarb sulfone and nitrile sulfone (Figure 7.50, IV) present. About 40% of the urinary ^{14}C-metabolites were water-soluble, and their identity was not ascertained. When carbonyl-^{14}C-labeled aldicarb sulfoxide (Figure 7.49, I) was administered orally to rats, the compound was hydrolyzed much more slowly than was aldicarb, but a greater percentage of radioactivity was eliminated in the urine. The metabolites were, however, identical to those from animals treated with aldicarb. In another study with various ^{14}C-labeled samples of aldicarb,[210] the same general metabolic pathway of aldicarb in rats was observed. However, aldicarb sulfoxide was not present at detectable levels.

Single oral doses of aldicarb to laying hens were rapidly metabolized and excreted, with 75% of the administered equivalents appearing in the feces after 24 hr.[211] Almost 50% of the metabolites were water-soluble in nature and 10% remained in the feces solids after extraction. Small amounts of aldicarb sulfoxide, aldicarb sulfone, and the N-hydroxymethyl derivative of aldicarb sulfone were present in the feces, but hydrolysis products

(III) (IV)

FIGURE 7.50. Nitrile metabolites of aldicarb.

were much more abundant. Feeding aldicarb to laying hens in the diet for 21 days did not change the nature of the metabolites in the excreta. Of the total radioactive aldicarb consumed during the 21-day feeding period, 80% was eliminated in the feces. By 7 days after the birds had been returned to a normal diet, the total excretion approached 90% of the consumed radioactivity. Carbamate-type metabolites in the tissues of eggs were essentially nonexistent; the residues consisted primarily of unextractable materials and polar unknown metabolites, as well as of typical hydrolysis products of aldicarb.

In a lactating cow, ^{35}S-aldicarb administered orally was essentially gone after 24 hr.[212] More than 80% of the administered radioactivity was in the urine and only 0.5% appeared in the feces. Total radiocarbon in milk collected over a 540-hr period was equivalent to 3% of the administered dose. Metabolites in the urine and milk were similar to those which had been isolated from rat urine. Continuous ingestion of the carbamate insecticide by lactating cows as part of the diet for 14 days did not alter the magnitude or nature of the residues eliminated daily via milk, urine, or feces.[213] Aldicarb equivalents in the milk were approximately 1/100 of the insecticide level in the feed, with about 15% of the residues present as aldicarb and about 4% as aldicarb sulfoxide. Total elimination of the consumed radiocarbon was 1% in the milk, 92% in the urine, and 3% in the feces.

Resistant strains of houseflies absorbed almost 80% of topically applied N-methyl-^{14}C aldicarb within 24 hr; susceptible flies absorbed 10% less.[155] The major difference between the susceptible and the resistant insects was in the amount of radioactivity excreted – 40% of the absorbed radiocarbon after 24 hr by resistant insects and only 7% by susceptible insects. The amount of the treatment evolved as ^{14}C-carbon dioxide by both strains was 7 to 10% within a 24-hr period.[214] Aldicarb sulfoxide represented 75% of the residues within the insect body, whereas aldicarb sulfone accounted for only 6%. Internal water-soluble metabolites constituted 20% of the total radiocarbon, and there was no evidence that the houseflies hydrolyzed aldicarb.

The carbamate insecticide penetrated boll weevil cuticle at about half the rate observed for houseflies.[215] Although aldicarb sulfoxide was a major nonconjugated metabolite in both species, the formation of water-soluble metabolites was much greater in houseflies than in boll weevils. About 26% of the aldicarb-^{14}C equivalents excreted by flies was in the form of water-soluble metabolites after 6 hr, whereas they constituted less than 4% in the excreta of the boll weevil. Absorption of topically applied aldicarb was also slower by boll weevils than by third-instar tobacco budworms,[216] but the metabolism of the absorbed material was the same for both species, with hydrolysis accounting for less than 5% of the applied dose.

Mosquito microsome preparations metabolized 30% of ^{35}S-labeled aldicarb added to the enzyme system.[106] Aldicarb sulfoxide was the major metabolite, constituting 25% of the added radiocarbon, whereas aldicarb sulfone accounted for only 1% and hydrolytic products equaled about 3%.

Metabolism of aldicarb in cotton plants involves oxidation to the sulfoxide and sulfone analogs, hydrolysis of these carbamate materials, and the formation of water-soluble metabolites.[214,217-220] One investigation showed that there was 56% loss of the ^{14}C-carbonyl aldicarb 14 days after treatment of the plants.[217] With the radiocarbon on the oxime moiety, however, only 20 to 30% disappeared, indicating that approximately 20% of the compound had been hydrolyzed to form ^{14}C-carbon dioxide.

Cotton plants grown in soils treated at 1 lb/acre with S-methyl-^{14}C aldicarb harbored a total of 209 ppm equivalents in the foliage after 9 days.[219] Aldicarb sulfoxide was present at 148 ppm, sulfone at 15 ppm, and the water-soluble metabolites constituted approximately 44 ppm aldicarb equivalents. Aldicarb *per se* was present in the foliage at 2 ppm. After 146 days, there were only trace amounts of aldicarb and approximately 0.5 ppm each of aldicarb sulfoxide, sulfone, and the hydrolysis product of aldicarb sulfoxide. The remainder of the residues, 7.5 ppm, consisted of unknown water-soluble metabolites. Seeds from cotton plants grown in soils treated with radioactive aldicarb at 1 lb/acre at planting and 2 lb/acre side-dress 58 days later contained 2.8 ppm total ^{14}C-aldicarb equivalents when harvested 88 days after the last treatment. Only 14% was present as carbamate-type metabolites, consisting of equal quantities of aldicarb sulfoxide and aldicarb sulfone. About 70% of the radiocarbon in the seeds was water-soluble in nature.

An examination of the water-soluble metabolites in cotton plants at harvest showed that they numbered up to ten, six of which were identified.[218] Approximately 80% existed as sugar conjugates of 2-methyl-2(methylsulfinyl)propanol (Figure 7.51, V). Other products in small quantities were 2-methyl-2-(methylsulfonyl)propanol, 2-methyl-2-(methylsulfinyl)propionaldehyde oxime (Figure 7.51, VI), and 2-methyl-2-(methylsulfonyl)propionaldehyde oxime. Nonconjugated metabolites soluble in water included 2-methyl-2-(methylsulfinyl)propionamide (Figure 7.51, VII), 2-methyl-2-(methylsulfinyl)propionic acid (Figure 7.51, VIII), and 2-methyl-2-(methylsulfonyl)propionic acid. There was no evidence that any of the water-soluble metabolites contained the carbamate moiety.

Radioactive residues in the foliage of potatoes grown in soils containing S-methyl-^{14}C aldicarb

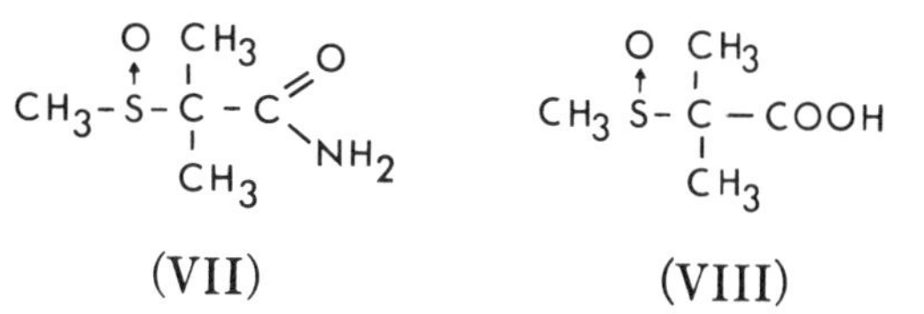

FIGURE 7.51. Water-soluble metabolites of aldicarb formed in the cotton plant. Metabolites V and VI existed as glycoside conjugates, and metabolites VII and VIII were in the free form. With the exception of metabolite VII, the corresponding sulfone analogs (O←S→O) were also detected.

were composed of 56% aldicarb sulfone, 7% aldicarb sulfoxide, and 30% water-soluble metabolites when harvested 90 days after applying the insecticide.[221] ^{14}C-Residues in the tubers were made up of 66% water-soluble materials, 10% aldicarb sulfone, and 5% aldicarb sulfoxide; the remaining residues were hydrolysis products.

The half-life of intact aldicarb in soils is generally between 7 and 10 days.[217,222] As in animals and plants, the compound is converted to the sulfoxide analog and, to a lesser degree, to the sulfone derivative. There is rapid binding of aldicarb to soil particles, indicated by the fact that 14% of the radioactive aldicarb added to soils could not be recovered after only 30 min.[220] During this time, 8% was already oxidized to the sulfoxide. After 2 weeks, 60% loss of the added radiocarbon was observed; after 8 weeks, almost 90% loss had been encountered. Ten metabolites were detected in the soil samples, including three unidentified products not formed in the cotton plant. Aldicarb sulfoxide and sulfone were the major metabolites extractable from the soil, although after 8 weeks the residues consisted almost entirely of bound ^{14}C-materials. Extractability of aldicarb and its metabolites from soils varied with the soil types, removal being more efficient from sandy soils and less efficient from clay soils.[217]

High soil moisture enhanced the rate of degradation of aldicarb and also increased volatilization of aldicarb and its toxic metabolites.[223] Laboratory studies have shown that the rate of loss of aldicarb may not be influenced as much by moisture content as by the drying process of the soil.[224]

METHOMYL
(Lannate, Nudrin, DuPont 1179)

Methomyl (S-methyl N-[(methylcarbamoyl)oxy] thioacetimidate) is a commercial insecticide registered in the United States by the Environmental Protection Agency for insect control on 13 different crops. These include tobacco, sweet corn, potatoes, cabbage, cauliflower, broccoli, and head lettuce.[225] Although methomyl metabolism studies have obviously been conducted for some time, the original data submitted to the Environmental Protection Agency in request for a tolerance first appeared in the literature in 1973. Previously, however, portions of the data had been presented by the IUPAC Commission on Terminal Residues in 1971.[226]

Very simple metabolic products for methomyl in rats and in plants have been presented.[225,227] The pathway involves degradation of ^{14}C-methomyl (S-methyl-1-^{14}C N-[(methylcarbamoyl) oxy] thioacetimidate) to ^{14}C-carbon dioxide and ^{14}C-acetonitrile (Figure 7.52). Neither the hydrolysis product of methomyl, S-methyl-N-hydroxythioacetimidate, nor the sulfoxide and sulfone of methomyl were detected as metabolites. In soils, the degradative pathway is even simpler; ^{14}C-carbon dioxide is the only significant labeled metabolite formed from radioactive methomyl. Polar metabolites are present in all of the substrates, but these have been postulated, and in part confirmed, to consist of naturally occurring components resulting from incorporation of the radioactive carbon after total decomposition of the methomyl molecule.

A metabolism study involving three rats constitutes the total information reported on the degradation of this carbamate in mammals.[225] Treatment of one of these rodents with ^{14}C-methomyl yielded the results summarized in Figure 7.53. Most of the administered radioactivity was recovered in the exhaled air and was identified as carbon dioxide and acetonitrile. The feces contained only trace amounts of radioactivity, but the urine accumulated about 12% of the dose 24 hr after treatment. Residues in the body tissues at this time totaled a maximum of 10% of the administered radiocarbon. Accountability of the applied ^{14}C-methomyl was slightly greater than

FIGURE 7.52. General metabolic pathway of methomyl proposed for animals and plants.

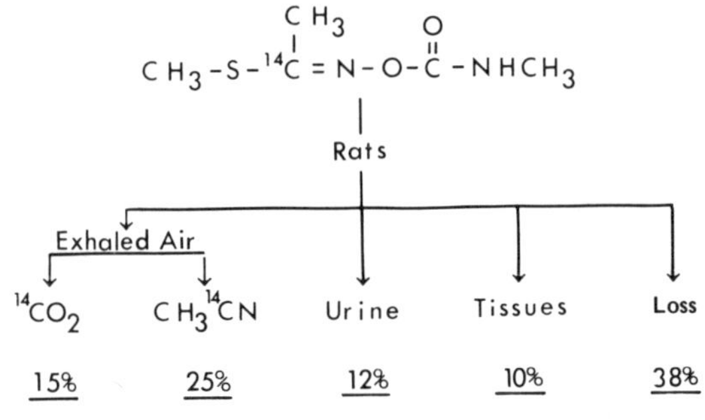

FIGURE 7.53. Disposition of methomyl (percent of dose) 24 hr after oral treatment of rats.[225]

60%. By 72 hr, the percentages as ^{14}C-carbon dioxide, ^{14}C-acetonitrile, and urinary ^{14}C-metabolites were 18, 34, and 15 respectively, making the total recovery of ^{14}C-methomyl after 3 days approximately 70%.

Treatment of a second rat with ^{14}C-methomyl resulted in greater elimination of ^{14}C-carbon dioxide and ^{14}C-urinary metabolites. In this test, however, the radioactive acetonitrile was not trapped and quantitated. The third animal was treated in the same manner as the others and sacrificed after 24 hr to determine the levels of methomyl-^{14}C equivalents in the tissues.

The chemical identity of the urinary ^{14}C-metabolites was not established, although it was confirmed that methomyl, its hydrolysis product, and its sulfoxide and sulfone were not present. β-Glucuronidase-aryl sulfatase treatment of the conjugates in the urine was ineffective in converting the polar metabolites to apolar ones. The result of acid treatment of the urinary metabolites was not reported.

Methomyl was metabolized in tobacco, corn, and cabbage in much the same manner as it was in rats.[227] Tobacco plants grown in a nutrient solution containing ^{14}C-methomyl removed up to 25% of the added radiocarbon over a 4-week period. About 14% of the absorbed dose was characterized as methomyl and about 3% as metabolites of unknown identity. Volatile components accounted for the remainder of the absorbed radioactivity consisting of ^{14}C-carbon dioxide and ^{14}C-acetonitrile in equal amounts.

One week after foliar treatment of cabbage and corn plants with ^{14}C-methomyl, 74% remained in or on the cabbage and 49% persisted in the corn plants. The volatile components, ^{14}C-carbon dioxide and ^{14}C-acetonitrile, accounted for all of the radiocarbon lost from the leaves. In cabbage, these metabolites were present in equal amounts, whereas in treated corn plants four times more ^{14}C-acetonitrile was produced than ^{14}C-carbon dioxide. Of the material remaining in the plants, only 6 to 7% was present as intact methomyl; the balance characterized as polar and nonpolar metabolites. The latter, 30% and 11% of the total ^{14}C-residues in cabbage and corn respectively, were subsequently identified as fatty acids naturally occurring in these plants. Polar products, 62% and 82% of the radiocarbon in cabbage and corn, were not identified. However, β-glucosidase failed to liberate any S-methyl N-hydroxythioacetimidate or other compounds which might have been present as sugar conjugates.

With the exception that ^{14}C-acetonitrile was not formed, the metabolism of ^{14}C-methomyl in soils was not unlike that in plants.[228] Under field conditions, approximately 71% of the applied radioactivity was lost as volatile components after 1 month, with 85% after 12 months. ^{14}C-Residues not extracted with methanol constituted 26% of the applied dose after 1 month, but had declined to 15% by 12 months. Distribution of these residual materials among different soil fractions was as follows: hymatomelanic acid, 3%; fulvic acid, 26%; alpha-humus, 35%; beta-humus, 8%; and a volatile component lost during concentration, 28%. Treatment of the unextracted ^{14}C-residues with hot 0.1*N* sodium hydroxide released 62% of the radiocarbon.

The metabolism of injected ^{14}C-methomyl by cabbage loopers produced volatile components and unknown polar materials which could not be converted to organo-soluble products by either acid or enzymic treatment.[229] However, volatile metabolites were not formed when the insects were injected with the hydrolysis product of methomyl, S-methyl-N-hydroxythioacetimidate. This established that the metabolic pathway of methomyl does not proceed through the oxime route in the production of ^{14}C-carbon dioxide and ^{14}C-acetonitrile. The same situation is probably true for other organisms. Since metabolites containing the intact carbamate moiety may thus serve as intermediates in the formation of carbon dioxide and acetonitrile, attempts should be made to more completely characterize these intermediates. The significance of such studies is enhanced by the fact that a major portion of the terminal residues in plants and soils have yet to be identified. Similarly, the metabolic pathway of methomyl in mammals is incomplete in that 25% to 30% of the applied insecticide was not recovered. Nevertheless, the virtual explosion of methomyl into natural products or simple organic molecules makes this compound very appealing from an environmental toxicology viewpoint.

THIOCARBOXIME
(Talcord, WL 21959, SD 17250-W)

Male and female rats dosed orally (2.5 mg/kg) with radioactive thiocarboxime (S-2-cyanoethyl-N-[(methylcarbamoyl)oxy]thio [1,2-^{14}C] acetimidate) metabolized and excreted the carbamate insecticide in a similar manner.[230] A summary of the data is presented in Table 7.5. The exhaled air and urine provided the major routes of elimination of the administered radiocarbon; only small quantities were excreted in the feces.

Elimination via the urine was virtually complete after 24 hr, but expiration of radioactivity continued throughout the 3-day collection period; i.e., approximately 12% of the administered carbon-14 was exhaled during the second day and 4% during the third day. Analysis of the radioactive components expired during the first 24 hr showed that they consisted of carbon dioxide and acetonitrile (Figure 7.54, top), with the latter comprising 81% of the total radiocarbon in the air. By the third day, 87% of the exhaled radioactivity was acetonitrile.

TABLE 7.5

Excretion and Distribution of Radioactivity in Rats after Oral Treatment with Thiocarboxime[230]

	Percent of dose (hr after treatment)	
Material	24	96
Urine	34	38
Feces	1.1	2.8
Exhaled air	26	44[a]
Gut	–	1.4
Carcass	–	2.6
Skin	–	1.5
Total	61	90

[a]Analysis performed 72 hr after treatment.

Treatment with the hydrolysis product of thiocarboxime, S-2-cyanoethyl-N-hydroxythio-(1,2-^{14}C)acetimidate, also yielded ^{14}C-carbon dioxide and ^{14}C-acetonitrile in the exhaled air of treated rats, but their production relative to the thiocarboxime treatments varied sufficiently to suggest a different pathway of formation. In the case of the hydrolysis product, only 13% of the dose was exhaled during the first 24 hr, and this consisted of ^{14}C-carbon dioxide and ^{14}C-acetonitrile in almost equal quantities (Figure 7.54, bottom). Although this facet was not considered by the authors,[230] the data indicate that thiocarboxime is metabolized to a compound other than its hydrolysis product prior to degradation to acetonitrile. This would be possible if the thio group was removed without loss of the carbamate moiety, for example.

Further support for the formation of an intermediate other than the parent oxime comes from the fact that radioactivity persisted in the exhaled air at relatively high levels for three days, but elimination of radiocarbon in the urine occurred at significant levels for only 24 hrs. Furthermore, characterization of the urinary metabolites of thiocarboxime indicated that any oxime formed would be readily conjugated and/or excreted, as shown in Figure 7.55. Thus, the ^{14}C-materials in the urine of treated rats included the parent compound, its hydrolysis product (Figure 7.55, I), and the O-glucuronide (Figure 7.55, II) and O-sulfate (Figure 7.55, III) of the hydrolysis product, which all together accounted for 95% of the total radiocarbon in the urine. Compounds resulting from hydrolysis of the carbamates (Figure 7.55, I, II, and III) were equivalent to 32% of the administered dose. No radioactive material

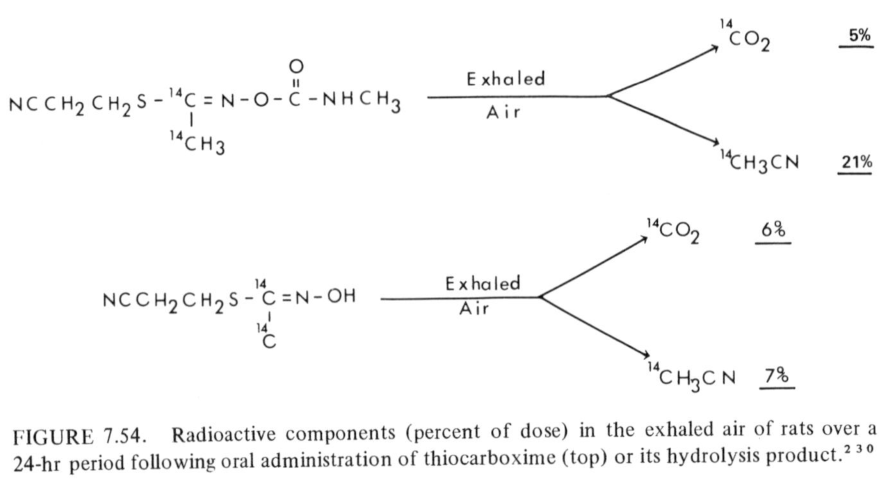

FIGURE 7.54. Radioactive components (percent of dose) in the exhaled air of rats over a 24-hr period following oral administration of thiocarboxime (top) or its hydrolysis product.[230]

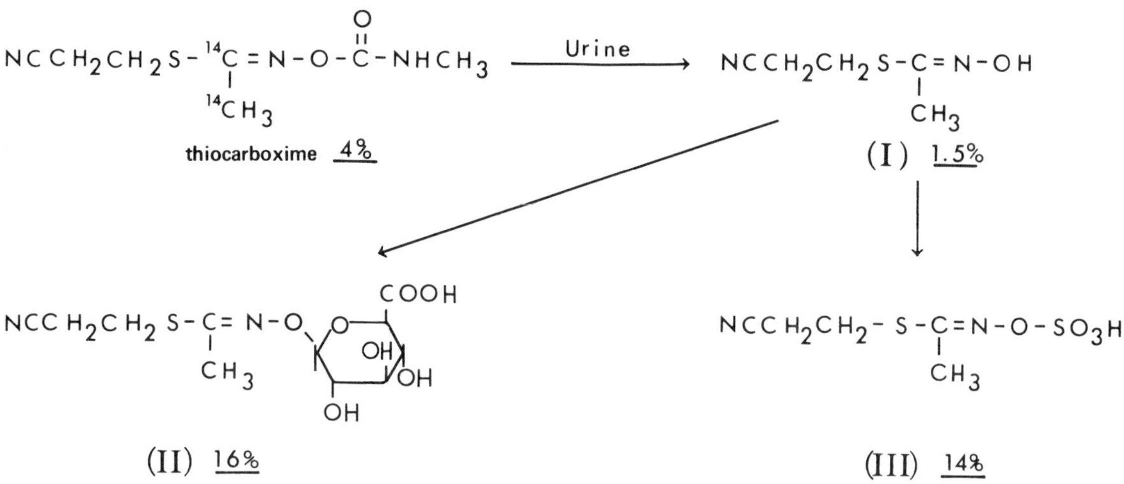

FIGURE 7.55. Urinary radiocarbon (percent of dose) from rats 96 hr after oral treatment with thiocarboxime.[230]

lacking the $-SCH_2CH_2CN$ group was detected in the urine. Thus, if such a metabolite did exist, it was retained in the body until it was metabolized to acetonitrile and carbon dioxide.

DS-15647

A compound structurally like aldicarb, DS-15647 [3,3-dimethyl-1-(methylthio)-2-butanone O-(methylcarbamoyl)oxime] has also shown similar insecticidal properties. It has demonstrated both systemic and contact activity against various phytophagous pests, especially certain species of those attacking cotton.[231]

Unlike methomyl and thiocarboxime, the metabolic fate of DS-15647 is essentially the same as the oxime carbamate aldicarb in plants and soils.[231] Sulfur oxidation to the sulfoxide derivative (Figure 7.56, I) occurs rapidly in plants, followed by a slower conversion of the sulfoxide to the sulfone (Figure 7.56, II). Other metabolites formed included the hydrolysis products of the three carbamates as well as some compounds, carbamates and noncarbamates, of a very polar nature. The identity of the latter products was not determined.

In soils, the same degradation pattern as that observed in plants was evident. However, radioactivity added to the soil as ^{35}S-labeled DS-15647 dissipated quickly. After 3 days there was 38% loss of radioactivity, and by 56 days only trace amounts could be detected.

Contrary to aldicarb, where the sulfoxide derivative is a more potent anticholinesterase agent than the parent compound or the sulfone analog, DS-15647 sulfone demonstrated the greatest inhibition of cholinesterase. Also, it was more toxic to boll weevils and lady beetle adults than DS-15647 sulfoxide, but not as toxic as DS-15647. The toxicity of aldicarb sulfone to boll weevils, on the other hand, was at least $\frac{1}{100}$ that of aldicarb or its sulfoxide.

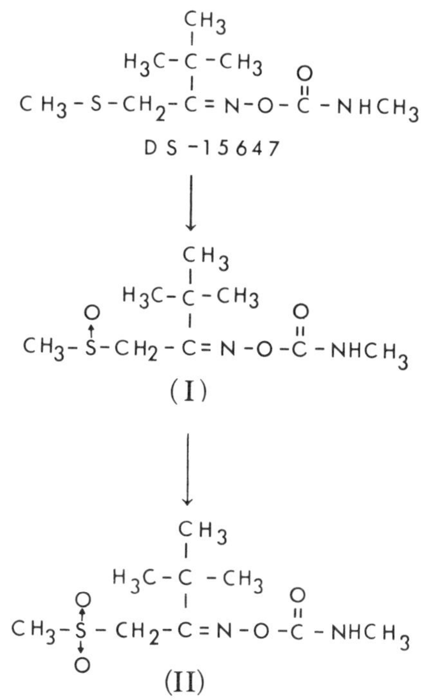

FIGURE 7.56. Oxidative metabolism of 3,3-dimethyl-1-(methylthio)-2-butanone O-(methylcarbamoyl)oxime (DS-15647) in cotton plants and soil. (Adapted from Reference 231, with permission.)

REFERENCES

1. **Egan, H.**, IUPAC Commission on Terminal Residues, *J. Assoc. Off. Anal. Chem.*, 50, 1071, 1967.
2. **Fukuto, T. R. and Sims, J. J.**, Metabolism of insecticides and fungicides, in *Pesticides in the Environment*, White-Stevens, R., Ed., Marcel Dekker, Inc., New York, 1971, 145.
3. **Ernst, W.**, Der Stoffwechsel von Pestiziden in Säugetieren, *Residue Rev.*, 18, 131, 1967.
4. **Lykken, L. and Casida, J. E.**, Metabolism of organic insecticide chemicals, *Can. Med. Assoc. J.*, 100, 145, 1969.
5. **O'Brien, R. D.**, *Insecticides, Action and Metabolism*, Academic Press Inc., New York, 1967, 83.
6. **Terriere, L. C.**, Insecticide-cytoplasmic interactions in insects and vertebrates, in *Annual Review of Entomology*, Annual Reviews, Inc., Palo Alto, Calif., 1968, 75.
7. **Dahm, P. A.**, Chemistry and metabolism of insecticides, in *Agricultural Practices and Water Quality*, Willrich, T. L. and Smith, G. E., Eds., Iowa State University Press, Ames, Iowa, 1970, 167.
8. **Casida, J. E. and Lykken, L.**, Metabolism of organic pesticide chemicals in higher plants, *Annu. Rev. Plant Physiol.*, 20, 607, 1969.
9. **Esser, H. O.**, The biodegradation of pesticides in the soil, *Meded. Rijksfac. Landbouwwet Gent*, 35, 753, 1970.
10. **Casida, J. E.**, Chemistry and metabolism of terminal residues of organophosphorus compounds and carbamates, in *Fate of Pesticides in Environment*, Tahori, A. S., Ed., Gordon and Breach, London, 1972, 295.
11. **Ryan, A. J.**, The metabolism of pesticidal carbamates, *Crit. Rev. Toxicol.*, 1, 33, 1971.
12. **Schlagbauer, B. and Schlagbauer, A. W.**, Metabolism of carbamate pesticides, *Residue Rev.*, 42, 1, 1972.
13. **Kuhr, R. J.**, Metabolism of methylcarbamate insecticide chemicals in plants, *J. Sci. Food Agric. Suppl.*, p. 44, 1968.
14. **Kuhr, R. J.**, Metabolism of carbamate insecticide chemicals in plants and insects, *J. Agric. Food Chem.*, 18, 1023, 1970.
15. **Kuhr, R. J.**, The formation and importance of carbamate insecticide metabolites as terminal residues, in *Pesticide Terminal Residues*, Tahori, A. S., Ed., Butterworth, New York, 1971, 199.
16. **Boecker, E. and Draber, W.**, Carbamate (insecticides), *Chem. Pflanzenschutz Schaedlingsbekaempfungsm.*, p. 219, 1970.
17. **Rivière, J. L.**, Les carbamates insecticides, *Ann. Zool. Ecol. Anim.*, 2, 479, 1970.
18. **Dorough, H. W.**, Metabolism of insecticidal methylcarbamates in animals, *J. Agric. Food Chem.*, 18, 1015, 1970.
19. **Fukunaga, K.**, Metabolism and degradation of carbamate insecticides, in *Pesticide Terminal Residues*, Tahori, A. S., Ed., Butterworth, New York, 1971, 163.
20. **Knaak, J. B.**, Biological and nonbiological modifications of carbamates, *Bull. W.H.O.*, 44, 121, 1971.
21. **Asakawa, M.**, Carbamate insecticides, *Noyaku Seisan Gijutsu*, 24, 1, 1971.
22. **Fukuto, T. R.**, Metabolism of carbamate insecticides, *Drug Metab. Rev.*, 1, 117, 1972.
23. **Hodgson, E. and Casida, J. E.**, Biological oxidation of N,N-dialkyl carbamates, *Biochim. Biophys. Acta*, 42, 184, 1960.
24. **Hodgson, E. and Casida, J. E.**, Metabolism of N,N-dialkyl carbamates and related compounds by rat liver, *Biochem. Pharmacol.*, 8, 179, 1961.
25. **Oonnithan, E. S. and Casida, J. E.**, Oxidation of methyl- and dimethylcarbamate insecticide chemicals by microsomal enzymes and anticholinesterase activity of the metabolites, *J. Agric. Food Chem.*, 16, 28, 1968.
26. **Krishna, J. G. and Casida, J. E.**, Fate in rats of the radiocarbon from ten variously labeled methyl- and dimethylcarbamate-^{14}C insecticide chemicals and their hydrolysis products, *J. Agric. Food Chem.*, 14, 98, 1966.
27. **Kuhr, R. J. and Casida, J. E.**, Persistent glycosides of metabolites of methylcarbamate insecticide chemicals formed by hydroxylation in bean plants, *J. Agric. Food Chem.*, 15, 814, 1967.
28. **Rahn, H. W.**, Über den Nachweis insectizidwirkender Carbaminate und ihrer Umwandlungsprodukte, *Arch. Exp. Pathol. Pharmakol.*, 241, 157, 1961.
29. **Zubairi, M. Y. and Casida, J. E.**, Detoxication of dimetilan in cockroaches and houseflies, *J. Econ. Entomol.*, 58, 403, 1965.
30. **Lambrech, J. A. (Union Carbide Corp.)**, U.S. Patent 2,903,478, September, 8, 1959.
31. **Back, R. C.**, Significant developments in eight years with Sevin insecticide, *J. Agric. Food Chem.*, 13, 198, 1965.
32. Union Carbide Corporation, Sevin Carbaryl Insecticide for Abundant Food and Fiber, Technical Information Bulletin, 1974.
33. **Casida, J. E.**, Mode of action of carbamates, *Annu. Rev. Entomol.*, 8, 39, 1963.
34. **Carpenter, C. P., Weil, C-S., Palm, P. E., Woodside, M. M., Nair, J. H., and Smyth, H. F.**, Mammalian toxicity of 1-naphthyl-N-methylcarbamate (Sevin insecticide), *J. Agric. Food Chem.*, 9, 30, 1961.
35. **Casida, J. E. and Augustinsson, J.**, Reaction of plasma albumin with 1-naphthyl-N-methylcarbamate and certain other esters, *Biochim. Biophys. Acta*, 36, 411, 1959.
36. **Casida, J. E., Augustinsson, J., and Jonsson, G.**, Stability and reaction mechanisms with certain esterases with certain carbamate insecticides, *J. Econ. Entomol.*, 53, 205, 1960.
37. **Kolbezen, M. J., Metcalf, R. L., and Fukuto, T. R.**, Insecticidal activity of carbamate cholinesterase inhibitors, *J. Agric. Food Chem.*, 2, 864, 1954.
38. **Miskus, R., Gordon, H. T., and George, E. A.**, Colorimetric determination of 1-naphthyl N-methylcarbamate in agricultural crops, *J. Agric. Food Chem.*, 7, 612, 1959.

39. **Johnson, D. P. and Critchfield, F. E.,** Determination of Sevin insecticide and its metabolites in poultry, tissues and eggs, *J. Agric. Food Chem.,* 11, 77, 1963.
40. **Johnson, D. P. and Stansbury, H. A.,** Adaptation of Sevin insecticide (carbaryl) residue methods to various crops, *J. Agric. Food Chem.,* 13, 235, 1965.
41. **Casida, J. E.,** *Radiation and Radioisotopes Applied to Insects of Agricultural Importance,* Internationnal Atomic Energy Agency, Vienna, 1963, 223.
42. **Skraba, W. J. and Young, F. G.,** Radioactive Sevin (1-naphthyl-1-carbon^{14}N-methylcarbamate), a convenient synthesis, *J. Agric. Food Chem.,* 7, 612, 1959.
43. **Krishna, J. G., Dorough, H. W., and Casida, J. E.,** Synthesis of N-methylcarbamates via methyl isocyanate-^{14}C and chromatographic purification, *J. Agric. Food Chem.,* 10, 462, 1962.
44. **Dorough, H. W., Leeling, N. C., and Casida, J. E.,** Non-hydrolytic pathway in metabolism of N-methylcarbamate insecticides, *Science,* 140, 170, 1963.
45. **Dorough, H. W. and Casida, J. E.,** Nature of certain carbamate metabolites of the insecticide Sevin, *J. Agric. Food Chem.,* 12, 294, 1964.
46. **Leeling, N. C. and Casida, J. E.,** Metabolites of carbaryl (1-naphthyl N-methylcarbamate) in mammals and enzymatic systems for their formation, *J. Agric. Food Chem.,* 14, 281, 1966.
47. **Richey, F. A., Bartley, W. J., Fitzpatrick, J. T., and Kurtz, A. P.,** Chemical synthesis of the carbaryl metabolite *trans*-5, 6-dihydro-5, 6-dihydroxy-1-naphthyl methylcarbamate, *J. Agric. Food Chem.,* 20, 825, 1972.
48. **Knaak, J. B., Tallant, M. J., Bartley, W. J., and Sullivan, L. J.,** The metabolism of carbaryl in the rat, guinea pig and man, *J. Agric. Food Chem.,* 13, 537, 1965.
49. **Hassan, A., Zayed, S. M. A. D., and Abdul-Hamid, F. M.,** Metabolism of 1-naphthyl N-methylcarbamate (Sevin) in the rat, *Biochem. Pharmacol.,* 15, 2045, 1966.
50. **Sullivan, L. J., Eldridge, J. M., Knaak, J. B., and Tallant, M. J.,** 5,6-Dihydro-5, 6-dihydroxycarbaryl glucuronide as a significant metabolite of carbaryl in the rat, *J. Agric. Food Chem.,* 20, 980, 1972.
51. **Abdel-Wahab, A. M., Kuhr, R. J., and Casida, J. E.,** Fate of C^{14}-carbonyl-labeled aryl methylcarbamate insecticide chemicals in and on bean plants, *J. Agric. Food Chem.,* 14, 290, 1966.
52. **Mumma, R. O., Khalifa, S., and Hamilton, R. H.,** Spectroscopic identification of metabolites of carbaryl in plants, *J. Agric. Food Chem.,* 19, 445, 1971.
53. **Knaak, J. B., Tallant, M. J., Kozbelt, S. J., and Sullivan, L. J.,** Metabolism of carbaryl in man, monkey, pig and sheep, *J. Agric. Food Chem.,* 16, 465, 1968.
54. **Mostaffa, I. Y., Hassan, A., and Zayed, S. M. A. D.,** Translocation and degradation of ^{14}C-labeled Sevin in cotton plants, *Z. Naturforsch.,* 218, 1060, 1966.
55. **Mehendale, H. M. and Dorough, H. W.,** Glucuronidation mechanisms in the rat and their significance in metabolism of insecticides, *Pestic. Biochem. Physiol.,* 1, 307, 1971.
56. **Bend, J. R., Holder, G. M., Protos, E., and Ryan, A. J.,** Water-soluble metabolites of carbaryl (1-naphthyl N-methylcarbamate) in mouse liver preparations and in the rat, *Aust. J. Biol. Sci.,* 24, 535, 1971.
57. **Strother, A.,** Comparative metabolism of selected N-methylcarbamates by human and rat liver fractions, *Biochem. Pharmacol.,* 19, 2525, 1970.
58. **Strother, A.,** In-vitro metabolism of methylcarbamate insecticides by human and rat liver fraction, *Toxicol. Appl. Pharmacol.,* 21, 112, 1972.
59. **Matsumura, F. and Ward, C. T.,** Degradation of insecticides by the human and rat liver, *Arch. Environ. Health,* 13, 257, 1966.
60. **Chin, B. H., Eldridge, J. M., Anderson, J., and Sullivan, L. J.,** The Metabolism of Carbaryl in the Intact Rat, Perfused Liver, and Liver Explants, paper presented at the 165th American Chemical Society Meeting, Chicago, August 1973.
61. **Sullivan, L. J., Chin, B. H., and Eldridge, J. E.,** The In-vitro Metabolism of Carbaryl by Liver Explants of Bluegill, Catfish, Perch, Goldfish and Kissing Gourmai, paper presented at the 165th American Chemical Society Meeting, Chicago, August 1973.
62. **Sullivan, L. J., Chin, B. H., and Carpenter, C. P.,** In-vitro vs. in-vivo chromatographic profiles of carbaryl anionic metabolites in man and lower animals, *Toxicol. Appl. Pharmacol.,* 22, 161, 1972.
63. **Chin, B. H., Eldridge, J. M., and Sullivan, L. J.,** Metabolism of carbaryl by selected human tissues using an organ-maintenance technique, *Clin. Toxicol.,* 7, 37, 1974.
64. **Baron, R. L. and Locke, R. K.,** Utilization of cell culture techniques in carbaryl metabolism studies, *Bull. Environ. Contam. Toxicol.,* 5, 287, 1970.
65. **Locke, R. K., Bastone, V. B., and Baron, R. L.,** Studies of carbamate pesticide metabolites utilizing plant and mammalian cells in cultures, *J. Agric. Food Chem.,* 19, 1205, 1971.
66. **Locke, R. K.,** Thin-layer chromatography of 1-naphthyl N-hydroxy,N-methylcarbamate and its application in two in vitro studies involving carbaryl, *J. Agric. Food Chem.,* 20, 1078, 1972.
67. **Locke, R. K.,** Possible in-vitro production of N-O-conjugates of N-hydroxycarbaryl from carbaryl by tobacco cells in culture, *Fed. Proc. Fed. Am. Soc. Exp. Biol.,* 31, 520, 1972.
68. **North, H. H.,** Pesticide metabolism in mammalian cell cultures, *Diss. Abst. Int. B,* 33, 2135, 1972.
69. **Oonnithan, E. S. and Casida, J. E.,** Metabolites of methyl- and dimethylcarbamate insecticide chemicals as formed by rat liver microsomes, *Bull. Environ. Contam. Toxicol.,* 1, 59, 1966.

70. **Reiner, E. and Skrinjaric-Spoljar, M.**, Hydrolysis of some monomethylcarbamates in human sera, *Croat. Chem. Acta,* 40, 87, 1968.
71. **Skrinjaric-Spoljar, M. and Reiner, E.**, Enzymatic degradation of carbamates, *Arh. Hig. Rada Toksikol.,* 19, 251, 1968.
72. **Sakai, K. and Matsumura, F.**, Degradation of certain organophosphate and carbamate insecticides by human brain esterases, *Toxicol. Appl. Pharmacol.,* 19, 660, 1971.
73. **Sakai, K. and Matsumura, F.**, Esterases of mouse brain active in hydrolyzing organophosphate and carbamate insecticides, *J. Agric. Food Chem.,* 16, 803, 1968.
74. **Shah, A. H. and Guthrie, F. E.**, Penetration of insecticides through the isolated midgut of insects and mammals, *Comp. Gen. Pharmacol.,* 1, 391, 1970.
75. **Shah, A. H. and Guthrie, F. E.**, In-vitro metabolism of insecticides during midgut penetration, *Pestic. Biochem. Physiol.,* 1, 1, 1971.
76. **Shah, P. V. and Guthrie, F. E.**, Penetration of insecticides through isolated sections of the mouse digestive system: effects of age and region of the intestine, *Toxicol. Appl. Pharmacol.,* 25, 621, 1973.
77. **Pekas, J. C.**, Intestinal metabolism and transport of naphthyl N-methylcarbamate in vitro (rat), *Am. J. Physiol.,* 220, 2008, 1971.
78. **Pekas, J. C.**, Intestinal hydrolysis, metabolism, and transport of a pesticidal carbamate in pH 6.5 medium, *Toxicol. Appl. Pharmacol.,* 23, 62, 1972.
79. **Pekas, J. C. and Paulson, G. D.**, Intestinal hydrolysis and conjugation of a pesticidal carbamate in vitro, *Science,* 170, 77, 1970.
80. **Casper, H. H., Pekas, J. C., and Dinusson, W. E.**, Gastric absorption of a pesticide (1-naphthyl N-methylcarbamate) in the fasted rat, *Pestic. Biochem. Physiol.,* 2, 391, 1973.
81. **Casper, H. H. and Pekas, J. C.**, Absorption and excretion of radiolabeled 1-naphthyl-N-methylcarbamate (carbaryl) by the rat, *Proc. N. Dak. Acad. Sci.,* 24, 160, 1971.
82. **Baron, R. L.**, Radioactive lactose in skim milk following administration of carbonyl-^{14}C carbaryl to a lactating cow, *J. Assoc. Off. Anal. Chem.,* 51, 1046, 1968.
83. **Dorough, H. W. and Ivie, G. W.**, Carbon-14 milk constituents from cows fed carbamate labeled with carbon-14 on the carbonyl, *Science,* 159, 732, 1968.
84. **Lucier, G. W., McDaniel, O. S., Williams, C., and Klein, R.**, Effects of chlordane and methylmercury on the metabolism of carbaryl and carbofuran in rats, *Pestic. Biochem. Physiol.,* 2, 244, 1972.
85. **Dorough, H. W.**, Carbaryl residues in milk and meat of dairy animals, in *Pesticide Terminal Residues,* Tahori, A. S., Ed., Butterworth, New York, 1971, 173.
86. **Knaak, J. B. and Sullivan, L. J.**, Metabolism of carbaryl in the dog, *J. Agric. Food Chem.,* 15, 1125, 1967.
87. **Paulson, G. D., Zaylskie, R. G., Zehr, M. V., Portnoy, C. E., and Feil, V. J.**, Metabolites of carbaryl (1-naphthyl methylcarbamate) in chicken urine, *J. Agric. Food Chem.,* 18, 110, 1970.
88. **Dorough, H. W.**, Carbaryl-C^{14} metabolism in a lactating cow, *J. Agric. Food Chem.,* 15, 261, 1967.
89. **Baron, R. L., Palmer, N. J., Ross, R., Doherty, J., and Jacobson, W. C.**, Distribution of radioactivity in milk resulting from oral administration of ^{14}C-labeled carbaryl, *J. Assoc. Off. Anal. Chem.,* 51, 32, 1968.
90. **Baron, R. L., Sphon, J. A., Chen, J. T., Lustig, E., Doherty, J. D., Hansen, E. A., and Kolbye, S. M.**, Confirmatory isolation and identification of a metabolite of carbaryl in urine and milk, *J. Agric. Food Chem.,* 17, 883, 1969.
90a. **Paulson, D. G. and Feil, V. J.**, The fate of a single oral dose of carbaryl (1-naphthyl N-methylcarbamate) in the chicken, *Poultry Sci.,* 48, 1593, 1969.
90b. **Best, E. M. Jr. and Murray, B. L.**, Observations on workers exposed to Sevin insecticide: a preliminary report, *J. Occup. Med.,* 4, 507, 1962.
91. **Whitehurst, W. E., Bishop, E. T., and Critchfield, F. E.**, The metabolism of Sevin in dairy cows, *J. Agric. Food Chem.,* 11, 167, 1963.
92. **Andrawes, N. R., Chancey, E. L., Crabtree, R. J., Herrett, R. A., and Weiden, M. H. J.**, Fate of naphthyl-1-^{14}C carbaryl in laying chickens, *J. Agric. Food Chem.,* 20, 608, 1972.
93. **Dorough, H. W., Mehendale, H. M., and Lin, T.**, Modification of carbaryl metabolism in rats with monoamine oxidase inhibitors, *J. Econ. Entomol.,* 65, 958, 1972.
94. **Culver, D. J., Lin, T., and Dorough, H. W.**, Effect of monoamine oxidase inhibitors on carbaryl metabolism, *J. Econ. Entomol.,* 63, 1369, 1970.
95. **Lin, T. and Dorough, H. W.**, Influence of insecticide exposure of rats, on in vivo and in vitro metabolic activity, *Arch. Environ. Contam. Toxicol.,* 2, 364, 1974.
96. **Eldefrawi, M. E. and Hoskins, W. M.**, Relation of the rate of penetration and metabolism to the toxicity of Sevin to three insect species, *J. Econ. Entomol.,* 54, 401, 1961.
97. **Andrawes, N. R. and Dorough, H. W.**, Metabolic fate of carbaryl naphthyl-^{14}C in boll weevils and bollworms, *J. Econ. Entomol.,* 60, 453, 1967.
98. **Camp, H. B. and Arthur, B. W.**, Absorption and metabolism of carbaryl by several insect species, *J. Econ. Entomol.,* 60, 803, 1967.
99. **Zayed, S. M. A. D., Hassan, A., and Hussein, P. M.**, Metabolism of carbamate drugs. II. Degradation of 1-naphthyl N-methylcarbamate (Sevin) in adult larvae of the adult cotton leafworm, *(Prodenia lipura), Biochem. Pharmacol.,* 15, 2057, 1966.

100. **Sugiyama, H., Kanki, T., and Shigematsu, H.,** Difference in the metabolic fate of carbaryl between the silkworm, *Bombyx mori* L. (Lepidoptera: Bombycidae), and the black-tipped leafhopper, *Bothrogonia japonica* Ishihara (Hemiptera: Tittigellidae), *Appl. Entomol. Zool.*, 6, 57, 1971.
101. **Kuhr, R. J.,** Comparative metabolism of carbaryl by resistant and susceptible strains of the cabbage looper, *J. Econ. Entomol.*, 64, 1373, 1971.
102. **Kuhr, R. J. and Davis, A. C.,** Toxicity and metabolism of carbaryl in the European corn borer, *Pestic. Biochem. Physiol.*, 5, 330, 1975.
103. **Ku, T.-Y. and Bishop, J. L.,** Penetration, excretion and metabolism of carbaryl in susceptible and resistant German cockroaches, *J. Econ. Entomol.*, 60, 1328, 1967.
104. **Tsukamoto, M. and Casida, J. E.,** Albumin enhancement of oxidative metabolism of methylcarbamate insecticide chemicals by the housefly microsome-NADPH system, *J. Econ. Entomol.*, 60, 617, 1967.
105. **Tsukamoto, M. and Casida, J. E.,** Metabolism of methylcarbamate insecticides by the $NADPH_2$-requiring enzyme system from houseflies, *Nature*, 213, 49, 1967.
106. **Shrivastava, S. P., Georghiou, G., and Fukuto, T. R.,** Metabolism of N-methylcarbamate insecticides by mosquito larval enzyme system requiring $NADPH_2$, *Entomol. Exp. Appl.*, 14, 333, 1971.
107. **Kuhr, R. J.,** Possible role of tyrosinase and cytochrome P-450 in the metabolism of 1-naphthyl methylcarbamate (carbaryl) and phenyl methylcarbamate by houseflies, *J. Agric. Food Chem.*, 17, 112, 1969.
108. **Price, G. M. and Kuhr, R. J.,** Metabolism of the insecticide carbaryl (1-naphthyl N-methylcarbamate) by fat body of the blowfly larva, *Calliphora erythrocephala*, *Biochem. J.*, 112, 133, 1969.
109. **Wilkinson, C. F. and Brattsten, L. B.,** Microsomal drug-metabolizing enzymes in insects, *Drug Metab. Rev.*, 1, 153, 1972.
110. **Matsumura, F. and Sakaki, K.,** Degradation of insecticides by esterases of the American cockroach, *J. Econ. Entomol.*, 61, 598, 1968.
111. **Hanna, M. A. and Attallah, Y. H.,** Penetration and biodegradation of carbaryl in susceptible and resistant strains of the Egyptian cotton leafworm, *J. Econ. Entomol.*, 64, 1391, 1971.
112. **Szeicz, F. M., Plapp, F. W., and Vinson, S. B.,** Tobacco budworm: penetration of several insecticides into the larvae, *J. Econ. Entomol.*, 66, 9, 1973.
113. **Plapp, F. W., Jr.,** Comparison of insecticide absorption and detoxification in larvae of the bollworm, *Heliothis zea*, and the tobacco budworm, *H. virescens*, *Pestic. Biochem. Physiol.*, 2, 447, 1973.
114. **Moriyama, H., Sugiyama, H., and Shigematsu, H.,** Hydroxy metabolite derived from carbaryl in the silkworm, *Bombyx mori*, *Pestic. Biochem. Physiol.*, 2, 1, 1972.
115. **Bend, J. R., Holder, G. M., Protos, E., and Ryan, A. J.,** The metabolism of carbaryl in the cattle tick, *Boophilus microplus* (Canestrini), *Aust. J. Biol. Sci.*, 23, 361, 1970.
116. **Schuntner, C. A., Schnitzerling, H. J., and Roulston, W. J.,** Carbaryl metabolism in larvae of organophosphorus- and carbamate-susceptible and resistant strains of cattle tick, *Boophilus microplus*, *Pestic. Biochem. Physiol.*, 1, 424, 1971.
117. **Stewart, N. E., Millemann, R. E. and Breese, W. P.,** Acute toxicity of the insecticide Sevin and its hydrolytic product 1-naphthol to some marine organism, *Trans. Am. Fish. Soc.*, 96, 25, 1967.
118. **Berenbom, M. and Young, L.,** The isolation of 1- and 2-naphthyl sulfuric acid and 1- and 2-naphthyl glucuronide from the urine of rats dosed with 1- and 2-naphthol, *Biochem. J.*, 49, 165, 1951.
119. **Boyland, E. and Wiltshire, G. H.,** The metabolism of naphthalene, 1-naphthol and 1:2-dihydroxy-1:2-dihydroxynaphthalene by rats injected with naphthalene, *Biochem. J.*, 50, 37, 1953.
120. **Butler, G. C. and Packham, M. A.,** Reinvestigation of the fate of injected glucuronic acid, glucurone and naphthol glucuronide in the rat, *Arch. Biochem.*, 56, 551, 1955.
121. **Packham, M. A. and Butler, G. C.,** Studies on the biological synthesis of glucuronides, *J. Biol. Chem.*, 194, 349, 1952.
122. **Packham, M. A. and Butler, G. C.,** The fate of injected sodium glucuronate and glucurone in the rat, *J. Biol. Chem.*, 207, 639, 1954.
123. **Dorough, H. W., McManus, J. P., Kumar, S. S., and Cardona, R. A.,** Chemical and metabolic characteristics of 1-naphthyl β-D-glucoside, *J. Agric. Food Chem.*, 22, 642, 1974.
124. **Corner, E. D. S. and Young, L.,** 1:2-Dihydronaphthalene-1:2-diol and its role in the metabolism of naphthalene, *Biochem. J.*, 61, 132, 1955.
125. **Umezawa, S. and Masamune, H.,** On detoxication of naphthalene ingested with special reference to a new uronic acid excreted, *J. Biochem.* (Tokyo), 28, 487, 1938.
126. **Sperber, I.,** Some experiments on the formation of glucuronic acid derivatives in the chicken, *K. Lantsbrukshoegkolan Ann.*, 15, 108, 1948.
127. **Boose, R. B. and Terriere, L. C.,** Quantitative aspects of the detoxication of naphthalene by resistant and susceptible houseflies, *J. Econ. Entomol.*, 6, 580, 1967.
128. **Binning, A. F., Darby, J. J., Heenan, M. P., and Smith, J. N.,** The conjugation of phenols with phosphate in grass grubs and flies, *Biochem. J.*, 103, 42, 1967.
129. **Terriere, L. C., Boose, R. B., and Roubal, W. T.,** The metabolism of naphthalene and 1-naphthol by houseflies and rats, *Biochem. J.*, 79, 620, 1961.

130. **Hansen, L. G. and Hodgson, E.,** Metabolism of naphthalene and 1-naphthol by enzyme preparations from the housefly, *Musca domestica, Pestic. Biochem. Physiol.*, 1, 464, 1971.
131. **Dorough, H. W. and Wiggins, O. G.,** Nature of the water-soluble metabolites of carbaryl in bean plants and their fate in rats, *J. Econ. Entomol.*, 62, 49, 1969.
132. **Cardona, R. A. and Dorough, H. W.,** Synthesis of the β-D-O-glucosides of 4- and 5-hydroxy-1-naphthyl N-methylcarbamate, *J. Agric. Food Chem.*, 21, 1065, 1973.
133. **Andrawes, N. R. and Chancey, E. L.,** The Metabolism of 1-Naphthyl-^{14}C Carbaryl in Several Plant Species, paper presented at the 163rd American Chemical Society Meeting, Boston, April 1972.
134. **Sud, R. K., Sud, A. K. and Gupta, K. G.,** Degradation of Seven (1-naphthyl-N-methylcarbamate) by *Achromobacter* sp., *Arch. Mikrobiol.*, 87, 353, 1972.
135. **Tewfik, M. S. and Hamdi, Y. A.,** Decomposition of Sevin by a soil bacterium, *Acta Microbiol. Pol.*, 2, 133, 1970.
136. **Davies, J. F. and Evans, W. C.,** Ring fission of the naphthalene nucleus by certain soil *Pseudomonas, Biochem. J.*, 85, 21, 1962.
137. **Davies, J. F. and Evans, W. C.,** Oxidative metabolism of naphthalene by soil *Pseudomonas, Biochem. J.*, 91, 251, 1964.
138. **Zuberi, R. and Zubairi, M. Y.,** Carbaryl degradation by *Pseudomonas phaseolicola* and *Aspergillus niger, Pak. J. Sci. Ind. Res.*, 14, 383, 1971.
139. **Boush, G. M. and Matsumura, F.,** Insecticidal degradation by *Pseudomonas melophthora,* the bacterial symbiote of the apple maggot, *J. Econ. Entomol.*, 60, 918, 1967.
140. **Matsumura, F. and Boush, G. M.,** Metabolism of insecticides by microorganisms, *Soil Biol. Biochem.*, 2, 320, 1971.
141. **Bollag, J.-M. and Liu, S.-Y.,** Degradation of Sevin by soil microorganisms, *Soil Biol. Biochem.*, 3, 337, 1971.
142. **Bollag, J.-M. and Liu, S.-Y.,** Hydroxylations of carbaryl by soil fungi, *Nature,* 236, 177, 1972.
143. **Bollag, J.-M. and Liu, S.-Y.,** Fungal degradation of 1-naphthol, *Can. J. Microbiol.*, 18, 1113, 1972.
144. **Matsumura, F. and Boush, G. M.,** Degradation of insecticides by a soil fungus, *Trichoderma viride, J. Econ. Entomol.*, 61, 610, 1968.
145. **Liu, S.-Y. and Bollag, J.-M.,** Metabolism of carbaryl by a soil fungus, *J. Agric. Food Chem.*, 19, 487, 1971.
146. **Liu, S.-Y. and Bollag, J.-M.,** Carbaryl decomposition to 1-naphthyl carbamate by *Aspergillus terreus, Pestic. Biochem. Physiol.*, 1, 366, 1971.
147. **Rodriguez, L. D. and Dorough, H. W.,** Metabolism of carbaryl in soils, *Arch. Environ, Contam. Toxicol.*, in press, 1975.
148. **Kazano, H., Kearney, P. C., and Kaufman, D. D.,** Metabolism of methylcarbamate insecticides in soils, *J. Agric. Food Chem.*, 20, 975, 1972.
149. **Ohkawa, H., Kohara, T., Yoshihara, R., and Miyamoto, J.,** Metabolism of N-methylcarbamate insecticides, *Noyaku Kagaku,* 1, 86, 1973.
150. **Ohkawa, H., Yoshihara, R., Kohara, T., and Miyamoto, J.,** Metabolism of *m*-tolyl N-methylcarbamate (Tsumacide) in rats, houseflies and bean plants, *Agric. Biol. Chem.*, 38, 1035, 1974.
151. **Miyamoto, J., Yamamoto, K., and Matsumoto, T.,** Metabolism of 3,4-dimethylphenyl N-methylcarbamate in white rats, *Agric. Biol. Chem.*, 33, 1060, 1969.
152. **Miyamoto, J.,** A feature of detoxication of carbamate insecticides in mammals, in *Biochemical Toxicology of Insecticides,* O'Brien, R. D. and Yamamoto, I., Eds., Academic Press, New York, 1970.
153. **Miyamoto, J.,** Metabolism of organophosphorus compounds and carbamate insecticides, in *Fate of Pesticides in Environment,* Tahori, A. S., Ed., Gordon and Breach, London, 1972, 319.
154. **Baron, R. L. and Doherty, J. D.,** Metabolism and excretion of an insecticide (6-chloro-3,4-dimethylphenyl N-methylcarbamate) in the rat, *J. Agric. Food Chem.*, 15, 830, 1967.
155. **Metcalf, R. L., Osman, M. F., and Fukuto, T. R.,** Metabolism of ^{14}C-labeled carbamate insecticides to $^{14}CO_2$ in the housefly, *J. Econ. Entomol.*, 60, 445, 1967.
156. **Gemrich, E. G.,** Enzymatic degradation of 2-chloro-4-dimethylphenyl N-methylcarbamate by the fat bodies of *Blaberus gigantus, J. Agric. Food Chem.*, 15, 617, 1967.
157. **Friedman, A. R. and Lemin, A. J.,** Metabolism of 2-chloro-4,5-dimethylphenyl N-methylcarbamate in bean plants, *J. Agric. Food Chem.*, 15, 642, 1967.
158. **Slade, M. and Casida, J. E.,** Metabolic fate of 3,4,5- and 2,3,5-trimethylphenyl methylcarbamates, the major constituents in Landrin insecticide, *J. Agric. Food Chem.*, 18, 467, 1970.
159. **Parke, D. V.,** *The Biochemistry of Foreign Compounds,* Pergamon Press, New York, 1969, Ch. 7.
160. **Georghiou, G. P. and Metcalf, R. L.,** The absorption and metabolism of 3-isopropylphenyl N-methylcarbamate by susceptible and carbamate-selected strains of houseflies, *J. Econ. Entomol.*, 54, 231, 1961.
161. Anon., *m*-(1-Methylbutyl)phenyl methylcarbamate and *m*-(1-ethylpropyl)phenyl methylcarbamate: tolerances for residues, *Fed. Regist.*, 34, 1589, 1969.
162. **Sutherland, G. L. and Cook, J. W.,** Terminal carbamate residues, *J. Assoc. Off. Anal. Chem.*, 53, 992, 1970.
163. **Tucker, B. V. and Pack, D. E.,** Bux insecticide soil metabolism, *J. Agric. Food Chem.*, 20, 412, 1972.
164. **Cheng, H.-M. and Casida, J. E.,** Metabolites and photoproducts of 3-(2-butyl)phenyl N-methylcarbamate and N-benzenesulfenyl-N-methylcarbamate, *J. Agric. Food Chem.*, 21, 1037, 1973.
165. **Miskus, P. R., Andrews, T. L., and Look, M.,** Metabolic pathways affecting toxicity of N-acetyl Zectran, *J. Agric. Food Chem.*, 17, 842, 1969.

166. **Black, A. L., Chiu, Y. C., Fukuto, T. R., and Miller, T. A.**, Metabolism of 2,2-dimethyl-2, 3-dihydrobenzofuranyl-7 N-methyl-N-(2-toluenesulfenyl) carbamate in the housefly and white mouse, *Pestic. Biochem. Physiol.*, 3, 435, 1973.
167. **Douch, P. G. C. and Smith, J. N.**, Metabolism of *m-tert*-butylphenyl N-methylcarbamate in insects and mice, *Biochem. J.*, 125, 385, 1971.
168. **Douch, P. G. C. and Smith, J. N.**, The metabolism of 3,5,-di-*tert*-butylphenyl N-methylcarbamate in insects and by mouse liver enzymes, *Biochem. J.*, 125, 395, 1971.
169. **Dawson, J. A., Heath, D. F., Rose, J. A., Thain, E. M., and Ward, J. B.**, The excretion by humans of the phenol derived in vivo from 2-isopropoxyphenyl N-methylcarbamate, *Bull. W.H.O.*, 30, 127, 1964.
170. **Casida, J. E., Shrivastava, S. P., and Esaac, E. G.**, Selective recovery of volatile products from houseflies treated with radioactive insecticide chemicals and synergists, *J. Econ. Entomol.*, 61, 1339, 1968.
171. **Shrivastava, S. P., Tsukamoto, M., and Casida, J. E.**, Oxidative metabolism of ^{14}C-labeled Baygon by living houseflies and by housefly enzyme preparations, *J. Econ. Entomol.*, 62, 483, 1969.
172. **Shrivastava, S. P., Georghiou, G. P., Metcalf, R. L., and Fukuto, T. R.**, Carbamate resistance in mosquitoes. Metabolism of propoxur by susceptible and resistant larvae of *Culex pipiens fatigans, Bull. W.H.O.*, 42, 931, 1970.
173. **Tsukamoto, M., Shrivastava, S., and Casida, J. E.**, Biochemical genetics of housefly resistance to carbamate insecticide chemicals, *J. Econ. Entomol.*, 61, 50, 1968.
174. **Gromberg, R. R.**, Metabolism of Baygon in Corn Plants, Chemagro Report No. 29233, 1970.
175. **Williams, E., Meikle, R. W., and Redemann, C. T.**, Identification of metabolites of Zectran insecticide in dog urine, *J. Agric. Food Chem.*, 12, 457, 1964.
176. **Wheeler, L. and Strother, A.**, In vitro metabolism of the N-methylcarbamates, Zectran and Mesurol, by liver, kidney and blood of dogs and rats, *J. Pharmacol. Exp. Ther.*, 178, 371, 1971.
177. **Roberts, R. B., Miskus, R. P., Duckles, C. K., and Sakai, T. T.**, In vivo fate of the insecticide Zectran in spruce budworm, tobacco budworm and housefly larvae, *J. Agric. Food Chem.*, 17, 107, 1969.
178. **Williams, E., Mielke, R. W., and Redemann, C. T.**, The identification of metabolites of Zectran insecticide in broccoli, *J. Agric. Food Chem.*, 12, 453, 1964.
179. **Robertson, W. A. H., Fraser, J., and Clinck, G. C.** (Bootes Pure Drug Co.). British Patent 982,235, February 3, 1965.
180. **Miskus, R. P., Look, M., Andrews, T. L., and Lyon, R. L.**, Biological activity as an effect of structural changes in aryl N-methylcarbamate, *J. Agric. Food Chem.*, 16, 605, 1968.
181. **Church, D. D. and Flint, D. R.**, The Fate of Mesurol [4-(methylthio)-3, 5-xylylmethylcarbamate] in Soil, Chemagro Report No. 29591, 1971.
182. **Starr, R. I. and Cunningham, D. J.**, Degradation of ^{14}C-Labeled Mesurol in Soil and Water, paper presented at the 167th American Chemical Society Meeting, Los Angeles, April 1974.
183. **Dorough, H. W.**, Metabolism of Furadan (NIA-10242) in rats and houseflies, *J. Agric. Food Chem.*, 16, 319, 1968.
184. **Metcalf, R. L., Fukuto, T. R., Collins, C., Borack, K., El-Aziz, A., Munoz, R., and Cassil, C. C.**, Metabolism of 2,2-dimethyl-2,3-dihydrobenzofuranyl-7-N-methylcarbamate (Furadan) in plants, insects, and mammals, *J. Agric. Food Chem.*, 16, 300, 1968.
185. **Black, A. L., Chiu, L., Fahmy, M. A. H., and Fukuto, T. R.**, Selective toxicity of N-sulfenylated derivatives of insecticidal methylcarbamate esters, *J. Agric. Food Chem.*, 21, 747, 1973.
186. **Hicks, B. W., Dorough, H. W., and Davis, R. B.**, Fate of carbofuran in laying hens, *J. Econ. Entomol.*, 63, 1108, 1970.
187. **Ivie, G. W. and Dorough, H. W.**, Furadan C-14 metabolism in lactating cows, *J. Agric. Food Chem.*, 16, 849, 1968.
188. **Knaak, J. B., Munger, D. M., McCarthy, J. F., and Stater, L. D.**, Metabolism of carbofuran alfalfa residues in the dairy cow, *J. Agric. Food Chem.*, 18, 832, 1970.
189. **Knaak, J. B., Munger, D. M., and McCarthy, J. F.**, Metabolism of carbofuran in alfalfa and bean plants, *J. Agric. Food Chem.*, 18, 827, 1970.
190. **Dorough, H. W.**, Fate of Furadan insecticide in susceptible and carbamate-resistant houseflies, *Tex. Agric. Exp. Stn. Prog. Rep.*, PR-2832, July, 1970.
191. **Stenersen, J., Gilman, A., and Vardanis, A.**, Carbofuran: its toxicity to and metabolism by earthworm (*Lumbricus terrestris*), *J. Agric. Food Chem.*, 21, 166, 1973.
192. **Dorough, H. W.**, Fate of Furadan (NIA-10242) in bean plants, *Bull. Environ. Contam. Toxicol.*, 3, 164, 1968.
193. **Pree, D. J. and Saunders, J. L.**, Metabolism of carbofuran in mugho pine, *J. Agric. Food Chem.*, 22, 620, 1974.
194. **Ashworth, R. J. and Sheets, T. J.**, Uptake and translocation of carbofuran in tobacco plants, *J. Econ. Entomol.*, 63, 1301, 1970.
195. **Ashworth, R. J. and Sheets, T. J.**, Metabolism of carbofuran in tobacco, *J. Agric. Food Chem.*, 20, 407, 1972.
196. **Penner, D. and Early, R. W.**, Effect of alachlor, butylate, and chlorobromuron on carbofuran distribution and metabolism in barley and corn., *Weed Sci.*, 21, 360, 1973.
197. **Hamill, A. S. and Penner, D.**, Chlorobromuron–carbofuran interactions in corn and barley, *Weed Sci.*, 21, 363, 1973.
198. **Robbins, J. E., Bakke, J. E., and Feil, V. J.**, Metabolism of 4-benzothienyl N-methylcarbamate (Mobam) in rats: balance study and urinary-metabolite separation, *J. Agric. Food Chem.*, 17, 236, 1969.
199. **Robbins, J. D., Bakke, J. E., and Feil, V. J.**, Metabolism of benzo[b]thienyl-4-yl methylcarbamate (Mobam) in dairy goats and a lactating cow, *J. Agric. Food Chem.*, 18, 130, 1970.

200. **Morallo, B. D.,** The metabolism of Mobam (4-benzo[b]thienyl N-methylcarbamate) in resistant and susceptible house flies, *Diss. Abstr. Int. B,* 30, 3226, 1970.
201. **Williams, P. P. and Stolzenberg, R. L.,** Ruminal bacterial degradation of benzo(b)thien-4-yl methylcarbamate (Mobam) and effect of Mobam on ruminal bacteria, *Appl. Microbiol.,* 23, 745, 1972.
202. **Kaufman, D. D. and Kearney, P. C.,** Microbial degradation of isopropyl-N-3-chlorophenylcarbamate and 2-chloroethyl N-3-chlorophenylcarbamate, *Appl. Microbiol.,* 13, 443, 1965.
203. **Knowles, C. O.,** Metabolism of two acaricidal chemicals, N′-(4-chloro-*o*-tolyl)-N, N-dimethylformamidine (chlorphenamidine) and *m*-[(dimethylamino)methylene] amino phenyl methylcarbamate hydrochloride (formetanate), *J. Agric. Food Chem.,* 18, 1038, 1970.
204. **Sen Gupta, A. K. and Knowles, C. O.,** Fate of formetanate-^{14}C acaricide in the rat, *J. Econ. Entomol.,* 63, 10, 1970.
205. **Ahmad, S. and Knowles, C. O.,** Degradation of formetanate acaricide by rat liver preparations, *J. Econ. Entomol.,* 63, 1690, 1970.
206. **Ahmad, S. and Knowles, C. O.,** Formamidase involvement in the metabolism of chlorphenamidine and formetanate acaricides, *J. Econ. Entomol.,* 64, 792, 1971.
207. **Sen Gupta, A. K. and Knowles, C. O.,** Galecron-^{14}C [N′-(4-chloro-*o*-tolyl)-N, N-dimethylformamidine] metabolism in the dog and goat, *J. Econ. Entomol.,* 63, 951, 1970.
208. **Knowles, C. O. and Sen Gupta, A. K.,** Photodecomposition of the acaricide N′-(4-chloro-*o*-tolyl)-N, N-dimethylformamidine, *J. Econ. Entomol.,* 62, 344, 1969.
209. **Andrawes, N. R., Dorough, H. W., and Lindquist, D. A.,** Degradation and elimination of Temik in rats, *J. Econ. Entomol.,* 60, 979, 1967.
210. **Knaak, J. B., Tallant, M. J., and Sullivan, L. J.,** The metabolism of 2-methyl-2-(methylthio)propionaldehyde O-(methylcarbamoyl)oxime in the rat, *J. Agric. Food Chem.,* 14, 573, 1966.
211. **Hicks, B. W., Dorough, H. W., and Mehendale, H. M.,** Metabolism of aldicarb pesticide in laying hens, *J. Agric. Food Chem.,* 20, 151, 1972.
212. **Dorough, H. W. and Ivie, G. W.,** Temik-S^{35} metabolism in a lactating cow. *J. Agric. Food Chem.,* 16, 460, 1968.
213. **Dorough, H. W., Davis, R. B., and Ivie, G. W.,** Fate of Temik-carbon-14 in lactating cows during a 14-day feeding period, *J. Agric. Food Chem.,* 18, 135, 1970.
214. **Metcalf, R. L., Fukuto, T. R., Collins, C., Brock, K., Burke, J., Reynolds, H. T., and Osman, M. F.,** Metabolism of 2-methyl-2-(methylthio)propionaldehyde O-(methylcarbamoyl)oxime (Temik) in plants and insects, *J. Agric. Food Chem.,* 14, 579, 1966.
215. **Andrawes, N. R. and Dorough, H. W.,** Metabolism of Temik in boll weevils and houseflies, *Tex. Agric. Exp. Stn. Prog. Rep.,* PR-2833, 8, 1970.
216. **Bull, D. L., Lindquist, D. A., and Coppedge, J. R.,** Metabolism of 2-methyl-2-(methylthio)propionaldehyde O-(methylcarbamoyl)oxime (Temik, UC-21149) in insects, *J. Agric. Food Chem.,* 15, 610, 1967.
217. **Coppedge, J. R., Lindquist, D. A., Bull, D. L., and Dorough, H. W.,** Fate of 2-methyl-2-(methylthio)propionaldehyde O-(methylcarbamoyl)oxime (Temik) in cotton plants and soil, *J. Agric. Food Chem.,* 15, 902, 1967.
218. **Bartley, W. J., Andrawes, N. R., Chancey, E. L., Bagley, W. P., and Spurr, H. W.,** The metabolism of Temik aldicarb pesticide [2-methyl-2-(methylthio)propionaldehyde O-(methylcarbamoyl)oxime] in cotton plants, *J. Agric. Food Chem.,* 18, 446, 1970.
219. **Andrawes, N. R., Romine, R. R., and Bagley, W. P.,** Metabolism and residues of Temik aldicarb pesticide in cotton foliage and seed under field conditions, *J. Agric. Food Chem.,* 21, 379, 1973.
220. **Bull, D. L.,** Metabolism of UC-21149 [2-methyl-2-(methylthio)propionaldehyde O-(methylcarbamoyl)oxime] in cotton plants and soil in the field, *J. Econ. Entomol.,* 61, 1598, 1968.
221. **Andrawes, N. R., Bagley, W. P., and Herrett, R. A.,** Metabolism of 2-methyl-2-(methylthio)propionaldehyde O-(methylcarbamoyl)oxime (Temik aldicarb pesticide) in potato plants, *J. Agric. Food Chem.,* 19, 731, 1971.
222. **Andrawes, N. R., Bagley, W. P., and Herrett, R. A.,** Fate and carryover properties of Temik aldicarb pesticide [2-methyl-2-(methylthio)propionaldehyde O-(methylcarbamoyl)oxime] in soil, *J. Agric. Food Chem.,* 19, 727, 1971.
223. **Bull, D. L., Stokes, R. A., Coppedge, J. R., and Ridgway, R. L.,** Further studies of the fate of aldicarb in soil, *J. Econ. Entomol.,* 63, 1283, 1970.
224. **Supak, J. R.,** The volatilization, degradation, absorption, and desorption characteristics of aldicarb [2-methyl-2-(methylthio)propionaldehyde O-(methylcarbamoyl)oxime] in soils and clays, *Diss. Abstr. Int. B,* 33, 982, 1972.
225. **Harvey, J., Jr., Jelinek, A. G., and Sherman, H.,** Metabolism of methomyl in the rat, *J. Agric. Food Chem.,* 21, 769, 1973.
226. **Baron, R. L.,** Toxicological considerations of metabolism of carbamate insecticides, in *Pesticide Terminal Residues,* Tahori, A. S., Ed., Butterworth, New York, 1971, 185.
227. **Harvey, J., Jr. and Reiser, R. W.,** Metabolism of methomyl in tobacco, corn and cabbage, *J. Agric. Food Chem.,* 21, 775, 1973.
228. **Harvey, J., Jr. and Pease, H. L.,** Decomposition of methomyl in soil, *J. Agric. Food Chem.,* 21, 784, 1973.
229. **Kuhr, R. J.,** Metabolism of methomyl in the cabbage looper, *Pestic. Biochem. Physiol.,* 3, 113, 1973.
230. **Hutson, D. H., Hoadley, E. C., and Pickering, B. A.,** Metabolism of S-2-cyanoethyl-N-[(methylcarbamoyl)oxy]thioacetimidate, an insecticidal carbamate in the rat, *Xenobiotica,* 1, 179, 1971.
231. **Whitten, C. J. and Bull, D.,** Fate of 3,3-dimethyl-1-(methylthio)-2-butanone O-(methylcarbamoyl)oxime (Diamond Shamrock DS-15647) in cotton plants and soil, *J. Agric. Food Chem.,* 22, 234, 1974.

Chapter 8

BIOLOGICAL AND ENVIRONMENTAL STABILITY

INTRODUCTION

Titles relating to pesticides in the environment have been embedded on the covers of a few dozen books and documents over the last 15 years. A number of these volumes describe the intricate chemical reactions which occur in various sectors of our surroundings, others examine the interaction of pesticides with a particular environmental compartment, such as soil or water, and still others take aim at the effects of these toxins on nontarget living organisms. Most approach the subject objectively, some are partial, a few are unscientific, but all of them were created out of genuine concern for the potential hazards associated with the widespread use of pesticide chemicals.

These publications together with periodic reports in the news media have succeeded in creating public concern and considerable controversy regarding the benefits vs. the risks associated with the use of pesticides. However, the above monographs only touch on the environmental implications of carbamate usage, partly because much of the information regarding stability, translocation, and product characterization have only recently appeared in the literature, and also because many consider these compounds to be inherently unstable and thus of little concern when compared to more persistent pesticides. In addition, there is a lack of extremely refined analytical techniques for detection of carbamates (see Chapter 9), particularly techniques such as those which were responsible for the discovery of minute amounts of organochlorine materials in almost every segment of the environment. Thus, this chapter is an essential part of our description of carbamate insecticides.

An attempt has been made on the following pages to present a composite of most of the published studies dealing with the environmental stability of methylcarbamate insecticides. It will be clear that much of the work has been conducted in the laboratory under simulated field conditions, but there have been several field projects which generally confirm laboratory findings. There is a definite need for more experiments which generate results under actual conditions encountered in the real world.

As might be expected, the carbamates vary widely in persistence, and each compound's lifetime in a particular situation depends on a large number of physical and chemical parameters. However, as a class, the carbamates are not long-term contaminants and, under similar circumstances, do not persist in our environment nearly as long as many of the organochlorine insecticides. Paradoxically, it is this apparent environmental instability which has restricted many of the carbamate candidates to the laboratory and prohibited their commercial use.

In some cases, release of a carbamate insecticide into the environment is followed by production of a metabolite or degradation product which still possesses biological activity. Where these transformations are known to take place, experiments have been conducted to ensure that these products also have limited existence in our surroundings.

From the standpoint of human exposure, commercial carbamates, when applied according to instructions on the label, have a good safety record thus far. They have not become a significant residue in our diet, nor are they found in minute quantities in our body fat. But carbamates are toxicants, nonetheless, and we should be concerned about their use and the possible side effects which potentially accompany their application. Thus, it is imperative that as much as possible be learned about this family of insecticides, particularly as they grow in importance over the coming years of increasing demand for food and fibers.

ANIMALS

Considerable quantities of insecticides are applied to domestic animals as dusts, water suspensions, or dietary supplements for the control of noxious insects. In most cases, studies related to this application are performed to assure total safety to the animal receiving the treatment and, in the case of farm animals, to guarantee minimum residues of toxicants in tissues and other products, such as milk and eggs.

When applied topically, the insecticide is subject to volatilization, atmospheric decomposition, run-off, and rub-off, similar to that encountered with plant foliage, although the physical chemistry of the surfaces could cause very different responses. This aspect of carbamate degradation by animals has received little attention, however. Once the compound penetrates the skin, or after its oral ingestion, the insecticide becomes available to the metabolic enzymes within the animal and, thus, is open to attack via oxidation, hydrolysis, and conjugation. Parent compound and/or metabolites are then either stored in various tissues or excreted from the animal body.

Although in many instances insecticides are applied to animals intentionally, there are circumstances where animals eat grains, forage, or other foods which have been previously treated with insecticides. In this event, the animal may contact not only the parent material, but may also ingest various degradation products formed in or on the feed.

When animals that have been treated with, or have consumed, insecticides become part of our diet, it is very important that the nature of the residues be known. In this regard, a substantial degree of research has been conducted relative to the fate of carbamates in dairy cattle, particularly with respect to the levels of products eliminated in milk, where residues must be zero.

Carbaryl in Dairy Animals

Early in the history of carbaryl's use, reports appeared pertaining to the effect of carbaryl-treated feed on the flavor and odor of milk from dairy cattle, and, more precisely, on the appearance of residues in the milk. One of the first studies included feeding cows technical carbaryl for 2 weeks at levels of 50, 150, and 450 ppm of their daily roughage.[1] Samples of milk taken at regular intervals during and after treatment showed no detectable carbaryl following analysis with a colorimetric method of 0.01 ppm sensitivity. In addition, no off-flavor or odors due to the presence of carbaryl were noted, and the cattle appeared perfectly normal physiologically and exhibited no symptoms of poisoning.

Another report appearing at the same time indicated that after carbaryl was fed to dairy cows, or after repeated spraying of cattle with 0.5% suspensions of carbaryl, no residues appeared in the milk.[2] Also, spraying beef cattle repeatedly with 0.5% carbaryl suspensions yielded no detectable residues in animal body fat. Later, it was discovered that increasing the active ingredient of the suspension to 1% resulted in some carbaryl in the milk for the first 24 hr following application, but gave no detectable residues thereafter.[3] Even when cattle were dusted with a 50% carbaryl formulation, no residue could be isolated from milk except for 1 hr post-treatment.[3,4] Thus, it was deemed safe to dust in the morning and milk at night.

These studies convinced most workers that carbaryl, if present in cow's milk, was there in very insignificant amounts; consequently, more attention was given to an examination of tissue residues. Cattle, sheep, goats, and hogs were sprayed several times with 1% carbaryl and were slaughtered 1 or 7 days later.[5] Though some residues were found within 24 hr of the last application, by 7 days no detectable carbaryl, 1-naphthol, or 1-naphthol conjugates could be found by colorimetric analysis of fat, muscle, liver, heart, kidney, or brain of any animal except for fat and brain of one goat. It was also demonstrated that small quantities of unchanged carbaryl were excreted in cattle feces.[6]

In 1964, a report appeared in the literature which prompted reconsideration of carbaryl metabolism, and consequently of carbaryl excretion in milk.[7] Capsules containing carbonyl-^{14}C carbaryl were fed to a goat, and all excretory products were analyzed radiometrically for 96 hr following treatment. Although very little actual carbaryl was recovered in milk, as much as 0.9 ppm carbaryl equivalents were present 8 hr after ingestion; these had disappeared by 72 hr. Thus, even though carbaryl or 1-naphthol *per se* was not in the goat's milk, other metabolites, missed by previous colorimetric analysis, were present. Subsequent experiments utilizing 1-naphthyl-^{14}C carbaryl fed to dairy cows at 0.25 and 3.05 mg/kg proved that carbaryl and free 1-naphthol were excreted in milk in minute quantities for only a short period after treatment. Also detected were a number of hydroxylated metabolites which did not respond to the colorimetric method of analysis.[8]

These later radiocarbaryl studies also elucidated the nature of the carbamate and hydrolytic metabolites which were excreted in urine and feces; more important, they revealed the extreme biological instability of carbaryl in dairy animals. Joined with the rapid metabolism was the virtual absence of residues in a wide variety of the treated animal's tissues.[8] Such rapid biotransformation of carbaryl into oxidative, hydrolytic, and conjugate products in dairy cows has been further elaborated by many other laboratories, as described in detail in Chapter 7.

One drawback to the radiotracer studies was that the insecticide was applied as a single dose. It was believed by some that this may not reflect the situation where an animal consumes low levels of insecticide daily. Thus, a study was initiated in which cows were fed 1-naphthyl-^{14}C carbaryl at 10, 30, or 100 ppm for 14 days.[9] At all dose levels, rapid equilibration was established between intake, metabolism, and excretion of products, so that the amounts of recovered radiocarbon were similar over the 14-day period. In milk, a very small percentage of carbaryl equivalent, representing approximately 0.2% of dose, was found at each level. Little of this was carbaryl itself; mostly it consisted of oxidative and hydrolytic products. *In toto*, the picture drawn from the results of the 14-day feeding was very much like that depicted in the single-dose experiments.

However, there have been reports in recent years, based on colorimetric methods, of persistent and higher milk residues than those discussed above. For example, cattle sprayed with 0.3% carbaryl excreted the parent compound in milk for 69 hr post-treatment.[10] Repeated spraying of cows with 1% carbaryl suspensions yielded residues of 0.1 to 0.5 ppm in milk.[11,12] An oral dose of 0.6 or 0.3 mg/kg gave carbaryl residues in milk which did not reach a peak until 6 days after treatment;[13] significant residues were found even 9 days after dosing. The reasons for these higher, persistent milk loads are not known, though they may be due in part to different experimental conditions and differences in the test animals.

Other Carbamates in Farm Animals

Other carbamates administered to cows yield different results, but all compounds are characterized by low production of radioactive materials in milk, as shown in Figure 8.1. Of the insecticide equivalents given, only 0.6% was eliminated from the animals as parent carbofuran,[14] and no parent aldicarb[15,16] or Mobam[17] could be found. In addition, no residues of Mobam were found in any cow tissue except the brain, where 0.1 ppm Mobam equivalents resided after 24 hr. A few other tissues from Mobam-treated goats had minor quantities of radioactivity, with a maximum of 1.3

	MILK ^{14}C	URINE ^{14}C	FECES ^{14}C
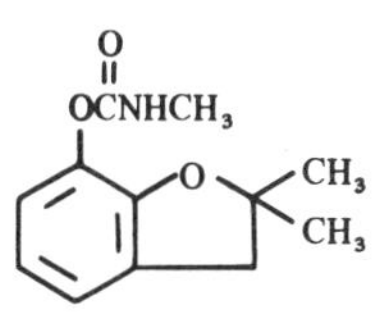 carbofuran	0.16%	90%	0.5%
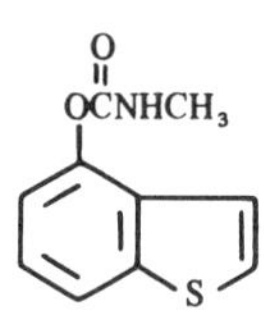 mobam	1.0%	87%	12%
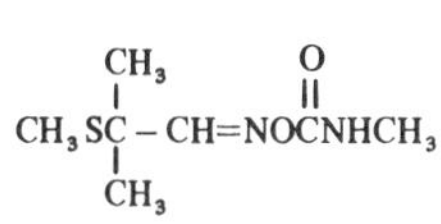 aldicarb	1.2%	83%	0.6%

FIGURE 8.1. Fate of radiocarbon 24 hr after ingestion of radiolabeled methylcarbamate insecticides by dairy cows.[14,15,17]

ppm in the liver 24 hr post-treatment. Once again, an exception to the rule appears in a report stating that, after oral ingestion of Baygon (0.5 to 0.85 mg/kg), parent compound residues persisted in milk for as long as 24 days after exposure.[13]

In a study designed to mimic actual farm conditions, cattle were fed silage composed of Bermuda grass which had been sprayed with Mobam.[18] As part of the investigation, it was demonstrated that Mobam in ensilage was stable for at least 30 days. While consuming equivalent levels of 0.25 to 0.90 mg/kg, the dairy cows released no Mobam or its phenol in any of the milk examined. Various physiological parameters were monitored during the study, but the cattle showed no ill effects from feeding on the Mobam-treated silage.

In a similar program, carbofuran was mixed with silage at different levels just before feeding to dairy cows.[19] Interestingly, the concentration of insecticide in the silage regulated the intake by the cows, i.e., the more carbofuran present, the slower the cows ate. In this way, clinical symptoms of poisoning were barely visible compared to cows receiving a large single oral dose of the insecticide. Cows consuming 197 ppm carbofuran for 11 days excreted a maximum of 0.13% of this in milk, most of which was 3-hydroxy carbofuran with no detectable parent compound.

Almost identical results were obtained when carbofuran-treated corn was made into silage and fed to cattle.[20] Although insecticide equivalents of 33 to 74 mg per cow were consumed, no carbofuran residues could be detected in milk at any time. Also, there were no differences between control and carbofuran-treated cows with respect to weight increase, milk production, food intake, blood cholinesterase activity, and other physiological parameters.

Bypassing the silage and introducing either carbonyl-^{14}C or ring-^{14}C carbofuran into lactating Holstein cows via rumen fistula also gave similar results.[21] After 4 days, 2.5% of the dose from both labeled preparations entered the milk, but less than 2% of this was parent compound. When cattle were fed alfalfa contaminated with radioactive carbofuran metabolites, they excreted about 2.5% of the radiocarbon from carbonyl-^{14}C residues, little of which was carbofuran. Plant residues originating from ring-^{14}C carbofuran yielded only 0.7% of the administered radioactivity in milk after ingestion by the cows. Examination of the radioactive products excreted from the cattle indicated that the plant glycosides fed to the cows were metabolized and conjugated as glucuronides and sulfates.

The same type of experiment has been conducted with rats. Water-soluble products recovered from bean plants which had previously been injected with ^{14}C-carbaryl were fed to rats.[22] After 24 hr, 70% of the radiocarbon had been voided in the urine. By 4 days, 93% of the radioactivity was recovered in the urine and feces. Most of the products retained their water-soluble properties.

The biostability of Butacarb in sheep was examined because this material is used for control of sheep blowflies on animals destined for human consumption.[23] Lambs were dipped in a 0.5% suspension of Butacarb and slaughtered at weekly intervals up to 6 weeks. The suspension contained Butacarb labeled with ^{3}H in the ring and ^{14}C on the N-methyl group or N-methyl-^{14}C Butacarb. In all experiments, no residues greater than 0.2 ppm were found in liver, kidney, fat, or muscle. It was suggested that, under normal practices, these levels would probably never exceed 0.1 ppm in tissues of sheep treated with Butacarb.

Carbaryl, Aldicarb, and Carbofuran in Poultry and Fish

Just as there exists a potential for transfer of insecticide residues from feeds or surface treatments into meat and milk of dairy and beef cattle, there is also a possibility that such movement occurs into the meat and eggs of poultry. Accordingly, numerous studies concerning the fate of ingested or topically applied carbaryl in chickens have been undertaken.[24-28]

In general, no serious ill effects were noted after feeding or dusting chickens with relatively high levels of carbaryl. Deposits of carbaryl plus 1-naphthol were occasionally detected in a few tissues, such as muscle (1 ppm maximum) and fat (1 to 2 ppm), during and shortly after treatment, but by 7 days tissues were free of carbaryl. Some residues appeared in eggs, but usually only after a large oral dose of insecticide had been administered. Likewise, residues in a wide variety of tissues could be detected shortly after a high oral dose (1,000 to 2,500 mg/kg), but it was thought that these dissipated rapidly to near zero within 7 to 10 days.[29] Experiments using radiocarbaryl indicated that carbaryl deposits in bird tissues and eggs existed primarily in the parts-per-billion range.[30,31] Also, most of the radioactive residues in eggs were in the form of typical hydrolytic, oxidative, and conjugate carbaryl metabolites.

Residues in eggs and tissues of chickens given a single oral dose of radioaldicarb were found to be very low.[32] A maximum of 0.18 ppm aldicarb equivalents appeared in eggs within 24 hr, but by 10 days it dropped to 0.01 ppm, none of which was aldicarb itself. The same rapid decline was evident in residues located in muscle, liver, and kidney, where again no aldicarb *per se* was present. Repeating the study by feeding the oxime carbamate as part of the chickens' diet did not yield considerable differences in tissue and egg deposits.

Carbofuran was also administered to chickens as a single oral treatment.[33] Because of differences in specific activities, the carbonyl-^{14}C preparation was given at a rate of 0.3 mg/kg and the ring-^{14}C carbofuran at 2.7 mg/kg. The lower dose caused egg residues no greater than 0.02 ppm, which disappeared after 6 days. The higher level yielded maximum deposits of 0.13 ppm in the eggs, which diminished to 0.01 ppm after 10 days. Examination of chicken tissues revealed that the largest residues were in liver within 6 hr of treatment (0.42 to 2.73 ppm). However, after 1 day most tissues were nearly free of residues.

Another marketable animal which could possibly encounter and accumulate carbaryl from its feed or surroundings is fish. Although carbaryl is not generally applied directly to water, it is possible that low levels of the insecticide could enter the fish environment. Therefore, catfish were exposed to water concentrations of 1-naphthyl-^{14}C carbaryl at 0.05 or 0.25 mg/l, or were fed 0.28 or 2.8 mg/kg/week, for 50 days.[34] Maximum retained radioactivity totaled only 9 ppb after feeding and 11 ppb after water exposure to the higher doses. Thus, carbaryl's demonstrated instability in animals was extended to include the catfish.

The above investigations, together with numerous metabolism studies in other species, leave little doubt that methylcarbamate insecticides are very unstable in living animals. The low levels of carbamate which most animals would likely encounter in their environment should be rapidly degraded and eliminated without great harm. With respect to farm animals, methylcarbamate insecticides applied according to label directions should not appear in meat, milk, or eggs. However, this does not imply that all metabolic products formed from carbamates in animals are without problems, although the evidence to date would support this correlation.

PLANTS

The growing of plants for food and fiber products constitutes a large portion of agriculture. This task, which increases in importance each year as the world population continues to expand, can be lightened in many instances by the judicious use of herbicides, fungicides, nematicides, and insecticides. The latter offer plants protection from damage by insect feeding or by transmission of fatal disease organisms. Protection is usually accomplished through coating the plant surfaces with a poison barrier, or by feeding the plant through the roots and/or foliage with a poison which effects its action after insect ingestion.

Such foliar treatments or root feedings demand a certain persistence of insecticide on or within the plant in order to be effective. Persistence, however, must not result in unduly high residues in the harvestable portion of the crop. Although it is almost impossible to treat a plant with any insecticide and find absolutely no residues at any time thereafter, dissipation rates on various crops can be obtained with enough precision to allow accurate predictions of post-treatment levels. Thus, tolerances can be established which permit low quantities of certain insecticides to be present on crops at the time of harvest. Obviously, these tolerances are set at levels which will ensure maximum safety to the consumer. In summary, the stability of an insecticide on or in an agricultural plant must be such that it persists long enough to effectively control insect pests, but not so long that harmful deposits remain at harvest time.

Of course, many plants exist in our environment which are not treated with pesticides, but become contaminated by spray drift and subterranean movement of the chemicals. In those instances where such plants serve as food for other species, it is important that the insecticide rapidly decompose within the nontarget plant.

In Chapter 7 it was demonstrated that carbamate insecticides are not very stable in plants, being rapidly decomposed by oxidation and conjugation. However, the majority of metabolism studies were performed in the laboratory using stem injections or root feedings, which may not always reflect actual field conditions. Thus, the biological stability discussed in this chapter centers around residue data collected from crop treatments under commercial growing conditions or from laboratory and greenhouse tests where field conditions were emulated. In most cases, only the parent compound and other known organo-soluble products were assayed.

Carbaryl

Since carbaryl is registered for use on approximately 100 different crops, a considerable amount of data pertaining to the biostability of this compound in and on plants have been obtained. Most of the published information is concerned with residues on fruits and vegetables. With carbaryl, as with most insecticides, a considerable amount of residue data are generated by universities, experiment stations, government laboratories, and industry for registration purposes. Much of this information does not reach publication in scientific journals and thus is not available for review.

The loss of carbaryl from orange and lemon peels following conventional high-pressure spray treatment of the respective trees at two different rates is shown in Figure 8.2. Although the same dissipation line is drawn for both citrus fruits, there were slight differences; in lemon peels carbaryl had a half-life of 28 days compared to 42 days in Valencia oranges.[35] It is important to note that the concentrations reported are for peels, and not for whole fruit. Since the peels alone represented 20 to 30% of the fruit weight, total fruit residues would be considerably lower. In no case was any carbaryl or naphthol detected in edible portions of the fruits.

As illustrated in Figure 8.2, doubling the rate of application approximately doubled the deposit throughout the dissipation curve. It can also be seen that residue disappearance follows first-order reaction kinetics. There is usually a rapid initial loss of insecticide shortly after application, due to volatilization and weathering, and then a slower

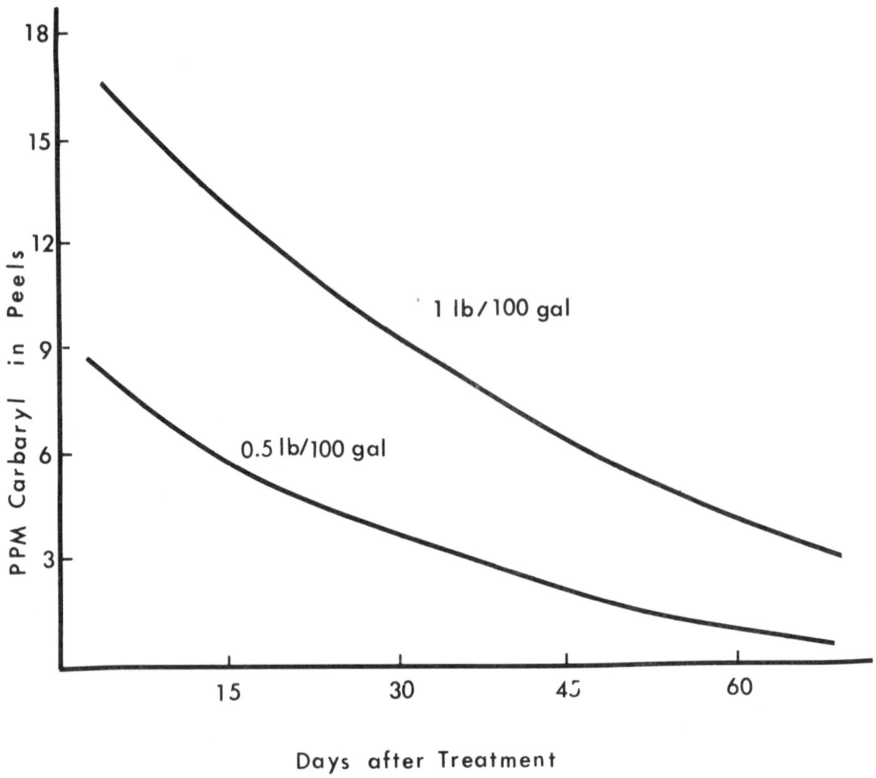

FIGURE 8.2. Dissipation of carbaryl from peels of field-sprayed lemons and oranges treated at two different rates.[35]

loss over a longer period of time, due to degradation of the parent compound. A third change in slope may follow, which is sometimes thought of as the true persistence curve. From such a multislope curve it is obvious that the half-life of a residue is the time it takes for only the first half of the residue to disappear, and this time will almost always be considerably shorter than the time it takes for the last half of the residue to completely dissipate.

Although carbaryl residues behaved similarly on lemons and on oranges, the dynamics of loss for the same insecticide on other fruits was shown to be quite different.[36] Treatment of trees with 0.15% carbaryl sprays gave varied initial deposits on three different fruits (Figure 8.3). Coverage on peaches averaged 13 ppm; much lower levels, 3 ppm and 1.5 ppm respectively, were detected on apples and plums shortly after application. Undoubtedly some of these differences were due to variations in the surface of the fruits. It was recommended that, in order to reach Romania's tolerance of 2 ppm (carbaryl tolerance on most fruits in the United States in 10 ppm), growers would have to harvest plums, apples, and peaches, respectively, 2 days, 7 days, and 21 days after application of 0.15% carbaryl sprays.

Based on initial deposits, half-lives for carbaryl residues on the three fruits ranged from 7 to 10 days. Almost identical half-lives were found for carbaryl on strawberries and on gooseberries.[37] The latter study also showed that carbaryl residues present on apples during storage did not decrease appreciably over 4 months time, and that there was some movement of carbaryl from the apple peel into the pulp.

When carbaryl was sprayed on three different apple varieties, similar initial deposits were found with similar dissipation curves.[38] Once again, the half-lives were calculated to be around 10 days, which is fairly consistent from studies conducted in Canada,[38] Russia,[37] and Romania.[36] Although difficult to compare, initial deposits on the fruits may not have been the same. Interestingly, each country had different tolerance limits for carbaryl on apples.

Residue data for vegetables are often gathered from various parts of the plants. For example, after spraying brinjal plants (eggplant) with an 0.15% carbaryl emulsion, leaves, stems, and fruit were shown to have 52, 36, and 40 ppm of carbaryl, respectively.[39] The disappearance of insecticide from all three surfaces varied little, with a reported half-life of about 6.5 days. Within 12 to 13 days, EPA tolerances of 10 ppm had been achieved.

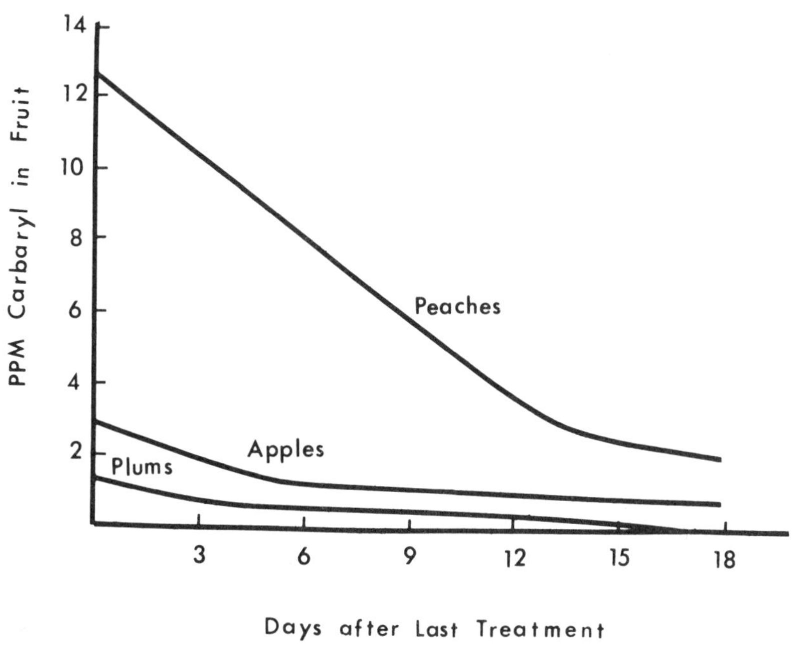

FIGURE 8.3. Deposition and dissipation of residues from fruit on trees sprayed with carbaryl.[36]

Considerably smaller residues of carbaryl resulting from similar application rates on eggplant were found by other workers.[40] Even at a higher rate with a 0.4% suspension (2.2 lb/acre), initial residues on eggplant reached only 17 ppm, although the same treatment level delivered twice the insecticide to cabbage (Table 8.1). At lower dosages, cabbage also contained about twice as much carbaryl, but doubling the application rate did not double the initial deposit for either plant. In fact, it required an almost fourfold increase in dose level to achieve a twofold increase in surface residues. This phenomenon is well known to field entomologists and is related, in part, to the fact that doubling the amount of formulated insecticide also doubles the amount of wetting agents, suspension agents, etc. The result is that a proportionately larger amount of insecticide runs off the treated surface at the higher level.

Half of the carbaryl residues had dissipated from eggplant and cabbage in about 1 day, regardless of application rate (Table 8.2). Thus, the dissipation curves for initial carbaryl deposits on cabbage were very similar, even though the quantities of insecticide involved were different. Washing the vegetables before residue analysis showed that most of the carbaryl deposits had not penetrated beyond the outer surfaces.

An aerial spray of 0.15% carbaryl to tomatoes at 210 gal/acre left fruit residues of only 0.1 ppm after 6 days, with no detectable levels present after 9 days.[41] This rapid loss also was demonstrated to take place on tomato leaves following ground spraying with carbaryl suspensions.[42] But, as seen in Table 8.3, dissipation was considerably prolonged by adding Pinolene, a β-pinene polymer, to the spray mixture (1.56 lb carbaryl + 1 gal Pinolene/100 gal spray/acre). Residue analyses, in this case, were accomplished with a thin-layer cholinesterase-inhibition technique (see Chapter 9), so that some of the large differences may have been due to interactions with Pinolene. It is unfortunate that the results were not substantiated by a method more specific for carbaryl. Another study, using apples and a different spreader-sticker, indicated that the additive had little effect on carbaryl deposition or dissipation.[38]

The action of Pinolene was attributed to an occlusion and dissolution of the insecticide within the Pinolene film which formed on the foliage. This afforded protection from rain and wind erosion, volatilization, and photodecomposition.

TABLE 8.1

Differences in Initial Deposits on Cabbage and Eggplant Sprayed with Various Rates of Carbaryl[40]

Dose (lb/acre)	Initial deposit (ppm)	
	Cabbage	Eggplant
0.55	15	8
1.1	24	12
2.2	34	17

TABLE 8.2

Percent of Initial Carbaryl Deposit Recovered from Cabbage Treated at Three Different Rates

Days after treatment	Percent carbaryl recovered		
	0.55 lb/acre	1.1 lb/acre	2.2 lb/acre
1	61	56	55
3	39	34	31
5	26	25	21
7	17	17	15

Adapted from Reference 40, with permission.

TABLE 8.3

Effect of Pinolene on Carbaryl Persistence on Tomato Leaves[42]

Days after treatment	Average carbaryl recovered (ppm)	
	No Pinolene	Pinolene added
0	48	500
1	11	600
3	8	350
5	2	220
7	0	175
22	0	15

Note again that the treatment without Pinolene resulted in very rapid initial loss of carbaryl during the first 24 hr, followed by slower disappearance over the next few days.

An average compilation of carbaryl residues on vegetables from several sources sprayed at the rate of 1 lb/acre was reported early in carbaryl's history.[43] Broccoli was shown to collect and retain more carbaryl on a parts-per-million basis than tomatoes, cucumbers, or snap beans. However, even with an initial deposit of 8.6 ppm on broccoli, residues had fallen to 0.67 ppm by 7 days post-treatment. The spinach group of vegetables accumulated higher initial carbaryl deposits than did most other vegetables, probably due to larger leaf surface area.[44] An initial average deposit of 52 ppm rapidly degraded to about 9 ppm within 1 week. Lettuce residues were usually lower and also dissipated shortly after application. With these plants, a larger percentage of the decrease in residues was due to dilution by growth.

Although generally not thought of as a good soil insecticide, carbaryl has been used as such for control of the carrot maggot. Soil treated with 5 to 16 g/running meter left 0.4 to 0.5 ppm of carbaryl in carrots 12 days after drenching.[45] By harvest, however, all residues had disappeared, despite the fact that some residual carbaryl was still present in the soil. Actual movement of carbaryl from the soil into the carrots was not established. When the insecticide was applied to soil as a 10% granular formulation, it was shown to migrate through the root system into the aerial parts of corn plants.[46] Colorimetric analysis exposed residues as high as 28 ppm in leaves 35 days after treatment with 3.3 lb/acre.

Systemic activity also has been demonstrated following topical carbaryl treatments. An average of 1.75% of the radioactivity from leaf-applied carbonyl-^{14}C carbaryl was translocated to active growing centers of cacao seedlings.[47] However, transferred radiocarbon was not identified as carbaryl *per se* and may have been mostly metabolites. Absorption of the applied insecticide into the leaf took place for up to 35 days, whereupon a drastic drop in rate occurred, probably because of exhaustion of external residues. It was shown that the amount of carbaryl absorbed and translocated to other parts of the seedling varied little with different seasonal conditions.

Foliar application of 1-naphthyl-^{14}C carbaryl to wheat, alfalfa, peanuts, and tomatoes under field conditions was followed by a rapid loss of surface radioactivity to very low levels by 14 days.[48] The small percentages that were absorbed by each species were immediately metabolized to the expected oxidative and conjugate products.

As discussed in the previous section, forage crops are often treated with insecticides prior to feeding to farm and dairy animals. Thus, rapid degradation of applied materials is essential to minimize transfer of residues to these animals. Aerial application of carbaryl at 1 lb/acre to red clover and grasses left residues of 30 ppm, which dissipated by 50% in 1 day, and to only 0.1 ppm by 28 days.[49] When dusted on red clover at 1.5 lb/acre, carbaryl disappeared somewhat slower, leaving deposits of 0.2 to 0.5 ppm after 49 days, and less than 0.1 ppm at 65 days.[50] A dose of 4 oz/acre delivered to range grasses by airplane or ground sprayer produced initial deposits of 34 ppm. However, when the rate was increased to 1 lb/acre, more than 100 ppm were detected. With either rate, more than 80% of the residues had disappeared from the grass within 21 days.

Stability of carbaryl on alfalfa depended on dose, mowing time, and drying procedure used to obtain the feed.[51] Sun-dried hay that had been cut within 2 weeks of treatment contained residues from a trace to more than 100 ppm after 78 to 105 days in storage. None could be found, on the other hand, in hay mowed 3 weeks after treatment and dried in the dark.

Spraying 0.2% emulsions of carbaryl on field corn resulted in initial residue levels of 72 ppm on leaves, but these levels had dropped exponentially to 1 ppm by 3 weeks.[46] If rain fell shortly after treatment, residues were markedly reduced, as expected. But with aged residues, rain washed off considerably smaller percentages of persisting materials. Dramatic effects of precipitation, and probably of sunlight, on carbaryl dissipation are illustrated in Figure 8.4, where plants grown in open air are compared to plants grown under a

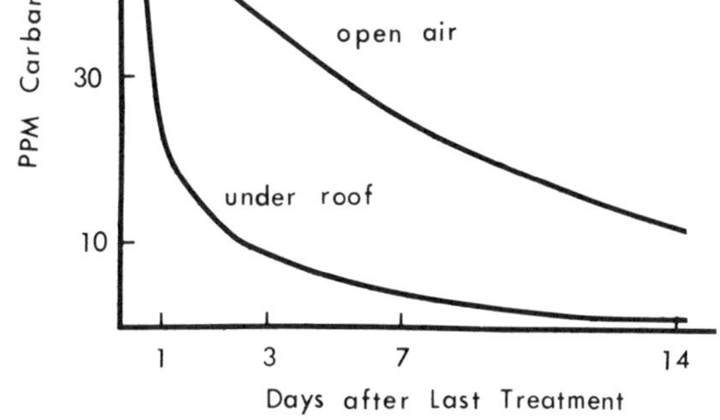

FIGURE 8.4. Dissipation of carbaryl residues from leaves of corn grown under cover and in an open field.[46]

roof. Low temperatures were also found to slow losses of residues from corn leaves. The husks also contained carbaryl residues, but only in the outer wrapper leaves; no residues were detected in the grain, stover, silk, or even in the inner husk leaves.

Carbofuran

Residue studies of other methylcarbamate insecticides appear less frequently in the literature, but the success of carbofuran and aldicarb as systemic agents has prompted several plant studies with these insecticides. For example, the uptake of carbofuran into field corn was examined following a granular broadcast application to the soil and after an in-furrow granular treatment.[52] As anticipated, concentration of the pesticide in the furrow resulted in larger absorption and translocation into leaves and stalks. After 4 months, carbofuran equivalents in the corn plants totaled 11.7 ppm, compared to only 5.5 ppm from broadcast applications. However, no accumulation of residues was found in the kernels with either practice.

Carbofuran metabolism took place primarily in the leaves, where more than 90% conversion to 3-hydroxy and 3-keto carbofuran occurred at the silage stage. By contrast, only 5 to 15% degradation took place in stalks and cobs, although total residues were much lower than in the leaves. It was noted that while the leaves were desiccating, 3-hydroxy carbofuran was oxidized to the 3-keto derivative, whereas drying stalks hydrolyzed 3-hydroxy carbofuran to the corresponding phenol. The researchers[52] concluded that, even though they had worked with higher-than-recommended rates, corn plant residues easily fell within established tolerances for carbofuran in corn fodder and forage.

Carbofuran deteriorated more rapidly in alfalfa sprayed with carbofuran emulsions at 0.5 and 1 lb/acre.[53] This particular study was one of the few where doubling of the application rate resulted in more than doubling of the initial insecticide deposit, from 48 ppm to 136 ppm. Within 1 week, in both plots, more than half of the deposit was not recovered, and no measurable quantities (greater than 0.2 ppm) persisted beyond 2 weeks. Dehydration of alfalfa at 870°C removed about 65% of carbofuran contaminants.

Coastal Bermuda grass contained 24 ppm of carbofuran after treatment at 1 lb/acre.[20] In only 7 days, a mere 4% of this remained, and by 14 days less than 0.04 ppm were recovered. Alfalfa cut 1 day after spraying and placed in silage for 30 days at 30°C did not show any decline in the 16 ppm of residue content. Likewise, field-sprayed corn which was cut and silo-cured for 55 days contained carbofuran residues ranging from 1.1 to 3.6 ppm during the entire period. Carbofuran residues persisted even through feeding the silage to cattle for another 56 days. However, the authors[20] made it clear that the high application rates and almost immediate harvest were not typical and that under normal practices only very low residues would be present in silage.

Carrots containing carbofuran residues stored at 4.4°C tended to increase their residue content with time.[54] Apparently this was due to loss of water during storage. In-furrow treatment of carrot seeds with subsequent foliage sprays later in the season left harvest residues of 0.1 to 0.5 ppm carbofuran plus 3-hydroxy carbofuran 10 to 50 days after the last application.

Soil placement of carbofuran for a study of intake into cabbage plants was performed on a small scale with carbonyl-^{14}C insecticide.[55] Within 1 week post-application, less than 10% of the radioactivity recovered from cabbage leaves was in the form of the parent compound; 80% was 3-hydroxy carbofuran. A steady decrease in recoverable carbofuran followed, so that less than 1% of the leaf residue was present as the parent material by 56 days. Meanwhile, the 3-hydroxy metabolite increased to 90% after 2 weeks, but dropped to 75% by 8 weeks.

Using somewhat different formulations and techniques, another worker applied 3% sand-core granules of carbofuran to rice plants in a greenhouse study.[56] Various manipulations of experimental treatments were conducted to follow carbofuran movement and stability in the rice environment. Results, based on an insect bioassay, revealed that the insecticide could move laterally through soil and water, that movement into the plant was more rapid through stem and roots than through leaves, and that large amounts of water were required for most efficient transport. Carbofuran was retained primarily in the soil, and not in the rice plant; consequently, the soil was more important in determining the stability of the insecticide.

Insect bioassay of mugho pine trees grown in soil treated with carbofuran granules indicated that insecticidally active residues persisted in needles 2 years after application.[57] Initial treatment with 2 g per tree yielded peak needle deposits of 282 ppm in 2 years; new needles had 1.3 ppm, and old needles contained 9.7 ppm. This apparent slow metabolic rate and long-term persistence were confirmed in a later study using radiocarbofuran, where more intermediates were identified.[58]

Aldicarb

In studies where an insecticide is applied to soil or to water, the detection of bioactivity in plants grown therein does not necessarily imply that the parent insecticide has been translocated within the plant. It is always possible that a metabolite, formed in the soil or in the plant, is the moving toxicant. For instance, when aldicarb was applied in-row at a rate of 2 lb/acre with bean seed, no aldicarb was present in bean seeds or pods when harvested 90 days later.[59] However, 0.09 ppm of aldicarb sulfoxide and sulfone were found. No parent compound was in the aerial portion of field-grown beans, but about 50 ppm of sulfoxide and sulfone persisted after 47 days, with 7 ppm still present after 103 days.

A similar aldicarb stability pattern has been established for other plants and is illustrated in Figure 8.5. Within 9 days after in-furrow application of S-methyl-^{14}C aldicarb to cotton plants, most of the 209 ppm total equivalents in the foliage was present as aldicarb sulfoxide.[60] A small concentration of sulfone was also found, but very little of the parent insecticide. Presumably, over the first few days there was a very rapid loss of aldicarb with a corresponding build-up of sulfoxide. As can be seen in Figure 8.5, the sulfoxide concentration remained constant for a few days, and then it too quickly disappeared. Meanwhile, the sulfone concentration peaked and remained level for about 1 week, followed by slower dissipation. Careful scrutiny of the curves indicates that all of the sulfone did not result from *in situ* oxidation of the sulfoxide. As suggested in the report, some of the sulfone was probably transported into the leaves from the soil, where it had already been formed from aldicarb. Movement of the residues from leaves into cotton seeds was negligible. Many of these results were confirmed by residue analysis of field-treated cotton grown by commercial procedures.

Other research has shown that aldicarb persistence depends on soil moisture. Cotton seedlings exposed to in-furrow placement of 10% aldicarb granules at 15 lb/acre were compared in dry land and irrigated land.[61] Post-harvest cotton from dry land plots averaged 0.07 ppm residues in seed and 0.05 ppm in the lint. Similar samples from irrigated fields contained a residue level of 0.01 ppm in both seed and lint. The particular residue analysis procedure used in the study required oxidation of recovered products, so that the distribution of aldicarb, sulfoxide, and sulfone

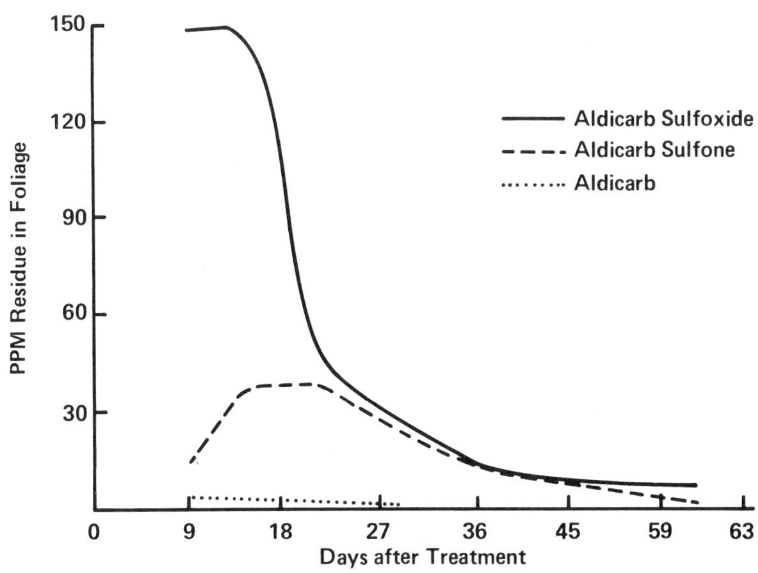

FIGURE 8.5. Presence of aldicarb and its sulfoxide and sulfone in cotton foliage after soil treatment.[60]

was not known. However, since all three of these chemicals maintain considerable toxicity, the biostability of all should be understood.

Lower application rates (1 to 4 lb/acre) placed either in-row or as a side-dress to cotton seedlings produced much lower residues in cottonseed.[62] After 60 to 183 days, seed contamination averaged less than 0.007 ppm of aldicarb equivalent, with only a few exceptions containing 0.008 to 0.016 ppm of sulfone.

Translocation of ^{35}S-aldicarb in cotton plants resulted in concentration of most of the residues in older leaves.[63] Maximum leaf accumulation from soil applications took place within 2 weeks, and apparently little or no uptake occurred after this time. However, volatilization of aldicarb or metabolites from the plants or soil may have masked any further build-up. The behavior of aldicarb and its sulfoxide was similar to that shown in Figure 8.5, but the sulfone accumulated at a slow rate throughout the 6-week test period to a final concentration of 30 to 40 ppm.

Foliar application of ^{35}S-aldicarb to cotton plants in the field yielded results similar to those found for soil treatments.[64] However, metabolism was more rapid, and very little translocation from leaves to other plant parts took place.

Uses of aldicarb on fruits and vegetables are restricted to soil applications. When distributed at the base of apple trees at rates of 0.1 to 0.4 lb per tree, residues of 0.08 to 0.83 ppm sulfoxide and 0.04 to 0.14 ppm sulfone were detected in the fruit 143 days later.[62] When pear trees were subjected to 1 to 4 oz of insecticide, the fruit had residues of 0.25 to 0.80 ppm at day 40 and of 0.07 to 1.0 ppm at day 68. A single application of 2 oz per tree resulted in higher fruit contamination than two 1-oz-per-tree treatments spaced 50 days apart. This was attributed to more complete absorption of insecticide earlier in the season. Pears were stored for more than 1 year at 0°C without a significant reduction in residues.

Citrus trees also absorb and translocate aldicarb residues. Applications of 1.4 oz to individual navel orange trees resulted in sulfoxide residues of 12.8 ppm and 2.63 ppm in fruit peel and pulp respectively after 100 days.[65] Smaller deposits of 2.24 ppm and 0.58 ppm were identified as aldicarb sulfone, but no aldicarb (less than 0.01 ppm) *per se* was found. Higher application rates at the base of Temple orange trees (4 oz per tree), Pineapple orange trees (4.8 or 9.6 oz per tree), and Marsh grapefruit trees (3.2 oz per tree) resulted in much lower fruit residues, although samples were taken much later after treatment.[66] For example, in Temple oranges the residues reached a maximum of only 0.046 ppm 33 weeks after application. The contamination was distributed about equally between juice and peels of the fruit. By 38 weeks, levels had dropped to 0.019 ppm, at which time tree leaves contained 0.007 ppm. Similar results with somewhat higher residue levels were found with grapefruit trees, but Pineapple orange trees absorbed and translocated a much greater amount of insecticide. The latter plants had been subjected to heavy irrigation shortly after treatment, which probably accounts for the differences. Even so, maximum levels in fruit at 25 weeks were only 3.7 ppm.

Repeated foliar applications of aldicarb to sugarbeets did not result in substantial downward movement of the systemic insecticide.[62] Leaves contained 0.58 ppm sulfoxide and 0.86 ppm sulfone, whereas crowns and roots held only about 0.03 ppm of each metabolite 126 days after treatment ceased. The low residues in all parts of the plant are important, because cattle are often fed waste products from sugar production.

Other forage is also protected from insects through the use of aldicarb. Bands of 10% granules (6.2 lb/acre) on both sides of alfalfa rows gave very steady plant residues of about 4.5 ppm from day 12 to day 140 post-treatment.[67] But when alfalfa was cut and dried in the greenhouse, residues were considerably higher through the early post-treatment period. Lower application rates, of course, gave lower residues.

Although parent compound was not separated from oxidized products, it is unlikely that very much aldicarb was present in the alfalfa based on greenhouse studies.[68] At several sampling dates, however, it was found that the sulfone predominated over the sulfoxide. When pea aphids were placed on treated plants, mortality rose and fell with the rise and fall of sulfoxide residues.

Aldicarb has also been demonstrated to be unstable in field-grown potatoes exposed to in-furrow treatments of S-methyl-^{14}C insecticide.[69] The expected sulfoxidation products were found, but at no time could any parent material be extracted from any portion of the plants. Analysis of individual metabolites showed that tubers contained many more water-soluble and hydrolytic products than did leaves.

A recent series of papers reported on the effects of various external factors on aldicarb residues in treated plants. For example, lettuce grown in greenhouses is often subjected to a carbon dioxide-enriched atmosphere. Under these circumstances, plants drenched with aldicarb solutions exhibited no differences in uptake or metabolism and accumulated residues in the same manner as control plants.[70] Greenhouse-grown tomatoes treated with aldicarb by soil-drill showed symptoms of phytotoxicity, but subsequent plant growth and the weight and yield of fruit were not affected by the treatment.[71] Surprisingly, average residues in tomatoes of aldicarb plus sulfoxidation products were five times higher after a dose of 0.5 g per plant than after a dose of 0.25 g per plant. Finally, the effect of several systemic fungicides on the uptake and metabolism of aldicarb in soybeans was examined.[72] None of the compounds altered uptake or degradation to nontoxic products, but, because of analytical procedures, it was not possible to examine the effects on oxidative metabolism.

Methomyl

Because of the metabolism of methomyl to small chemical fragments (see Chapter 7), the only residue which appears to bear watching is the parent compound itself. When applied as a 0.03% suspension to greenhouse tomatoes, a build-up of methomyl occurred on the fruit with repeated applications[73] (Figure 8.6). As indicated, 1 week after the first spray residues were about 0.1 ppm, but 1 week after the third spray residues had reached 0.35 ppm.

In a comparative study of field-grown crops, methomyl was shown to be quite unstable.[74] Residues in and on unpolished rice 60 to 83 days after the last application were always below 0.001 ppm. In apples, residues ranged from 0.023 ppm 45 days after the last treatment to 0.164 ppm 30 days after the third application. Oranges contained less than 0.001 ppm methomyl 78 days after the sixth spray had been applied. Green tea and Chinese cabbage contained small quantities of residues within 1 week of treatment, which virtually disappeared after 2 weeks.

Repetitive season-long treatments of cabbage and corn with methomyl under field conditions support the above findings.[75] Only the outer waste-leaves of cabbage had substantial residues (0.8 to 0.9 ppm) 8 days after the final application. At the same time, corn fodder contained 2 ppm methomyl, but no residues were detected in the kernels.

Other Carbamates

With other methylcarbamate insecticides, environmental-persistence data are limited. On forest tree foilage, as judged by insect bioassay, Zectran was stable for about 10 days on white spruce and on European larch.[76,77] Zectran and Matacil persisted for 10 days on balsam fir. Carbaryl was rapidly lost from white spruce.

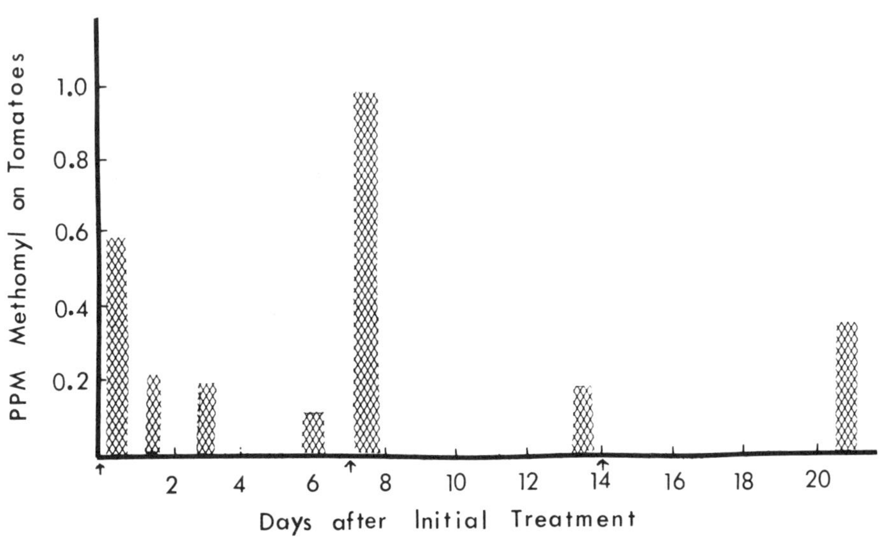

FIGURE 8.6. Residues of methomyl on tomatoes after repeated treatments. Arrows indicate spray dates.[73]

Baygon appeared to be fairly stable on tamarack, but not on jack pine. Further studies with isopropoxy-^{14}C Baygon in young balsam fir trees revealed that penetration of the insecticide through tree needles varied considerably with formulation.[78] Mixtures containing sugars or urea gave much better penetration and had the added advantage of overcoming the phytotoxic properties of certain oil-base formulations. However, following penetration into the needles, there was very little transport via phloem to other parts of the tree.

One of the carbamates most resistant to plant degradation may be formetanate. After topical application of the ring-^{14}C preparation to orange seedlings, as much as 75% of the radioactivity was found to have penetrated the foliage by 12 days.[79] Of this, 22% was present as unchanged parent carbamate. Even after injection into plant stems, considerable quantities of the parent compound persisted. Assuming that similar degradation occurs with field-applied materials, it would appear that formetanate is relatively stable in plants.

It is evident from the above examples, and from metabolism data presented in Chapter 7, that methylcarbamate insecticides are not very stable in or on plants. With aldicarb, where toxic metabolites are formed immediately after absorption, residues tend to be more persistent. Penetration of applied carbamates through leaf surfaces with subsequent translocation is limited, but those possessing systemic properties are absorbed through roots and transported to other parts of the plant. Since most systemic insecticides are applied to the soil, it is not always possible to ascertain whether metabolites are formed after entry and translocation in the plant or whether they are formed in the soil. However, even if penetration occurs, metabolism is rapid, thus ensuring a limited lifetime of the toxic parent carbamate within plants.

SOILS

For many years the fate of pesticides in the environment was considered to be synonymous with the fate of pesticides in soils. This concept has changed in recent times and now includes interactions with air, water, nontarget plants, animals, and microorganisms. However, soils still constitute a major environmental compartment, and persistence and degradation of insecticides in soils have been the subjects of many research projects.

Like studies with plants and animals, soil studies have often been confined to the laboratory under artificial conditions of light, temperature, and humidity. Although these procedures have divulged considerable information on general metabolic pathways and comparative degrees of stability, they cannot always be directly translated into field situations. Thus, when considering the stability of insecticide chemicals in the environment, it is important to examine data which originated from experiments performed in the field or from laboratory and greenhouse studies closely approximating field conditions.

As previously described, the stability of an insecticide within an animal is dependent mainly on the enzymic makeup or biochemical characteristics of the animal itself. In plants, enzyme systems are also operative, but other external factors, such as rain, sunlight, and atmospheric degradation, enter the picture. The complexity increases even more when the behavior of a chemical in soil is considered. A wide range of external and internal factors, including a diverse microbial population, interact with the compound and its metabolites. Some of these factors, such as hydrogen ion concentration (pH), can cause rapid chemical breakdown unparalleled in plants or animals. On the other hand, the ability of soil particles to adsorb and bind chemicals for long periods of time can result in very slow degradation rates, compared to rates in plants or animals. Of course, soil residues are also subject to many of the same weathering conditions that contribute to pesticide degradation on plants.

One of the most important factors related to the persistence of a chemical in soils is the microflora and microfauna that are almost always present. Some of these organisms can catalyze chemical and biochemical transformations of the most complex molecules. However, their aura of infallibility has, in certain instances, led to a false sense of security, giving rise to the assumption that, once in the soil, a pesticide can be forgotten. Thus, when an insecticide reaches the soil, its fate is dependent on a host of conditions, including soil type, pH, organic content, mineral ion content, and moisture content, the nature of the soil colloids, the flow of liquid and air through the soil, the amount of cultivation and plant growth present, and the exposure to environmental parameters, such as wind, sunlight, rain, temperature, humidity, etc.

The perpetration of soils with insecticides can be intentional or accidental. Intentional dosing includes application for control of soil-inhabiting insect pests or for systemic movement into plants for management of foliage feeders. The dusting and spraying of plants, animals, or other surfaces with insecticides always results in some contamination of supporting soils, so that this can be considered intentional. Accidental exposure occurs via movement of the insecticide in air or water away from the application site to a nontarget area. The soil, then, constitutes a reservoir for these chemicals where they may be ultimately degraded or removed by air, water, or plant and animal uptake.

Carbaryl

Despite carbaryl's long-standing use as an insecticide, there are few reports in the literature pertaining to its stability in soil. Part of the reason is that carbaryl is not widely used as a soil insecticide, and thus its persistence in soils relative to its effectiveness as a toxicant has been of little interest. Also, one paper appeared in 1965 which indicated that the insecticide was very rapidly degraded in the soil environment, as shown in Figure 8.7.[44] Three rates of the insecticide tilled into the top 6" of sandy loam soil dissipated with almost equal half-lives of 8 days. By 40 days, total destruction of the parent molecule had been completed, regardless of the application rate. A few years later, another study, where carbaryl was applied to shallow mud flats for oyster pest control, also established that by 42 days carbaryl had been almost completely degraded.[80] On the other hand, following treatment at very high rates in carrot plots (5 to 16 g/running meter), carbaryl residues after 12 days ranged from 20 to 118 ppm, and some of the insecticide persisted at high levels for as long as 82 days.[45]

In the 1970's, carbaryl's persistence in soils was reexamined by several investigators. One study attempted to quantitate the available data on the interactions of carbaryl residues in soil with other aspects of the environment, particularly plants.[81] It was concluded that a level of carbaryl in the tillage layer of soil below 1 ppm would not lead to "impermissible" contamination of plants and water coming in contact with the soil. The factors exerting the greatest influence on carbaryl levels in soil were the rate of application, chemical formulation, and frequency of application.

Carbaryl residues of 1 ppm should be fairly easy to obtain in most soils shortly after treatment. For example, spraying of carbaryl at the rate of 3 lb/acre on apple trees left deposits in the soil beneath the tree canopy of 14 ppm.[82] By 1 to 2 days, residues had fallen to 3 ppm, and at the next sampling date, day 14, carbaryl levels

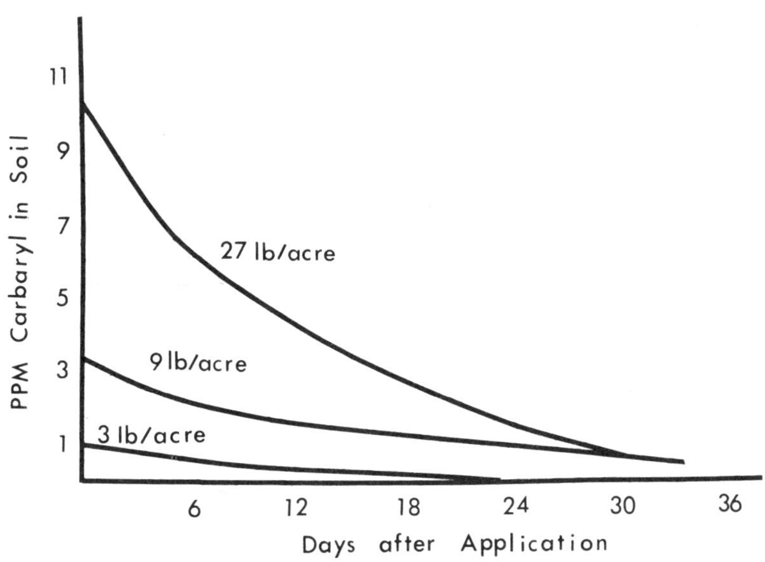

FIGURE 8.7. Degradation of carbaryl mixed at different rates with sandy loam soil. (Adapted from Reference 44, with permission.)

were well below 1 ppm. In another orchard, two cover sprays of carbaryl spaced 2 weeks apart distributed a total of 6 lb/acre, but no soil residue above 1 ppm could be detected within 8 days of the last application. A watershed area treated with a 50% wettable powder suspension of carbaryl at 3 lb/acre contained less than 1 ppm soil residues at all sampling dates thereafter.[83] Analysis of another watershed soil treated by in-furrow placement of carbaryl with corn seed showed that a lag period of 25 to 116 days occurred before carbaryl began to degrade.[84] However, decomposition then proceeded rapidly, so that by 135 days 95% of the insecticide had disappeared from the soil. None of the associated waterways in the orchards or watershed areas contained appreciable carbaryl residues.[82-84] Run-off water from the watersheds occasionally contained carbaryl, but the total amount leaving the area over an entire season was only 0.14% of the applied insecticide.[84]

Somewhat greater stability of carbaryl in soils has been reported from laboratory studies. Mixing ^{14}C-carbonyl carbaryl with five different soils at rates of 2 ppm and 200 ppm showed that the metabolic rate was dependent on soil type.[85] Maximum degradation occurred in clay loam, and minimum degradation took place in loamy sand, although dissipation was not complete from any soil after 32 days. A portion of the radiocarbon was always unextractable with ethyl ether or ethanol, the amount being in rough proportion to the amount of soil organic matter. Though the loss pattern of carbaryl was similar at 2 ppm and 200 ppm, degradation of 3,5-dimethylphenyl N-methylcarbamate was influenced by initial soil concentration. Approximately the same results were obtained in another study using clay loam soil and 1-naphthyl-^{14}C-labeled carbaryl.[86] Other indications also point to some degree of carbaryl persistence in soils.[87]

However, in no research reported thus far have there been any indications that carbaryl persists as long in soils as do certain of the organochlorine insecticides.[88,89] In case of the latter, one of the contributing factors to their increased stability is adsorbance onto soil particles. Although the physical chemistry of this process has been considered in some detail with organochlorines, only one in-depth study has been completed with carbaryl. It was shown that the carbamate insecticide adsorbed more readily to acid soils than to neutral or alkaline soils.[83,90] This is probably due to decreased displacement of carbaryl from the active sites by water as the pH drops. In a similar manner, dry soils bind more carbamate because of less competition with water.

Humified organic matter from three different soil types had about equal capacity to adsorb, although chemical analysis indicated that silty clay loam was in a higher state of oxidation.[90] When carbaryl was mixed with various isolated soil adsorbents, equilibrium was attained in 2 hr or less, and the amount adsorbed decreased as temperature decreased. The binding was found to be completely reversible, however. The overall conclusions with respect to carbaryl binding in a soil aqueous system were that the sorption was not chemisorption, but rather that it was physical attraction via van der Waal bonds between the hydrophobic portions of carbaryl and the adsorbent surface.

In one study using kaolinite and bentonite clays, the adsorptive isotherms for four carbamate insecticides dissolved in water were compared.[91] The capacity of both clays to bind the carbamates was in the order Pyrolan > dimetilan > Baygon > carbaryl. However, bentonite adsorbed considerably more of each compound than did kaolinite. Thus, when introduced into an aqueous carbaryl solution of 1 mg/l concentration, 163 g of kaolinite, but only 93 g of bentonite, were needed to remove 90% of the insecticide. On the other hand, with Pyrolan at 1 mg/l, 9 g of kaolinite and 0.65 of bentonite precipitated 90% of the insecticide.

Under normal agricultural practices, the binding sites of a particular soil would probably not become saturated. But what happens in those areas where large quantities of pesticides and their containers are disposed of? To simulate this condition, a study was conducted where a number of pesticides were mixed with a loam soil at the extremely high rate of 5 tons/acre.[92] One of the test materials was carbaryl, and it was shown that 56 days after application most of the carbaryl still remained in the soil. However, degradation was based on the amount of carbon dioxide evolved, which could be misleading, since this would not include nonhydrolytic metabolism of carbaryl, nor would it discount any change in normal carbon dioxide production due to interaction of carbaryl with organic matter in the soil.

The effects of carbaryl on soil microbes was not

as great as one might have expected at such high concentrations. However, the unformulated insecticide did reduce bacteria populations and fungal growth, although it increased *Streptomyces* production. Interestingly, formulated carbaryl treatment resulted in an increase in bacteria populations. Whether these population fluctuations occur at normal agricultural levels of carbaryl is not known, but it would seem unlikely. Although changes in carbaryl formulation did not alter degradation, other pesticides did produce large variations in biodegradation rates with changes in formulation carriers.

Undoubtedly, one of the major components of soils responsible for carbaryl dissipation is the microorganism complex. As described in detail in Chapter 7, different microbes metabolize carbaryl and 1-naphthol at various rates, using many biochemical reactions. What is perhaps most important with respect to environmental stability is the ability of certain microorganisms to split the naphthalene ring, which ultimately results in total conversion into natural soil constituents.

Carbofuran

The systemic methylcarbamates are by design more persistent in soils than carbaryl. For example, field microplots of sandy soil mixed with carbofuran at a rate of 10 ppm continued to kill at least 97% of added cabbage maggots for 90 days after treatment.[93] Even by 120 days 63% mortality was still found. Lower levels of 3 ppm carbofuran were degraded more rapidly. In greenhouse tests with the same soil, carbofuran at concentrations as low as 2 ppm persisted for at least 100 days.

Many other insect bioassays have been used for determining carbofuran's stabiltiy in a variety of soils. When compared at different temperatures in the laboratory, 0.5 ppm carbofuran mixed with sandy soil retained 100% bioactivity for a minimum of 16 weeks at 13°C.[94] As expected, raising the temperature to 24°C decreased carbamate persistence. The Collembola species used for the bioassay was found to be very sensitive to carbofuran, and thus an examination of insecticide suppression of field populations of this insect complex was undertaken. It was discovered that as long as 2 months after carbofuran application the number of Collembola in the soil was significantly reduced. A call was made for more careful examination of changes in nontarget invertebrate populations caused by soil application of pesticides such as carbofuran.

In a study using the common field cricket as the assay agent, it was revealed that carbofuran mixed at 9.4 ppm was only moderately persistent, with a half-life in sandy loam soil of about 35 days and a total degradation time of approximately 16 weeks.[95] A different insect, the wireworm, in the same type of soil containing a lower concentration of insecticide (1.85 ppm) yielded an almost identical half-life of 36 days.[96] In muck soil, carbofuran's half-life was extended to 10 to 12 weeks, but by 25 weeks bioactivity had mostly disappeared.[95]

Information on soil stability can be obtained second-hand from studies on the effectiveness of systemic insecticides.[97,98] For instance, following in-furrow application of carbofuran at 1 lb/acre with corn plants, leaves that were removed and exposed to leafhoppers continued to kill all insects for 34 days.[97] Various soil conditions caused differences in carbofuran effectiveness, as judged by bioassay. Corn grown on soils with low moisture and/or heavy makeup was more susceptible to leafhoppers. Of course, a loss in effectiveness does not necessarily imply that more carbofuran was degraded; it may reflect an increase in the retention of active ingredients in the soil, thus making carbofuran less available for transport into the plant.

Taken together, all of these bioassay studies point to moderate stability of carbofuran in soils. Variations in rates from different investigations can be partially explained by differences in soil types, formulation of toxicant, experimental conditions, and insect species employed. However, it is important that the amount of insecticide added to the soil in this type of project be carefully considered. If, for example, one applies 20 times the level needed to kill 100% of the insects, then as much as 95% degradation will have to occur before any change in insect mortality is noted. Similarly, if a large percentage of the applied material is bound to soil colloids or is unavailable to the insects in some other manner, mortality figures will imply a much faster rate of degradation. Thus, in a bioassay system of analysis, what is being measured is the availabilty of an active form of molecule, be it parent or metabolite, under the soil and weather conditions of the experiment. Indeed, this type of assay gives more valuable data in regard to the environmental availability of toxic molecules, but it should not be interpreted to coincide with actual disappearance of the insecticide from the soil.

Analysis of carbofuran by gas chromatography showed that dissipation rates of the insecticide from silt loam soil were dependent on application methods.[99] A broadcast of 10% granules at a rate of 4.8 lb/acre was followed by a degradation half-life of 46 days. When the granules were drilled into the furrow with seeds at rates of 2.8 and 3.7 lb/acre, half-lives increased to 94 and 117 days respectively. The decrease in rate following burial of the insecticide was due, in part, to decreased losses from evaporation, atmospheric decomposition, and photodegradation.

Certain areas of the field plots were characterized as "rapid disappearance" spots, with carbofuran half-lives of only 33.5 days. All of these areas had a higher pH, more clay in the soil texture, and higher moisture content. After application to the soil, carbofuran was slowly converted to 3-keto carbofuran, which accounted for a maximum of 5 to 10% of existing residues. The metabolite concentration in the soil peaked 29 days after broadcast of the parent compound, but it did not peak until 60 to 80 days after in-furrow placement.

Carbofuran radiolabeled with ^{14}C in the ring position and in the carbonyl position was applied to four types of field-collected soil.[100] The carbonyl-labeled preparation reached 50% degradation in 4 weeks in one type of silt loam soil (R), but half of the insecticide persisted for more than 54 weeks in another type of silt loam (S) and in organic soils (Figure 8.8). Sterilization of the soils had little effect on degradation rates, except for the clay loam soil, which showed a profile after sterilization like that of the silt loam-S soil.

Most of the radiolabeled products from ^{14}C-carbonyl carbofuran consisted of expired radioactive carbon dioxide, and since the amount recovered from three of the soils did not change after sterilization, it is apparent that degradation was not due to soil microbes. More likely, carbofuran was hydrolyzed in the soils, and the released methylcarbamic acid decomposed to yield ^{14}C-carbon dioxide. With ring-^{14}C carbofuran, an increase of soil-bound radioactivity with time was found. After reaching peak concentration, the residues decreased, producing a corresponding evolution of ^{14}C-carbon dioxide. A similar pattern followed treatment of soils with ^{14}C-carbofuran phenol. Various failures to remove the phenol from soil particles led to the conclusion that the molecule was chemically bound rather than merely adsorbed via van der Waals forces, hydrogen bonding, or simple cation-exchange mechanisms.

In part, the difference in stability of carbofuran in the two silt loam soils (see Figure 8.8) was probably due to variations in hydrogen ion concentration. When the natural pH of 6.0 in the silt loam-S sample was adjusted to pH 7.8, degradation was appreciably increased (Figure 8.9).

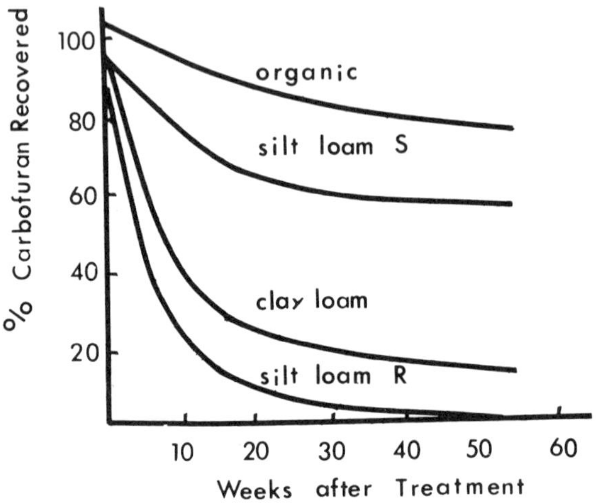

FIGURE 8.8. Degradation of carbonyl-^{14}C carbofuran in four different soils.[100]

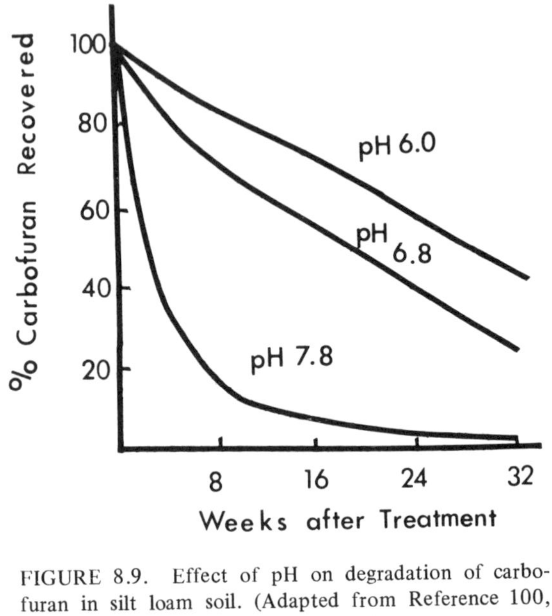

FIGURE 8.9. Effect of pH on degradation of carbofuran in silt loam soil. (Adapted from Reference 100, with permission.)

Aldicarb

Another systemic carbamate which has been the subject of many soil studies is aldicarb. An early investigation using a colorimetric analysis procedure, which did not distinguish aldicarb from its oxime and sulfoxides, reported on the stability of the molecules in acid sandy soils of citrus groves.[66] Within 30 to 40 days after application of 10% granular insecticide at a rate of approximately 15 lb/acre, total residues in three different orchards averaged 5, 150, and 1,960 ppb. Although treatment methods, rates, and the moisture content of each soil were not the same, these experimental variables fail to explain the large differences in soil deposits. Closer examination of soil characteristics might have provided some reasons for the divergent results.

Irrigation of one plot with 4″ of water shortly after treatment moved aldicarb down to between 3′ and 4′. There was also evidence for translocation of insecticide to soil and trees adjacent to the test area. This was probably due to a combination of lateral movement of pesticide through soil and air, and to some intergrowth of root systems among neighboring trees. More recent documentation confirmed the ease with which aldicarb can be leached through wet soil shortly after application.[101,102] In drier soils, aldicarb and metabolites crept up as well as down, and the upward movement eventually led to loss of aldicarb from the soil via codistillation with escaping water.

Since movement away from the site of application may in some instances cause inefficient use of aldicarb, various formulation materials have been tested in an effort to impede rapid translocation. In this regard, a range of granular aldicarb formulations were prepared from a variety of base materials and binders in order to examine their effect on movement of the insecticide out of the granules into water.[103] Results summarized in Table 8.4 point out that changing the base material while holding the mesh size constant gave 0.5-hr release percentages which ranged from 100 for corn cobs down to 15 for cellulose acetate. Altering the mesh size of calcium sulfate from 30/60 to 10/20 affected release rates only slightly, but a change in the mesh size of petroleum charcoal WIS 20 formulations resulted in considerable variation of release rates. Finally, reducing the amount of binder in the charcoal formulations also appreciably affected extraction into water. Correlation of aldicarb release from granules into water with uptake and translocation from soil into

TABLE 8.4

Effect of Base Material, Mesh Size, and Binders on Release of Aldicarb from 10% Granular Formulations into Water[103]

Base material	Mesh size	Binder (%)	% Aldicarb released after 0.5 hr	% Aldicarb released after 24 hr
Corn cobs	16/30	–	100	100
Wheat flour	16/30	–	80	100
Charcoal WS[a]	16/30	20	70	100
Cellulose acetate	20/40	–	15	53
Calcium sulfate	30/60	–	81	90
	10/20	–	60	81
Charcoal WIS[b]	+ 60	20	79	100
	8/16	20	17	31
	16/30	20	23	48
	16/30	10	42	100
	16/30	5	73	100

[a]Water-soluble binder added to petroleum charcoal.
[b]Water-insoluble binder added to petroleum charcoal.

cotton plants was not always good; however, the formulations best able to retain the insecticide in the water immersion tests generally offered extended bioactivity in greenhouse plants.

In the field, a 10% granular aldicarb formulation was applied to soil at a rate of 2 lb/acre for control of pests of broad beans. Thin-layer chromatographic analysis of soil samples collected after 2 months demonstrated that approximately 22% of the applied aldicarb equivalents were still present, mostly as aldicarb sulfoxide.[59] By 4 months the total had dropped to 13%, but this time the majority of the residue was aldicarb sulfone. At neither sampling date was any parent compound detected.

Several other field and greenhouse studies portray a more rapid degradation of aldicarb in soils. For example, an S-methyl-^{14}C aldicarb formulation placed in-furrow at a rate of 3 lb/acre produced an initial soil load of 13 ppm aldicarb equivalent.[104] In only 1 week this had fallen to 3.5 ppm, and by 2 months only 0.07 ppm were detected. Rapid formation of the sulfoxide metabolite followed by a slower build-up of sulfone was again evident. A broadcast treatment of 10% granular aldicarb at the base of Scotch pine trees (0.5 to 2 lb) gave similar dissipation results.[105] The fine loam soil destroyed half of all toxic residues in about 2 weeks, with total degradation complete in 1 to 2 months. A third study, where ^{35}S-aldicarb was mixed with field soil and replaced for dissipation under field conditions, also supported these results.[64]

Altering the pH of several different soils before addition of aldicarb had little effect on degradation rates.[101] However, under conditions of low moisture, the type of soil was a controlling factor. In construction sand the half-life for all toxic aldicarb materials was only 1 day, but in sandy loam soils 50% degradation was complete only after 4 to 7 days.

The importance of the above field studies in determining aldicarb's stability in soil can be appreciated by comparing the results with those obtained under laboratory conditions. Three different soil types were taken from the field and mixed with ^{35}S-aldicarb at rates of 20 ppm.[63] The mixtures were stored in tightly capped amber bottles in the dark at 25 to 28°C until analysis. The resulting dissipation curves for the fine sand, silty clay loam, and clay soils are pictured in Figure 8.10. There was fairly rapid conversion of

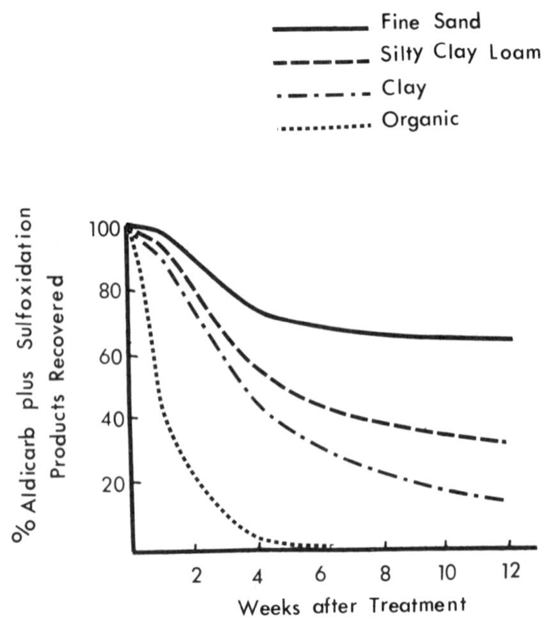

FIGURE 8.10. Degradation of ^{35}S-labeled aldicarb in different soils.[63]

aldicarb to its sulfoxide in all three soils, and half of the parent material was gone in 7 to 12 days. The sulfoxide and sulfone persisted much longer, so that degradation rates of the total toxic complex were appreciably slower than those obtained in field studies. When the insecticide was drilled into an organic soil and placed in the greenhouse, a loss curve much like that found in field plots resulted (Figure 8.10). Presumably, this was due to contributions from leaching, volatilization, greater absorption onto organic matter, uptake by plants growing in the soil, higher temperatures, and more moisture. However, when comparing laboratory and field experiments, the quantity of insecticide added to the soil must be considered, since this can significantly alter the rate of decay. For instance, equal treatment of clay loam soil mixtures containing 5, 20, or 50 ppm of aldicarb yielded respective half-lives for the total toxic complex of 1 week, 3 weeks, and 5 weeks.[102]

Bux and Landrin

Although not a systemic agent, Bux is an effective toxin for certain soil insect pests, and thus is often applied to soils. The insecticide is composed of a 3-to-1 mixture of 3-(1-methylbutyl)phenyl N-methylcarbamate and 3-(1-ethylpropyl)phenyl N-methylcarbamate. Following application of the mixed isomers in bands with corn plants at 2 lb/acre, soil samples removed at harvest time 5 to 6 months later averaged 0.08 ppm residue.[106] None of this was identified as the parent phenols. When the phenols themselves were mixed with soil, 50% degradation took place in only 1 day. This rapid degradation coupled with low inherent toxicity make the phenolic portions of Bux rather innocuous. It is tempting to translate these findings to other phenols of methylcarbamates, but this must await further study.

Working with ^{14}C-carbonyl Bux, the same researchers[106] applied acetone or hexane solutions of the insecticide to field-collected silt and silt loam soils. Incubation at the 1-ppm level revealed that both isomers of Bux were reduced to about half in 1 week. Both produced one major metabolite which also had a soil half-life of approximately 1 week at the 1-ppm level. However, neither metabolite was formed in sterile sand, and after 1 week only 5 to 15% degradation had occurred, indicating the importance of microbial attack in the degradation of this particular methylcarbamate.

An attempt at approaching field or actual-use conditions brought about the formulation of radiolabeled Bux granules. In this form, both isomers were decomposed at a slower rate, with half-lives in silt soil of about 2 weeks at the 10-ppm level. However, the metabolic route was the same, with production of a single metabolite which reached a maximum of only 1% of the total residue. Thus, with this insecticide, the only apparent entity of any concern would be the parent compound itself.

Another investigation, where Bux was incorporated into sandy soil at 10 ppm and subsequently bioassayed with insects, indicated that half of the killing power of the carbamate was lost in about 6 weeks.[107] However, the initial soil load was twice the LD_{100} dose, so that the actual half-life of the parent insecticide was probably closer to 2 to 3 weeks, as found in the other Bux study. Results also showed that Bux was more easily inactivated in dry sand and in moist organic soils than it was in moist sand. Another methylcarbamate included in the same series of tests was Landrin, which is also a 3-to-1 mixture of isomers, namely, 2,4,5- and 2,3,5-trimethylphenyl N-methylcarbamate. In moist sand, where bioactivity persisted for the longest time, Landrin's half-life was about 2 weeks.

Other Carbamates

The stability of other carbamate insecticides in soils has received only cursory attention. Based on insect bioassay, Isolan was 100% depleted from sandy soil in 100 days.[108] At the same time, 85% had disappeared from muck soil, and 75% was lost from silt loam soil. Introduced into the same test, Baygon killed most insects in muck and silt loam for as long as 6 months after application. Such extreme stability of this carbamate has not been demonstrated in various living organisms or in other areas of the environment.

Mesurol half-lives in a number of soils ranged from only 4 days to almost 2 months, with the most rapid decay, due mainly to hydrolysis, taking place in more alkaline soils.[109] In acidic soils, considerably more sulfoxidation occurred prior to hydrolysis. This was also true in sandy loam soil as compared to silt loam or highly organic silt loam soils.[110] Alkaline soils also efficiently degraded formetanate, so that half of the parent compound was lost in less than 2 days.[111,112] There was speculation that some of the formetanate may have complexed directly with soil components, but no concrete proof was presented. Storage of ^{14}C-carbonyl 3,5-dimethylphenyl N-methylcarbamate with five different soils at 25°C for 32 days revealed moderate differences in degradation rate.[85] The degree of hydrolysis, based on ^{14}C-carbon dioxide evolution, was highest in a clay loam soil and lowest in a loamy sand soil. The amount of residual radioactivity in the soils was roughly proportional to the organic content, and the only organo-soluble compound recovered was the parent insecticide.

Finally, two newer oxime carbamates have been subjected to field degradation in soils, although the application methods were artificial. After mixing ^{35}S-labeled 3,3-dimethyl-1-(methylthio)-2-butanone O-(methylcarbamoyl)oxime with a fine sandy loam soil at a rate of about 13 ppm, the sample was placed in a steel-screen pocket and buried in the same soil.[113] Retrieval and analysis showed the insecticide half-life to be 3 days, and by 1 week only about 30% of the applied radiosulfur could be accounted for. Part of the loss was due to downward movement of the material. There was some build-up of the expected sulfinyl and sulfonyl derivatives, but these were virtually nonexistent after 56 days.

The other oxime, methomyl, was pipetted onto a silt loam soil encased in a cylinder and lightly covered with the same soil.[114] Only 1.8% of the applied dose (equivalent to 4.5 lb/acre) was present as parent compound after 1 month. Even more rapid dissipation took place in fine sand and loamy sand soils. Unlike the results from other systemic carbamates, there was very little downward or lateral movement of methomyl, despite its water-soluble nature. Unextractable radioactivity remaining in soils after treatment was attributed to incorporation of a metabolite, ^{14}C-carbon dioxide, into natural soil components.

In summary, the available evidence points to slight to moderate persistence of carbamate insecticides and their known toxic metabolites in soils of many types. Obviously, there are a myriad of factors which govern the ultimate fate of a carbamate, be they microbial makeup, physical and chemical characteristics of the soil, or weather conditions of light, temperature, and moisture. However, through these variables a certain pattern emerges. Of the carbamates studied to date, almost all have had half-lives of short duration in soils, ranging from only a few days to a few weeks, with virtually total destruction of parent compound and known toxic derivatives completed by 1 to 4 months. Exceptions to this would be carbofuran, where reported half-lives have ranged from 18 to 378 days, and, subject to further investigation, perhaps Baygon.

There is proof for vertical and lateral movement of some of the more water-soluble systemic carbamates through soils, and also upward movement out of the soil into the air, so that loss of parent molecule from soils may not mean loss from the total environment. Some drastic differences from experiment to experiment or from compound to compound emphasize once again the need to investigate each individual situation to assure that the particular compound and soil in question do not represent one of the exceptions to the general trend. For example, soil pH in the range of 6 to 8 did not greatly alter the degradation rate of aldicarb in several different soils,[101] but half of carbofuran was destroyed in only 3 weeks at pH 7.8, compared to about 28 weeks at pH 6.0 in the same soil.[100] In another instance, microbial attack was primarily responsible for Bux destruction,[106] whereas carbofuran degradation in certain soils was independent of microorganisms.[100]

Thus, with normal agricultural and sanitation practices, commercially available methylcarbamate insecticides applied to the soil intentionally or accidentally should not carry over from one year to the next, and many will persist for only a few months. This does not mean that degradation products cannot accumulate, but those that have been identified as toxic thus far will also easily dissipate within a year's time. A limited collection of data indicate that, under normal-use practices, carbamate effects on soil microbes and other soil nontarget invertebrates are minimal, although temporary imbalances may result. Further comment on this topic, however, must await further research.

WATER AND AIR

The insecticidal contamination of very remote areas of our environment occurs primarily via dispersal of the compounds through air and water. Although the chemicals seldom are intentionally added to water or air, the insecticides enter these media as a by-product of spray and dust applications, as industrial effluents during their manufacture, or as chemical spills. Once an insecticide is deposited in the environment, it is continuously subject to volatilization, codistillation, condensation, dissolution, and adsorption, any combination of which may result in redeposition of the material far from the site of release.

More often than not, movement of the insecticide in air and water depends on the concomitant movement of dust, small bits of soil, and other particles which act as carriers for the insecticide molecules. Depending on the chemical nature of the molecule, there may be considerable exchange between adsorbance onto particulates and vaporization or dissolution into the free state. Those insecticides which do not bind tightly are not as susceptible to settling out or precipitation, and thus tend to move farther.

Of course, there are instances where insecticides are purposely added to water or air for insect control, particularly for insects affecting the health of man and of domestic animals. These applications are usually limited to small bodies of water or to air confined within a dwelling, but movement away from the site of application certainly takes place. Perhaps the worst offenders in this particular case are the household aerosol sprays.

Once the insecticide has found its way into air or waterways, it is not immune to decomposition.

In water, the rate of degradation depends on factors such as pH, temperature, light, salt content, amount of suspended solids, and presence of microorganisms, particularly microflora. In air, temperature, light, humidity, and particulate matter are important considerations.

How large a role air and water degradation play in the total destruction of insecticides, especially carbamates, is mostly speculation, due to the limited information available from field experiments and because of the difficulty in tracing the fate of a chemical through the total environment. Often it is felt that, once transported to air or large bodies of water, the chemical is subject to such great dilution that its presence can be ignored. The fallacy of this judgment can be illustrated by considering the biomagnification of residues in an aquatic environment. It is not uncommon for fish and invertebrates living in water contaminated with parts per billion of organochlorine insecticide to filter out and concentrate the residue more than 1,000-fold.[88] Although this is primarily due to the extreme preference of these compounds for lipid rather than water, there is evidence for some accumulation of the more water-soluble carbamates by river and pond inhabitants.

When assessing the environmental stability of insecticides in air and water, it is important that experiments be conducted under conditions most like those actually occurring. Carbamate studies under such conditions have been few in number, but fortunately the investigations conducted have been rather complete and have included a number of different chemical structures. Studies of carbamate interactions with air are minimal and can only be used as indices to real situations.

Fresh Water

One extensive study examined the fate and persistence of carbaryl, Baygon, Pyrolan, and dimetilan in Nile river water.[91,115] After mixing 40 mg of each insecticide with 10 l of river water, phosphate buffer was added to maintain a pH of 7.2. Air was gently bubbled through the tanks, which were held constant at 25°C. The resulting degradation curves are pictured in Figure 8.11. Carbaryl and Baygon had essentially disappeared in 1 week, but the dimethylcarbamates continued to persist unchanged for at least 3 months. With carbaryl, 1-naphthol was produced, with a peak concentration evident after 3 days, followed by

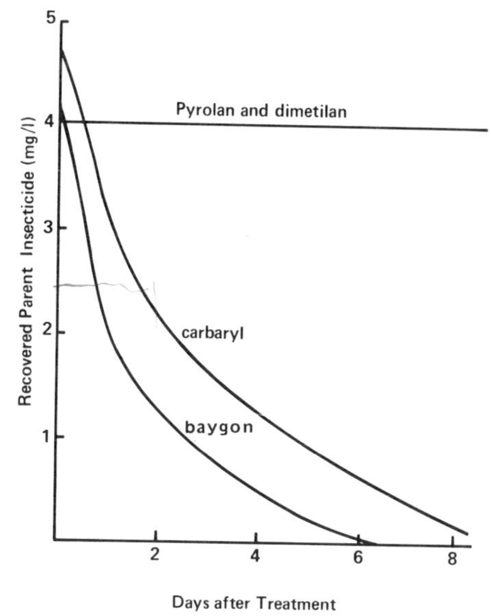

FIGURE 8.11. Degradation of carbamate insecticides in river water.[115]

total dissipation at the end of the week. Hydrolysis of the methylcarbamate ester was predominantly biological.

Consecutive additions of carbaryl to fresh river water demonstrated that there was some acclimatization of the microbes to carbaryl and its hydrolysis product. Thus, by the fourth addition, a carbaryl concentration of 17.5 mg/l was almost completely destroyed in only 1 day, with no recoverable 1-naphthol present. Similar findings resulted when Baygon was incubated with the river water, although microbial changes were not as great and there was a slight build-up of *o*-isopropoxy phenol after each addition of fresh water plus insecticide.

In a 50:50 mixture of sewage and river water the stability of carbaryl was even lower. However, dimetilan and Pyrolan were quite resistant to attack. On the other hand, the corresponding pyrazolyl hydrolysis products were almost immediately decomposed in river water.

Although other research has shown adverse effects of carbaryl on aquatic microorganisms (see Chapter 5), a recent study pointed to increased growth in the presence of carbaryl. Field-collected green algae placed in media made to simulate eutrophic lakes were mixed with carbaryl to give final insecticide concentrations of 0.1 to 1.0 ppm.[116] After 6 days at 21°C with a photoregime of 12 hr of light, the cell biomass had increased by 44 to 57%.

Sealed glass jars containing 20 l of Little Miami river water were contaminated with 10 ppm of various carbamates and stored in sunlight and in artificial light with occasional shaking.[117] Thin-layer chromatographic analysis revealed that 50% of Mesurol had disappeared in less than 3 days, with total destruction completed by 1 week. Decomposition was mainly via hydrolysis to 4-methylthio-3,5-dimethyl phenol, which in turn had dissipated at the end of 4 weeks. Carbaryl and Zectran had similar half-lives, but the parent hydrolysis product from either compound was not detected. Either they had not formed, or their appearance was followed by immediate degradation, as indicated in the Nile river study.[115] Matacil and Baygon half-lives were close to 1 week, and both methylcarbamates were more resistant to total destruction, Baygon residues persisting for as long as 8 weeks.

Farm pond waters have also demonstrated an ability to rapidly transform carbaryl into 1-naphthol, the reaction being catalyzed by organic and inorganic constitutents of the water.[118] Hydrolysis was enhanced considerably by agitation or by addition of glucose. The hydrolysis product, 1-naphthol, was easily destroyed when an isolate, possibly a *Flavobacterium*, was added to the pond water. As much as 20 ppm of 1-naphthol disappeared in 12 days. However, interpretation of these results is difficult, since sterile controls also exhibited rapid loss of 1-naphthol, suggesting chemical breakdown or volatilization. The pathway for 1-naphthol degradation based on identified products is shown in Figure 8.12. It was postulated that, since these products did not accumulate in the water, they were only intermediate to the formation of other unidentified metabolites. Evidence was also presented for toxicity of carbaryl to native pond bacteria.

With the exception of bacterial response to carbaryl, other pond water yielded similar results.[83] Samples of sterile or raw water containing 20 ppm carbaryl degraded the insecticide to 1-naphthol at an identical rate, although sterile water required shaking in order to equal raw water. Maximum formation of 1-naphthol took place after 21 to 28 days, during which time the number of bacteria nearly doubled, in contrast to control flasks without carbaryl. It was stated that the increased growth was probably due to released methylcarbamic acid, which served as a carbon and nitrogen source.

The study also provided further evidence for acclimatization of certain microorganisms to carbamate insecticides. When 0.5 ml of the pond water, which had been preincubated with 30 ppm of carbaryl for 1 year, was added to the test flasks, a dramatic increase in rate of carbamate hydrolysis ensued. Within 2 weeks, no 1-naphthol could be found. Of particular significance was a concurrent rapid build-up of primarily one bacterium population to the exclusion of all other colonies.

An oxime carbamate, aldicarb, was added at the high level of 100 mg/l to rain and overflow water which had collected in ditches near untreated fields.[102] Prior to aldicarb addition, the water had been filtered through paper into enameled tubs. With exposure to sunlight and open air whenever possible, it took about 46 weeks for water residues to become reduced to 0.4 ppm.

One of the few projects which did not involve removal of test water from its surroundings was conducted with a new carbamate, 3-*sec*-butylphenyl N-methyl-N-(phenylthio)carbamate or RE 11775.[119] An emulsifiable concentrate formulation of the insecticide was poured into a sewage-holding pond at a rate of 0.1 ppm for control of *Culex* mosquito larvae. Based on a mosquito bioassay, a mere 10% of the original toxicant remained intact after 8 hr, and by 24 hr all insecticide was gone. When the sewage water was brought into the laboratory and compared to tap water for degradation of 0.01 to 0.03 ppm RE 11775, all of the methylcarbamate disappeared from the sewage sample within 8 hr, whereas no loss occurred in sanitized water. In another experi-

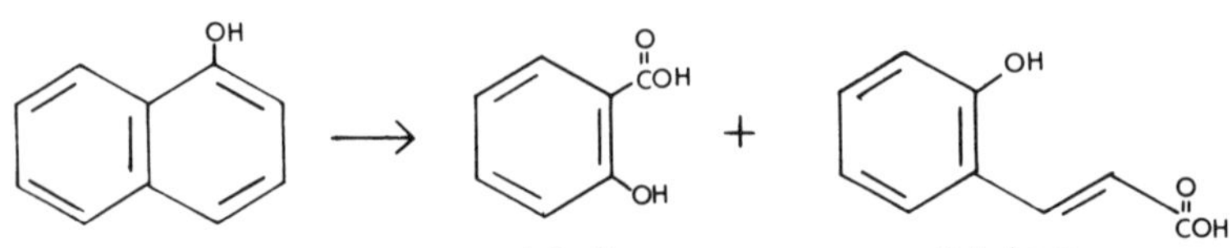

FIGURE 8.12. Decomposition of 1-naphthol in farm pond water.[118]

ment, assay of water 4 and 8 hr after treatment of a flooded alfalfa field and pasture respectively showed no residue of the applied 0.1 lb/acre.

Sea Water

Carbaryl and 1-naphthol stability in sea water has also been examined.[80,120,121] Aquaria containing 4 l of sea water, with and without 2″ of bay mud on the bottom, were dosed with 40 mg of either ring- or carbonyl-^{14}C-labeled carbaryl.[80] The tanks were covered and held at 8°C and at 20°C, with continuous aeration and fluorescent lighting for 10 hr per day. At 8°C, with no mud present, the carbamate ester was 50% destroyed in 38 days, mostly via hydrolysis to 1-naphthol. Under the same circumstances in the 20°C aquarium, carbaryl had almost completely dissipated in less than half this time. Further evaluation of temperature effects produced the relationship pictured in Figure 8.13. Similar curves were found with tanks held in the dark or under fluorescent light during the 4-day experiments. In addition, 1-naphthol was also more rapidly degraded at elevated temperatures.

The presence of mud in the 8°C test chambers caused rapid loss of carbaryl from the water, so that carbaryl plus 1-naphthol concentration was reduced by 90% after only 10 days. This was mostly due to adsorption onto the mud, where carbaryl was slowly degraded.

Some aquaria were incubated as closed systems containing a carbon dioxide trap. As expected, evolution of ^{14}C-carbon dioxide was demonstrated. The amounts were higher from the carbonyl-labeled insecticide than from the ring-labeled preparation, which also was not surprising. However, total radiocarbon recovery at the end of the experiments was only 40%. It was suggested that microbes converted the ^{14}C-carbon dioxide to labeled methane, which would not have been trapped, but it is doubtful that this accounted for all of the 60% loss. Levels of carbaryl from 5 to 25 ppm were toxic to small crustaceans, nemertean worms, and polychaetes.

Similar mini-ocean setups were used for further study of 1-^{14}C-1-naphthol degradation.[120] As shown in Figure 8.14, substrate destruction in water without light or bottom mud was 50% in 2 weeks. Exposure to light increased the rate of degradation about twofold, and addition of mud decreased the half-life to 2 days, regardless of whether incubation took place in the light or dark. In each case, if the "oceans" were sterilized before 1-naphthol treatment, or if the supply of air was limited, degradation was slowed appreciably, particularly in the aquaria containing bottom mud. Sterilization also completely stopped the evolution of ^{14}C-carbon dioxide. Finally, some of the most dramatic changes in degradation were effected by altering the pH (Figure 8.15). Unfortunately, these results were obtained with distilled water rather than sea water, but when the initial degradation rate is plotted against pH, a smooth bell-shaped curve results, with maximum 1-naphthol stability at pH 6.3. However, the pH of natural sea water was found to be 8.2, which is coincident with minimum stability.

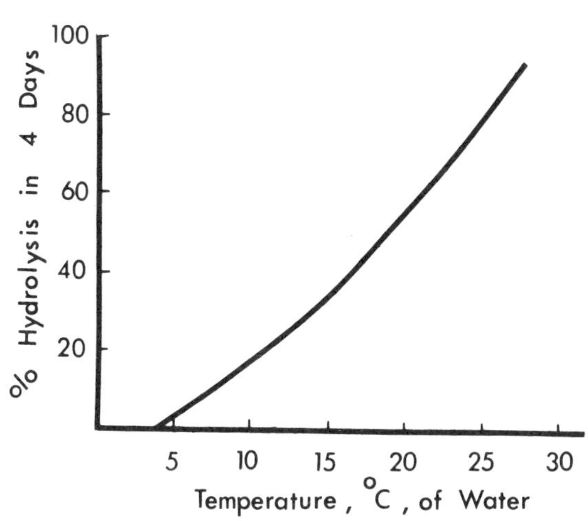

FIGURE 8.13. Effect of temperature on hydrolysis of carbaryl in sea water.[80]

Model Ecosystems

Recent attempts to emulate environmental conditions have involved the use of fresh-water aquaria complete with land mass, plants, and food-chain animals. In one instance, ring- or carbonyl-^{14}C-labeled carbofuran was applied to sorghum plants at the equivalent rate of 1 lb/acre.[122] Radiocarbon moved into the water, with peak concentrations appearing after 7 days; the levels from the ring-labeled insecticide (0.3 ppm) were much higher than those from the carbonyl-labeled preparation (0.04 ppm). This would imply hydrolysis of the methylcarbamate with subsequent release of ^{14}C-carbon dioxide. In the following 3 weeks, the concentration of ^{14}C in the water decreased by about 50%. Metabolites isolated from the water included N-hydroxymethyl carbofuran, 3-hydroxy carbofuran, 3-keto

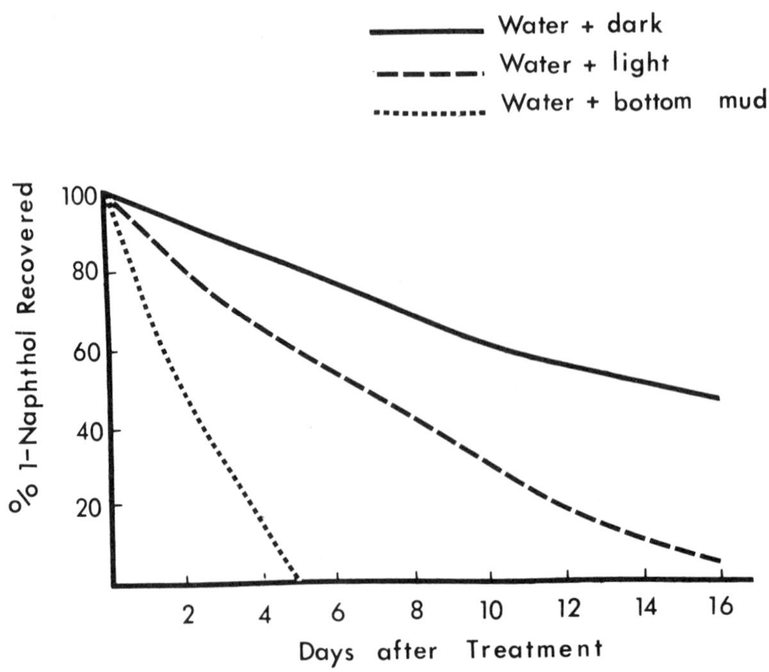

FIGURE 8.14. Degradation of 1-naphthol in sea water. (Adapted from Reference 120, with permission.

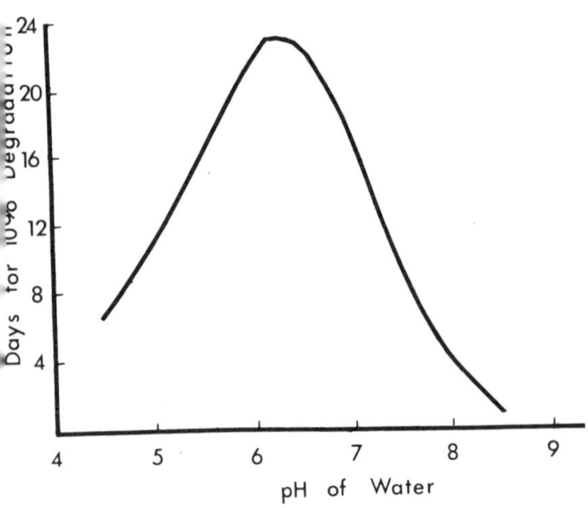

FIGURE 8.15. Effect of pH on degradation of 1-naphthol in water in the presence or absence of light or bottom mud.[120]

carbofuran, 3-hydroxy carbofuran phenol, and carbofuran phenol.

Metabolites of carbofuran were also present in all of the microbes, invertebrates, and fish present in the ecosystem. Presumably, they were formed, at least in part, within the animals, since dead crabs removed shortly after treatment contained large amounts of unchanged carbofuran. Biomagnifications of the total radiocarbon compared to water were around a 1,000-fold for algae, *Elodea,* and snails, and 100- to 300-fold for clams, crabs, *Daphnia,* fish, and mosquito larvae. However, this does not necessarily mean that carbofuran itself was biomagnified.

Ring-^{14}C carbaryl and carbonyl-^{14}C 3,5-dimethylphenyl N-methylcarbamate were added to soil at a rate of 3.0 ppm in a similar ecosystem.[123] Maximum concentration of carbaryl radioactivity in water was 300 ppb after 10 days; this fell to 7.6 ppb by 22 days. The other carbamate released a steady flow of radiocarbon into the water up to a maximum level of 35.8 ppb at day 34. Bioaccumulation ratios reached several thousand for algae and duckweed, and a few hundred for snails, catfish, and crayfish. About 60% of the radioactivity from each insecticide was tightly bound to soil and never became available to the water.

Much more is known about carbamate ester hydrolysis in water from results of laboratory studies with buffered distilled water, but this has been covered in detail in Chapters 2 to 4. Suffice it to say that at higher temperatures and more alkaline pH methylcarbamate hydrolysis is generally most rapid.

There are some indications that small amounts of carbamate residues are transported through the environment by movement in water. Three watershed studies have shown that carbaryl[83,84] and carbofuran[99] may be washed from soil by surface run-off following rainfall. In case of the latter compound, from 0.5 to 2.0% of the applied insecticide was removed from the land by run-off but only 0.14% of carbaryl was transported in run-off waters. Most of the insecticides were dissolved in the water itself rather than attached to associated solid materials.

Air

Data on the decomposition of methylcarbamate insecticides in air are virtually nonexistent. One study shed some light on the persistence of Baygon within the air of a closed structure. The carbamate was applied as a wettable powder on the wall/roof junction of houses in 2-m swaths at the rate of 2 g/m^2 every 35 days.[124] Air was sampled in the houses for 1 hr every week with an aqueous sodium hydroxide trap. On the day following the first treatment, 7 ng of Baygon was recovered per liter of air. After 14 days, no insecticide was trapped. With repeated treatments and high humidity, increased air concentrations of insecticide persisted from one spray to the next. When the humidity dropped, the air levels decreased accordingly. Throughout the tests there was no evidence for intoxication of any residents living in the dwellings.

In an open-air situation, air was sampled in orchards by bubbling it through fritted-glass impingers containing alcohol.[125] During the spraying operation, average carbaryl residues of 0.6 mg/m^3 air were present. By inserting this value into a mathematical equation, it is possible to calculate a "percent toxic dose" of 0.025 for carbaryl. However, interpretation of this value is difficult. What is most important is that air trapped at the time of treatment in residential areas within 1,000 to 2,000 feet of the sprayed orchards contained almost as much carbaryl (up to 0.5 mg/m^3) as the air at the site of application.

Certainly the presence of carbamates in air depends to a large degree on their inherent volatility. In this regard, movement into air has been shown to vary considerably with chemical structure, as illustrated in Table 8.5.[126] When distributed on glass plates at a rate of 0.5 μg/cm^2, radiocarbon from ^{14}C-labeled preparations of various aryl methylcarbamates was 50% dissipated at the end of periods ranging from 18 min to 18 hr. On plant foliage, where the original deposit was at a higher level (8.3 μg/cm^2) and was exposed to wind and sunlight, decomposition was qualitatively similar, although the rates for 50% loss ranged from 3 hr to more than 3 days. The one main exception to this was Zectran, where radioactivity on plant foliage was much more persistent. This was probably caused by the formation of less volatile degradation products.

TABLE 8.5

Volatilization of Carbonyl-^{14}C Aryl Methylcarbamates from Glass and Plant Foliage

	50% Loss (hr)	
Compound	Glass[a]	Plant foliage[b]
Mesurol	18	72
Carbaryl	14	68
HRS-1422	5.2	15
Baygon	1.8	8
Matacil	1.6	8
Zectran	1.2	61
Banol	0.7	16
UC 10854	0.3	3

[a]Applied at 0.5 μg/cm^2 and held under fluorescent light at 25°C.
[b]Applied at 8.3 μg/cm^2 and held in open air.

Adapted from Reference 126, with permission.

Formulation materials have also been shown to change the evaporation rates of carbamates, particularly Baygon.[78] Using isopropoxy-^{14}C insecticide, the following mixtures were prepared: 1% Baygon in 4:3:2:1 xylene:acetone:AR60: Tween 80; 1% Baygon in 8:2 ethylene glycol:20% Dowanol EPH; and 1% Baygon in 80% aqueous ethanol. Aliquots of each formulation were pipetted onto 2.5-cm circles of parafilm, which were then exposed to a greenhouse environment. In only 3 hr, half of the alcoholic Baygon had disappeared from the circle. After 48 hr, 50% of the xylene-base-formulated insecticide was gone, but only 16% of the Baygon in the ethylene glycol mixture was lost.

Another study with the same insecticide indicated that formulation materials had less effect on Baygon evaporation.[127] Painted boards were dosed with 100 mg/ft^2 carbamate in the forms of a 1% oil spray, a 13.9% spray concentrate, a 75%

wettable powder, and technical material dissolved in ethanol. Half-lives in the laboratory for the active ingredient of each mixture ranged from 13 to 17 days, with the greatest variations occurring in the first few days after application. This was thought to be due to losses of the formulation materials prior to loss of insecticide.

In both air and water, perhaps more than anywhere else, photodegradation significantly hastens the breakdown of carbamate residues. This aspect will be discussed in the next section. However, the contribution from this mechanism, compared to chemical hydrolysis and to microbial metabolism, appears to be minor.

Putting all of these scattered bits of information together does not create a dependable predictive theory. However, barring large-scale spills, most methylcarbamates and their phenols entering natural waterways would probably be rapidly destroyed with total persistence in the water *per se* of only a few days, or at most a few weeks. The dimethylcarbamates would be expected to persist somewhat longer. Based on knowledge of organochlorine behavior in air,[88,125,128] carbamate insecticides undoubtedly exist in air at extremely low concentrations, although the lifetime of any one particular molecule would probably be very short.

What is the biological significance of these minute quantities of toxins? Will long-term exposure to these very low levels of carbamates have adverse cumulative effects on living organisms? These questions cannot be answered with the available scientific information, and they may never be answered for species with generation times of many years, such as man. Certainly exposure from air and water is less than that encountered during ingestion of pesticide-contaminated food products.

PHOTODECOMPOSITION

In addition to the multitude of chemical and biological reactions which contribute to the cleansing of pesticide residues from the environment, there are reactions catalyzed by the presence of sunlight. Often the degradation products from all three mechanisms are identical, but photodecomposition has produced some unique structures. Obviously, the role of photolysis in the overall destruction of an applied pesticide depends on the ability of the sun's rays to reach the deposit. More precisely, it appears that light with wavelengths in the ultraviolet spectrum must come in contact with the pesticide.

Since penetration of ultraviolet light into solid matter is limited, photodecomposition of insecticides in soil is restricted to residues on or very near the surface. The same is probably true of plants, although it is not necessary for the residue to be exposed to direct sunlight, since ultraviolet reflection occurs. Perhaps the most vulnerable insecticides are those persisting in air and water. In the case of the latter, ultraviolet light has been shown to penetrate relatively thick layers of clear water.[129]

Once exposure has occurred, the extent of photodecomposition depends on the duration of exposure, the intensity and wavelength of the light, the surrounding media, temperature, humidity, and the inherent chemistry of the insecticide molecule. Since only the last of these can be controlled under field conditions, most of the experimental data on photodecomposition originate from laboratory studies. This is unfortunate, because the "erratic" results obtained from field experiments reflect the natural variation in intensity and wavelength distribution of the sun together with dramatic weather changes from day to day and from one location to the next.

However, in the laboratory it is important to use conditions which approximate reality, particularly with respect to the wavelength and intensity of irradiation. Much of the sunlight responsible for pesticide degradation falls into the wavelength range of 300 to 450 nm.[130] There are "black lights" available, which offer light of intensity similar to the sun's through this spectral region (Figure 8.16). Normal white fluorescent laboratory lights also emit ultraviolet light through the required range, but the energy is considerably lower. The intensity of fluorescent sun lamps, on the other hand, is high, but the wavelength distribution covers only part of the sun's spectrum. One approach which perhaps yields a spectrum nearest to natural light involves the use of a mercury lamp equipped with a borosilicate glass filter,[130] but this has not been employed in many experiments. Most often, insecticides are exposed to germicidal ultraviolet lamps, which emit essentially monochromatic light at 254 nm with intensities similar to the sun's.

Like the study of carbamates, photolysis of pesticide chemicals is a relatively new area of research. Ten to 15 years ago, only a few people

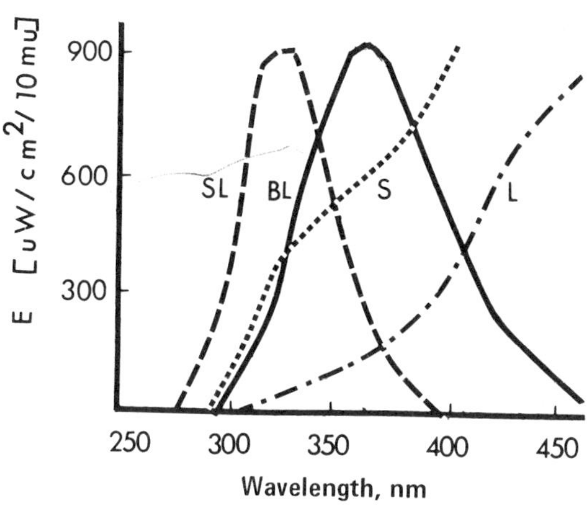

FIGURE 8.16. Emission of spectra of various light sources. SL = fluorescent sunlamp; BL = black light; S = normal sunlight; L = normal white fluorescent lamp. (Adapted from Reference 130, with permission.)

were working on photoalteration of pesticides, but as the methods, interest, and potential importance of this mode of degradation became established, more researchers began to examine the photostability of pesticides. Today it is often a routine part of a total metabolic study.

The first extensive look at ultraviolet degradation which included carbamates appeared in 1961.[131] Of the 141 pesticide chemicals exposed to ultraviolet light (254 nm), three were carbamate insecticides. Results showed that 10 mg of Isolan and of Pyrolan spotted on filter paper were moderately susceptible to photodecomposition; results with carbaryl were inconclusive, however, although degradation was evident.

In 1965, photolysis of carbamates became an important topic, following a report involving six aryl methylcarbamate insecticides.[132] Solutions (1 mg/ml) of the chemicals in absolute ethanol or in hexane were irradiated in open petri dishes for 1 to 3 hr with high- and low-intensity ultraviolet sources (254 nm) and with sunlight. Decomposition products were separated and located by a thin-layer cholinesterase assay.

Baygon and UC 10854 did not form any additional cholinesterase inhibitors after exposure to sunlight and were only slightly affected by artificial light. Mesurol decomposed to the same few products in sunlight and in ultraviolet light, although the degree of breakdown was greater in the laboratory. Photolysis of Matacil was extensive after treatment with any light. This was also true for Zectran, although most of the cholinesterase inhibitors were formed from impurities in the technical Zectran samples.

Carbaryl formed only one minor inhibitory product after exposure to sunlight or weak ultraviolet light, but intense ultraviolet irradiation caused the formation of several more products. Changing the solvent from ethanol to methanol or acetone did not alter the results, but exposure in hexane yielded much greater photodecomposition into compounds which moved differently on thin-layer plates. Of greatest significance was the finding that exposure of crystalline carbaryl or 50% wettable powder formulations of carbaryl for long periods to high-intensity ultraviolet lamps caused little, if any, degradation to other anticholinesterase agents. However, this does not agree entirely with an earlier study, where pure carbaryl was shown to be immune to photolysis, but dust, powder, and emulsifiable concentrate formulations of the insecticide were extensively decomposed after only 8 hr of exposure to sunlight or to ultraviolet lamps.[133]

The photostability of unformulated carbaryl has also been demonstrated after its application to silica gel thin-layer plates.[126] A spot of 100 μg of insecticide was irradiated for 3 hr with normal room fluorescent light, a 254-nm ultraviolet lamp, and a 366-nm lamp. In only one case did decomposition occur, and that was with short-wavelength irradiation, where a minor degradation product was detected with ninhydrin reagent at the origin of the developed plate.

Seven other methylcarbamates also resisted attack by room light and long-wave ultraviolet light, but the 254-nm rays partially converted Mesurol to at least seven products, two of which were the corresponding sulfoxide and sulfone. In addition, Baygon, UC 10854, HRS-1422, and Banol formed two to four unidentified photoproducts after exposure to the short-wave lamp. As indicated in the solvent studies described above,[132] Zectran and Matacil were again demonstrated to be very susceptible to photolysis by ultraviolet light of wavelength 254 nm.

When the same eight insecticides were placed on plant foliage, recovered degradation products were similar, at least qualitatively, to those found on thin-layer plates. Once again, Zectran and Matacil were most vulnerable to attack, and this led to more extensive examination of the photodecomposition of these two carbamates. Carbonyl-^{14}C samples of both compounds were irradiated (254 nm) for 20 hr to yield at least 12 photoproducts; the majority of these involved modifications of the dimethylamino group,[134] as illustrated in Figure 8.17. The same products were also formed on plant foliage exposed to sunlight.

These studies, together with an earlier report based on intensive ultraviolet irradiation of Zectran on bean leaves,[135] suggest that environmental residues of these two aryl methylcarbamates would be very susceptible to decomposition by photolysis. In fact, this susceptibility has prompted a report that Zectran was probably photodecomposed on the cuticle of insects.[136] However, insect cuticle or plant surfaces are not biologically inert, so that it is difficult to prove whether photoalteration or biological metabolism produced the recovered products. But despite the ease of breakdown, the derivatives depicted in Figure 8.17 all retain considerable toxicity; thus, in the end it is the stability of these products which determines the environmental safety of Zectran and Matacil. With Zectran, there is evidence that the analogs are more stable on foliage.[126]

Considerable attention has been focused on the photodecomposition of carbamates in aqueous solution. For example, insecticide solutions of 20 mg/l contained in petri dishes were irradiated with germicidal tubes (254 nm maximum wavelength) while standing in a cabinet lined with aluminum foil.[91] Altering the pH of the solution produced the striking differences listed in Table 8.6. Half of the carbaryl was destroyed after 8 min of illumination at pH 8, but at pH 5 the half-life was extended to 1 hr. In only 5 min 1-naphthol could be detected in the water; this was also photolyzed, but at a slower rate. Four other photoproducts were recovered, but ester hydrolysis was the major degradative pathway. Hydrolysis to 1-naphthol has been demonstrated in other photodecomposition studies,[130,132,137] and the photoinstability of 1-naphthol has also been confirmed.[138]

The photolytic decomposition of Baygon responded similarly to pH changes (Table 8.6), although the variation was not as great.[91] Formation of the parent phenol occurred, and it tended to accumulate with time. Such rapid and complete decay in a matter of minutes is in contrast to results from Baygon irradiation in organic solvent[132] or on thin-layer plates.[126] Thus, the environment of the substrate can influence decomposition rate.

Another important parameter is chemical structure, as pointed out by the rapid loss of both dimethylcarbamates, Pyrolan and dimetilan, regardless of pH[91] (Table 8.6). Both of the parent

TABLE 8.6

Effect of pH on Photodecomposition of Carbamate Insecticides in Aqueous Solution[91]

Compound	pH	Minutes required for 50% degradation
Carbaryl	5	60
	7	39
	8	8
Baygon	5	19
	7	17.5
	9	9
Pyrolan	5	7.5
	7, 9	6
Dimetilan	5, 7, 9	15

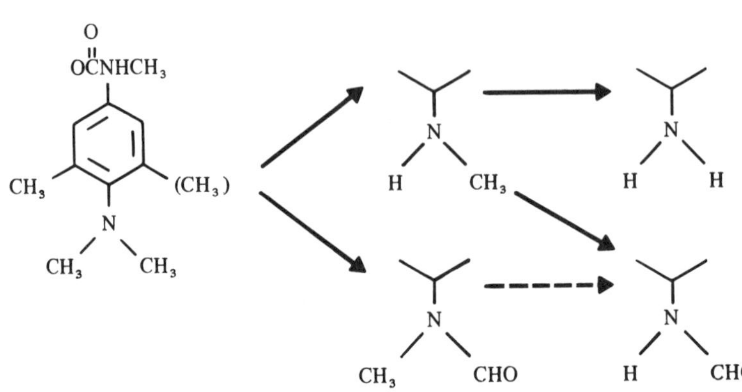

FIGURE 8.17. Photodecomposition of the dimethylamino group of Zectran and Matacil.[134]

enols were detected, and they too photodecomposed in less than 1 hr. The almost immediate destruction of Pyrolan may have been due to the fact that its ultraviolet absorption spectrum exhibits a maximum (248 nm) which is closest to the 254 nm used to photolyze. However, this cannot be interpreted to mean that Pyrolan will explode at the first hint of sunlight, since the sun's wavelengths are considerably higher than 254 nm (Figure 8.16). In fact, it has been demonstrated that irradiation of an ethanolic or hexane solution of Pyrolan with sunlight did not result in photodecomposition, even though extensive breakdown took place under a 254-nm ultraviolet tube.[139] On the other hand, dimetilan photodecomposed in solvent during treatment with either artificial or natural light.

A laboratory system which may more closely approximate sun exposure was designed for a study of formetanate degradation.[140] In a round-bottom flask containing an aqueous solution of the insecticide (250 ppm) a high-pressure mercury lamp was immersed. The emission spectrum included wavelengths of 286 nm and above, with intensity similar to that of the sun. After 4 days, the same photoproducts were formed in distilled water (pH 3) and in river water (pH 7.1), although the extent of breakdown was not given. The major reactions involved the conversion of the amidine group to a formamide with subsequent hydrolysis of the carbamate ester.

Photolysis of C-8353 [2-(1,3-dioxolane-2-yl) phenyl N-methylcarbamate] in water (0.025 to 1.25%) and in methanol (0.5%) at 25°C apparently resulted in different modes of degradation.[141] Using a quartz reaction vessel, the solutions were irradiated with an ultraviolet lamp having a primary emission at 360 nm. In the aqueous system the carbamate was hydrolyzed to its parent phenol and the dioxolane ring was cleaved to give 2-methylcarbamyl benzaldehyde (Figure 8.18), but in alcohol no photochemical reaction took place. However, with lamps producing light of 254 nm or 300 nm wavelength, methanolic solutions yielded more than 85% production of a single compound, 3-methyl-2H-1,3-benzoxazine-2,4-(3H)-dione (Figure 8.18). This likely resulted from cyclization of the benzaldehyde derivative, which, for some reason, was not possible in an aqueous medium.

A more novel photoreaction series has been found for 4-methylphenyl N-methylcarbamate.[142] Mercury-arc illumination of a 0.21*M* solution of the carbamate in *t*-butanol resulted in 25% hydrolysis to *p*-cresol and 57% conversion plus rearrangement to N-methyl 2-hydroxy-5-methylbenzamide (Figure 8.19).

Sunlight, a 275-W sun lamp, and a germicidal tube (254 nm) were all used to irradiate RE 5365 and its N-benzenesulfenyl derivative RE 11775.[143] The 3-(2-butyl)phenyl N-methylcarbamate, RE 5365, was more resistant than its derivative to degradation when irradiated as a 2-mg/ml solution in methanol. Only the ultraviolet lamp produced extensive photodecomposition to at least six products, one of which was the parent phenol and another the N-hydroxymethyl analog. On the other hand, the 3-(2-butyl)phenyl N-benzenesulfenyl-N-methylcarbamate, RE 11775, was converted to six or more photoproducts after exposure to any of the light sources. This is difficult to understand, since the products were partly the same as those formed from RE 5365, i.e., the benzenesulfenyl group was cleaved to give RE 5365, which in turn degraded to its phenol and N-hydroxymethyl derivative. However, the latter two products were found in only minor amounts, if at all, after sunlight or sun lamp irradiation of RE 5365. Also, N-demethyl RE 5365 was formed from RE 11775, but not from RE 5365.

Exposure of carbonyl-^{14}C RE 11775 on glass (0.5 $\mu g/cm^2$) and on silica gel plates (5 $\mu g/cm^2$) to ultraviolet light gave 99% and 50% degradation respectively in only 5 min. Similar exposure to sunlight gave 20% and 30% photodestruction in 30

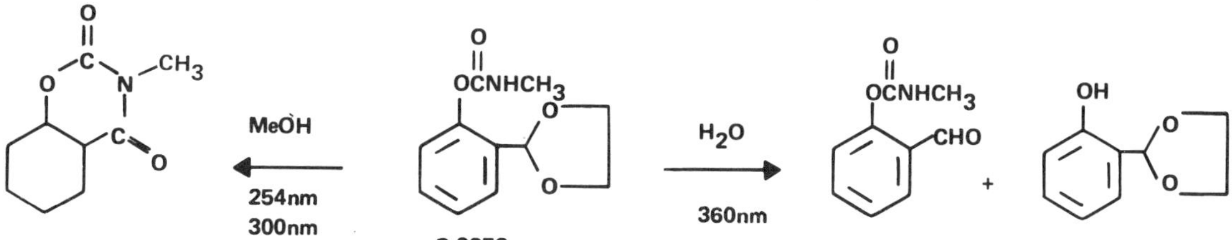

FIGURE 8.18. Photodecomposition of C-8353 in water and methanol.[141]

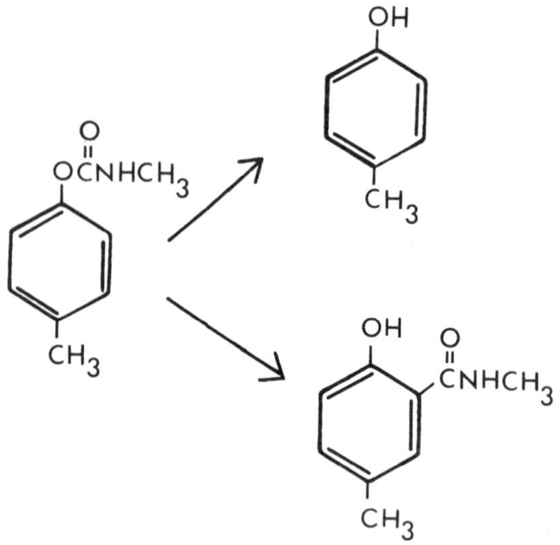

FIGURE 8.19. Photodecomposition of 4-methylphenyl N-methylcarbamate in *t*-butanol.[142]

min. A large number of products resulted from either experiment, with the following general photopathways evident: the sulfur atom was oxidized to sulfoxide; the N–S bond of RE 11775 was very labile, yielding RE 5365; there was ester cleavage of RE 11775; and modification of the N-methyl group of RE 11775 was more extensive than on RE 5365. Finally, the susceptibility of RE 11775 to ultraviolet degradation also extends to aqueous solutions, where a concentration of 0.02 to 0.1 ppm was shown to photodecompose by more than 60% after 8 hr in direct sunlight.[119]

Room light and sunlight have been compared with respect to their ability to catalyze destruction of Landrin[144] and carbofuran.[145] As expected, both insecticides were decomposed much faster outdoors, but in each case there was also some alteration in the laboratory. Crystalline carbofuran formed its 3-hydroxy analog after 2 days of sunlight, and by 21 days three other products appeared, one of which was probably the parent phenol.[145] Radiolabeled Landrin on bean leaves at 1 $\mu g/cm^2$ was almost completely destroyed after 4 days in the sun.[144] At that time only 11 to 20% of the applied ^{14}C was recovered. The major photoproducts resulted from hydroxylation of ring and N-methyl groups, with very little hydrolysis evident.

When all of these data are digested in an attempt to offer some conclusions regarding the contribution of photodecomposition to the environmental stability of carbamates, the result is disappointing. Certainly, there exists a potential for light-catalyzed destruction of methylcarbamate molecules, particularly for certain compounds, such as Zectran and Matacil. There is also evidence which implies that carbamates dissolved in water are sensitive to rather low intensities of ultraviolet light through a broad spectral range. Finally, laboratory experiments have shown that parameters such as pH, exposure medium, duration of exposure, and wavelength and intensity of light can greatly influence the degree of photodecomposition.

However, in too few instances have laboratory experiments produced the same results as exposure of the insecticides to sunlight. In most cases carbamates were irradiated with ultraviolet lamps while dissolved in an organic solvent, a situation which is believed by some to simulate exposure on a plant surface. The value of this information is limited because solvent illumination was generally accomplished with essentially monochromatic light at a wavelength of 254 nm, whereas illumination from the sun does not involve light of 254 nm, but includes a spectral range of wavelengths from 300 to 450 nm. In addition, dissolution of the insecticide in a solvent makes practically every molecule a potential target, whereas insecticides coated onto silica gel, plant foliage, or other surfaces are not as completely available. Thus, the very rapid and complete destruction in solvents is not repeated when the compound is applied to thin-layer plates or bean leaves. Therefore, the inference of extensive degradation of methylcarbamate insecticides by light made by some researchers may be unjustified, and further investigation under conditions nearer those actually existing in the field are considered essential.

HUMAN EXPOSURE

In the previous sections of this chapter, the stability of carbamate insecticides in various portions of the environment has been examined. The conclusions were that the persistence of these chemicals, with a few exceptions, is limited, and thus exposure of the general public to harmful quantities of carbamates would be very unlikely. Of course, those individuals involved with the manufacture and formulation of the insecticides, and with insecticide application and harvesting of sprayed crops, are exposed to higher levels of toxicant. Although the accident record with methylcarbamates and people in these categories of work has been relatively good, more studies are needed to ensure the health of those in daily close contact with carbamates.

With regard to the general population, there is undoubtedly some exposure to minute quantities of methylcarbamate insecticides during their use as control agents for public health and forest pests, as house and garden sprays, and as protectants for crops grown for fiber and animal feed. Concomitant with all of these uses is release of some insecticide into air, water, and soil, where the chemical is diluted and transported away from the site of application for potential contact with even more humans.

However, the most obvious source of carbamate exposure is through the consumption of insecticide-contaminated food. Because of the need to protect food crops from insects, and because of the almost impossible task of removing all residues before harvest, certain quantities of insecticide are allowed to be present on raw agricultural commodities. For example, with carbaryl these legal tolerances range from near zero to as high as 10 ppm. Although preliminary experiments have ensured that these levels are perfectly safe for human consumption, the most commonly used insecticides are subject to continuous examination.

Thus, in cooperation with the Food and Agricultural Organization of the United Nations and the World Health Organization, an international group of experts convenes periodically to assess the degree and significance of pesticide residues in food. After assimilating the available information for the most commonly used pesticides on a world-wide basis, the group publishes a list of "acceptable daily intake" (ADI) levels to be used as a guide for potential hazards from human consumption of pesticide-contaminated food.[146] Input into the calculations includes toxicology data, persistence data, level of residue present, loss of residue through processing, etc.

The only carbamate insecticide to be included in the FAO-WHO list thus far is carbaryl. The established acceptable daily intake was initially set at 0.02 mg insecticide/kg body weight, based on tolerance levels of 0.2 ppm for potatoes, and up to 10 ppm for some fruits, small berries, some vegetables, nuts, and olives. However, this has recently been changed to 0.01 mg/kg.[146] Part of the reason for the decrease is due to some information suggesting that carbaryl may have adverse effects on reproductive physiology and/or kidney function in animals. The panel also indicated that there is a potential for the daily intake level to be exceeded with carbaryl.

Levels of Carbaryl in Food

Even with the lower value, food surveys which have included carbaryl as a potential contaminant indicate that levels of the carbamate are far below the accepted daily intake. For instance, examination of 27 food products from 1967 through 1969 showed only 0.1 mg carbaryl in the total daily food ration of residents living in rural areas of Russia where 207 to 218 tons of pesticides were used annually.[147] In another area, where the yearly use of pesticides was only one fourth as much, daily intake levels were about 0.05 mg per person. Using a conservative weight estimate of 50 kg per person, this would yield average daily intake values of 0.002 and 0.001 mg/kg respectively.

Market basket surveys conducted by the Food and Drug Administration in the United States rarely turn up any significant quantities of methylcarbamate insecticides, with the exception of carbaryl. Samples taken consist of 82 food and beverage items selected from retail stores in cities of various populations across the country.[148] The amount of food analyzed was equal to that consumed by a 16- to 19-year-old male over a 2-week period. Thus, the intake levels (4 kg/day) probably represent maximum amounts.

Recoveries of carbaryl from food composites collected during the period of June 1964 through April 1970 are summarized in Table 8.7.[149] The greatest number of samples containing carbaryl appeared in 1965, when 7.4% of the composites yielded a daily intake of 0.15 mg. However, when the average daily intake was calculated, it was shown to fall below that established as safe by the WHO-FAO committee. In 1968 and 1970, no carbaryl was detected in any composite, and in the other years, average daily intake levels were very low.

Most of the carbaryl residues were usually found in beverages, with raw fruits serving as the largest source in edible commodities.[150,151] However, of the 6,763 fruit samples analyzed from 1964 to 1969, only 4.1% contained detectable amounts of carbaryl, giving an average residue of 0.02 ppm. The insecticide appeared in only 1.4% of more than 8,000 grain samples (average less than 0.005 ppm) and in 3% of products made from grains (average of 0.009 ppm). Of the 134 composites of ready-to-eat food, only 6% contained carbaryl at an average residue level of 0.011 ppm.

The relatively rare appearance of carbaryl in the market basket can be appreciated when compared to the total pesticide residues in food. About half of all fruits, vegetables, dairy products, and animal feeds contained one or more pesticide chemicals, although the residues were below 0.51 ppm 95% of the time.[151] Residues occurred even more frequently in fish, red meat, and poultry, but methylcarbamates were very rarely discovered in this group of foods. Using the years from 1964 to 1966, when carbaryl was most prevalent (Table 8.7), it can be seen that the frequency of occurrence was far less than that found for organochlorine insecticides, as shown in Table 8.8.[152] The organophosphorus compounds and the herbicide 2,4-D appeared about as often as carbaryl.

Interestingly, the total daily intake of 0.082 mg per person was highest with carbaryl. However, this amount, and those shown in Table 8.7, did not approach the average daily intake of 0.01 mg/kg established by the World Health Organization. Part of the reason for this is that the WHO-FAO committee assumes residue levels on food at the time of eating to be the same as the maximum allowed by legal tolerances. Also, in the case of carbaryl, residue losses during processing were not considered, since the amount of information available at the time was limited.

TABLE 8.7

Recovery of Carbaryl from Food Composites Taken from Retail Food Stores in the United States[149]

Year	Number of food composites	Percentage of composites with carbaryl	Daily intake (mg)	ADI[a] (mg/kg)
1965	216	7.4	0.15	0.002
1966	312	2.7	0.026	0.0005
1967	360	1.1	0.007	0.0001
1968	360	0	0	0
1969	360	0.8	0.003	0.00004
1970	360	0	0	0

[a]Suggested maximum safe average daily intake for carbaryl by WHO is 0.01 mg/kg.

Removal of Residues by Processing

Recent studies, and a few earlier reports, have shown that carbaryl residues on foods are greatly diminished by washing and processing. For example, washing apricots, lemons, or oranges with water removed as much as 75% of the surface deposits.[35,153] The degree of removal from cabbage and eggplant was shown to be dependent on the length of aging of the residue.[40] Thus, shortly after carbaryl was sprayed on cabbage, residues of 15 to 34 ppm were cut down to about 1 ppm with a simple wash. More carbaryl was recovered at later times, but it represented only about half of the total residue after 3 to 7 days.

Commercial canning and freezing of vegetables has been shown to essentially eliminate many insecticide residues from these foods.[154] This is particularly true when the processing involves husking, peeling, or shelling of the edible portion, leaving most of the residue behind. Washing by immersion or spraying can also remove considerable quantities, although the amount lost depends on the chemical nature of the insecticide and on surface characteristics of the crop (Table 8.9). Most insecticides easily rinse off tomatoes; leafy vegetables, such as spinach, tend to retain their residues, but even from these carbaryl is usually removed by a simple wash. The addition of detergents to the water often increases the amount removed, and blanching, especially water-blanching, reduces the contamination even more. Filling, closing, and processing in sealed containers generally has little effect on carbaryl residues.

TABLE 8.8

The Appearance of Insecticide Residues in 516 Food Composites from FDA Surveys between 1964 and 1966

Insecticide	Percentage of positive composites	Daily intake (mg)
DDT	37.4	0.037
Dieldrin	20.2	0.006
Lindane	13.6	0.004
BHC	6.2	0.003
Carbaryl	4.7	0.082
Malathion	4.5	0.009
2,4-D	3.5	0.004
Diazinon	3.0	0.001

Adapted from Reference 152, with permission.

TABLE 8.9

Water-washing of Insecticide Residues off Vegetables[154]

Compound	Percent removed from Spinach	Tomato	Broccoli
Carbaryl	87	97	77
DDT	48	91	–
Malathion	–	95	–
Parathion	9	–	0
Diazinon	0	88	–

The waste water from these operations usually contains some carbaryl, but the relative quantity of carbaryl discharged is generally much smaller than that found during the processing of foods contaminated with other insecticides. Solids from peels, shells, and pulp are often buried in sanitary land fills, but some are used for animal feeding, so that it is important that these wastes be low in insecticide content.

In almost all studies where the effects of processing on residues are examined, insecticides are applied to the crop at higher than recommended rates and/or the crops are harvested at intervals closer to the last application than allowed. For example, green beans were sprayed with carbaryl at 2 lb/acre two times and were harvested immediately after the last application.[155] An initial residue of 7.6 ppm was reduced by 70% with commercial water-blanching at 85°C for 1.5 to 3 min. Further processing did not remove any more insecticide. Home blanching of beans carrying 11 ppm carbaryl cut the residue down to 2 ppm, and after freezing or canning the contamination was lowered to less than 1 ppm.

Tomatoes sprayed with a carbaryl dose of 6 lb/acre at 10-day intervals were sampled on the day of the last spray.[156] An average residue of 5.2 ppm carbaryl was reduced by 82% with a cold-water wash without detergent, and by 99% with a warm-water wash with detergent. Removal of peels by two commercial procedures reduced the residues to almost zero. A simple hand wash removed 77%, and home peeling lowered residues by 92%. The waste material after production of tomato juice contained only 0.2 ppm carbaryl.

With broccoli, carbaryl residues were allowed to remain on the plant for 2 days prior to harvest.[157] Then, commercial preparation resulted in almost total removal of the 12-ppm contamination. Again, the majority was taken off by 30-sec water immersion followed by a 15-sec spray. Steam-blanching at 100°C for 4 to 8 min did little harm to the residue, but blanching by immersion in 85°C water for 6 min reduced the residue considerably. Overall removal by home washing, blanching, and freezing totaled 90%, but simple washing and home cooking lowered the level by only 55%.

A longer spray-to-harvest period of 7 days was used in a spinach study.[158] Washing lowered the initial deposit of 21 ppm by 66 to 88%, and, although steam-blanching had no effect on carbaryl, water-blanching brought the residue down to only 5% of the original deposit. After the spinach reached the final canning procedure, carbaryl residues fell below detectable limits (0.05 ppm). On the other hand, home preparation of spinach using a hand wash removed 70% of the residue, but no further reduction occurred after cooking.

The effects of canning and storage on carbaryl residues were studied by fortifying spinach and apricots before processing.[159] Initial carbaryl concentrations of 10.5 ppm in spinach and of 11.4 ppm in apricots were lowered by 44% and 12% respectively following 1-hr processing at temperatures above 100°C. After standing for 1 year at ambient temperature, there was little change in the insecticide level in the canned fruit or vegetable. However, increasing the storage temperature to 38°C destroyed 67% of the spinach residues and 17% of the apricot residues. Thus, it appears that the amount of carbaryl present in canned products is dependent on temperature of processing and storage and on the nature of the food product, with acid foods causing more prolonged stability.

Although residues of carbaryl and other carbamates do not often appear in dairy and other animal products, it is probable that processing of these commodities would not lead to reduced contamination.[160] One animal product which may be a source of carbaryl intake is honey. When bees were fed diets containing 2 ppm carbaryl, the insects transferred the insecticide to bee bread and to honey.[161] Open-comb honey residues of about 0.025 ppm were shown to accumulate to about 0.040 ppm after 56 days. This was probably due to residue concentration by evaporation of water. In bee bread, carbaryl levels attained a maximum at 0.1 ppm after 1 month, but decreased to 0.05

ppm by 56 days. Thus, it appears that carbaryl is relatively stable in honey, although its presence in commercial honey has not been established.

From the above results, it is apparent that any raw food product which contains carbaryl residues will have a reduced insecticide content after the most simple commercial operation. Thus, assuming that the initial raw product was within established tolerances for carbaryl, the food reaching the dinner plate would have an additional margin of safety because of the lowering of residues by processing.

REFERENCES

1. **Gyrisco, G. G., Lisk, D. J., Fertig, S. N., Huddleston, E. W., Fox, F. H., Holland, R. F., and Trimberger, G. W.,** The effects of feeding high levels of Sevin on residue, flavor and odor of the milk of dairy cattle, *J. Agric. Food Chem.*, 8, 409, 1960.
2. **Roberts, R. H., Jackson, J. B., Westlake, W. E., Ackerman, A. J., and Claborn, H. V.,** Residue studies of livestock sprays containing Sevin, *J. Econ. Entomol.*, 53, 326, 1960.
3. **Camp, H. B., Buttram, J. R., Hays, K. L., and Arthur, B. W.,** Sevin residues in milk from dairy cows following dermal applications, *J. Econ. Entomol.*, 56, 402, 1963.
4. **Eheart, J. E., Turner, E. C., and Dickinson, J.,** Residues of Sevin in whole milk from sprayed and dusted cows, *J. Econ. Entomol.*, 55, 504, 1962.
5. **Claborn, H. V., Roberts, R. H., Mann, H. D., Bowman, M. C., Ivey, M. C., Weidenbach, C. P., and Radeleff, R. D.,** Residues in body tissues of livestock sprayed with Sevin or given Sevin in the diet, *J. Agric. Food Chem.*, 11, 74, 1963.
6. **Whitehurst, W. E., Bishop, E. T., and Critchfield, F. E.,** The metabolism of Sevin in dairy cows, *J. Agric. Food Chem.*, 11, 167, 1963.
7. **Dorough, H. W. and Casida, J. E.,** Nature of certain carbamate metabolites of the insecticide Sevin, *J. Agric. Food Chem.*, 12, 294, 1964.
8. **Dorough, H. W.,** Carbaryl-C^{14} metabolism in a lactating cow, *J. Agric. Food Chem.*, 15, 261, 1967.
9. **Dorough, H. W.,** Carbaryl residues in milk and meat of dairy animals, in *Pesticide Terminal Residues*, Tahori, A. S., Ed., Butterworth, New York, 1971, 173.
10. **Hurwood, I. S.,** Studies on pesticide residues. II. Carbaryl residues in the body tissues and milk of cattle following dermal application, *Queensl. J. Agric. Anim. Sci.*, 24, 69, 1967.
11. **Petrovskii, V. V.,** Detection of Sevin and 1-naphthol residues in cattle blood and milk, *Veterinariya*, 43, 85, 1966.
12. **Petrovskii, V. V.,** Level and duration of Sevin occurrence in cow's milk, *Veterinariya*, 47, 39, 1970.
13. **Gajduskova, V.,** Excretion of 1-naphthyl N-methylcarbamate (carbaryl) and 2-isopropoxyphenyl N-methylcarbamate (aprocarb) with milk, *Vet. Med.*, 17, 335, 1972.
14. **Ivie, G. W. and Dorough, H. W.,** Furadan-C-14 metabolism in a lactating cow, *J. Agric. Food Chem.*, 16, 849, 1968.
15. **Dorough, H. W. and Ivie, G. W.,** Temik-^{35}S metabolism in a lactating cow, *J. Agric. Food Chem.*, 16, 460, 1968.
16. **Dorough, H. W., Davis, R. B., and Ivie, G. W.,** Fate of Temik-carbon-14 in lactating cows during a 14-day feeding period, *J. Agric. Food Chem.*, 18, 135, 1970.
17. **Robbins, J. D., Bakke, J. E., and Feil, V. J.,** Metabolism of benzo[b′] thien-4-yl methylcarbamate (Mobam) in dairy goats and a lactating cow, *J. Agric. Food Chem.*, 18, 130, 1970.
18. **Bowman, M. C., Lowery, R. S., Leuck, D. B., and Beroza, M.,** Effects of feeding cattle forage treated with Mobam, *J. Econ. Entomol.*, 61, 1495, 1968.
19. **Miles, J. T., Demott, B. J., Hinton, S. A., Montgomery, M. J., and Bennett, S. E.,** Effect of feeding carbofuran on the physiology of the dairy cow and on pesticide residues in milk, *J. Dairy Sci.*, 54, 478, 1971.
20. **Leuck, D. B., Bowman, M. C., Johnson, J. C., Jr., Beck, E. W., and Lowery, R. S.,** Niagara NIA-10242 insecticide residues: their persistence in coastal Bermuda grass, corn, and silages, and their effect, fed in silage, on lactating cows, *J. Econ. Entomol.*, 61, 1349, 1968.
21. **Knaak, J. B., Munger, D. M., McCarthy, J. F., and Sater, L. D.,** Metabolism of carbofuran alfalfa residues in the dairy cow, *J. Agric. Food Chem.*, 18, 832, 1970.
22. **Dorough, H. W. and Wiggins, O. G.,** Nature of the water-soluble metabolites of carbaryl in bean plants and their fate in rats, *J. Econ. Entomol.*, 62, 49, 1969.
23. **Baker, M. W., Crampton, E. L., Harrison, I. R., and Watkins, T. I.,** Residues of Butacarb in sheep muscle, fat, liver, and kidney following dipping for the control of sheep blowfly (*Lucilia* spp.), *J. Sci. Food Agric.*, 18, 392, 1967.

24. **Furman, D. P. and Pieper, G. R.,** Systemic acaricidal effects of Sevin in poultry, *J. Econ. Entomol.,* 55, 355, 1962.
25. **Johnson, D. P., Critchfield, F. E., and Arthur, B. W.,** Determination of Sevin insecticides and its metabolites in poultry tissues and eggs, *J. Agric. Food Chem.,* 11, 77, 1963.
26. **McCay, C. F. and Arthur, B. W.,** Sevin residues in poultry products, *J. Econ. Entomol.,* 55, 936, 1962.
27. **Nir, I., Weisenberg, F., Hadani, A., and Egyed, M.,** Studies of the toxicity excretion and residues of Sevin in poultry, *Poult. Sci.,* 45, 720, 1966.
28. **Vulchovski, Y., Petrov, D., and Stoimenov, V.,** Residual amounts of Sevin in the eggs, organs, and tissues of birds, *Vet. Med. Nauki,* 9, 83, 1972.
29. **Burdeinaia, R. V.,** Veterinary and sanitary evaluation of chicken carcasses after Sevin poisoning, *Veterinariya,* 48, 96, 1969.
30. **Paulson, G. D. and Feil, V. J.,** The fate of a single oral dose of carbaryl (1-naphthyl N-methylcarbamate) in the chicken, *Poult. Sci.,* 48, 1593, 1969.
31. **Andrawes, N. R., Chancey, E. L., Crabtree, R. J., Herrett, R. A., and Weiden, M. H. J.,** Fate of naphthyl-1-^{14}C carbaryl in laying chickens, *J. Agric. Food Chem.,* 20, 608, 1972.
32. **Hicks, B. W., Dorough, H. W., and Mehendale, H. M.,** Metabolism of aldicarb pesticide in laying hens, *J. Agric. Food Chem.,* 20, 151, 1972.
33. **Hicks, B. W., Dorough, H. W., and Davis, R. B.,** Fate of carbofuran in laying hens, *J. Econ. Entomol.,* 63, 1108, 1970.
34. **Korn, S.,** The uptake and persistence of carbaryl in channel catfish, *Trans. Am. Fish. Soc.,* 102, 137, 1973.
35. **Gunther, F. A., Blinn, R. C., and Carman, G. E.,** Residues of Sevin on and in lemons and oranges, *J. Agric. Food Chem.,* 10, 222, 1962.
36. **Polizu, A., Greger, H., and Alexandri, A. V.,** Studies on the persistency of carbaryl residues in fruit, *Qual. Plant. Mater. Veg.,* 20, 215, 1971.
37. **Bogomolova, Z. N.,** Sevin stability in fruit and berry crops during vegetation and prolonged storage, *Vopr. Pitan.,* 27, 55, 1968.
38. **Williams, K.,** Persistence of Sevin and Diazinon residues on fruits, *Proc. Entomol. Soc. B.C.,* 56, 57, 1959.
39. **Deshmukh, S. N. and Lal, R.,** Dissipation of carbaryl residues from brinjal (*Solanum melogena* L.) plants, *Indian J. Entomol.,* 32, 201, 1970.
40. **Mann, G. S. and Chopra, S. L.,** Residues of carbaryl on crops, *Pestic. Monit. J.,* 2, 163, 1969.
41. **Cuant, P., Carrasco, J. M., and Martinez, R.M.,** Contamination of agricultural produce by pesticides. II. Pesticide residues in tomatoes, *Rev. Agroquim. Tecnol. Aliment.,* 8, 472, 1968.
42. **Blazquez, C. H., Vidyarthi, A. D., Sheehan, T. D., Bennett, M. J., and McGrew, G. T.,** Effect of Pinolene (*beta*-pinene polymer) on carbaryl foliar residues, *J. Agric. Food Chem.,* 18, 681, 1970.
43. **Granett, P. and Reed, J. P.,** Field evaluation of Sevin as an insecticide for pests of vegetables in New Jersey, *J. Econ. Entomol.,* 53, 388, 1960.
44. **Johnson, D. P. and Stansbury, H. A.,** Adaptation of Sevin insecticide (carbaryl) residue method to various crops, *J. Agric. Food Chem.,* 13, 235, 1965.
45. **Stobwasser, H.,** The residues of some esters of phosphoric acid and of Sevin in carrots after field applications, *Z. Pflanzenschutz,* 70, 459, 1963.
46. **Dhal, P. and Lal, R.,** Colorimetric determination of carbaryl residues in field-treated corn, *J. Assoc. Off. Anal. Chem.,* 54, 1095, 1971.
47. **Sundaram, A. and Sundaram, K. M. S.,** Penetration, translocation, and persistence of the insecticides 1-naphthyl N-methyl-carbamate (Sevin) and O,O-dimethyl-O-(3-methyl-4-nitrophenyl) phosphorothioate (Sumithion) in cacao seedlings *(Theobroma cacao), Ghana J. Sci.,* 7, 1, 1967.
48. **Andrawes, N. R., Chancey, E. L., and Bartley, W. J.,** The Metabolism of 1-Naphthyl-^{14}C Carbaryl in Several Plant Species, paper presented at the 163rd American Chemical Society Meeting, Boston, April 1972.
49. **Huddleston, E. W. and Gyrisco, G. G.,** Residues of Sevin and 1-naphthol on forage from aerial applications of Sevin in oil, *J. Econ. Entomol.,* 53, 484, 1960.
50. **George, D. A., Butler, L. I., Maitlen, J. C., Rusk, H. W., and Walker, K. C.,** Insecticide Residues on Forage Crops, ARS 33-112, U.S. Department of Agriculture, Washington, D.C., 1967.
51. **Dimitrov, S. and Vuleva, R.,** Studying the Sevin (carbaryl) retention time in alfalfa, *Vet. Med. Nauki,* 8, 51, 1971.
52. **Turner, B. C. and Caro, J. H.,** Uptake and distribution of carbofuran and its metabolites in field-grown corn plants, *J. Environ. Qual.,* 2, 245, 1973.
53. **Fahey, J. E., Wilson, M. C., and Armbrust, E. J.,** Residues of Supracide and carbofuran in green and dehydrated alfalfa, *J. Econ. Entomol.,* 63, 589, 1970.
54. **Finlayson, D. G., Brown, M. J., Campbell, C. J., and Williams, I. H.,** Rates, methods and persistence of insecticides used for preventing carrot maggot damage, *J. Entomol. Soc. B.C.,* 69, 14, 1972.
55. **Van Middelem, C. H. and Peplow, A. J.,** Evaluation of extraction procedures for the removal of ^{14}C-carbofuran and its toxic metabolites from cabbage leaves, *J. Agric. Food Chem.,* 21, 100, 1973.
56. **Bowling, C. C.,** Lateral movement, uptake, and retention of carbofuran applied to flooded rice plants, *J. Econ. Entomol.,* 63, 239, 1970.

57. **Pree, D. J. and Saunders, J. L.,** Bioactivity and translocation of carbofuran residues in mugho pine, *Environ. Entomol.,* 2, 262, 1973.
58. **Pree, D. J. and Saunders, J. L.,** Metabolism of carbofuran in mugho pine, *J. Agric. Food Chem.,* 22, 620, 1974.
59. **Skrentny, R. F. and Ellis, J. A.,** Control of *Aphis fabae* on broad beans *(Vicia faba)* by the systemic action of *gamma*-BHC, thionazin and aldicarb, *Pestic. Sci.,* 1, 45, 1970.
60. **Andrawes, N. R., Romine, R. R., and Bagley, W. P.,** Metabolism and residues of Temik aldicarb pesticide in cotton foliage and seed under field conditions, *J. Agric. Food Chem.,* 21, 379, 1973.
61. **Woodham, D. W., Edwards, R. R., Reeves, R. G., and Schutzmann, R. L.,** Total toxic aldicarb residues in soil, cottonseed, and cotton lint following a soil treatment with the insecticide on the Texas high plains, *J. Agric. Food Chem.,* 21, 303, 1973.
62. **Maitlen, J. C., McDonough, L. M., Dean, F., Butt, B. A., and Landis, B. J.,** Aldicarb Residues in Apples, Pears, Sugarbeets and Cottonseed: Performance in Apples and Pears, ARS 33-135, U.S. Department of Agriculture, Washington, D.C., 1970.
63. **Coppedge, J. R., Lindquist, D. A., Bull, D. L., and Dorough, H. W.,** Fate of 2-methyl-2-(methylthio)propionaldehyde O-(methylcarbamoyl)oxime (Temik) in cotton plants and soil, *J. Agric. Food Chem.,* 15, 902, 1967. 15, 902, 1967.
64. **Bull, D. L.,** Metabolism of UC-21149 [2-methyl-2-(methylthio)propionaldehyde O-(methylcarbamoyl)oxime] in cotton plants and soil in the field, *J. Econ. Entomol.,* 61, 1598, 1968.
65. **Tashiro, H., Chambers, D. L., Shaw, J. G., Beavers, J. B., and Maitlen, J. C.,** Systemic activity of UC-21149 against the citrus red mite, citrus thrips, California red scale and spirea aphid on nonbearing orange trees, *J. Econ. Entomol.,* 62, 443, 1969.
66. **Hendrickson, R. and Meagher, W. R.,** Residues of 2-methyl-2-(methylthio)propionaldehyde O-(methylcarbamoyl)-oxime (Temik) in citrus following soil applications, *J. Agric. Food Chem.,* 16, 903, 1968.
67. **Halfhill, J. E. and Maitlen, J. C.,** Residues of aldicarb in fresh and dried alfalfa, *J. Econ. Entomol.,* 66, 557, 1973.
68. **Tamaki, G., Halfhill, J. E., and Maitlen, J. C.,** The influence of UC-21149 on the aphidiid parasite, *Aphidius smithi* on populations of the pea aphid, *J. Econ. Entomol.,* 62, 678, 1969.
69. **Andrawes, N. R., Bagley, W. P., and Herrett, R. A.,** Metabolism of 2-methyl-2-(methylthio)propionaldehyde O-(methylcarbamoyl)oxime (Temik aldicarb pesticide) in potato plants, *J. Agric. Food Chem.,* 19, 731, 1971.
70. **Krueger, H. R., Lindquist, R. K., and Bauerle, W. L.,** Phorate and aldicarb in greenhouse lettuce: effect of CO_2 on residues, *J. Econ. Entomol.,* 66, 573, 1973.
71. **Lindquist, R. K., Krueger, H. R., Spadafora, R. R., and Mason, J. F.,** Application of aldicarb to greenhouse tomatoes: plant growth, fruit yields, greenhouse whitefly control, and residues in fruits, *J. Econ. Entomol.,* 65, 862, 1972.
72. **Krueger, H. R. and Mason, J. F.,** Phorate and aldicarb: effects of systemic fungicides on their levels in soybeans, *J. Econ. Entomol.,* 66, 815, 1973.
73. **Krueger, H. R., Lindquist, R. K., Mason, J. F., and Spadafora, R. R.,** Application of methomyl to greenhouse tomatoes: greenhouse whitefly control and residues in foliage and fruits, *J. Econ. Entomol.,* 66, 1223, 1973.
74. **Kawahara, T. and Maeda, H.,** Residue analysis of mesomil in crops, *Noyaku Kensasho Hokoku,* 11, 89, 1971.
75. **Harvey, J., Jr. and Reiser, R. W.,** Metabolism of methomyl in tobacco, corn and cabbage, *J. Agric. Food Chem.,* 21, 775, 1973.
76. **Nigam, P. C.,** Contact and Residual Toxicity Studies of Zectran against Eighteen Species of Forest Insect Pests, Information Report CC-X-29, Chemical Control Research Institute, Ottawa, Ontario, December 1972.
77. **Nigam, P. C.,** Summary of Contact, Stomach and Residual Toxicity of Insecticides against Forest Insect Pests during 1973, Information Report CC-X-46, Chemical Control Research Institute, Ottawa, Ontario, October 1973.
78. **Webb, J. A. and Nigam, P. C.,** Penetration and Translocation of Baygon Formulations in Balsam Fir, Information Report CC-X-45, Chemical Control Research Institute, Ottawa, Ontario, September 1973.
79. **Knowles, C. O. and Sen Gupta, A. K.,** Metabolism of formetanate acaricide in orange seedlings, *J. Econ. Entomol.,* 63, 615, 1970.
80. **Karinen, J. F., Lamberton, J. G., Stewart, N. E., and Terriere, L. C.,** Persistence of carbaryl in the marine estuarine environment. Chemical and biological stability in aquarium systems, *J. Agric. Food Chem.,* 15, 148, 1967.
81. **Spynu, E. I., Molozanova, L. G., and Stefanskii, K. S.,** Hygienic criteria for assessing the extent of soil contamination with pesticides, *Gig. Sanit.,* 35, 79, 1970.
82. **Kuhr, R. J., Davis, A. C., and Bourke, J. B.,** Dissipation of Guthion, Sevin, Polyram, Phygon and Systox from apple orchard soil, *Bull. Environ. Contam. Toxicol.,* 11, 224, 1974.
83. **Ahlrichs, J. L., Chandler, L., Monke, E. J., and Reuszer, H. W.,** Effect of Pesticide Residues and other Organo-toxicants on the Quality of Surface and Ground Water Resources, PB Report No. 211080, U.S. National Technical Information Service, Springfield, Va., 1970.
84. **Caro, J. H., Freeman, H. P., and Turner, B. C.,** Persistence in soil and losses in runoff of soil-incorporated carbaryl in a small watershed, *J. Agric. Food Chem.,* 22, 860, 1974.
85. **Kazano, H., Kearney, P. C., and Kaufman, D. D.,** Metabolism of methylcarbamate insecticides in soils, *J. Agric. Food Chem.,* 20, 975, 1972.
86. **Rodriguez, L. D. and Dorough, H. W.,** Metabolism of carbaryl in soils, *Arch. Environ. Contam. Toxicol.,* in press, 1975.

87. **Foschi, S., Cesari, A., Ponti, I., Bentivogli, P. G., and Bencivelli, A.,** Study of the degradation and vertical movement of pesticides in soil, *Not. Mal. Piante,* 82, 37, 1970.
88. **Edwards, C. A.,** *Persistent Pesticides in the Environment,* CRC Press, Cleveland, 1970.
89. **Brooks, G. T.,** *Chlorinated Insecticides,* CRC Press, Cleveland, 1974.
90. **Leenheer, J. A. and Ahlrichs, J. L.,** A kinetic and equilibrium study of the adsorption of carbaryl and parathion upon soil organic matter surfaces, *Soil Sci. Soc. Am. Proc.,* 35, 700, 1971.
91. **Aly, O. M. and El-Dib, M. A.,** Studies of the persistence of some carbamate insecticides in the aquatic environment, in *Fate of Organic Pesticides in the Aquatic Environment,* (Advances in Chemistry Series III), American Chemical Society, Washington, D.C., 1972, 210.
92. **Stojanovic, B. J., Kennedy, M. V., and Shuman, F. L., Jr.,** Edaphic aspects of the disposal of unused pesticides, pesticide wastes, and pesticide containers, *J. Environ. Qual.,* 1, 54, 1972.
93. **Read, D. C.,** Persistence of some newer insecticides in mineral soils measured by bioassay, *J. Econ. Entomol.,* 62, 1338, 1969.
94. **Thompson, A. R.,** Persistence of biological activity of seven insecticides in soil assayed with *Folsomia candida, J. Econ. Entomol.,* 66, 855, 1973.
95. **Harris, C. R.,** Laboratory studies on the persistence of biological activity of some insecticides in soils, *J. Econ. Entomol.,* 62, 1437, 1969.
96. **Onsager, J. A. and Rusk, H. W.,** Potency of the residues of some nonpersistent insecticides in soil against wireworms, *J. Econ. Entomol.,* 62, 1060, 1969.
97. **Bhirud, K. M. and Pitre, H. N.,** Influence of soil class and soil moisture on bioactivity of carbofuran and disulfoton in corn in greenhouse tests. Relationship to leafhopper vector control and corn stunt disease incidence, *J. Econ. Entomol.,* 65, 324, 1972.
98. **Abdellatif, M. A., Hermanson, H. P., and Reynolds, H. T.,** Effect of soil clay and organic matter content upon systemic efficacy of two carbamate insecticides, *J. Econ. Entomol.,* 60, 1445, 1967.
99. **Caro, J. H., Freeman, H. P., Glotfelty, D. E., Turner, B. C., and Edwards, W. M.,** Dissipation of soil-incorporated carbofuran in the field, *J. Agric. Food Chem.,* 21, 1010, 1973.
100. **Getzin, L. W.,** Persistence and degradation of carbofuran in soil, *Environ. Entomol.,* 2, 461, 1973.
101. **Bull, D. L., Stokes, R. A., Coppedge, J. R., and Ridgway, R. L.,** Further studies of the fate of aldicarb in soil, *J. Econ. Entomol.,* 63, 1283, 1970.
102. **Quaraishi, M. S.,** Edaphic and water relationships of aldicarb and its metabolites, *Can. Entomol.,* 104, 1191, 1972.
103. **Stokes, R. A., Coppedge, J. R., and Ridgway, R. L.,** Chemical and biological evaluation of the release of aldicarb from granular formulations, *J. Agric. Food Chem.,* 18, 195, 1970.
104. **Andrawes, N. R., Bagley, W. P., and Herrett, R. A.,** Fate and carryover properties of Temik aldicarb pesticide [2-methyl-2-(methylthio)propionaldehyde O-(methylcarbamoyl)oxime] in soil, *J. Agric. Food Chem.,* 19, 727, 1971.
105. **Kearby, W. H., Ercegovich, C. D., and Bliss, M., Jr.,** Residue studies on aldicarb in soil and Scotch pine, *J. Econ. Entomol.,* 63, 1317, 1970.
106. **Tucker, B. V. and Pack, D. E.,** Bux insecticide soil metabolism, *J. Agric. Food Chem.,* 20, 412, 1972.
107. **Harris, C. R.,** Laboratory evaluation of candidate materials as potential soil insecticides, IV., *J. Econ. Entomol.,* 66, 216, 1973.
108. **Gupta, D. S. and Rawlins, W. A.,** Persistence of two systemic carbamate insecticides in three types of soils, *Indian J. Entomol.,* 28, 482, 1966.
109. **Starr, R. I. and Cunningham, D. J.,** Degradation of ^{14}C-Labeled Mesurol in Soil and Water, paper presented at the 167th American Chemical Society Meeting, Los Angeles, April 1974.
110. **Church, D. D. and Flint, D. R.,** The Fate of Mesurol [4-(Methylthio)-3,5-xylylmethylcarbamate] in Soil, Chemagro Report No. 29591, 1971.
111. **Knowles, C. O.,** Metabolism of two acaricidal chemicals, N′-(4-chloro-*o*-tolyl)-N,N-dimethylformamidine (chlorphenamidine) and *m*-{[(dimethylamino)methylene]amino}phenyl methylcarbamate hydrochloride (formetanate), *J. Agric. Food Chem.,* 18, 1038, 1970.
112. **Arurkar, S. K. and Knowles, C. O.,** Decomposition of formetanate acaracide in soil, *Bull. Environ. Contam. Toxicol.,* 5, 324, 1970.
113. **Whitten, C. J. and Bull, D. L.,** Fate of 3,3-dimethyl-1-(methylthio)-2-butanone O-(methylcarbamoyl)oxime (Diamond Shamrock DS-15647) in cotton plants and soil, *J. Agric. Food Chem.,* 22, 234, 1974.
114. **Harvey, J., Jr. and Pease, H. L.,** Decomposition of methomyl in soil, *J. Agric. Food Chem.,* 21, 784, 1973.
115. **Aly, O. M. and El-Dib, M. A.,** Studies of the persistence of some carbamate insecticides in the aquatic environment. 1. Hydrolysis of Sevin, Baygon, Pyrolan and dimetilan in waters, *Water Res.,* 5, 1191, 1971.
116. **Stadnyk, L., Campbell, R. S., and Johnson, B. T.,** Pesticide effect on growth and ^{14}C assimilation in a fresh-water alga, *Bull. Environ. Contam. Toxicol.,* 6, 1, 1971.
117. **Eichelberger, J. W. and Lichtenberg, J. J.,** Persistence of pesticides in river water, *Environ. Sci. Technol.,* 5, 541, 1971.
118. **Hughes, L. B., Jr.,** Fate of carbaryl insecticide in surface waters, *Diss. Abstr. Int. B,* 32, 3108, 1971.
119. **Schaefer, C. H. and Dupras, E. F., Jr.,** Factors affecting the stability of the carbamate insecticide RE 11775, *Mosq. News,* 32, 201, 1972.

120. **Lamberton, J. G. and Claeys, R. R.**, Degradation of 1-naphthol in sea water, *J. Agric. Food Chem.*, 18, 92, 1970.
121. **Paris, D. F. and Lewis, D. L.**, Chemical and microbial degradation of ten selected pesticides in aquatic systems, *Residue Rev.*, 45, 95, 1973.
122. **Yu, C. C., Booth, G. M., Hansen, D. J., and Larsen, J. R.**, Fate of carbofuran in a model ecosystem, *J. Agric. Food Chem.*, 22, 431, 1974.
123. **Kanazawa, J., Isensee, A. R., and Kearney, P. C.**, Distribution of Two Methylcarbamate Insecticides in an Aquatic Microecosystem, Pesticide Paper No. 66, presented at the 167th American Chemical Society Meeting, Los Angeles, April 1974.
124. **Miller, C. W. and Shafik, M. T.**, Occurrence of Propoxur in Air Following Intradomiciliary Spraying, Pesticide Paper No. 26, presented at the 163rd American Chemical Society Meeting, Boston, April 1972.
125. **Jegier, Z.**, Occurrence and distribution of pesticide residues. Pesticide residues in the atmosphere, *Ann. N.Y. Acad. Sci.*, 160, 143, 1969.
126. **Abdel-Wahab, A. M., Kuhr, R. J., and Casida, J. E.**, Fate of ^{14}C-carbonyl-labeled aryl methylcarbamate insecticide chemicals in and on bean plants, *J. Agric. Food Chem.*, 14, 290, 1966.
127. **Dorough, H. W. and Crouch, G. W.**, Persistence and residual effectiveness of various formulations of Baygon (Bayer 39007) against the housefly, *J. Econ. Entomol.*, 59, 1188, 1966.
128. **Seiber, J. N. and Woodrow, J. E.**, Determination of Pesticides and Their Transformation Products in Air, Pesticide Paper No. 3, presented at the 167th American Chemical Society Meeting, Los Angeles, April 1974.
129. **Koller, L. R.**, *Ultraviolet Radiation*, 2nd ed., Wiley & Sons, New York, 1965.
130. **Crosby, D. G.**,Experimental approaches to pesticide photodecomposition, *Residue Rev.*, 25, 1, 1969.
131. **Mitchell, L. C.**, The effect of ultraviolet light (2,537 Å) on 141 pesticide chemicals by paper chromatography, *J. Assoc. Off. Agric. Chem.*, 44, 643, 1961.
132. **Crosby, D. G., Leitis, E., and Winterlin, W. L.**, Photodecomposition of carbamate insecticides, *J. Agric. Food Chem.*, 13, 204, 1965.
133. **Okada, K., Nomura, K., and Yamamoto, S.**, Chemistry of stability of pesticides. I. The deterioration of formulated 1-naphthyl-N-methylcarbamate by ultraviolet irradiation, *Nippon Nogei Kagaku Kaishi*, 35, 739, 1961.
134. **Abdel-Wahab, A. M. and Casida, J. E.** Photooxidation of two 4-dimethylaminoaryl methylcarbamate insecticides (Zectran and Matacil) on bean foliage and of alkylaminophenyl methylcarbamates on silica gel chromatoplates, *J. Agric. Food Chem.*, 15, 479, 1967.
135. **Kenaga, E. E., Doty, A. E., and Hardy, J. L.**, Laboratory insecticidal tests with 4-dimethylamino-3,5-xylyl methylcarbamate, *J. Econ. Entomol.*, 55, 466, 1962.
136. **Roberts, R. B., Miskus, R. P., Duckles, C. K., and Sakai, T. T.**, In-vivo fate of the insecticide Zectran in spruce budworm, tobacco budworm, and housefly larvae, *J. Agric. Food Chem.*, 17, 107, 1969.
137. **Fedorova, Y. N. and Karchik, O. N.**, Effect of external factors on the decomposition of Sevin, *Khim. Sel'sk. Khoz.*, 8, 762, 1970.
138. **Wauchope, R. D. and Haque, R.**, Nonbiological Degradation of Carbaryl in Aqueous Solution, Pesticide Paper No. 33, presented at the 163rd American Chemical Society Meeting, Boston, April 1972.
139. **Eberle, D. O. and Gunther, F. A.**, Chromatographic, spectrometric and irradiation behavior of five carbamate insecticides, *J. Assoc. Off. Agric. Chem.*, 48, 927, 1965.
140. **Su, G. C. C. and Zabik, M. J.**, Photochemistry of bioactive compounds. Photolysis of *m*-(N,N-dimethylformamidine)-phenyl N-methylcarbamate hydrochloride in water, *J. Agric. Food Chem.*, 20, 642, 1972.
141. **Pape, B. E., Para, M. F., and Zabik, M. J.**, Photochemistry of bioactive compounds: photodecomposition of 2-(1,3-dioxolane-2-yl)phenyl-N-methyl carbamate, *J. Agric. Food Chem.*, 18, 490, 1970.
142. **Trecker, D. J., Foote, R. S., and Osborn, C. L.**, Photochemistry of aryl carbamates, *Chem. Commun.*, 1034, 1968.
143. **Cheng, H.-M. and Casida, J. E.**, Metabolites and photoproducts of 3-(2-butyl)phenyl N-methylcarbamate and N-benzenesulfenyl-N-methyl-carbamate, *J. Agric. Food Chem.*, 21, 1037, 1973.
144. **Slade, M. and Casida, J. E.**, Metabolic fate of 3,4,5- and 2,3,5-trimethylphenyl methylcarbamates, the major constituents in Landrin insecticide, *J. Agric. Food Chem.*, 18, 467, 1970.
145. **Metcalf, R. L., Fukuto, T. R., Collins, C., Borck, K., El-Aziz, S. A., Muñoz, R., and Cassil, C. C.**, Metabolism of 2,2-dimethyl-2,3-dihydrobenzofuranyl-7-N-methylcarbamate (Furadan) in plants, insects, and mammals, *J. Agric. Food Chem.*, 16, 300, 1968.
146. Pesticide Residues in Food, World Health Organization Technical Report Series No. 458, FAO Agricultural Series No. 84, Rome, December 1969.
147. **Gontovaya, N. A. and Shtenberg, A. I.**, Residual amounts of some pesticides occurring in daily food rations of residents in two rural microdistrics of the Moldavian SSR, *Vopr. Pitan.*, 30, 28, 1971.
148. **Duggan, R. E. and Lipscomb, G. Q.**, Dietary intake of pesticide chemicals in the United States (II), *Pestic. Monit. J.*, 2, 153, 1969.
149. **Duggan, R. E. and Corneliussen, P. E.**, Dietary intake of pesticide chemicals in the United States (III), June 1968 to April 1970, *Pestic. Monit. J.*, 5, 331, 1972.
150. **Duggan, R. E. and Weatherwax, J. R.**, Dietary intake of pesticide chemicals, *Science*, 157, 1006, 1967.
151. **Duggan, R. E., Lipscomb, G. Q., Cox, E. L., Heatwole, R. E., and Kling, R. C.**, Pesticide residue levels in foods in the United States from July 1, 1963 to June 30, 1969, *Pestic. Monit. J.*, 5, 73, 1971.

152. **Duggan, R. E.,** Pesticide residues in foods, *Ann. N.Y. Acad. Sci.,* 160, 174, 1969.
153. **Viel, G., Hascoet, M., and Dubroca, G.,** The elimination of pesticide residues on apricots by washing before canning, *Phytiatr. Phytopharm.,* 15, 41, 1966.
154. **Farrow, R. P., Elkins, E. R., Rose, W. W., Lamb, F. C., Ralls, J. W., and Mercer, W. A.,** Canning operations that reduce insecticide levels in prepared foods and in solid food wastes, *Residue Rev.,* 29, 73, 1969.
155. **Elkins, E. R., Lamb, F. C., Farrow, R. P., Cook, R. W., Kawai, M., and Kimball, J. R.,** Removal of DDT, malathion and carbaryl from green beans by commercial and home preparative procedures, *J. Agric. Food Chem.,* 16, 962, 1968.
156. **Farrow, R. P., Lamb, F. C., Cook, R. W., Kimball, J. R., and Elkins, E. R.,** Removal of DDT, malathion and carbaryl from tomatoes by commercial and home preparative methods, *J. Agric. Food Chem.,* 16, 65, 1968.
157. **Farrow, R. P., Lamb, F. C., Elkins, E. R., Cook, R. W., Kawai, M., and Cortes, A.,** Effect of commercial and home preparative procedures on parathion and carbaryl residues in broccoli, *J. Agric. Food Chem.,* 17, 75, 1969.
158. **Lamb, F. C., Farrow, R. P., Elkins, E. R., Kimball, J. R., and Cook, R. W.,** Removal of DDT, parathion, and carbaryl from spinach by commercial and home preparative methods, *J. Agric. Food Chem.,* 16, 967, 1968.
159. **Elkins, R. E., Farrow, R. P., and Kim, E. S.,** The effect of heat processing and storage on pesticide residues in spinach and apricots, *J. Agric. Food Chem.,* 20, 286, 1972.
160. **Street, J. C.,** Methods of removal of pesticide residues, *Can. Med. Assoc. J.,* 100, 154, 1969.
161. **Winterlin, W., Walker, G., and Luce, A.,** Carbaryl residues in bees, honey, and bee bread following exposure to carbaryl via the food supply, *Arch. Environ. Contam. Toxicol.,* 1, 362, 1973.

Chapter 9

RESIDUE METHODOLOGY

INTRODUCTION

Residue chemists involved in the analysis of methylcarbamate insecticides must possess a very high degree of skill. Although there is generally no lack of technically adequate detection systems, many of the carbamates cannot be monitored by the same techniques that are employed for other pesticides. For example, the analysis of most pesticides (e.g., chlorinated hydrocarbons) by gas chromatography (GLC) is a relatively simple matter once suitable extraction and cleanup methods have been perfected. Operating parameters effecting different retention times for the toxicant and for various interfering materials may be developed, thus allowing some extraneous peaks to be tolerated, provided that they do not coincide with the pesticide(s) under consideration. However, even if extraction efficiency and cleanup are found to be excellent, carbamates often degrade during gas chromatography. This may be partially prevented by special column preparations and by selecting just the right GLC parameters, but even then the results are not always satisfactory.

One approach used to circumvent thermal degradation of the intact carbamate is its derivatization to a "stable" form. In the case of phenyl N-methylcarbamate insecticides, derivatization often involves hydrolysis of the carbamate ester and reaction of the released phenol with a reagent to yield a product which can be detected by GLC analysis. This technique is not without certain inherent problems, however. For example, one can never be sure whether the phenol originated from chemical or metabolic degradation of the insecticide or if it occurred naturally in the substrate being analyzed. Since phenols are usually less toxic to animals than the parent insecticides, it is essential that they be separated from the intact carbamates prior to derivatization. Otherwise, the final results cannot be properly interpreted.

Another problem results from the fact that hydroxylation is a common metabolic pathway for carbamate insecticides, and one type of product is a ring-hydroxylated analog of the parent compound (see Chapter 7). Since this metabolite is a phenol, it is susceptible to derivatization with the test reagent and may indeed interfere with analysis of the parent phenol. In a similar manner, the very reactive derivatizing reagents may complex with one or more of the many phenols which occur ubiquitiously in nature to form interfering materials not otherwise encountered in residue analysis. The presence of such products often contributes significantly to the already laborious task of preparing samples for analysis by gas chromatography or by other methods of detection. The same considerations apply to methods based on methylamine derivatization, where carbamate-type metabolites constitute a portion of the residues.

Although there are many other problems associated- with the methodological aspects of carbamate residue analysis, it is usually difficult to anticipate these problems based on the literature available. Most authors claim good sensitivity, high reproducibility of results, and a rather straightforward procedural format. However, the reported methods do not always adapt easily from laboratory to laboratory, so standard residue analysis procedures are difficult to obtain. Several review articles[1-4] have dealt with the technical problems associated with carbamate insecticide analysis, helping to explain why simple, repeatable methods are difficult to establish. The following discussion will elaborate on some of these problems and suggest possible remedial procedures.

COLORIMETRIC METHODS

Colorimetric procedures have long been used for analysis of methylcarbamate insecticide residues because of the relative simplicity of these methods and, for the most part, their reliability and sensitivity, which ranges from submicrogram to microgram quantities. However, certain difficulties are frequently encountered, including particularly the stability of the color complex and interference from coextractives.[5] In addition, most complex formation depends on initial hydrolysis of the methylcarbamate residue into parent phenol and methylamine, and it is one of the latter products that actually combines with the colorimetric reagent. *p*-Nitrobenzenediazonium fluoborate, one of the most commonly used coupling agents, has been employed for detection of a variety of carbamates on several different substrates.[6-9] In the case of carbaryl analyses on crops, for example, the lower limit of detection approaches the 0.01-ppm level. The reagent is also effective for determination of carbaryl in formulation, where it was found that phosphoric acid in the coupling solution stablized the color of the azo dye.[10] With this reagent, it is the phenolic portion of the molecule that forms the colored species.

Another coupler, diazotized 3-nitroaniline-4-sulfonic acid, has been shown to be suitable for the colorimetric determination of eight N-methylcarbamates.[11] Maximum color formation was obtained in 20 min, and the color was stable for at least 24 hr. An excess of methanol (10 to 15%) in the test solution severely limited the range of Beer-Lambert plots. At lower concentrations, however, linearity was observed for insecticide levels up to 8 ng.

Total aldicarb residues in soil and water were determined by using a method specific for detection of the carbamoyloxime group.[12-15] Residues were hydrolyzed with base to yield aldicarb oxime, which in turn was subjected to acid hydrolysis to release hydroxylamine. The latter was oxidized with iodine to nitrous acid and quantitated by using a standard nitrite-diazo dye procedure. The color stability of the diazo compound was limited, but removing the excess iodine through extraction with bromobenzene lengthened its life considerably. Thus, less than 1% decay of color intensity was observed 10 min after extraction of the iodine.[16]

Diazotized 2,5-dichloroaniline sulfate forms a ruby-red complex directly with carbaryl, whereas 1-naphthol yields a distinctly different yellow-red color with much less sensitivity.[17] Several other carbamate insecticides and herbicides did not produce a colored product under the prescribed conditions. When the technique was applied to carbaryl formulations and to carbaryl-fortified crops, recoveries averaged 95% at the 4-ppm to 32-ppm level.[18] Sensitivity was about 0.1 μg of carbaryl, and little interference from crop extracts or from 1-naphthol was noted. Similar color formation and sensitivity for carbaryl formulations were observed when using diazotized *o*-toluidine.[19]

ENZYMIC METHODS

Enzyme inhibition techniques have been used for the detection of organophosphorus and carbamate insecticides since their introduction as toxicants. These insecticides are active, at least in part, because of their ability to inhibit the enzyme acetylcholinesterase (Chapter 3). Therefore, it is not surprising that attempts have been made to utilize this action for analytical purposes. The fact is, however, that, although an extremely useful screening tool, enzyme inhibition techniques offer little in the way of a specific routine method for residue analysis.

Enzyme assays of any type are highly variable. Even when ultrapure compounds and enzymes are used, it is not unusual that I_{50} values vary two- to tenfold when determined in the same laboratory, and even more when the same assays are performed in different laboratories. This does not suggest that poor technique is at fault, but simply that the experimental conditions (temperature, enzyme and substrate source and concentration, reaction time, etc.) are so exacting that high reproducibility would hardly be expected. Add to these basic parameters such extraneous variables as type of supports (paper, thin-layer adsorbents), mixture of pesticides, solvent residues, naturally occurring inhibitors, etc., and it becomes apparent that it is difficult to devise a standardized enzyme inhibition method. Although instrumental methods of analysis also present their share of problems, they are far more likely to be resolved than those associated with enzymic techniques.

As might be expected, the first attempts to quantitate carbaryl by enzyme inhibition were made by using a cholinesterase (ChE) procedure originally developed for organophosphorus insecticides.[20,21] The sensitivity of the method was approximately 0.05 μg of pesticide. Many enzymic methods for carbaryl and other carbamates have since been reported, and the topic has been reviewed a number of times.[1,4,22-25]

The enzyme procedure preferred in the late 1950s for carbamate analysis utilized the Δ pH technique for estimating the level of residues.[26] Acetylcholine was incubated with cholinesterase in buffered solution with and without the insecticide residue. In the presence of a pesticide or inhibitor, there was a decrease in acetic acid production due to inhibition of the enzymic cleavage of acetylcholine. This reduction in enzyme activity (Δ pH) was correlated to the inhibitor concentration, and a calibration curve was constructed.[21] Although housefly head ChE was 100-fold more sensitive to carbaryl than ChE from horse plasma, bovine erythrocytes, or human plasma, the mammalian sources were used routinely because of their ready availability.

In addition to measuring change in pH as a means of quantitating residues of insecticides with antiesterase activity, a Δ fluorescence method has been developed.[27,28] In this case, a substrate is selected (i.e., N-methylindoxyl acetate) which is hydrolyzed by ChE to a highly fluorescent material (i.e., N-methylindoxyl). Quantitation is assessed fluorometrically, and the decrease in fluorescence is related to inhibitor concentration.

A survey of the more recent reports on enzymic residue methods for detecting methylcarbamates indicates that thin-layer-chromatographic techniques are receiving much more attention than other methods. This approach allows rapid separation of pesticide mixtures and offers sensitivity often in the parts-per-billion range.[25] Paper chromatography may be used in a similar manner,[29] but generally it does not provide the resolution and rapidity obtained with TLC.

Chromatoplates used for separation and detection of anticholinesterase insecticides in the early 1960s had to undergo special, time-consuming preparation. Consequently, TLC was not extensively utilized until the method was perfected for use with standard TLC plates and procedures around 1968. In the initial study,[30] the pesticides were applied to aluminum oxide or silica gel chromatoplates. After development in an appropriate solvent mixture, the plates were sprayed with a solution of enzyme (bee brain ChE) and buffer, incubated for 20 min at room temperature, and then sprayed again with indophenyl acetate solution. After 10 min, inhibitors appeared as white spots on an intense blue background. Quantitation was achieved by copying the chromatogram, cutting out the spot, weighing, and plotting on semilogarithmic paper. Limits of detectability were at the subnanogram level, and the total assay took about 40 min.

Many modifications of this procedure have appeared in the literature since 1968. Variations include TLC supports, enzyme source, substrates, and quantitative considerations. Bee brain ChE is still used often, and sensitivity in the picogram range is obtained with substrates such as β-naphthyl acetate.[31-34] In addition, nonspecific mammalian tissue esterases, particularly those of the liver, are preferred by a number of investigators.[35-39] Pig liver extracts in combination with the substrates indophenyl acetate and 5-bromoindoxyl acetate have proved to be very effective for the analysis of a variety of methylcarbamate insecticides, with sensitivity at the nanogram to picogram levels.[40-42] Horse serum is also used as a ChE source, but does not appear to be as sensitive as the liver preparation.[43]

Enzymes other than ChE are less promising for use in TLC enzyme residue techniques. Carbaryl showed only slight activity against β-amylase and none against α-amylase.[44] Alkaline and acid phosphatases were demonstrated in one study to be unsuitable for carbaryl detection,[45] whereas another study[46] reported that acid phosphatase was sensitive to 1 μg and alkaline phosphatase to 30 μg of the carbamate.

Occasionally, enzymic procedures that do not involve TLC or paper chromatography have been used for the analysis of carbamate insecticides. One such method calls for the pesticide residue to be placed on agar gel containing ChE.[47] Zones of inhibition are visualized using acetylcholine and bromothymol blue, and the limit of detection with carbaryl was 0.25 μg.

Perhaps the most ingenious enzymic method for tracing carbamate and organophosphorus insecticides is one which was designed to monitor air and water for the presence of ChE inhibitors. The system, designated as the Cholinesterase Antagonist Monitor, CAM-1, is automated and equipped with both visible and audible alarms.[48,49] The heart of CAM-1 is an electrochemical cell consisting of two platinum electrodes separated by an enzyme pad holder. The latter is prepared from horse serum ChE adsorbed onto aluminum hydroxide gel and applied to urethane foam.

When in operation, a constant current is applied to the cell, and water or air together with a solution of butyrylthiocholine iodide substrate is pumped over the immobilized enzyme. If no inhibitor is present, the enzyme hydrolyzes the substrate, releasing thiocholine iodide. The free thiol group causes a continuous low electrochemical voltage to flow between the electrodes. In this state, no alarm is given. However, if the enzyme is inhibited, less free thiol is formed, and the voltage increases. The increase in voltage is transformed into a visual and/or audible alarm system. Although no claims are made that the CAM-1 in its present form is suitable for residue analysis, the sensitivity is reportedly adequate to prevent accidental poisoning by ChE inhibitors in water supplies. No minimum detectable limits were given, but carbaryl in water at 50 ppm caused a response, and as little as 0.2 ppm of DDVP, an organophosphate, was also detected. Changing the enzyme source from horse serum to bee brain would probably increase the sensitivity of the system to carbaryl and probably to other methylcarbamate insecticides.

The future of enzyme inhibition techniques for carbamate residue analysis may be predicted on the basis of the history of their use with organophosphorus insecticides. In case of the latter, enzymic methods were widely employed until gas-chromatographic procedures were developed for the phosphorus-containing materials. Then, little use was made of analysis via enzyme inhibition. This probably will also happen with carbamates, although the final preferred method may be something other than gas chromatography. However, this point has yet to be achieved, and enzymes will continue to provide an interim method for estimating residual levels of methylcarbamate insecticides.

FLUORESCENCE

Spectrophotofluorometry has been especially suitable for analysis of carbaryl residues because of the strongly fluorescent nature of the naphtholate anion. The method has been applied to residues in honeybees, tomato and bean leaves, fish, and fruits and to the carbaryl content of mixed formulations.[50-54] Detection of methylcarbamate residues on thin-layer chromatograms by spectrofluorometric means has involved a variety of methods. Carbaryl, for example, was located by *in situ* fluorometry in the low nanogram range after spraying the plate with dilute base.[55] Mesurol and several other sulfur-containing pesticides were resolved at the 100-300-ng level by exposing the TLC plate to bromine and then spraying with solutions of metal ions and chelating agents or pH-sensitive fluorescent indicators.[56,57]

Several 5-unsubstituted-3-hydroxy flavones have been employed as fluorogenic spray reagents for location of polar carbamate materials separated by cellulose TLC.[58] The complexed flavones fluoresced brightly in media of high dielectric constants or mild basicity, but exhibited negligible fluorescence in nonpolar environments. With reversed-phase chromatography on a low-polarity prepurified cellulose support, polar carbamate insecticides and other pesticides containing polar functions appeared as strongly fluorescent spots on a weakly fluorescent or nonfluorescent background. The procedure, initially applied to Baygon, has been extended to other carbamates.[59] Detection limits are between 0.01 and 0.06 μg, which represents a five- to tenfold increase in sensitivity over colorimetric visualization techniques. Organophosphates and *s*-triazine herbicides were also detected in the same range, whereas chlorinated hydrocarbons were observed at the 0.1-μg level. Although the method appears suitable for rapid screening programs and has potential for quantitative determination, its application to actual methylcarbamate insecticide residue analysis has not been reported.

Other fluorogenic labeling techniques used for the detection of pesticide residues have recently been summarized.[60] Many procedures require hydrolysis of the methylcarbamate insecticides prior to derivatization of the hydrolytic products with fluorescent or potentially fluorescent compounds. Subsequent cleanup and separation of the adducts followed by *in situ* fluorometry allows detection and quantitation of residues at the nanogram level. For example, conditions for methylcarbamate hydrolysis and labeling of the resulting phenols and methylamine with dansyl chloride (1-dimethylaminonaphthalene-5-sulfonyl chloride) have been established, and TLC systems have been developed for the separation of the dansylated derivatives of nine carbamate insecticides.[61-63] Peak areas vs. concentration plots were linear for as much as 300 to 400 ng per spot, and reproducibility was satisfactory for concentrations above 10 ng per spot. Total carbamate recovery from water samples fortified with Landrin, Mesurol, or Baygon by measurement of the dansylmethylamine derivatives ranged from 86 to 104%.[64] Zectran and Matacil, upon hydrolysis and derivatization with dansyl chloride or 4-chloro-7-nitrobenzo-2,1,3-oxadiazole (NBD-Cl), yielded fluorescent indicators derived from both methylamine (from the methylcarbamate group) and dimethylamine (from the phenyl dimethylamino group).[65]

Aqueous alkaline hydrolysis of Landrin and

carbofuran with concomitant derivatization of the released methylamine with NBD-Cl in an upper organic layer was sensitive to 10 to 40 ppb of the aryl methylcarbamates both in water and in soil samples.[66] Interferences observed during *in situ* fluorometric measurements were reduced because of the selectivity of the reaction and because of the nonfluorescent nature of the NBD-Cl hydrolysis product.

GAS CHROMATOGRAPHY

Even though considerable problems are involved in gas-chromatographic analysis of methylcarbamate insecticide residues, it is obviously the method preferred by most residue chemists, provided that these problems can be circumvented. Numerous reports have appeared in the literature describing attempts to separate carbamates by GLC, either as the parent molecule and metabolites or as a derivatized form of the pesticide. Few of these methods are entirely adequate for routine analysis of the carbamates, but, by and large, each contributes something to the future success of gas chromatography as a tool for the detection of methylcarbamate insecticide residues.

The most desirable method for GLC analysis is direct injection and quantitation of the intact insecticide and toxic metabolites. This is relatively easy to do with organochlorine insecticides because they are usually eluted unchanged and are detected at extremely low quantities by using electron-capture (EC) techniques. Unfortunately, methylcarbamate insecticides do not behave in such a simple and direct manner. Most are thermally decomposed and/or retained on the solid support of the GLC column.[67-69] Moreover, with the exception of a few halogenated carbamates, they do not respond to electron capture.

Careful preparation of columns, using low-polarity methyl and phenyl silicone polymers, modest temperatures, and special column-conditioning procedures, has partially solved some of the chromatographic problems.[70-72] This, however, relates more to the parent insecticides than to their metabolites. Quantitation has been improved by using a nitrogen-specific detector[2] or a flame-photometric detector for sulfur-containing compounds.[73] Even so, derivatization remains an essential feature of methylcarbamate analaysis, and the salient points will be presented following a brief discussion of GLC analysis of nonderivatized carbamates and their metabolites.

Methods for Intact Carbamates

Some methylcarbamate insecticides can be analyzed with little difficulty using conventional GLC parameters. For example, Baygon, Zectran, and Matacil gave responses equivalent to approximately 90% of the injected compound when chromatographed on an OV-17 column operated at 180°C.[69] The method was effective in assaying tissue homogenates for Zectran and Mesurol,[74] although the latter compound and carbaryl were degraded by about 70%.[69] This variation in stability of the carbamates to GLC analysis is quite typical of the results obtained in numerous similar studies.

Carbaryl, as well as other methylcarbamates and certain metabolites, showed only slight decomposition on a DC-200 column at 180°C.[75] As little as 20 ng of some of the insecticides gave 50% full-scale deflection with a Coulson electrolytic-conductivity detector. It was pointed out that degradation was most apparent when the injector port became contaminated from repeated injection of crop extracts; however, the performance was restored by changing the glass wool plug and resilylation of the column. Even with these precautions, methomyl degraded to a much greater degree than the other compounds that were evaluated.

Nanogram amounts of carbaryl apparently chromatograph without detectable degradation on a 3% SE-30 column at 145°C.[76,77] However, the column had to be equilibrated with microgram quantities of carbaryl during conditioning and with 50- to 100-ng amounts immediately preceding analysis. Although slight decomposition to 1-naphthol may have occurred, the results agreed well with wet analytical methods. Carbofuran and 3-hydroxy carbofuran residues in fortified berries were similarly quantitated by direct GLC following acid hydrolysis of water-soluble conjugates, methylene chloride extraction, and Florisil column cleanup.[78] The N-specific electrolytic-conductivity detector showed that recoveries of both compounds averaged 86% over a range of 0.2 to 5.0 ppm.

Deactivation of column supports may prove useful for GLC determinations of thermally unstable compounds like the carbamates.[79] Acid-treated Chromosorb W coated with 6% Carbowax

20M in the usual manner, then heated to 260 to 280°C overnight under minimum nitrogen flow, and finally exhaustively extracted with methanol resulted in a support which, though highly deactivated, still retained chromatographic capabilities. Carbaryl, Mesurol, promecarb, and Zectran were recovered intact in greater than 90% yield from such a column when analyzed in the presence of crop extracts.[80] In addition, a mixture of the insecticides separated into detectable peaks on a 1.75-m column at 163°C.

Methomyl residues in water, soil, sediment, and tobacco have been determined on a 10% DC-200 column at 140°C using a flame-photometric detector.[81] Although the intact carbamate residue was injected, the structure of the eluted material was not ascertained. Recoveries ranged from 71 to 91% for samples fortified at the 0.01- to 0.05-ppm levels. This method effectively detects both the sulfoxide and sulfone metabolites of methomyl and has been used to analyze several field-treated crops for total methomyl residues.[82,83] The oxime carbamate also chromatographs well when injected onto a 6% OV-210/4% OV-101 mixed-phase column at 160°C.[84] A microcoulometric detector in the nitrogen mode produced a good standard curve for 25 to 125 ng of the compound.

Many procedures call for hydrolysis of methomyl to its corresponding oxime prior to GLC analysis.[85-88] Using this approach and a 4% FFAP column, a microcoulometric detector in the sulfur mode yielded a linear response for up to 120 ng of the oxime.[85] Approximately 75% of the theoretical sulfur was detected, but significant interference was observed in extracts of crops having a high sulfur content. Flame-photometric detector sensitivity for 50% full-scale deflection was 41 ng of the methomyl oxime.[86]

Aldicarb is metabolized to the toxic metabolites, aldicarb sulfoxide and sulfone. Therefore, any suitable residue procedure must take into consideration the total bioactive complex either by summation of individual compound analyses or through conversion of the residues to a common material followed by GLC quantitation. In an earlier method, aldicarb sulfoxide and sulfone were each determined individually via direct GLC, using a flame-photometric detector.[89] Aldicarb *per se*, which had previously been isolated from the metabolites, was also analyzed separately after oxidation to the sulfone. In each case, the material actually eluting from the column was the oxime of aldicarb sulfone. The necessary separation of oxime metabolites from the methylcarbamoyloximes was accomplished by column chromatography.

Deposits of aldicarb and its sulfoxide and sulfone on fortified crops, field-treated potatoes, and soil were separated from each other by Florisil-column and prepared for GLC analysis by peracetic acid oxidation to aldicarb sulfone.[90,91] During GLC, the aldicarb sulfone was thermally degraded to the corresponding nitrile sulfone, however. Although hydrolysis products were not separated from intact carbamate residues prior to GLC analysis, thermal degradation of the sulfoxidized oximes to the respective nitrile was apparently inhibited by using wet nitrogen as the carrier gas. The detection limit was 5 ppb.

Determination of aldicarb residues in field-treated cotton foliage and seed utilized a Florisil column to resolve the methylcarbamate residues, followed by sulfoxidation, and then a second Florisil-column separation to remove oxime and nitrile sulfone metabolites from aldicarb sulfone residues.[92] GLC analyses of the individual fractions, such as the sulfone nitrile, showed amounts of aldicarb sulfoxide and sulfone which agreed very well with results obtained by radiometric assays.

A similar procedure for aldicarb residue detection depended upon peracetic acid oxidation of a mixture of aldicarb and its metabolites without prior resolution.[73] The carbamate sulfones were then separated from the non-carbamate sulfones by Florisil-column chromatography before GLC analysis on a mixed-phase Carbowax/DC-200 column. The compound actually measured by the flame-photometric detector was probably the sulfone oxime. Sensitivity of the procedure was less than 10 ppb, and negligible interference was observed in extracts of five crops examined. The method has been applied to field-treated alfalfa for analysis of total aldicarb residues,[93] and various modifications of the "total aldicarb residue" technique have been very successful for monitoring other crops, soils, and animal tissue.[94-97]

Derivatization Methods

Phenols resulting from hydrolysis of aryl N-methylcarbamate insecticides usually do not behave well on GLC columns and/or do not respond to EC detection; the exceptions are some of the halogen- or sulfur-containing phenols.[98] However, conversion of phenols to derivatives with enhanced sensitivity to EC or alkali-flame detection may provide an efficient means of residue analysis. Alternatively, coupling of the methylamine portion of N-methylcarbamates to reagents sensitive to EC detection may also prepare certain carbamates for ready analysis. In fact, derivatization of some type appears to be the most promising approach to successful determination of carbamate insecticide residues by GLC techniques.

Acylation of eight methylcarbamate insecticides with various perfluoroacyl anhydrides (Figure 9.1) reportedly yields respective nitrogen derivatives of the intact carbamates which possess shorter retention times, improved peak symmetry, and increased stability to gas-chromatographic conditions as compared to the parent carbamates.[99] Actually, the exact nature of the derivative is seldom demonstrated, and one cannot rule out the possibility that hydrolysis occurs and that the resulting phenol is acylated.

Derivatization using trifluoroacetic (TFA) anhydride has proved to be of considerable help in the analysis of a variety of methylcarbamate insecticides.[100] Uncatalyzed acetylation in benzene at 100°C for 2 hr was shown to be quantitative for carbofuran and carbaryl at the nanogram level. Response of an EC detector to the TFA derivatives was linear over the range from 0.5 to 10 ng. Pentafluoropropionyl and heptafluorobutyryl analogs of the insecticides also gave excellent sensitivity, but offered no particular advantage over TFA compounds. Catalytic amounts of pyridine enabled the condensation of pentafluoropropionic anhydride with N-methylcarbamates to proceed at room temperature. Moreover, the lower reaction temperature significantly reduced the amount of interfering materials and permitted quantitation at the picogram level. Finally, GLC analysis of carbamate residues in air, using 5% SE-30 on Chromosorb W, an EC detector, and TFA derivatization allowed detection of as little as 20 ng of certain of the compounds.[101]

Among several reagents used in an attempt to effect gas-chromatographic separation and EC detection of aglycone metabolites of carbaryl, the chloromethyldimethylsilyl derivatives of 1-naphthol and 1,4-, 1,5-, and 1,6-naphthylenediols (Figure 9.2, Structure I) gave the best results.[102] Additionally, TFA derivatives of intact 4-hydroxy, 5-hydroxy, 6-hydroxy, and 7-hydroxy carbaryl

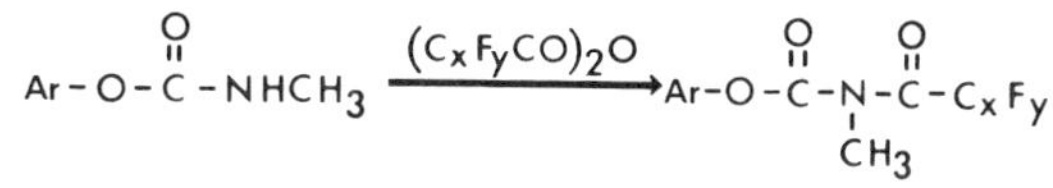

FIGURE 9.1. Proposed derivitization of aryl N-methylcarbamates by reaction with a perfluoroacyl anhydride.[99]

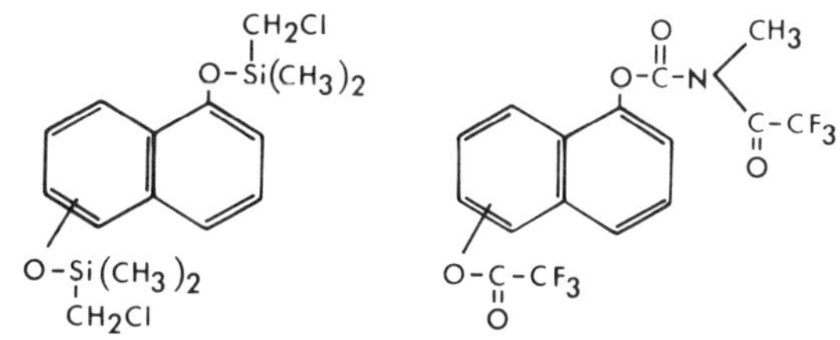

I II

FIGURE 9.2. Chloromethyldimethylsilyl derivatives (I) and trifluoracetyl derivatives (II) of carbaryl metabolites.

(Figure 9.2, Structure II) were extremely sensitive to ^{63}Ni detection; 35 pg of carbaryl gave 50% full-scale deflection. With N-hydroxymethyl carbaryl, 1-naphthol trifluoroacetate was formed upon reaction with trifluoroacetic anhydride. In another study, carbaryl and 12 other substituted phenyl N-methylcarbamates were analyzed using initial silica gel-TLC separation of the pesticides into four bands, followed by acetone extraction of the areas and by trifluoroacetylation of the residues in ethyl acetate.[103] Without TLC separation, the derivatized mixture could not be resolved by GLC. The method, however, is a good one and has demonstrated merit as a general methylcarbamate insecticide residue procedure for rice.[104,105]

There are several methods available for the derivatization of phenols released after hydrolysis of parent carbamates. Trichloroacetylation of 1-naphthol, the hydrolysis product of carbaryl, resulted in the detection of 0.02 ng of the trichloroacetate (Figure 9.3, Structure III) using a ^{90}Sr EC detector.[106] The brominated 1-naphthyl acetate derivative (Figure 9.3, Structure IV) can be determined under the same conditions as the trichloroacetate compound.[107] Recoveries from milk, bees, and pollen fortified at 0.005 to 16 ppm averaged 89% and 95% for the trichloroacetylated and brominated derivatives respectively. Free 1-naphthol was separated from carbaryl prior to derivatization by partitioning the methylene chloride extract with a sodium hydroxide solution.

Carbofuran and its metabolites present in fortified crops and field-treated tomatoes have also been determined as trichloroacetates.[108] Alumina-column chromatography was employed to separate the phenols from the carbamates. The sensitivity of the method was 0.04 ppm for the 3-hydroxy metabolite and 0.01 ppm for carbofuran.

Trichloroacetylation of residues of Baygon and its 2-hydroxy and N-hydroxymethyl metabolites in feed crops and field-treated pasture grass, animal tissues, and milk have been reported.[109,110] The procedure is comprehensive in scope, but somewhat lengthy for routine analysis. The inclusion of water-soluble conjugates was achieved by prior enzymic hydrolysis in order to free the aglycones which were included in the analysis. Following chloroform extraction and silica gel cleanup, a portion of the eluted residues was analyzed for 2-hydroxy Baygon by reaction of the respective phenol with trichloroacetyl chloride (Figure 9.4, Structure VI). The N-hydroxymethyl analog was analyzed as the trichloroacetylated Baygon phenol (Figure 9.4, Structure V). Limits of detection in oat grain were reported to be 0.02 ppm for Baygon and N-hydroxymethyl Baygon and 0.05 ppm for 2-hydroxy Baygon, Carbaryl, Baygon, and methomyl residues in tobacco were similarly analyzed as the trichloroacetylated derivatives on an OV-1/OV-17 column.[111] There were no interferences from the tobacco extracts evident by flamephotometric detection.

Acetylation of liberated phenols with monochloroacetic anhydride has been employed for the determination of Banol in Bermuda grass, milk, and foods, of Baygon in air, and of carbaryl residues in food crops.[50,112-114] Acetylation, in the presence of pyridine, is complete in 10 min at room temperature, and EC reaction to the chloroacetate is linear up to 200 pg. Recoveries of carbaryl in tomatoes, cucumber, lettuce, and unpolished rice at levels of 0.05 to 0.2 ppm have averaged between 70 and 96%. The procedure has been extended to include Meobal determinations,[115] confirmation of carbaryl residues in honey bees,[50] and quantitation of sulfate conjugates of carbaryl, Mobam, and 4-chlorophenyl N-methylcarbamate.[116-119]

A sensitive aryl methylcarbamate residue method that appears promising is based on conversion of parent phenols to 2,4-dinitrophenylaryl ethers by reaction with 1-fluoro-2,4-dinitrobenzene (Figure 9.5) and quantitation by EC detection. Minimum quantities detectable are approximately 0.1 to 0.2 ng. Since these adducts generally require short columns and high oven temperatures, interfering material tends to elute early, resulting in decreased backgrounds. The procedure, initially applied to herbicidal carbamates,[120] has been used for the determination of four N-methylcarbamate insecticides added to certain crops and river water.[121] Free phenols in the extracts were removed by ceric sulfate oxidation in 20% acetone prior to carbamate

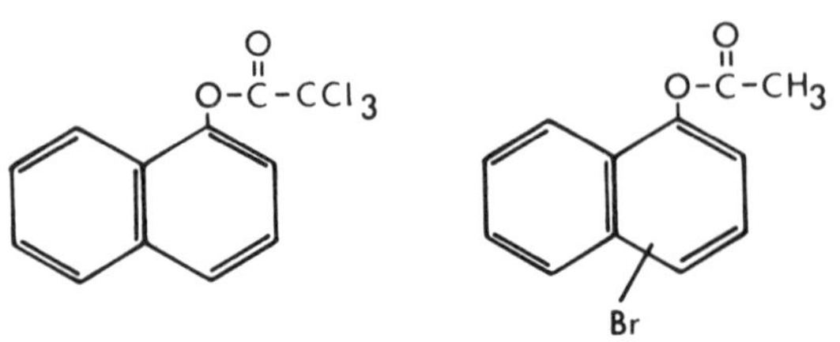

III IV

FIGURE 9.3. Products from trichloroacetylation (III) or acetylation plus bromination (IV) of 1-naphthol.

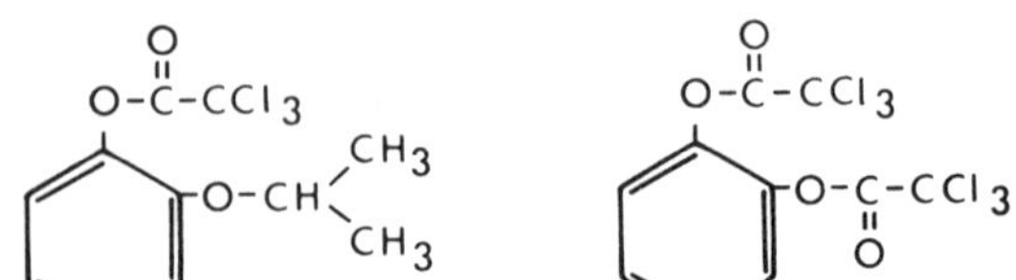

V VI

FIGURE 9.4. Products from trichloroacetylation of Baygon phenol (V) and 2-hydroxy Baygon phenol (VI).

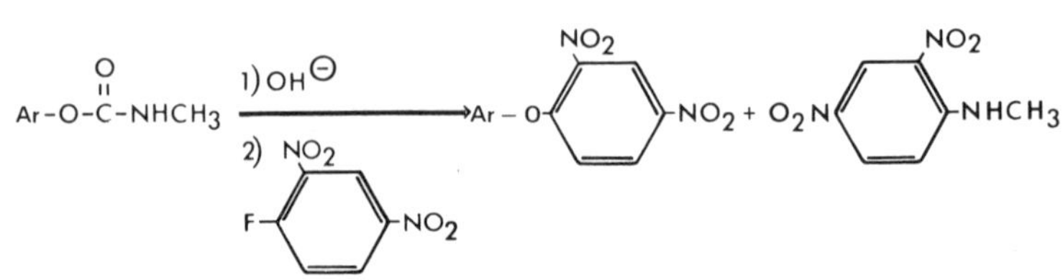

FIGURE 9.5. Derivitization of aryl N-methylcarbamates by hydrolysis and reaction with 1-fluoro-2,4-dinitrobenzene.

hydrolysis and derivatization. Recoveries at the 0.2-ppm level for carbaryl, Mesurol, and Baygon in peas, lettuce, and apples varied by only 0 to 9%.

A similar method combined hydrolysis and derivatization of the resulting phenols with 1-fluoro-2,4-dinitrobenzene (FDNB) in a single step at pH 11.[122-127] Recoveries of 13 methylcarbamates from crops fortified with 0.05-ppm of insecticide ranged from 88 to 115%. Excess FDNB, which may interfere during analysis, can be removed from the sample by converting it to water-soluble material by reaction with glycine.[128] The method has been used for recovery of carbofuran and its metabolites from soil,[125] water and plant material,[126] and corn.[127] Carbofuran phenol was not extracted under these conditions.

Dissipation rates of carbaryl[129] and Landrin[130] from various soils have also been determined by GLC analysis of the respective 2,4-dinitrophenyl ethers. Higher yields of 2,3,5- and 3,4,5-trimethylphenol derivatives were obtained from reaction in aqueous media at room temperature than under reflux conditions in acetone. Since acetone extraction of soil incubated for 3 days with Landrin phenol yielded less than 13% of the applied phenol, prior separation of the hydrolysis products from the intact methylcarbamate residues was not considered to be essential.

As indicated in Figure 9.5, FDNB can also react with the methylamine (or dimethylamine) that is released upon hydrolysis of carbamate insecticides. This reaction has been utilized for the EC detection of eight N-methyl- and two N,N-dimethylcarbamate pesticides.[131,132] Once again, interfering peaks from unreacted FDNB present in the final GLC solution were removed by conversion of the excess reagent into a water-soluble adduct by reaction with glycine prior to benzene extraction. The method has been modified for the determination of methomyl residues in rapeseed oil, where the most satisfactory conditions were basic hydrolysis at 82°C for 10 min, followed by coupling with FDNB for 10 min, and then cleanup with glycine for 10 min.[133] In this case, however, additional unreacted FDNB had to be removed by a microcolumn cleanup on silica gel.[134] The electron-capture detector responded linearly over a range of 0.2 to 5 ng of N-methyl-2,4-dinitroaniline. When the FDNB method was used for analyzing rice, straw, and soil for carbofuran and Baygon residues, the detection limit was 0.005 ppm.[135]

α-Bromo-2,3,4,5,6-pentafluorotoluene has been proven to be successful as a derivatizing agent for phenols (Figure 9.6) and other acidic nucleophiles in river water.[136-138] Almost quantitative conversion of microgram quantities of 1-naphthol and *sec*-amylphenol to the pentafluorobenzyl ethers

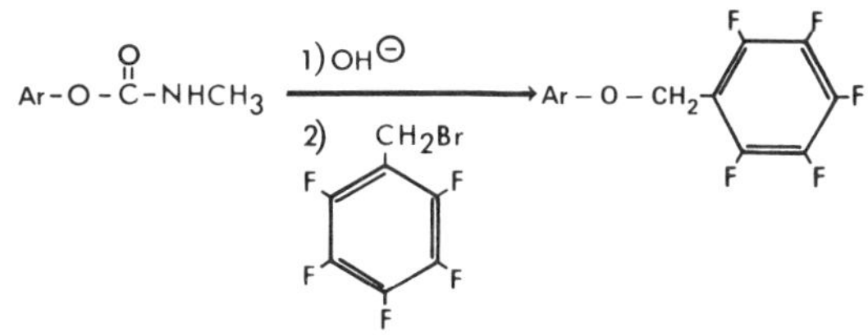

FIGURE 9.6. Derivitization of aryl N-methylcarbamates by hydrolysis and reaction with *α*-bromo-2,3,4,5,6-pentafluorotoluene.

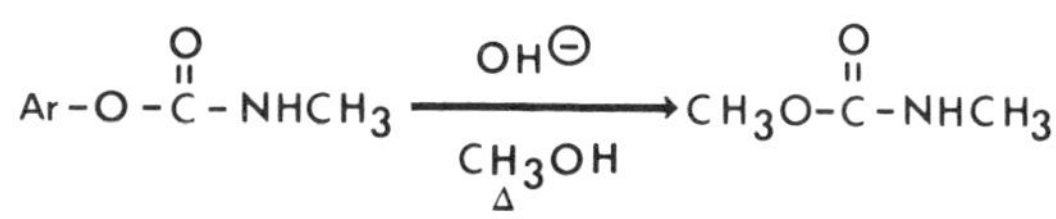

FIGURE 9.7. On-column transesterification of aryl N-methylcarbamates.[140]

was obtained using potassium carbonate in acetone. Excess pentafluorobenzyl bromide was separated from the derivatized phenols by column chromatography. Electron-capture-sensitive phenolic portions of several methylcarbamates were formed by preparing them as the pentafluorobenzyl as well as the 2,4-dinitrophenyl and 2,6-dinitro-4-trifluoromethylphenol ethers.[139] The response was of the same magnitude for all derivatives, and GLC separation was effected.

On-column transesterification of aromatic methylcarbamate insecticides to methyl N-methylcarbamate using methanol containing catalytic amounts of sodium hydroxide (Figure 9.7) resulted in a highly sensitive procedure for carbamate analysis with a Rb_2SO_4 detector.[140] The technique has been applied to estimation of total carbofuran residues on field-treated lettuce[141] and soil,[142] and of methomyl residues in tobacco.[143] Recoveries from spiked tobacco averaged about 80%, and the sensitivity was 0.02 ppm.

OTHER METHODS

Thin-layer chromatography (TLC) is used in a

number of residue procedures to separate compounds of interest from extraneous materials and also as a substrate upon which the insecticide residues can be detected by spectrophotometric methods, chromogenic and fluorogenic spray reagents, enzyme inhibition, or fluorogenic labeling.

p-Nitrobenzenediazonium fluoborate was found to be effective for visualizing 0.1-μg amounts of eight carbamate pesticides following their thin-layer chromatography on alumina and exposure to potassium hydroxide.[144] Some degree of selectivity was due to the fact that N-substituted O-alkyl carbamates and substituted urea herbicides were not sensitive to this reagent. A more recent study demonstrated that minimum detectable levels of 12 N-methylcarbamates on alumina plates range from 0.02 to 10 μg after treatment with the diazo compound.[145] A slight advantage in the use of alumina and polyamide layers over silica gel support for separation of a mixture of insecticides was reported. In an evaluation of polyamide TLC plates for pesticide analysis, from 0.02 to 0.05 μg of nine different N-methylcarbamates were detected following derivatization.[146] In addition, the plates provided more ready detection of UV-absorbing materials than did the silica gel layers.

A number of N-methyl- and N,N-dimethylcarbamate insecticides (2 to 5 μg) were exposed on silica gel after spraying with methanolic *p*-dimethylaminobenzaldehyde.[147] Although chlorinated hydrocarbon and organophosphorous insecticides were not visualized with this reagent, several substituted urea and N-phenylcarbamate herbicides were detected at the same level as the carbamate insecticides.

Treatment of electron-rich carbamate and urea compounds with 9-dicyanomethylene-2,4,7-trinitrofluorene or other pi-electron acceptor to form colored pi-complexes is a simple, rapid, nondestructive TLC-visualization technique.[148-150] The procedure is particularly useful for substrate location on thin-layer chromatograms when regeneration of the intact compound is needed for subsequent manipulations, i.e., mass spectroscopy. Quantitation by *in situ* reflectance spectroscopy was found to be linear from 1 to 8 μg of insecticide on either cellulose or silica gel plates. However, cellulose gave better reproducibility at lower concentrations, and color stability was more pronounced on cellulose than on silica gel.

High-pressure liquid chromatography (LC), though still relatively limited, has shown considerable potential for methylcarbamate insecticide analysis because of its mild operating conditions, specificity, and possibilities for automation. Already, the chromatographic characteristics of many carbamates have been reported.[151] For example, separation of carbofuran, Matacil, Landrin, 1-naphthol, and Mesurol is possible on a Permaphase ODS column using 6% methanol in water as the mobile phase. The 254-nm UV detector required 25 to 1500 ng for 25% full-scale deflection. Others have been successful in separating four N-methylcarbamates by using a polar bonded stationary phase with hexane-methylene chloride mixtures as solvents.[152] An increase in response to Baygon was observed when a 270-nm detector was substituted for a 254-nm detector. Other carbamates for which LC techniques have been perfected include Zectran, carbaryl, and methomyl.[153-156]

The interface of a fluorescent detector with liquid-chromatographic equipment offers a means for substantially increasing the sensitivity to suitably derivatized carbamate insecticides.[157] Recent work indicates that minimum detectable amounts of dansylated phenols range from 2 to 10 ng using a commercially available fluorometer.[158] Linear curves up to 600 ng were obtained for Mesurol and carbofuran derivatives. In a similar way, carbofuran residues on fortified lettuce were determined by LC following simple acetonitrile extraction and labeling with dansyl chloride.[159]

Mass spectroscopy (MS) has been of tremendous value in the identification of methylcarbamate insecticide residues and their metabolites. Thus, in 1969, mass spectral data were presented which supported the previously proposed structure of the carbaryl metabolite 5,6-dihydro-5,6-dihydroxy N-methylcarbamate.[160] Within the next few years, MS confirmation of carbaryl residues and of several other important metabolites were reported.[161-164] Even though electron-impact MS usually reveals little of the molecular ion, the fragmentation patterns often allow differentiation of very similar species. For example, 4-hydroxy-1-naphthyl N-methylcarbamate glucoside can be distinguished from the 5-hydroxy counterpart upon careful examination of the spectra.[164] The use of mass spectroscopy for identification and quantitation purposes has been applied to many other methylcarbamates;[165-167] indeed, this may prove to be the preferred technique for the future analysis of methycarbamate insecticide residues.

NMR[168] and IR[169-171] function to some

degree as analytical tools for carbamate insecticides, particularly for analysis of formulations. However, these techniques suffer from a lack of sensitivity and require careful cleanup and isolation of individual components prior to analysis. Despite these difficulties, IR procedures have been used for quantitation of methylcarbamate residues in grapes[172] and carbaryl residues in orchard soil.[173] For the most part, though, NMR and IR are mainly used for providing confirmatory identification of carbamate residues, and not as routine analytical tools for residue analysis.

REFERENCES

1. **Thornburg, W.,** Pesticide residues, *Anal. Chem.,* 43, 145R, 1971.
2. **Williams, I. H.,** Carbamate insecticide residues in plant material: determination by gas chromatography, *Residue Rev.,* 38, 1, 1971.
3. **Kanazawa, J.,** Annual review of analysis of pesticides for agriculture, *Bunseki Kagaku,* 22, 155R, 1973.
4. **Ruzicka, J. H. A.,** Organophosphorus and carbamate detection system, *Pestic. Sci.,* 4, 417, 1973.
5. **Benson, W. R. and Finocchiaro, J. M.,** Rapid procedure for carbaryl residues: modification of the official colorimetric method, *J. Assoc. Off. Anal. Chem.,* 48, 676, 1965.
6. **Fahey, J. E., Wilson, M. C., and Armbrust, E. J.,** Residues of Supracide and carbofuran in green and dehydrated alfalfa, *J. Econ. Entomol.,* 63, 589, 1970.
7. **George, D. A., Rusk, H. W., Powell, D. M., and Landis, B. J.,** An analytical method for *o*-isopropoxyphenyl methylcarbamate (Bayer 39007), its aphicidal value and persistence in potatoes and sugar beets, *J. Econ. Entomol.,* 60, 82, 1967.
8. **Takeda, M., Sekita, H., Otsuki, K., Takamatsu, C., and Tanage, H.,** Studies on analysis of pesticide residues in foods. Part 10. Analytical method for residues of carbaryl in agricultural products, *J. Food Hyg. Soc. Jap.,* 14, 467, 1973.
9. **Takagi, Y. and Tanida, K.,** Pretreatment for photometric determination of carbaryl in unpolished rice, *J. Food Hyg. Soc. Jap.,* 14, 482, 1973.
10. **McDermott, W. H. and DuVall, A. H.,** Carbaryl insecticide: analysis of formulations by colorimetry, *J. Assoc. Off. Anal. Chem.,* 53, 896, 1970.
11. **Ramasamy, M.,** A colorimetric method for the determination of eight carbamate insecticide residues, *Pestic. Sci.,* 5, 383, 1974.
12. **Blom, J.,** Sensitive and specific reaction for nitrate and for hydroxylamine, *Ber. Dtsch. Chem. Ges.,* 59, 121, 1926.
13. **Johnson, D. P. and Stansbury, H. A., Jr.,** Determination of Temik residues on fruits and vegetables, *J. Assoc. Off. Anal. Chem.,* 49, 399, 1966.
14. **Kearby, W. H., Ercegovich, C. D., and Bliss, M., Jr.,** Residue studies on aldicarb in soil and Scotch Pine, *J. Econ. Entomol.,* 63, 1317, 1970.
15. **Quraishi, M. S.,** Edaphic and water relationships of aldicarb and its metabolites, *Can. Entomol.,* 104, 1191, 1972.
16. **Lee, D. F. and Roughan, J. A.,** Improvements to the nitrite-diazo dye (Blom's) method of determining hydroxylamine as used in the determination of residues of aldicarb, *Analyst,* 96, 798, 1971.
17. **Vonesch, E. E. and DeRiveros, M. H. C. K.,** Colorimetric determination of carbaryl in wettable formulations, *J. Assoc. Off. Anal. Chem.,* 54, 128, 1971.
18. **DeRiveros, M. H. C. K. and Vonesch, E. E.,** Colorimetric determination of carbaryl in apple, lettuce, chard, and tomato, *J. Assoc. Off. Anal. Chem.,* 54, 1083, 1971.
19. **Rangaswamy, J. R. and Majumder, S. K.,** Colorimetric method for estimation of carbaryl and its residues on grains, *J. Assoc. Off. Anal. Chem.,* 57, 592, 1974.
20. **Moorefield, H. H. and Tefft, E. R.,** Application of cholinesterase assay to residue analysis of 1-naphthyl N-methylcarbamate (Sevin), *Contrib. Boyce Thompson Inst.,* 19, 295, 1958.
21. **Zweig, G. and Archer, T. E.,** Residue determination of Sevin (1-naphthyl N-methylcarbamate) in wine by cholinesterase inhibition and paper chromatography, *J. Agric. Food Chem.,* 6, 910, 1958.
22. **Villeneuve, D. C.,** A review of enzymatic techniques used for pesticide residue analysis, *Adv. Chem. Ser.,* 104, 27, 1971.
23. **Mendoza, C. E ,** Analysis of pesticides by the thin-layer chromatographic-enzyme inhibition technique, *Residue Rev.,* 43, 105, 1972.
24. **Mendoza, C. E.,** Thin-layer chromatography and enzyme inhibition techniques, *J. Chromatogr.,* 78, 29, 1973.
25. **Mendoza, C. E.,** Analysis of pesticides by the thin-layer chromatographic-enzyme inhibition technique. Part II, *Residue Rev.,* 50, 43, 1974.

26. **Giang, P. A. and Hall, S. A.**, Enzymatic determination of organic phosphorus insecticides, *Anal. Chem.*, 23, 1830, 1951.
27. **Guilbault, G. G., Sadar, M. H., Kuan, S. S., and Casey, D.**, Enzymatic methods of analysis. Trace analysis of various pesticides with insect cholinesterases, *Anal. Chim. Acta*, 52, 75, 1970.
28. **Guilbault, G. G., Lozes, R. L., Moore, W., and Kuan, S. S.**, Effect of pesticides on cholinesterase from aquatic species: crayfish, trout and fiddler crab, *Environ. Lett.*, 3, 235, 1972.
29. **Gamson, R. M., Robinson, D. W., and Goodman, A.**, Test for anticholinesterase materials in water, *Environ. Sci. Technol.*, 7, 1137, 1973.
30. **Winterlin, W., Walker, G., and Frank, H.**, Detection of cholinesterase-inhibiting pesticides following separation on thin-layer chromatograms, *J. Agric. Food Chem.*, 16, 808, 1968.
31. **Mueller, B.**, Determination of pesticide residues in bee honey. Part I. Semiquantitative thin-layer-chromatographic determination of insecticide residues in bee honey, *Nahrung*, 17, 381, 1973.
32. **Mueller, B.**, Chemical toxicological tests on poisoned bees. 2. Review of methods for the chemical toxicological examination of poisoned bees, *Arch. Exp. Veterinaermed.*, 27, 145, 1973.
33. **Mueller, B. and Worseck, M.**, Thin-layer chromatographic/enzymic method, using enzymes of bee brain, for semiquantitative determination of carbaryl in dead bees, *Monatsh. Veterinaermed.*, 25, 558, 1970.
34. **Winterlin, W., Walker, G., and Luce, A.**, Carbaryl residues in bees, honey, and bee bread following exposure to carbaryl via the food supply, *Arch. Environ. Contam. Toxicol.*, 1, 362, 1973.
35. **Wales, P. J., McLeod, H. A., and McKinley, W. P.**, TLC-enzyme inhibition procedure to detect some carbamate standards and carbaryl in food extracts, *J. Assoc. Off. Anal. Chem.*, 51, 1239, 1968.
36. **Mendoza, C. E., Grant, D. L., Braceland, B., and McCully, K. A.**, Evaluation of esterases from livers of beef, pig, sheep, monkey, and chicken for detection of some pesticides by thin-layer chromatographic-enzyme inhibition technique, *Analyst*, 94, 805, 1969.
37. **Geike, F.**, Thin-layer chromatographic-enzymatic identification of carbamates. I. Identification of carbamate insecticides with bovine liver esterase, *J. Chromatogr.*, 53, 269, 1970.
38. **Anon.**, Methods for performing residue analysis of plant protection and pest control agents in foodstuffs. 6. Method for the identification and quantitative determination of insecticidal esterase inhibitors, *Nahrung*, 14, 671, 1970.
39. **Ohta, T. and Ishikura, S.**, Studies on the selectivity in detecting organophosphorous and carbamate pesticides by thin-layer chromotographic enzyme inhibition technique, *J. Hyg. Chem.*, 19, 219, 1973.
40. **Mendoza, C. E. and Shields, J. B.**, Sensitivity of pig liver esterase in detecting twelve carbamate pesticides on thin-layer chromatograms, *J. Chromatogr.*, 50, 92, 1970.
41. **Mendoza, C. E. and Shields, J. B.**, Determination of some carbamates by enzyme inhibition techniques using thin-layer chromatography and colorimetry, *J. Agric. Food Chem.*, 21, 178, 1973.
42. **Mendoza, C. E., McLeod, H. A., Shields, J. B., and Phillips, W. E. J.**, Determination of methomyl in rape seeds, oils, and meals, *Pestic. Sci.*, 5, 231, 1974.
43. **Filatov, G. V., Brytskov, V., and Kozin, V. A.**, Enzymic detection methods in chromatograms for chlorophos, Sevin, and their cholinesterase-inhibiting metabolites, *Tr. Vses. Nauchno Issled. Inst. Vet. Sanit.*, 39, 144, 1971.
44. **Geike, F.**, Thin-layer chromatographic-enzymatic detection of some pesticides by inhibition of amylases, *J. Chromatogr.*, 63, 343, 1971.
45. **Stoimenov, V. and Kunchev, K.**, Possibilities of demonstrating residual amounts of organic phosphorus and carbamate pesticides through the inhibition of some specific esterases and phosphatases, *Vet. Med. Nauki*, 10, 19, 1973.
46. **Geike, F.**, Thin-layer-chromatographic-enzymic detection of carbamates. III. Detection of insecticidal and herbicidal carbamates with phosphatases, *J. Chromatogr.*, 64, 383, 1972.
47. **Anon.**, Methods for performing residue analyses of plant protection and pest control agents in foodstuffs. 9. Method for the determination of insecticidal phosphoric, phosphonic and carbamic acid esters using agar diffusion, *Nahrung*, 14, 695, 1970.
48. **Goodson, L. H., Jacobs, W. B., and Davis, A. W.**, An immobilized cholinesterase product for use in the rapid detection of enzyme inhibitors in air or water, *Anal. Biochem.*, 51, 362, 1973.
49. **Goodson, L. H. and Jacobs, W. B.**, Rapid detection system for organophosphates and carbamate insecticides in water, *Environ. Prot. Technol. Ser.*, Rep. No. EPA-R2-72-010, p. 66, 1972.
50. **Argauer, R. J., Shimanuki, H., and Alvarez, C. C.**, Fluorometric determination of carbaryl and 1-naphthol in honeybees (*Aphis mellifera* L.) with confirmation by gas chromatography, *J. Agric. Food Chem.*, 18, 688, 1970.
51. **Argauer, R. J. and Bontoyan, W.**, Fluorometric analysis of carbaryl insecticide in mixed formulations, *J. Assoc. Off. Anal. Chem.*, 53, 1166, 1970.
52. **Ishii, Y.**, Determination of carbaryl (1-naphthyl N-methylcarbamate) in fish tissues by spectrophotofluorescence, *Bull. Agric. Chem. Insp. Stn.* (Tokyo), 10, 51, 1970.
53. **Ott, D. E., Ittig, M., and Friestad, H. O.**, Automated steam distillation and fluorometry for screening for carbaryl as 1-naphthol in fruits and vegetables, *J. Assoc. Off. Anal. Chem.*, 54, 160, 1971.
54. **Argauer, R. J. and Webb, R. E.**, Rapid fluorometric evaluation of the deposition and persistence of carbaryl in the presence of an adjuvant on bean and tomato leaves, *J. Agric. Food Chem.*, 20, 732, 1972.
55. **Frei, R. W., Lawrence J. F., and Belliveau, P. E.**, *In situ* fluorimetric method for the determination of Sevin and α-napthol on thin-layer chromatograms, *Fresenius Z. Anal. Chem.*, 254, 271, 1971.

56. **Belliveau, P. E., Mallet, V., and Frei, R. W.,** A spray method for the fluorescent detection of sulphur-containing organic compounds, *J. Chromatogr.*, 48, 478, 1970.
57. **Belliveau, P. E. and Frei, R. W.,** pH-sensitive fluorogenic spray reagents. I. General discussion and investigation of 1,2-dichloro-4,5-dicyanobenzoquinone, *Chromatographia,* 4, 189, 1971.
58. **Mallet, V. and Frei, R. W.,** An investigation of flavones as fluorogenic spray reagents for organic compounds on a cellulose matrix. Part I. General discussion of the method. *J. Chromatogr.*, 54, 251, 1971.
59. **Mallet, V. and Frei, R. W.,** An investigation of flavones as fluorogenic spray reagents for organic compounds on a cellulose matrix. Part II. Detection of pesticides, *J. Chromatogr.*, 56, 69, 1971.
60. **Lawrence, J. F. and Frei, R. W.,** Fluorimetric derivatization for pesticide residue analysis, *J. Chromatogr.*, 98, 253, 1974.
61. **Frei, R. W. and Lawrence, J. F.,** Fluorigenic labeling of carbamates with dansyl chloride. I. Study of reaction conditions, *J. Chromatogr.*, 61, 174, 1971.
62. **Lawrence, J. F., Legay, D. S., and Frei, R. W.,** Fluorigenic labelling of carbamates using dansyl chloride. III. Thin-layer chromatographic properties of the derivatives, *J. Chromatogr.*, 66, 295, 1972.
63. **Lawrence, J. F. and Frei, R. W.,** Fluorigenic labelling of carbamates using dansyl chloride. II. Fluorescence phenomena of the derivatives, *J. Chromatogr.*, 66, 93, 1972.
64. **Frei, R. W. and Lawrence, J. F.,** Fluorigenic labelling of carbamates with dansyl chloride. IV. *In situ* quantitation of N-methyl carbamate insecticides on thin-layer chromatograms, *J. Chromatogr.*, 67, 87, 1972.
65. **Frei, R. W. and Lawrence, J. F.,** Determination of Matacil and Zectran by fluorigenic labelling, thin-layer chromatography, and *in situ* fluorimetry, *J. Assoc. Off. Anal. Chem.*, 55, 1259, 1972.
66. **Lawrence, J. F. and Frei, R. W.,** Fluorigenic labelling of N-methyl- and N,N-dimethylcarbamates with 4-chloro-7-nitrobenzo-2,1,3-oxadiazole, *Anal. Chem.*, 44, 2046, 1972.
67. **Krishna, J. G., Dorough, H. W., and Casida, J. E.,** Synthesis of N-methylcarbamates via methyl isocyanate-C^{14} and chromotographic purification, *J. Agric. Food Chem.*, 10, 462, 1962.
68. **Strother, A.,** Gas chromatography of various phenyl N-methylcarbamates, *J. Gas Chromatogr.*, 6, 110, 1968.
69. **Wheeler, L. and Strother, A.,** Chromatography of N-methylcarbamates in the gaseous phase, *J. Chromatogr.*, 45, 362, 1969.
70. **Cook, R. F., Stanovick, R. P., and Cassil, C. C.,** Determination of carbofuran and its carbamate metabolite residues in corn using a nitrogen-specific gas chromatographic detector, *J. Agric. Food Chem.*, 17, 277, 1969.
71. **Cassil, C. C., Stanovick, R. P., and Cook, R. F.,** A specific gas chromatographic method for residues of organic nitrogen pesticides, *Residue Rev.*, 26, 63, 1969.
72. **Riva, M. and Carisano, A.,** Direct gas chromatographic determination of carbaryl, *J. Chromatogr.*, 42, 464, 1969.
73. **Maitlen, J. C., McDonough, L. M., and Beroza, M.,** Rapid method for the extraction, cleanup and GLC determination of toxic residues of Temik, *J. Assoc. Off. Anal. Chem.*, 52, 786, 1969.
74. **Wheeler, L. and Strother, A.,** In-vitro metabolism of the N-methylcarbamates, Zectran and Mesurol, by liver, kidney and blood of dogs and rats, *J. Pharmacol. Exp. Ther.*, 178, 371, 1971.
75. **Laski, R. and Watts, R. R.,** Gas chromatography of organonitrogen pesticides, using a nitrogen-specific detection system, *J. Assoc. Off. Anal. Chem.*, 56, 328, 1973.
76. **Lewis, D. L. and Paris, D. F.,** Direct determination of carbaryl by gas–liquid chromatography using electron capture detection, *J. Agric. Food Chem.*, 22, 148, 1974.
77. **Weyer, L. G.,** Gas-liquid chromatographic method for the analysis of carbaryl insecticide formulations, *J. Assoc. Off. Anal. Chem.*, 57, 778, 1974.
78. **Williams, I. H. and Brown, M. J.,** Determination of carbofuran and 3-hydroxycarbofuran residues in small fruits, *J. Agric. Food Chem.*, 21, 399, 1973.
79. **Aue, W. A., Hastings, C. R., and Kapila, S.,** On the unexpected behavior of a common gas chromatographic phase, *J. Chromatogr.*, 77, 299, 1973.
80. **Lorah, E. J. and Hemphill, D. D.,** Direct chromatography of some N-methylcarbamate pesticides, *J. Assoc. Off. Anal. Chem.*, 57, 570, 1974.
81. **Reeves, R. G. and Woodham, D. W.,** Gas chromatographic analysis of methomyl residues in soil, sediment, water, and tobacco utilizing the flame photometric detector, *J. Agric. Food Chem.*, 22, 76, 1974.
82. **Masuda, T., Kanazawa, J., and Inoue, H.,** Determination of methomyl residues on potatoes, *Pestic. Technol.*, 28, 22, 1972.
83. **Kawahara, T. and Maeda, H.,** Residue analysis of Mesomil in crops, *Noyaku Kensasho Hokoku,* 11, 89, 1971.
84. **Williams, I. H.,** Direct gas chromatographic determination of methomyl (Lannate) with microcoulometric nitrogen detection, *Pestic. Sci.*, 3, 179, 1972.
85. **Pease, H. L. and Kirkland, J. J.,** Determination of methomyl residues using microcoulometric gas chromatography, *J. Agric. Food Chem.*, 16, 554, 1968.
86. **Kiigemagi, U., Wellman, D., Cooley, E. J., and Terriere, L. C.,** Residues of the insecticides phorate and methomyl in mint hay and oil, *Pestic. Sci.*, 4, 89, 1973.
87. **Lee, Y. W., Ford, R. J., McDonald, H., McKinlay, K. S., Putnam, L. G., and Saha, J. G.,** Residues of methomyl in rape plant and seed following its applications for the control of bertha armyworm, *Mamestra configurata* (Lepidoptera: Noctuidae), *Can. Entomol.*, 104, 1745, 1972.

88. **Krueger, H. R., Lindquist, R. K., Mason, J. F., and Spadafora, R. R.,** Application of methomyl to greenhouse tomatoes: greenhouse whitefly control and residues in foliage and fruits, *J. Econ. Entomol.,* 66, 1223, 1973.
89. **Maitlen, J. C., McDonough, L. M., and Beroza, M.,** Determination of residues of 2-methyl-2-(methylthio)propionaldehyde O-(methylcarbamoyl)oxime (UC-21149, Temik), its sulfoxide, and its sulfone by gas chromatography, *J. Agric. Food Chem.,* 16, 549, 1968.
90. **Carey, W. F. and Helrich, K.,** Improved quantitative method for the determination of aldicarb and its oxidation products in plant materials, *J. Assoc. Off. Anal. Chem.,* 53, 1296, 1970.
91. **Bromilow, R. H.,** Breakdown and fate of oximecarbamate nematicides in crops and soils, *Ann. Appl. Biol.,* 75, 473, 1973.
92. **Andrawes, N. R., Romine, R. R., and Bagley, W. P.,** Metabolism and residues of Temik aldicarb pesticide in cotton foliage and seed under field conditions, *J. Agric. Food Chem.,* 21, 379, 1973.
93. **Halfhill, J. E. and Maitlen, J. C.,** Residues of aldicarb in fresh and dried alfalfa, *J. Econ. Entomol.,* 66, 557, 1973.
94. **Lindquist, R. K., Krueger, H. R., Spadafora, R. R., and Mason, J. F.,** Application of aldicarb to greenhouse tomatoes: plant growth, fruit yields, greenhouse whitefly control, and residues in fruits, *J. Econ. Entomol.,* 65, 862, 1972.
95. **Kreuger, H. R. and Mason, J. F.,** Effects of plant growth regulators on levels of phorate and aldicarb in soybeans, *J. Agric. Food Chem.,* 22, 338, 1974.
96. **Woodham, D. W., Edwards, R. R., Reeves, R. G., and Schutzmann, R. L.,** Total toxic aldicarb residues in soil, cottonseed, and cotton lint following a soil treatment with the insecticide on the Texas high plains, *J. Agric. Food Chem.,* 21, 303, 1973.
97. **Woodham, D. W., Reeves, R. G., and Edwards, R. R.,** Total toxic aldicarb residues in weeds, grasses, and wildlife from the Texas high plains following a soil treatment with the insecticide, *J. Agric. Food Chem.,* 21, 604, 1973.
98. **Bowman, M. C. and Beroza, M.,** Determination of residues of Mobil MC-A-600 (benzo[b]thien-4-yl methylcarbamate) and its hydrolyis product (benzo[b]thiophene-4-ol) in coastal Bermuda grass and milk, *J. Agric. Food Chem.,* 15, 894, 1967.
99. **Seiber, J. N.,** N-Perfluoroacyl derivatives for methylcarbamate analysis by gas chromatography, *J. Agric. Food Chem.,* 20, 443, 1972.
100. **Lau, S. C. and Marxmiller, R. L.,** Residue determination of Landrin insecticide by trifluoroacetylation and electron-capture gas chromatography, *J. Agric. Food Chem.,* 18, 413, 1970.
101. **Sherma, J. and Shafik, T. M.,** A multiclass, multiresidue analytical method for determining pesticide residues in air, *Arch. Environ. Contam. Toxicol.,* 3, 55, 1975.
102. **Khalifa, S. and Mumma, R. O.,** Gas chromatographic separation of the aglycone metabolites of carbaryl, *J. Agric. Food Chem.,* 20, 632, 1972.
103. **Suzuki, K., Nagayoshi, H., and Kashiwa, T.,** Systematic separation and identification of 13 carbamate pesticides in their mixture, *Agric. Biol. Chem.,* 37, 2181, 1973.
104. **Sekita, H., Suzuki, T., Takeda, M., and Tanabe, H.,** Studies on analysis of pesticide residues in foods. XV. Analytical method for carbamate insecticide residues in unpolished rice by gas chromatography with ECD, *J. Food Hyg. Soc. Jap.,* 15, 219, 1974.
105. **Ueji, M. and Kanazawa, J.,** Method for the residue determination of carbamate pesticides in crops, *Jap. Agric. Res. Q.,* 8, 67, 1974.
106. **Butler, L. I. and McDonough, L. M.,** Specific GLC method for determining residues of carbaryl by electron-capture detection after derivative formation, *J. Assoc. Off. Anal. Chem.,* 53, 495, 1970.
107. **Gutenmann, W. H. and Lisk, D. J.,** Gas chromatographic residue determination of Sevin as brominated 1-naphthyl acetate, *J. Agric. Food Chem.,* 13, 48, 1965.
108. **Butler, L. I. and McDonough, L. M.,** Determination of residues of carbofuran and its toxic metabolites by electron capture gas chromatography after derivative formation, *J. Assoc. Off. Anal. Chem.,* 54, 1357, 1971.
109. **Stanley, C. W., Thornton, J. S., and Katague, D. B.,** Gas chromatographic method for residues of Baygon and metabolites in plant tissues, *J. Agric. Food Chem.,* 20, 1265, 1972.
110. **Stanley, C. W. and Thorton, J. S.,** Gas chromatographic method for residues of Baygon and its major metabolite in animal tissues and milk, *J. Agric. Food Chem.,* 20, 1269, 1972.
111. **Nesemann, E. and Seehofer, F.,** Methods for the quantitative determination of biocides in tobacco and tobacco products, *Beitr. Tabakforsch.,* 7, 251, 1974.
112. **Argauer, R. J.,** Determination of residues of Banol and other carbamate pesticides after hydrolysis and chloroacetylation, *J. Agric. Food Chem.,* 17, 888, 1969.
113. **Miller, C. W., Shafik, M. T., and Biros, F. J.,** A method for sampling and analysis of propoxur in air, *Bull. Environ. Contam. Toxicol.,* 8, 339, 1972.
114. **Ishii, Y. and Yamashita, Y.,** Studies on the microanalysis of N-methylcarbamate insecticides. Report III. Residue analysis of carbaryl, *Bull. Agric. Chem. Insp. Stn.,* (Tokyo), 12, 63, 1972.
115. **Kawahara, T.,** Determination of 3,4-dimethylphenyl N-methylcarbamate (MPMC) by electron-capture gas–liquid chromatography, *Bull. Agric. Chem. Insp. Stn.,* (Tokyo), 10, 45, 1970.
116. **Mehendale, H. M. and Dorough, H. W.,** Conjugative metabolism and action of carbamate insecticides, in *Insecticides,* Tahori, A. S., Ed., Gordon and Breach, New York, 1971.

117. **Paulson, G. D. and Portnoy, C. E.,** Sulfate ester conjugates: a one-step method for replacing the sulfate with an acetyl group, *J. Agric. Food Chem.*, 18, 180, 1970.
118. **Paulson, G. D., Zaylskie, R. G., Zehr, M. V., Portnoy, C. E., and Feil, V. J.,** Metabolites of carbaryl (1-naphthyl methylcarbamate) in chicken urine, *J. Agric. Food Chem.*, 18, 110, 1970.
119. **Paulson, G. D., Zehr, M. V., Dockter, M. M., and Zaylskie, R. G.,** Metabolism of *p*-chlorophenyl N-methylcarbamate in the rat and goat, *J. Agric. Food Chem.*, 20, 33, 1972.
120. **Cohen, I. C. and Wheals, B. B.,** An electron-capture gas chromatographic method for the determination of substituted urea and carbamate herbicides as 2,4-dinitrophenyl derivatives of their amine moieties, *J. Chromatogr.*, 43, 233, 1969.
121. **Cohen, I., Norcup, J., Ruzicka, J. H. A., and Wheals, B. B.,** An electron-capture gas chromatographic method for the determination of some carbamate insecticides as 2,4-dinitrophenyl derivatives of their phenol moieties, *J. Chromatogr.*, 49, 215, 1970.
122. **Benson, W. R.,** Report on carbamate pesticides, *J. Assoc. Off. Anal. Chem.*, 53, 351, 1970.
123. **Storherr, R. W.,** Reports on carbamate pesticides and on fumigants and miscellaneous pesticides, *J. Assoc. Off. Anal. Chem.*, 54, 324, 1971.
124. **Holden, E. R.,** Gas chromatographic determination of residues of methylcarbamate insecticides in crops as their 2,4-dinitrophenyl ether derivatives, *J. Assoc. Off. Anal. Chem.*, 56, 713, 1973.
125. **Caro, J. H., Glotfelty, D. E., Freeman, H. P., and Taylor, A. W.,** Acid ammonium acetate extraction and electron capture gas chromatographic determination of carbofuran in soils, *J. Assoc. Off. Anal. Chem.*, 56, 1319, 1973.
126. **Caro, J. H., Freeman, H. P., Glotfelty, D. E., Turner, B. C., and Edwards, W. M.,** Dissipation of soil-incorporated carbofuran in the field, *J. Agric. Food Chem.*, 21, 1010, 1973.
127. **Turner, B. C. and Caro, J. H.,** Uptake and distribution of carbofuran and its metabolites in field-grown corn plants, *J. Environ. Qual.*, 2, 245, 1973.
128. **Takase, I. and Osuga, S.,** Gas-chromatographic micro-determination of pesticides. Part 6. Determination of some residual N-methylcarbamate insecticides as 2,4-dinitrophenyl ester derivatives, *Noyaku Seisan Gijutsu*, 30, 22, 1973.
129. **Caro, J. H., Freeman, H. P., and Turner, B. C.,** Persistence in soil and losses in runoff of soil-incorporated carbaryl in a small watershed, *J. Agric. Food Chem.*, 22, 860, 1974.
130. **Asai, R. I., Gunther, F. A., and Westlake, W. E.,** Influence of some soil characteristics on the dissipation rate of Landrin insecticide, *Bull. Environ. Contam. Toxicol.*, 11, 352, 1974.
131. **Holden, E. R., Jones, W. M., and Beroza, M.,** Determination of residues of methyl- and dimethylcarbamate insecticides by gas chromatography of their 2,4-dinitroaniline derivatives, *J. Agric. Food Chem.*, 17, 56, 1969.
132. **Sumida, S., Takaki, M., and Miyamoto, J.,** Rapid gas chromatographic determination of micro quantities of N-methyl carbamates as their 2,4-dinitrophenyl derivatives, *Agric. Biol. Chem.*, 34, 1576, 1970.
133. **Mendoza, C. E. and Shields, J. B.,** Determination of methomyl by using 1-fluoro-2,4-dinitrobenzene reaction and gas–liquid chromatography, *J. Agric. Food Chem.*, 22, 255, 1974.
134. **Mendoza, C. E. and Shields, J. B.,** A microcolumn apparatus for rapid cleanup of 2,4-dinitrophenylmethylamine extracts for carbamate pesticide analysis, *J. Agric. Food Chem.*, 22, 528, 1974.
135. **Aizawa, T., Kamiyama, I., Sawafuji, S., and Takase, I.,** Studies on microanalysis by gas chromatography. Part 8. Residue analysis of carbamate pesticides by Afid, *J. Pestic. Sci.*, 2, 60, 1974.
136. **Kawahara, F. K.,** Gas chromatographic analysis of mercaptans, phenols, and organic acids in surface waters with use of pentafluorobenzyl derivatives, *Environ. Sci. Technol.*, 5, 235, 1971.
137. **McCallum, N. K. and Armstrong, R. J.,** The derivatisation of phenols for gas chromatography using electron capture detection, *J. Chromatogr.*, 78, 303, 1973.
138. **Johnson, L. G.,** Formation of pentafluorobenzyl derivatives for the identification and quantitation of acid and phenol pesticide residues, *J. Assoc. Off. Anal. Chem.*, 56, 1503, 1973.
139. **Seiber, J. N., Crosby, D. G., Fouda, H., and Soderquist, C. J.,** Ether derivatives of the determination of phenols and phenol-generating pesticides by electron capture gas chromatography, *J. Chromatogr.*, 73, 89, 1972.
140. **Moye, H. A.,** Reaction gas chromatographic analysis of pesticides. I. On-column transesterification of N-methylcarbamates by methanol, *J. Agric. Food Chem.*, 19, 452, 1971.
141. **Van Middelem, C. H., Moye, H. A., and Janes, M. J.,** Carbofuran and 3-hydroxycarbofuran determination in lettuce by alkali-flame gas chromatography, *J. Agric. Food Chem.*, 19, 459, 1971.
142. **Hubbell, D. H., Rothwell, D. F., Wheeler, W. B., Tappan, W. B., and Rhoads, F. M.,** Microbiological effects and persistence of some pesticide combinations in soil, *J. Environ. Qual.*, 2, 96, 1973.
143. **Tappan, W. B., Wheeler, W. B., and Lundy, H. W.,** Methomyl residues on cigar-wrapper and flue-cured tobaccos in Florida, *J. Econ. Entomol.*, 66, 197, 1973.
144. **Finocchiaro, J. M. and Benson, W. R.,** Thin layer chromatography of some carbamate and phenylurea pesticides, *J. Assoc. Off. Anal. Chem.*, 50, 888, 1967.
145. **Ogamo, A., Kasahara, H., and Nagasawa, K.,** Thin-layer chromatography of carbamate insecticides, *J. Hyg. Chem.*, 19, 340, 1973.
146. **Nagasawa, K., Yoshidome, H., and Kamata, F.,** Separation and detection of carbamates and related compounds on polyamide layers, *J. Chromatogr.*, 52, 453, 1970.
147. **El-Dib, M. A.,** Thin layer chromatographic detection of carbamate and phenylurea pesticide residues in natural waters, *J. Assoc. Off. Anal. Chem.*, 53, 756, 1970.

148. **Hutzinger, O., Jamieson, W. D., MacNeil, J. D., and Frei, R. W.,** Electron-donor-acceptor complexing reagents for the analysis of pesticides. I. Survey of reagents and instrumental techniques, *J. Assoc. Off. Anal. Chem.*, 54, 1100, 1971.
149. **MacNeil, J. D., Frei, R. W., and Hutzinger, O.,** Electron-donor-acceptor complexing reagents for the detection of pesticides on TLC. II. Carbamates, anilides, and ureas. Spray reagents, chromatography, and instrumental techniques, *Int. J. Environ. Anal. Chem.*, 1, 205, 1972.
150. **MacNeil, J. D., Frei, R. W., Frei-Haeusler, M., and Hutzinger, O.,** Electron-donor-acceptor reagent in the analysis of pesticides. VII. A simple model system hydrolysis of some carbamate pesticides, *Int. J. Environ. Anal. Chem.*, 2, 323, 1973.
151. **Thruston, A. D.,** Liquid chromatography of carbamate pesticides, *Environ. Prot. Technol. Ser.*, Rep. No. EPA-R2-72-079, October 1972, p. 1.
152. **Ishii, Y. and Otake, T.,** Studies on the analysis of pesticides by high-performance liquid chromatography. Part I. Carbamate insecticides, *Bull. Agric. Chem. Insp. Stn.* (Tokyo), 13, 32, 1973.
153. **Hosler, C. F., Jr.,** Degradation of Zectran in alkaline water, *Bull. Environ. Contam. Toxicol.*, 12, 599, 1974.
154. **Colvin, B. M., Engdahl, B. S., and Hanks, A. R.,** Determination of carbaryl in pesticide formulations and fertilizers by high-pressure liquid chromatography, *J. Assoc. Off. Anal. Chem.*, 57, 648, 1974.
155. **Leitch, R. E.,** Precise quantitative analysis with a stable, high-speed liquid–liquid chromatography column, *J. Chromatogr. Sci.*, 9, 531, 1971.
156. **Seiber, J. N.,** Reversed-phase liquid chromatography of some pesticides and related compounds, *J. Chromatogr.*, 94, 151, 1974.
157. **Cassidy, R. M. and Frei, R. W.,** A fluorescence detector for high-speed liquid chromatography, *J. Chromatogr.*, 72, 293, 1972.
158. **Frei, R. W., Lawrence, J. F., Hope, J., and Cassidy, R. M.,** Analysis of carbamate insecticides by fluorigenic labelling and high-speed liquid chromatography, *J. Chromatogr. Sci.*, 12, 40, 1974.
159. **Frei, R. W. and Lawrence, J. F.,** Fluorigenic labelling in high-speed liquid chromatography, *J. Chromatogr.*, 83, 321, 1973.
160. **Baron, R. L., Sphon, J. A., Chen, J. T., Lustig, E., Doherty, J. D., Hansen, E. A., and Kolbye, S. M.,** Confirmatory isolation and identification of a metabolite of carbaryl in urine and milk, *J. Agric. Food Chem.*, 17, 883, 1969.
161. **Liu, S.-Y. and Bollag, J.-M.,** Metabolism of carbaryl by soil fungus, *J. Agric. Food Chem.*, 19, 487, 1971.
162. **Durden, J. A., Jr. and Bartley, W. J.,** The mass spectra of 4-, 5-, 6-, and 7-hydroxy-1-naphthyl methylcarbamates, *J. Agric. Food Chem.*, 19, 441, 1971.
163. **Mumma, R. O. and Khalifa, S.,** Mass spectra of trifluoroacetyl derivatives of carbaryl and its aglycone metabolites, *J. Agric. Food Chem.*, 20, 1090, 1972.
164. **Cardona, R. A. and Dorough, H. W.,** Syntheses of the β-D-glucosides of 4- and 5-hydroxy-1-naphthyl N-methylcarbamate, *J. Agric. Food Chem.*, 21, 1065, 1973.
165. **Bartley, W. J., Andrawes, N. R., Chancey, E. L., Bagley, W. P., and Spurr, H. W.,** The metabolism of Temik aldicarb pesticide [2-methyl-2-(methylthio)propionaldehyde O-(methylcarbamoyl)oxime] in the cotton plant, *J. Agric. Food Chem.*, 18, 446, 1970.
166. **Biros, F. J.,** Applications of combined gas chromatography-mass spectrometry to pesticide residue identifications, *Adv. Chem. Ser.*, 104, 132, 1971.
167. **Rankin, P. C.,** Negative ion mass spectra of some pesticidal compounds, *J. Assoc. Off. Anal. Chem.*, 54, 1340, 1971.
168. **Keith, L. H. and Alford, A. L.,** The high resolution NMR spectra of pesticides. III. The carbamates, *J. Assoc. Off. Anal. Chem.*, 53, 157, 1970.
169. **Kovacs, M. F., Jr.,** Improved infrared method for carbaryl (Sevin) in commercial formulations, *J. Assoc. Off. Anal. Chem.*, 50, 566, 1967.
170. **McDermott, W. H.,** Infrared analysis of carbaryl insecticide: modification of the extraction procedure to accommodate liquid suspension formulations, *J. Assoc. Off. Anal. Chem.*, 56, 576, 1973.
171. **McDermott, W. H.,** Collaborative study of the infrared spectrophotometric determination of aldicarb in formulations, *J. Assoc. Off. Anal. Chem.*, 57, 642, 1974.
172. **Broderick, E. J., Bourke, J. B., Mattick, L. R., Taschenberg, E. F., and Avens, A. W.,** Determination of methylcarbamate pesticides in the presence of methyl anthranilate in concord grapes, *J. Assoc. Off. Anal. Chem.*, 49, 982, 1966.
173. **Kuhr, R. J., Davis, A. C., and Bourke, J. B.,** Dissipation of Guthion, Sevin, Polyram, Phygon and Systox from apple orchard soil, *Bull. Environ. Contam. Toxicol.*, 11, 224, 1974.

Chapter 10

RESISTANCE AND SYNERGISM

INTRODUCTION

Insects possess a strategic weapon which places them at a definite advantage in the chemical warfare waged between pest and man. Under pressure of attack by a particular poison, certain insect strains are able to develop total immunity to the compound within only a few short generations. To make matters worse, the immunity or resistance may encompass other insecticides to which the enemy has never been exposed. Even the one-time miraculous DDT soon lost some of its "miraculousness" when various insect pests could no longer be controlled by the white powder. Other powders and liquids replaced DDT, but the marathon race between the synthetic-chemists' ingenuity and the insects' environmental adaptability had begun, and in some instances the arthropods are far ahead. Indeed, the winning of this race may be the single most important challenge facing plant protection scientists of today. Therefore, a thorough understanding of the biochemical and genetic mechanisms governing resistance is of paramount importance.

The most quoted definition of resistance is that of the World Health Organization, as follows: "Resistance to insecticides is the development of an ability in a strain of insects to tolerate doses of toxicant which would prove lethal to the majority of individuals in a normal population of the same species."[1] The operative word in the definition is "development" which rules out those cases where a particular insect species is naturally insensitive to a toxicant, i.e., it is tolerant to levels of insecticide which kill closely related species. Natural tolerance is common to all members of the species and does not require previous exposure to the chemical.

Unfortunately, all investigators do not subscribe to these differences between tolerance and resistance, so that some confusion exists. For example, a strain of insects which has developed only three- or fourfold resistance to a compound is labeled "tolerant," but if another strain becomes 10- or 20-fold resistant, it is dubbed "resistant." Where the line is drawn between tolerance and resistance varies with the investigator. It would be better to call this process simply a low level of resistance versus a high level of resistance, for in both instances selection by insecticide pressure was probably responsible for producing the strains. Also, in certain situations, the highest degree of resistance that can ever be achieved is only two- or threefold, since detrimental factors coselected by insecticide pressure inhibit any further concentration of resistance.

Another term used for low-level resistance is "vigor tolerance" which implies an overall enhanced ability to withstand conditions of stress, including exposure to insecticides.[2] The selected individuals are more apt to survive and reproduce, not because of any specific insecticide resistance mechanism, but because of lowered sensitivity to heat, desiccation, etc. Thus, protection quite often extends to include several different types of insecticides and involves a very complicated genetic inheritance.[3]

The development of resistance in the field is revealed by a steady, though sometimes unnoticed, decline in effectiveness of a chemical from one treatment to the next. In other words, the insecticide is functioning as a Darwinian sieve, allowing only resistant mutants to pass through and mate for propagation of future generations. Each year the number of resistant individuals swells, until suddenly no control is possible with the selecting insecticide. The first response of the pesticide applicator is often to double or triple the application rate. Besides insulting the environment and running the risk of high harvest residues or dangerous surface deposits, the increased dosages may further concentrate resistant genotypes and accentuate possible cross resistance to other compounds. Ultimately, it is necessary to concede the battle to the insect and switch to another product.

How do the resistant strains originate? Evidence supports a preadaptive hypothesis; that is, a normal susceptible population of insects contains a few individuals carrying genetic codes for a resistance mechanism to insecticide A.[4] These individuals exist in low numbers in nature because of a physiological handicap, such as reduced fertility or slow development, which makes them less competitive. However, when their food and surroundings become tainted with insecticide A, it is the "weaklings" that are best able to survive and multiply. If, on the other hand, the original susceptible population had inherently contained a high frequency of genotypes possessing the resistance mechanism, the insects would have been naturally tolerant to insecticide A, and control would not have been possible even with the first application.

The speed with which resistance develops in an insect population depends on many factors, but of major importance is the intensity of selection. If control can be maintained at a 95% or better level, resistance build-up is likely to be very slow, probably due to lack of genetic variability in the survivors. But when treatments kill only 70 to 90% of the insects, a sufficient gene pool is left behind for effective isolation and concentration of resistant mutants. Other parameters then become important, including the number of generations the pest completes in a year, the frequency of resistant genotypes in the population, the number of genes involved in conferring resistance, whether the gene expression is dominant or recessive, history of exposure to other toxicants, and counteracting effects of natural selection.

The known mechanisms responsible for resistance to insecticides are limited in number and can be divided into three categories: behavioral, physiological, and biochemical. Behavioral resistance is usually manifested as an avoidance of a treated area; i.e., there is a phenotype in the population which does not frequent a particular surface, and consequently, when this surface is treated, members of this phenotype are not harmed. It may also be that resistant mutants exist which contact the surface, but only for a very brief period of time, possibly because of an irritation or repellency elicited by the insecticide deposit. This type of resistance is rare and, thus far, has not been found to occur with methylcarbamate insecticides.

Physiological resistance is based on selection of genes which control reduced transport of insecticide to the site of action, slower penetration of insecticide through the cuticle, greater storage in nonsensitive tissues, more rapid excretion, reduced sensitivity of target tissues, or increased general vigor. Any one of these selections alone is seldom enough to cause significant protection from carbamate poisoning, but in combination they can be very effective.

Biochemical resistance is most common for all types of insecticides and centers around differences in enzyme activities between resistant and susceptible strains. For example, resistant individuals contain higher levels or more active enzymes for metabolism (detoxication) of the selecting toxicant, or the target enzyme in resistant insects is much less sensitive to inhibition by the insecticide. Again, one of these mechanisms may not be sufficient for producing immunity, but when coexistent with another mechanism, resistance can be substantial.

There are probably other factors contributing to the selection of resistant strains, but until the genetics regulating protection from toxicity are completely unraveled, the whole picture will not be seen. More needs to be learned about the number of genes conferring resistance, their chromosomal location, and how they interact and express themselves. Studies thus far suggest that only a small number of major genes are responsible for transmitting resistance, and in many cases one major gene has been correlated with a particular physiological or biochemical mechanism involved in resistance.[3,5] The expression of the major gene as a viable phenotype, however, hinges on the gradual accumulation and melding of the necessary modifying or ancillary genes needed for survival and propagation of the resistant strain.

When a resistant strain is available and purified, individuals can be mated with insects from the original susceptible strain under laboratory supervision, so that the offspring (F_1 generation) can be observed. Further parent-F_1 crosses or F_1 inbreeding clarify the genotype responsible for resistance by indicating whether inheritance is completely dominant or recessive or somewhere between; whether it is primarily monofactorial; whether it is transmitted through cytoplasm or chromosomes; whether it is sex-linked or autosomal; and whether it is heterogeneous or homogeneous.[6] With some insect species, marker genes can be used to determine which chromosomes carry resistance, and in a few cases linkage maps allow pinpointing the locus of the responsible gene(s).

All of the above procedures, from determining the level of resistance in the field to locating the mechanism and genes responsible for immunity, depend on a technique for quantitating differences in response of susceptible and resistant insects. This is done by comparing toxicity of the test chemical to populations of both strains under laboratory conditions. Insects from each group are treated by topical application, injection, spray, dip, ingestion, or exposure to a coated surface with a range of insecticide dosages necessary to cause from approximately 5% mortality to about 95% mortality during a standard time period (usually 24 hr). After sufficient replication, results are averaged and plotted as logarithm of the dose against percentage mortality in probits (units of standard deviation from the mean).[7] Assuming that the test population follows a normal homogeneous distribution pattern, a linear regression line is fitted through the data points, as shown on the left in Figure 10.1 (line A).

The assumption that the test population is homogeneous may not always be correct, however.[8] For example, the data plotted on the right of Figure 10.1 could be interpreted as line B (normal distribution) or as line C. In case of the latter, the abrupt change in slope indicates that the population is really heterogeneous, and results should be interpreted accordingly.

The most useful information extracted from the regression lines in Figure 10.1 are the slope and LD_{50} (lethal dose to kill 50% of the test population). A susceptible strain of insects would be expected to give results somewhat similar to line A, with an LD_{50} value in this case of 2.7 μg of insecticide/g of insect. Depending on the genetic makeup of the species, the slope of the log dosage–mortality (ld–m) line would vary from steep for a highly homogeneous population, with little difference in susceptibility of individuals, to shallow for a population exhibiting a wide range of

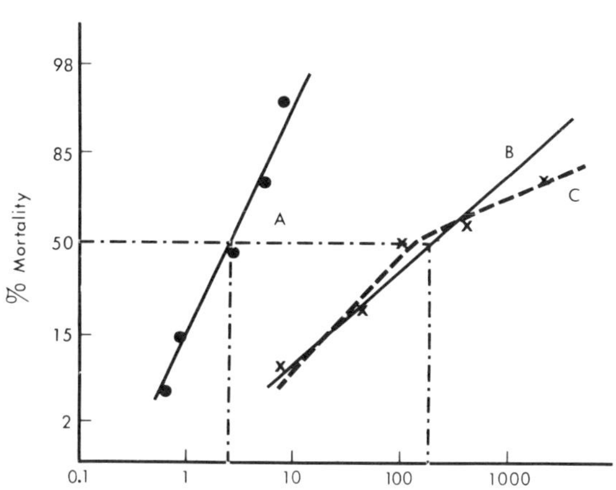

FIGURE 10.1. Log dosage–mortality (ld–m) curves for test populations of insects.

response to the toxicant. In general, field populations are inclined to yield shallower slopes than are found with laboratory strains, where genetic variability may be restricted.

As resistance develops, line A moves to the right, with a corresponding decrease in steepness of slope, to eventually give a response such as line B in Figure 10.1. The LD_{50} has increased to 180 μ/g, and the strain is said to be 67-fold resistant (LD_{50} from line B ÷ LD_{50} from line A). Note that, if LD_{95} values are compared, the resistance factor is even higher. The decline in slope is due to selection of resistant hybrids and homogeneous resistant genotypes which respond with greater variability to the insecticide. However, as resistance becomes more concentrated, the slope tends to steepen again.

A cross between resistant and susceptible insects yields a population which reacts somewhere between lines A and B, depending on the number of genes involved, degree of dominance, etc. Vigor tolerance is seen as a slight shift of the ld–m line to the right, but with no significant change in slope (or possibly an increase in slope). Since some of these differences may be minor, it is important to standardize test procedures, because the slope and LD_{50} for a particular population and insecticide can change considerably with application method (solvent used, whether dip or spray, etc.), sex ratio of insects, time to read death, temperature, age of insect, and so forth. Thus, great care must be taken to insure that these variables are kept to a minimum when comparing the response of different insect strains to an insecticide.

After establishing that a resistant strain has developed from pressure by a particular insecticide, the next chore is to determine whether the insects simultaneously acquired an insensitivity to the action of other insecticides. This umbrella protection is termed "cross resistance" or "multiple resistance." The difference between the two terms is genetic, though in outward appearance the result is the same. Cross resistance is operating when the same defense mechanism (usually governed by a single gene pair), such as increased oxidative detoxication, protects against several different poisons. Multiple resistance occurs when resistance to various chemicals is conferred by different, coexisting defense mechanisms in the same insect strain, with each mechanism apparently coded by a different gene rather than all mechanisms controlled by a pleiotropic expression of the same gene.[6] In any event, this ability of the insect to develop immunity to a compound before even "seeing" it has rendered many potential insecticides useless before they left the chemist's bench.

The opposite of multiple resistance is negatively correlated resistance, whereby selection of resistance by pressure from insecticide A leads to a corresponding selection of susceptibility to insecticide B, and vice versa. Therefore, when A is no longer effective in managing the insect, a switch to B gives nearly complete control again for a few years. Alternate use of insecticides A and B can then be repeated ad infinitum. Ideally, the genes controlling such a phenomenon should be allelic to ensure continuous back-and-forth selection. Unfortunately, negative correlation has been established for only a few pairs of insecticides, but the process continues to offer exciting possibilities for solving the problem of insect resistance.

There is some degree of reversion from resistance back to susceptibility when the pressure from an insecticide is released. The amount depends on the nature of controlling genetics and on the overall fitness of the resistant strain. For example, if resistance is mostly dominant and many individuals are homozygous, which is often the case in laboratory cultures, reversion may never take place. In the field, where interbreeding with more vigorous wild types is possible, the length of time needed for a return to susceptibility depends on the size and geographic distribution of the resistant population, on the number of "susceptible pockets" within this area, and on the extent of movement of both resistant and susceptible insects.[9] A complicated multifactorial inheritance of semidominant resistance is usually lost quite rapidly under these circumstances.

Of course, any return of insecticide pressure to the reverting population invariably results in a rapid recurrence of resistance. Presumably, the many, nonspecific ancillary genes which contributed to the build-up of a resistant strain are very slowly diluted. Thus, even though the frequency of resistance genes *per se* has returned to a level comparable to the original susceptible population, the redevelopment of resistance is an easy task because of the ready availability of the necessary background genetics.

The potential for insecticide resistance is present in every insect species man attempts to control. Some, such as the housefly, have responded by developing immunity to pyrethroids, nicotine, DDT, HCH, cyclodienes, organotins, organophosphates, carbamates, and even the microbial agent *Bacillus thuringiensis.*[5] Others, such as the European corn borer, have failed to produce a viable resistant strain under long and varied chemical pressures. The degree of resistance need not be great, for in some cases only a moderate increase in LD_{50} can render a compound useless under field conditions.[10]

A recent count catalogs 225 species of arthropods resistant to one or more insecticides.[11] Although the list includes some of the most devastating health and agricultural pests, it represents only a small percentage of the insects and mites which have been classified as detrimental to man's well-being. The following pages briefly summarize what is known about survival of insects under challenge by carbamate insecticides. Unfortunately, adequate documented studies of carbamate resistance are few in number, and the reader should consult several recent publications covering resistance to other groups of insecticides for a fuller picture of this fascinating and frightening adaptability by insects.[5,6,9-13]

OCCURRENCE OF RESISTANCE

Pressure under Field Conditions

Of the 225 documented cases of field resistance of arthropods to insecticides, only a handful involve carbamate compounds.[11] This is probably due mainly to the relatively short period that these materials have been available for large-scale commercial application. Early naive predictions that the new class of insect poisons would defy Darwin's theory were soon dashed when carbaryl resistance appeared in strains of the apple leaf roller[14] and the tobacco budworm.[15,16] More recently, carbaryl resistance has been measured to be near 40-fold in such pests as the sugar cane frog hopper,[17] green rice leafhopper,[18] Egyptian cotton leafworm,[19] and the cattle tick.[20]

Loss of carbamate sensitivity has not been limited to agricultural pests, nor to the compound carbaryl. Six field-collected strains of the German cockroach in Louisiana were shown to require 1.9 to 15 times more Baygon to obtain an LD_{50} comparable to an insecticide-susceptible strain.[21] The shallow slopes for the ld–m curves indicated considerable heterogeneity in the sample populations, and it was predicted that more selection pressure would yield rapid development of highly resistant genotypes. *Anopheles albimanus* mosquitoes in parts of Central America are resistant to carbaryl and Baygon, though the selection pressure probably did not originate from control measures aimed at the mosquitoes, but is more likely a side effect of intensive nearby spraying of cotton and rice fields.[22]

There are certainly many other reports of methylcarbamate failures in the field, but unless a comparison was made between the toxicities of the insecticide to susceptible (S) and to resistant (R) strains, it is impossible to ascertain whether loss of control was due to true resistance, natural tolerance, or faulty application techniques. In fact, the S strain should ideally be susceptible to all insecticides to ensure that tolerance of the methylcarbamate is not a result of cross resistance or multiple resistance from previous exposure to other insecticides. Thus, proving that resistance exists and proving that the resistance developed as a result of pressure from the carbamate are two different problems.

An example of this dilemma is the cabbage looper. For many years this pest of cole crops was controlled by DDT, but in the mid to late 1950's resistance developed to the point where DDT was no longer effective in New York State.[23] At that time, a resistant and a susceptible strain of loopers were collected and brought into the laboratory for culture. During the next several years, the organophosphate parathion served as a replacement, but once again the insect adapted, so that in 1968 the use of parathion for looper control was also curtailed.[24] Parathion-resistant larvae were saved for laboratory rearing, and the search for an alternative material began again. One of the more promising products uncovered was the oxime carbamate methomyl, which has since been widely recommended as a loopercide.

Throughout this period, loopers were exposed to only minor amounts of carbaryl, and yet a marked loss in sensitivity to the methylcarbamate took place. As shown in Figure 10.2, analysis revealed that the S and DDT-R strains were equally susceptible to carbaryl, giving a base-line LD_{50} value of 5 μg per 4th-instar larva. Development of parathion resistance, however, was accompanied by a simultaneous 15-fold increase in resistance to carbaryl, as seen by results with the parathion-R strain. Finally, loopers collected from cabbage plots in 1973 demonstrated total immunity to carbaryl, an immunity which had developed essentially in the absence of carbaryl.

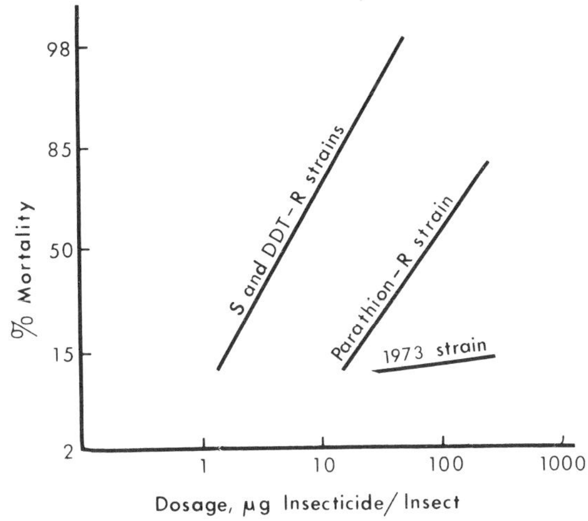

FIGURE 10.2. Toxicity of carbaryl to susceptible, DDT-resistant, parathion-resistant, and 1973 field strains of 4th-instar cabbage loopers.

In a similar way, DDT and parathion pressure were building cross resistance (10 X) to methomyl even before the compound was released for commercial application.[25] Then, methomyl selection from 1968 to 1973 enhanced the resistance factor to 27-fold. However, one very important difference between carbaryl and methomyl is the inherent toxicity of the compounds to S loopers; that is, carbaryl's LD_{50} is 5 μg per 4th-instar larva and methomyl's LD_{50} is 0.03 μg per 4th-instar larva. Therefore, even with 27-fold resistance among 1973 loopers, the toxicity of methomyl remained a low 0.8 μg per larva. In fact, at 5 μg per insect it is doubtful that carbaryl ever would have controlled loopers, and this natural tolerance to carbaryl was simply accentuated by DDT, parathion, and methomyl pressure.

Prior exposure to organochlorines and/or organophosphorus compounds does not automatically result in predisposed immunity to methylcarbamates, however. Bollworms up to 15-fold resistant to DDT showed no apparent cross resistance to carbaryl.[26] Field selection of endrin- and endosulfan-resistant strains of tobacco hornworms occurred without change in susceptibility to carbaryl.[27] *Anopheles albimanus* mosquitoes in areas heavily sprayed with DDT, dieldrin, and many organophosphates maintained their sensitivity to carbaryl and Baygon, although laboratory pressure of a field strain with the latter carbamate for only three generations produced more than 75-fold resistance to both structures.[28]

To remove some of the uncertain variables present in field studies, many researchers have turned to carefully controlled selection of insecticide-resistant strains in the laboratory. Unfortunately, the correlation between resistance development under laboratory pressure and that occurring in practice is not always good, though qualitative trends are evident. Thus, the need for careful documentation of carbamate-resistance outbreaks through prudent observations coupled with laboratory confirmation cannot be overemphasized.

Pressure Under Laboratory Conditions

Because of the problems caused by insect resistance to organochlorine and organophosphorus insecticides, one of the first studies conducted with carbamates included a test for potential resistance development. Laboratory pressure of houseflies at the LD_{70} level with UC 10854 (3-isopropylphenyl N-methylcarbamate) for 67 generations yielded only an eightfold change in sensitivity.[29] However, pressure with the very closely related 3-*t*-butylphenyl N-methylcarbamate produced almost instantaneous resistance, so that after only ten generations a 15-fold degree of protection was apparent (Figure 10.3). Selection with carbaryl gave a curve similar to the 3-*t*-butyl compound, whereas Pyrolan resistance developed at a rate between the two extremes.

The relative rates of resistance development to the four compounds could have been predicted by comparison of ld–m plots for susceptible flies, as shown in Figure 10.4. The very steeply sloped line through a narrow dosage range for UC 10854 indicates a fairly homogeneous response to this insecticide, making selection of resistant individuals difficult. A more heterogeneous response to Pyrolan allows for greater survival of resistant genotypes, which then breed immunity to this dimethylcarbamate. Finally, carbaryl and 3-*t*-butylphenyl N-methylcarbamate were virtually nontoxic to part of the population, and these individuals were rapidly selected at an LD_{70} level of pressure to yield highly resistant strains after only a few generations.

By increasing the selection dosage to give 90 to 95% mortality and by maintaining the pressure throughout larval and adult stages, a strain of flies resistant to UC 10854 was obtained after only seven generations.[30] However, tremendous differences in response between sex and life stage were observed. For instance, female adults were 50-fold resistant, but male flies were little more than

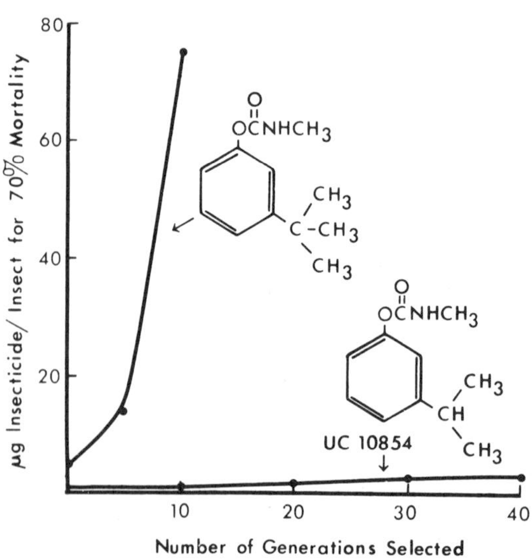

FIGURE 10.3. Development of resistance by houseflies to two closely related aryl methylcarbamate insecticides during laboratory pressure at the LD_{70} level. (Adapted from Reference 29, with permission.)

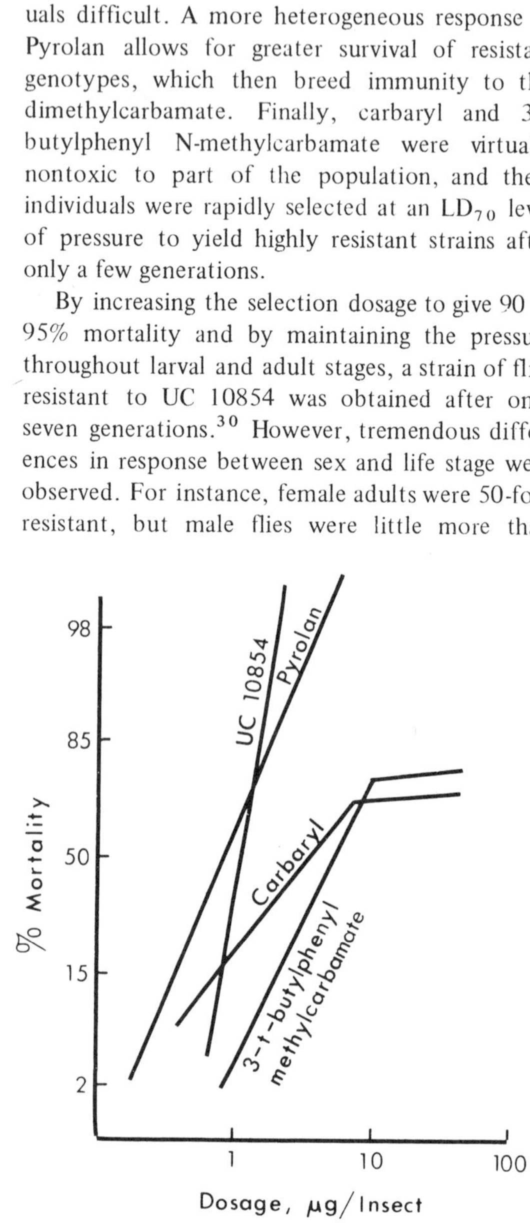

FIGURE 10.4. Log dosage–mortality curves for houseflies treated with various carbamate insecticides. (Adapted from Reference 29, with permission.)

threefold resistant, even after 22 generations of pressure. Meanwhile, the larvae also expressed a much lower build-up of insensitivity, being only 7.5 times more tolerant after 22 generations. With a dimethylcarbamate, Isolan, and the same strain of flies, the reverse was found; that is, more resistance occurred in larvae (20 X) than in adult females (7 X) after 20 generations of exposure. Continued pressure on the larvae for another 80 generations did not increase Isolan resistance in adults significantly.[31] Thus, life stage and sex must be considered when evaluating carbamate resistance in insects.

The importance of a standardized test method can also be illustrated from the above study. When the degree of resistance to UC 10854 by adult females was determined by using insecticide dissolved in oily dioctyl phthalate instead of acetone, the level changed from 50-fold to only sevenfold. In addition, when flies were injected instead of topically treated, the difference in LD_{50} between R and S strains dropped to only fivefold. Such results suggest that differences in cuticle permeability may contribute to carbamate resistance in the housefly.

The near inability of populations of malaria mosquitoes and southern house mosquitoes to develop resistance to UC 10854 after 50 and 30 generations, respectively, of larval selection at 80 to 95% mortality levels made this compound a promising candidate for use in public health.[22,32] The slight decreases in sensitivity were attributed to vigor tolerance, and no resistant genotype was thought to be present in the original populations. In addition, the carbamate-pressured strains lost some of their resistance to dieldrin and DDT, suggesting negatively correlated toxicity. Another study with three strains of *Culex* mosquito also revealed only slight inclination towards UC 10854 resistance.[33] However, UC 10854 was eventually ruled out as an agent for control of insect vectors because of mammalian-toxicity problems.

A more successful methylcarbamate insecticide in public-health programs is Baygon. Unfortunately, results from laboratory studies with houseflies and mosquitoes support potential resistance development in the field. For example, 2 years of pressure with Baygon yielded a 20-fold tolerant strain of houseflies.[34] With *Culex pipiens fatigans* mosquitoes, the degree of resistance depended on life stage and on method of exposure.[35] Larvae surviving 80 to 95% selection pressure for 35 generations were 25 times less sensitive to Baygon, whereas the adult forms were only threefold (contact toxicity) to fivefold (topical) resistant. Conversely, similar pressure on adults for 24 generations gave nearly the same fivefold (topical) to eightfold (contact) resistance in adults, but only threefold in larvae. From these results it would seem that Baygon's useful life as a mosquito larvicide would be much shorter than as an adulticide.

"Grooming" of mosquitoes by prior long-term exposure to other insecticides may hasten development of insensitivity to Baygon. Field-collected *Anopheles albimanus* resistant to DDT and dieldrin (and to some organophosphates), when subjected to rigorous Baygon pressure for only three generations, produced larvae almost immune to the carbamate, although the adults retained most of their susceptibility.[22,28] But when an insecticide-susceptible strain was reared in a Baygon-contaminated environment through both larval and adult stages for extended periods (14 to 19 generations), only minor resistance to Baygon occurred.[22]

Stored-product insects, such as the granary weevil, have also demonstrated an ability to resist Baygon's toxicity. Low-level pressure (50 to 70% mortality) through seven generations produced a fivefold resistant strain, and after 14 generations the difference in susceptibility between R and S individuals was more than sevenfold.[36]

The responses of insects to laboratory pressure with carbaryl vary considerably. Some species are either naturally tolerant or do not seem capable of developing any significant resistance. Thus, after 15 generations of exposure, Mexican bean beetles did not change in their sensitivity to carbaryl,[29] and European corn borers developed only a modest increase in tolerance after 12 generations of living on LC_{60}- to LC_{70}-contaminated diets.[37] An investigation indicated that potential mechanisms for resistance were present in borer populations,[38] but these were probably not exploited by the insects because of decreased fecundity, fertility, and general vigor which accompanied their selection. On the other hand, most studies indicate that selection of houseflies with carbaryl via topical application or exposure to a treated surface for a relatively low number of generations results in rapid build-up of resistance.[29,39,40]

Conflicting reports on resistance of lice to

carbaryl exist in the literature, although most of the differences in response by the test populations may have been due to previous insecticide exposure. One strain, resistant to DDT, was pressured for 30 generations with carbaryl, using an LD_{50} dose, and no significant change in sensitivity appeared.[41] However, the lice possessed some degree of natural tolerance to carbaryl and the selection pressure may have been too low to single out resistant mutants. Another strain, resistant to lindane, did yield a fivefold change in susceptibility to carbaryl following 25 lifetimes of carbamate exposure.[42] Continuous and more vigorous pressure brought a sudden burst in evolution so that by generation 85, greater than 285-fold resistance to carbaryl had developed.[43] In fact, most lice allowed to wallow in technical carbaryl powder for 24 hours survived apparently unharmed.

The development of resistance by insects is not totally undesirable. Possession of an ability to survive in an insecticide-contaminated environment would be an advantage for many parasitic and predaceous species, which are often "controlled" along with the pests during chemical treatments. Although only a few such species have been identified,[11] they are not generally investigated and may be much more numerous. Laboratory studies indicate that a Braconid parasite of boll weevils, when pressured with carbaryl at the LD_{50} level, developed fourfold resistance in seven generations.[44] Strains of predaceous mites collected from an orchard which had been sprayed for nine years with carbaryl were found to require 25 to 77 times more carbaryl for an LD_{50} dose than susceptible mites.[45] More cases of carbamate resistance undoubtedly exist among beneficial arthropods, and proposals for rearing resistant strains of parasites and predators in the laboratory for release into insecticide- sprayed areas continue to offer an interesting approach to integrated management of insect pests.

Multiple and Cross Resistance

Most information on multiple or cross resistance between different carbamate insecticides or between carbamates and other groups of insecticides is too meager to permit confident predictions, although some general trends are apparent. Knowing that carbamates are metabolized by a few major pathways in insects (Chapter 7), and that enhanced metabolic detoxication is the primary mechanism responsible for carbamate resistance (discussed later in this chapter), one would expect considerable cross resistance to occur among carbamate insecticides. Indeed, this has been shown to be true for houseflies.[29,30,39,46,47] An extreme example is the situation where complete resistance to UC 10854 or Isolan also causes virtual immunity to carbaryl, Baygon, Mesurol, Zectran, Pyrolan, Pyramat, and several others.[46] Note that in this case cross resistance extended from methylcarbamates to dimethylcarbamates (and vice versa), though in other studies methylcarbamate-immune strains of flies were found to be only moderately resistant to dimetilan, Isolan, and Pyrolan.[29,39] Finally, cross resistance from other carbamates to UC 10854 does not occur in housefly populations.[29,40]

Other species have also developed blanket resistance to several carbamate structures, though degrees of response vary with compound. Baygon-resistant (25 X) culicine mosquitoes displayed almost equal resistance to Baygon, carbofuran, and 2-isopropylphenyl *N*-methylcarbamate, with lower levels evident for many others, down to the oxime carbamate aldicarb (3.6 X).[35] Interestingly, cross resistance was shown to be uniformly greater among *ortho*-substituted phenyl *N*-methylcarbamates than among corresponding *meta*-substituted analogs, a situation which also appears in housefly resistance. Analysis of a field-collected strain of green rice leafhoppers from a heavily treated agricultural area revealed simultaneous resistance to Baygon (169 X), carbaryl (100 X), Banol (69 X), Meobal (24 X), Tsumacide (20 X), and several others, the lowest level once again occurring with an oxime carbamate, methomyl.[18] Baygon resistance in granary weevils crossed over to include Zectran and UC 10854,[36] and Baygon-resistant *Anopheles* larvae were almost equally tolerant of carbaryl.[28]

Examination of cross-resistance patterns in German cockroaches showed that the method of treatment is extremely important in determining resistance levels.[48] A strain selected by dipping nymphs in carbaryl-LC_{90} solutions developed 75-fold resistance to carbaryl when analyzed with a dip method. However, topical application of carbaryl to R and S roaches revealed no difference in carbamate susceptibility. On the other hand, resistance to Mobam was not apparent with the dip technique, but a difference in toxicity between

R and S strains emerged on topical application of Mobam. Cross resistance (topical) also extended to include dimetilan, Isolan, aldicarb, and particularly, Matacil, though not Baygon or UC 10854.

Since carbamates were developed in part for use on DDT- and cyclodiene-resistant species, the degree of overlap between organochlorine and carbamate resistance would, hopefully, be at a minimum. This is not an unreasonable expectation, since the mechanism imparting immunity in many DDT-resistant insects involves specific metabolic dehydrochlorination to the nontoxic DDE. Thus, DDT-R and S strains of these insects should respond similarly when challenged with a carbamate. Indeed, this has been the case for certain carbamates against houseflies,[29,49] honeybees,[50] and cabbage loopers (Figure 10.2). However, some DDT-resistant houseflies are also known to carry additional resistance to carbaryl, Pyrolan, Isolan, Baygon, UC 10854, and Zectran.[29,30,47,49,51] In addition, German cockroaches resistant to DDT were found to be 75 times more tolerant of carbaryl than susceptible roaches.[52] Such instances of reduced sensitivity to both DDT and a carbamate are probably due to multiple resistance rather than true cross resistance. In other words, the gene coding for DDTase is closely linked and carried through selection with the gene(s) governing resistance to carbamates.

Conversely, it is possible to develop DDT resistance during carbamate selection, as illustrated by the following examples. Long-term pressure (> 40 generations) of houseflies with UC 10854, Pyrolan, or Isolan yielded high levels of resistance to DDT, and sometimes to methoxychlor,[29,46] though other strains pressured by the same compounds retained their susceptibility to DDT.[29,30,53,54] Carbaryl-pressured flies[29,40,55] and Baygon- or UC 10854-pressured *Culex pipiens fatigans* mosquitoes[33,35] responded with slight multiple resistance to DDT, but *Anopheles albimanus* mosquitoes confined to a Baygon-contaminated environment increased their resistance to the organochlorine from fivefold to 20-fold.[28] Although body lice pressured with carbaryl lost their resistance to DDT initially, by the 78th generation the insects were immune to carbaryl and more than 458 times less sensitive to DDT.[43] Finally, granary weevils selected for Baygon resistance simultaneously displayed an increased tolerance for DDT.[36] Thus, it appears that selection for carbamate resistance usually leads to coselection for some degree of DDT resistance.

The expression of multiple resistance between carbamates and cyclodienes is totally unpredictable and is often different from that between carbamates and DDT. For example, pressure of a field strain of malaria mosquitoes with Baygon increased the insects' resistance to DDT, but did not change their response to dieldrin.[28] Conversely, selection of houseflies by treatment with phenyl N,N-dimethylcarbamate gave no change in DDT susceptibility, but produced total immunity to dieldrin.[54] Added to this, a UC 10854-resistant population of houseflies became resistant to both DDT and dieldrin,[46] whereas a carbaryl-resistant strain of flies showed no serious cross-resistance to either organochlorine.[40] To complete the confusion, a *Culex* mosquito strain moderately resistant to UC 10854 carried slight cross resistance to both DDT and dieldrin, but neither DDT- nor dieldrin-resistant strains were resistant to UC 10854.[33]

Information on other organochlorines is limited, but German cockroaches have been shown to develop multiple resistance to chlordane and to Mesurol, Baygon, or carbaryl.[52,56] Houseflies selected for resistance to phenyl N, N-dimethylcarbamate or UC 10854 also became resistant to chlordane or lindane.[30,46,54,57] However, selection for lindane insensitivity did not include resistance to phenyl N,N-dimethylcarbamate.[57] Until more is learned about carbamate-organochlorine interactions, it appears safe to say that, if an insect develops resistance to an organochlorine compound, it will probably remain sensitive to most carbamates, but insects developing resistance to carbamates will possibly also become more insensitive to the heavily chlorinated insecticides.

Because of a similar mode of action, organophosphorus and carbamate insecticides are often assumed to display general cross resistance in most insect species. This could only be true, however, if resistance involved a change at the site of action, i.e., in the central nerve synapses. Although there is some evidence for altered acetylcholinesterase in resistant strains (discussed later), most cross resistance between the two groups is probably due to a common type of metabolic detoxication. Unfortunately, the substrate specificity for the deactivation is not easy to define, so that cross-resistance patterns are not very predictable and certainly not mandatory.

Houseflies very resistant to UC 10854 were shown to be moderately cross resistant to a wide variety of organophosphates, with levels of 12 to 35 times for malathion down to three to four times for parathion.[46] Another UC 10854-R strain displayed less than a fivefold increase in LD_{50} for parathion, malathion, Chlorthion, diazinon, dicapthon, and ronnel.[30] High resistance to carbaryl or phenyl N,N-dimethylcarbamate[54,55] resulted in a slight carry-over to parathion or diazinon respectively, but flies modestly resistant to carbaryl were not cross resistant to diazinon, parathion, malathion, Sumithion, dimethoate, or Dipterex.[29,40] The only instance when houseflies exhibited a convincing level of cross resistance between a carbamate and organophosphate occurred when a diazinon-R strain was pressured with Pyrolan.[29] Eventually the flies became more than 40-fold resistant to both insecticides.

Only a three- or fourfold change in toxicity between susceptible and carbamate-resistant strains of culicine mosquitoes was observed when the insects were treated with malathion, diazinon, or bromophos, and even smaller differences were found with parathion, methyl parathion, fenthion, and Abate.[33,35,58] But Baygon-R (<100 X) anopheline mosquitoes were 36 times harder to kill with parathion and methyl parathion, and 20 times more tolerant to malathion and fenitrothion.[28] Finally, highly carbamate-resistant rice leafhoppers collected from the field were immune to malathion (580 X) and methyl parathion (400 X).[18,59]

Cross resistance from organophosphorus compounds to carbamates appears to evolve with ease and is not uncommon. Hence, several organophosphorus-R strains of houseflies proved to be almost impossible to kill with Baygon, UC 10854, or Zectran.[47] This is possibly true cross resistance via a common metabolic mechanism, since all carbamates were 100 times more toxic in the presence of piperonyl butoxide. Loss of carbaryl susceptibility (20 X) was more pronounced in flies pressured with parathion than for parathion itself (10 X), though cross resistance to the dimethylcarbamates, Isolan, Pyrolan, and dimetilan, was not as great.[49,60] Flies which had developed resistance (37 X) to diazinon through repeated exposures in a dairy barn were found to be equally opaque to carbaryl and also moderately resistant (10 to 14 X) to Isolan and to Pyrolan.[61] When the flies were placed under additional diazinon pressure in the laboratory resistance (6 X) to dimetilan also surfaced.[62] After 30 to 40 generations of rearing in the absence of diazinon, some regression of carbamate resistance took place, but the flies continued to be less sensitive than a susceptible strain.[63,64] Malathion-R houseflies[51] and green rice leafhoppers,[65] Chlorthion-R houseflies,[30] and parathion-R cabbage loopers[25] (Figure 10.2) have also displayed cross-resistance to certain methyl- and dimethylcarbamate insecticides.

Turning to public-health pests, there is fortunately much less multiple resistance evident. For example, *Aedes nigromaculis* field strains essentially immune to several organophosphorus insecticides were readily controlled by Baygon, carbofuran, Bux, and 3-(2-butyl)phenyl N-phenylthio-N-methylcarbamate.[66] The latter material was also very toxic to organophosphorus-R strains of *Culex tarsalis.*[67] Another *Culex* species, pressured in the laboratory with malathion for 33 generations, developed very little (2.2 X) cross resistance to carbaryl.[68] Lastly, bed bugs resistant to malathion and diazinon did not change in their susceptibility to carbamates.[69] Thus, methylcarbamate insecticides may prove to be very useful against disease vectors resistant to organophosphorus insecticides, though their use on OP-resistant agricultural pests may be more limited.

Negatively Correlated Resistance

The resistance researcher's dream of perfect negative correlation between resistance and susceptibility for two different compounds has not magically come true with the discovery of carbamate insecticides. However, some promising clues have been uncovered. For example, *Anopheles albimanus* mosquitoes became more susceptible to dieldrin and DDT when under pressure from the methylcarbamate UC 10854, whereas a corresponding strain reared for the same length of time in the absence of UC 10854 maintained a high level of organochlorine resistance.[32] Body lice developed almost 300-fold resistance to carbaryl through 85 generations of selection pressure, but at the same time the lice became 69 times more susceptible to lindane.[43] A strain of houseflies which appeared incapable of resisting phenyl N, N-dimethylcarbamate after many generations of pressure did, however, become less sensitive to organophosphorus and organochlorine insecticides.[54,57]

Thus, in these instances, the factors controlling resistance to carbamate esters appear to be independent of those imparting resistance to other compounds. However, to achieve complete negatively correlated toxicity, it should be possible to pressure with A and lose resistance to B, and then pressure with B while regaining susceptibility to A. In the above studies, only half of the picture has been completed. But the possibility of finding two compounds which operate as negative correlates continues to offer an intriguing and, hopefully, achievable solution to the problem of insect resistance to insecticides.

A recent discovery in the realm of negative correlation involves the use of two similar carbamate structures. When Egyptian cotton leafworms were pressured with carbaryl at the LD_{50} level for four generations, a strain resistant (10 X) to carbaryl resulted.[19] However, treatment of both R and S insects with the same dose of the closely related N-acetyl carbaryl resulted in greater kill among the resistant insects. Apparently, selection of carbaryl-resistance mechanisms was accompanied by coselection of an enhanced ability to deacetylate the carbaryl analog near the site of action. Thus, resistance to carbaryl imparted susceptibility to N-acetyl carbaryl.

MECHANISMS AND GENETICS

Enhanced Detoxication

Almost concomitant with the discovery of insect resistance to carbamates, a search was begun for biochemical or physiological differences between resistant and susceptible strains which would explain the protective mechanisms. Because of prior studies concerning resistance to other types of insecticides, it was no great surprise when the first reports appeared in 1961 indicating that carbaryl- and UC 10854-resistant houseflies were able to metabolize methylcarbamate insecticides at a faster rate than susceptible parent strains.[60,70] When metabolism was blocked by a synergist, the resistant insects regained susceptibility to the carbamates.

Since then, numerous other studies have shown that carbamate-resistant houseflies possess enhanced metabolic capabilities compared to their susceptible counterparts.[6,71-76] An example of the type of difference between strains is shown in Figure 10.5. Susceptible and Baygon-resistant flies

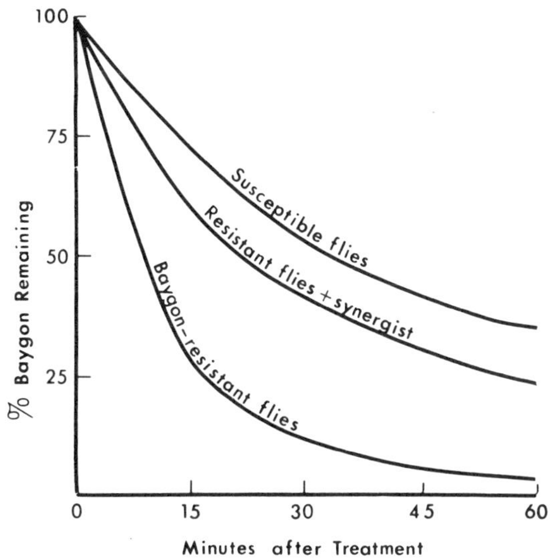

FIGURE 10.5. Metabolism of Baygon by susceptible and resistant houseflies in the presence or absence of synergist. (Adapted from Reference 75, with permission.)

were injected with 0.1 μg of radiolabeled insecticide, and recovery of parent material after various time periods gave respective rates of metabolism as shown.[75] The difference in rates is about three- to fourfold, which is substantially lower than the 50-fold difference between topical LD_{50} values for the S and R strains. However, the rate of metabolic destruction is probably sufficient in R insects to "keep up" with the rate of entry of insecticide through the cuticle, thus preventing sufficient toxicant from reaching the site of action. Note that pretreatment of the resistant flies with a synergist, which inhibited most metabolism, produced a degradation curve similar to that for the susceptible flies (Figure 10.5). Three different species of mosquito have been shown to resist Baygon's toxic action by a similar two- to threefold enhancement of in-vivo detoxication rate.[28,35,58,77] Likewise, the susceptibility of cabbage looper strains to carbaryl and methomyl is inversely related to the rate of metabolism of the carbamates.[25,78,79]

During several of these studies with S and carbamate-R insects, in-vitro subcellular preparations were used to determine the precise metabolic pathways involved in resistance. In almost all cases, detoxication occurred via oxidative conversions catalyzed by a mixed-function oxidase (mfo) enzyme system(s) requiring NADPH and molecular oxygen (see Chapter 6). Whether the same nonspecific enzyme system is responsible for resistance to all carbamates or whether closely related enzymes are operating on specific carbamate substrates has not been clearly resolved.

One component of the mfo system, namely cytochrome P-450, may be the critical factor in the resistance mechanism. Studies have shown that carbamate-resistant strains of cabbage loopers[79] and houseflies[80,81] possess elevated P-450 titers in comparison to susceptible strains, though the increased levels of cytochrome are not always sufficient to explain the increased metabolic rates. Perhaps the key lies in the recently discovered qualitative differences between binding sites of P-450 in microsomes from resistant and from susceptible strains of houseflies.[82] The fact that these differences are under genetic control implies that selection of cytochromes with greater specificity for certain insecticide structures is possible.

Metabolic pathways other than oxidative pathways may contribute more to resistance in some insect species. For example, carbaryl-resistant German cockroaches hydrolyzed more methylcarbamate than did susceptible roaches.[83] Conjugation of hydrolytic products also appeared to take place more readily in the resistant roaches, a situation similar to that occurring with carbofuran in carbamate-resistant houseflies.[84] Added to this, there have been indications of a synergist-sensitive mechanism other than oxidation affording protection from methylcarbamate toxicity in houseflies.[85] It is possible that the synergists act by interfering with glucose conjugation reactions, since this has been shown to occur in vitro with insect tissue preparations.[86,87]

One last possibility revolves around the mutant aliphatic esterase (aliesterase) theory, which plays a part in insect resistance to organophosphorus compounds.[88] The hypothesis of some researchers is that the activity of aliesterase is lower in resistant insects, the enzyme being partially replaced by a mutant aliesterase (e.g., phosphatase) which catalyzes the rapid detoxication (hydrolysis) of organophosphorus insecticides. In a similar way, Isolan-R (8 X) houseflies have been found to contain 50% less aliesterase activity than S flies.[71,89] It was demonstrated that, as resistance developed under Isolan pressure, aliesterase activity decreased and Isolan metabolism increased. Indeed, back-cross experiments implied that resistance to Isolan and low aliesterase activity were genetically linked and could have been transmitted by the same gene. In deference to these findings, however, it has been shown that green rice leafhoppers resistant to methylcarbamate or organophosphate insecticides had 6 to 14 times higher aliesterase activity than did susceptible leafhoppers,[90] and activity was also higher in other carbamate-resistant strains of houseflies.[3]

Regardless of the metabolic mechanism conferring resistance, the difference between metabolites found in R and S strains of insects is usually quantitative, and not qualitative. In other words, susceptible and resistant insects are able to form the same series of detoxication products, but resistant insects form them at a faster rate. Therefore, any insect showing an ability to metabolize a carbamate possesses the ability (genetic material) to become resistant to its toxicity. The fact that this does not necessarily follow *a priori* has been demonstrated recently with carbaryl and the European corn borer. Despite years of carbaryl pressure and an inherent ability of the borer to metabolize carbaryl at a reasonable rate, no significant resistance has evolved.[38]

Enhanced Penetration or Excretion

It is certainly conceivable that an insect could develop resistance to a carbamate by preventing or slowing down penetration of the toxicant through the cuticle or gut wall. Little has been done concerning examination of the latter possibility, though this could indeed be a factor contributing to resistance. With respect to cuticle penetration, results vary considerably. For example, carbaryl moved faster through susceptible German cockroach cuticle than through resistant cuticle,[83] and the degree of cuticle permeability among cabbage looper strains was qualitatively related to their susceptibility to methomyl.[25] Conversely, UC 10854-resistant houseflies actually absorbed methylcarbamates, such as aldicarb and UC 10854, at a faster rate than did susceptible flies.[70,72] Presumably, the enhanced absorption was compensated by much faster metabolic destruction of the toxicant, since much less intact carbamate was recovered from internal extracts of resistant strains than from those of susceptible insects. Lastly, several carbamate-resistant strains of houseflies were shown to absorb a wide variety of methylcarbamate insecticides at rates no different from those of susceptible insects.[75,85]

All of the above measurements of penetration

rates were conducted with topically treated insects; thus, conditions may not have been truly representative of actual insecticide exposures. A more realistic approach involved placing resistant and susceptible mosquito larvae in Baygon-contaminated water. Interestingly, cuticle from susceptible larvae removed more insecticide from the water than did resistant cuticle.[77] This phenomenon may be just as important as penetration rate through the integument.

Another potential mechanism contributing to carbamate resistance is enhanced excretion of insecticide from the insect's body, thereby reducing the amount of toxicant available for reaction with acetylcholinesterase. In insect metabolic studies performed with carbamates to date, this pathway does not appear to be very important, though considerably more evaluation is certainly needed to eliminate this hypothesis. What has been found in resistant houseflies and roaches is primarily an enhanced excretion of carbamate metabolites, with only modest amounts of unchanged parent compound recovered from either R or S insects.[72,75,83–85] The higher quantities of excreted products are probably simply a reflection of a higher metabolic rate in resistant strains, the primary function of the insects' defense mechanisms being conversion of lipid-soluble toxicants to water-soluble products for elimination from the body.

Modified Acetylcholinesterase

The toxic action of carbamate insecticides is dependent on the inhibition of acetylcholinesterase present in the synapses within the central nervous system of insects (see Chapter 3). It is evident that, if a population of insects could develop appropriate alterations in the rate (K_a) of carbamate-esterase complex formation, rate (k_2) of enzyme carbamylation, and/or rate (k_3) of decarbamylation (Chapter 3), their susceptibility to carbamates could be significantly reduced; that is, the population would be resistant to the insecticides. This hypothesis has been examined recently for two species of insects.

Houseflies 20-fold resistant to Baygon were compared with susceptible flies by analysis of acetylcholinesterase activity in head homogenates.[34] The concentration of carbaryl (9×10^{-8} *M*) or Baygon (4×10^{-9} *M*) required for 50% inhibition of nerve enzyme was nearly identical for R and S insects, indicating that little change in overall affinity or reactivity of the acetylcholinesterase had accompanied resistance development. In addition, the recovery rate (k_3) for carbamylated enzyme from both strains was identical, so that resistance had no effect on the sensitivity of AChE in this instance.

Green rice leafhoppers resistant to parathion or malathion, when compared with susceptible insects, showed no difference in cholinesterase inhibition by Baygon based on assays of whole-body homogenates.[90] However, one strain, resistant to organophosphates as well as to carbamates in the field, was found to have considerably lowered chlorinesterase sensitivity. Later studies demonstrated a parallel relationship between this loss in activity and a decrease in methylcarbamate toxicity, as illustrated in Table 10.1.[18] When toxicity ratios were plotted against cholinesterase inhibition ratios for several aryl methylcarbamates (Figure 10.6), a very good correlation (r = 0.96) indicated that resistance of the leafhoppers to methylcarbamate insecticides was probably due to insensitive acetylcholinesterase within the central nervous system.[91] A plausible explanation for the change in sensitivity is an altered anionic binding site on the enzyme, since the order of inhibition

TABLE 10.1

Methylcarbamate Insecticide Toxicity and Cholinesterase Inhibition in Resistant and Susceptible Strains of the Green Rice Leafhopper[18]

	Topical LD_{50} (µg/g)		Whole-body ChE I_{50} (M)	
Strain	Carbaryl	Baygon	Carbaryl	Baygon
Susceptible	0.71	2.6	1.4×10^{-6}	1.3×10^{-5}
Resistant	71	44	6.0×10^{-5}	1.5×10^{-3}
Ratio (R/S)	100	169	43	115

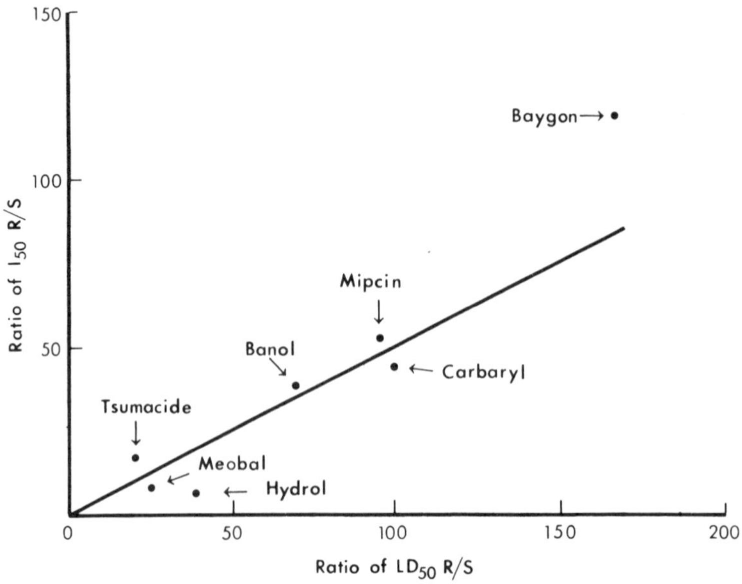

FIGURE 10.6. Relation of toxicity to cholinesterase inhibition for various aryl methylcarbamate insecticides against resistant and susceptible green rice leafhoppers. (Adapted from Reference 91, with permission.)

by different phenyl methylcarbamates in the S strain is dependent on ring substituents as follows: 2-i-C_3H_7 >2-i-C_3H_7O > 3-CH_3; by contrast, the order for inhibition of resistant acetylcholinesterase is 2-i-C_3H_7 = 3-CH_3 > 2-i-C_3H_7O.

Similar findings were uncovered in a recent comparative study of AChE in homogenates of susceptible and Baygon-immune *Anopheles albimanus* larvae.[91a] Although both enzymes were similar in their affinity (K_a) for the substrate acetylthiocholine, they responded quite differently to Baygon. Indeed, there was a decrease of 12,600-fold in the sensitivity (k_i) of the carbamate-resistant AChE as opposed to the susceptible enzyme. Though the altered sites on the insensitive AChE might be expected to confer resistance to all carbamate insecticides, evidence was presented to show that changes in sensitivity were specific for certain structures rather than generic.

Before leaving this section on resistance mechanisms it is important to reemphasize the fact that in most cases no one mechanism is singularly responsible for inferring resistance to a carbamate. In addition, it appears that different coexisting mechanisms operate on the toxicant in at least a multiplicative fashion rather than by simple addition.[92] A good example of this principle was recently illustrated with a housefly strain virtually immune to carbaryl.[9] A series of factors responsible for resistance were genetically isolated in substrains and recombined to give the results shown in Figure 10.7. Note that minor changes in detoxication or cuticle penetration, when put together, gave very large differences in susceptibility. This, of course, explains why many investigators examining only one resistance mechanism fail to uncover differences between R and S insects anywhere near the differences obtained for topical toxicity.

Genetics

The hereditary units which determine the physical and biochemical makeup of insects, including the presence or absence of resistance mechanisms, are the genes. Strung like beads on a necklace, they are arranged in precise order on lengths of chromosomal thread, and each thread has a duplicate. Thus, genes exist in pairs, or alleles, and although both members of the pair may be the same "shape" and "color" to distinguish them from other pairs, alleles may differ in their "message", that is, one allele confers susceptibility to a carbamate and another confers resistance. The frequency of the resistant allele and the degree to which its expression dominates over susceptibility determine the level of carbamate resistance in an insect population. This assumes, of course, that resistance is monofactorial, i.e., due to the action of a single gene pair.

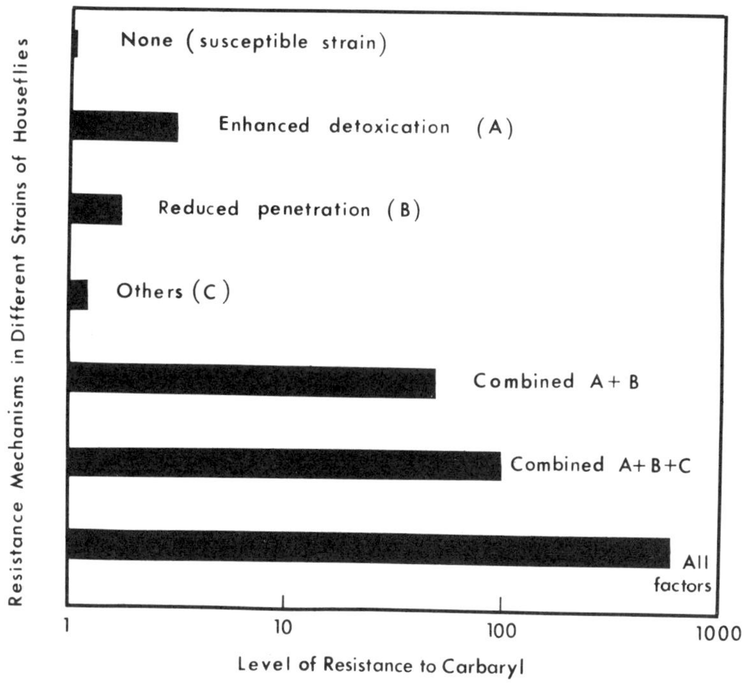

FIGURE 10.7. Combination of various resistance mechanisms contributing to carbaryl resistance in the housefly. (Adapted from Reference 9, with permission.)

Multifactorial inheritance, which appears to be the case for most carbamate resistance, as will be seen below, is more complicated.

Determining the exact location of each gene which contributes to resistance is extremely difficult, so that up to now attempts have been limited to finding the "necklaces" or chromosomes which carry the responsible genes. With respect to carbamates, genetic experiments have been conducted for a little more than 10 years and center almost exclusively around the housefly. This insect has six pairs of chromosomes; one pair of these are sex chromosomes, which are apparently not associated with inheritance of resistance. Thus, most studies deal with the five pairs of autosomes, which are simply referred to as chromosomes I to V (the Wagoner numbering system[93] will be used throughout this chapter).

The first report appeared in 1965 and concerned itself with the genetics of resistance to Isolan and UC 10854.[31] Reciprocal cross-experiments between Isolan-R or UC 10854-R and S houseflies demonstrated that the genes conferring resistance to the carbamates were located on autosomes, and not on sex chromosomes; that is, resistance was transmitted by either male or female flies. Both R strains inherited their enhanced tolerance for Isolan primarily through a partially dominant major R gene, though some minor cytoplasmic influence may also have been involved. A different organophosphorus-resistant strain of flies, which was cross-resistant (129 X) to carbaryl even in the presence of piperonyl butoxide, passed its protection from generation to generation via a more complicated polygenic system.

Later studies using flies with visible chromosome markers placed the location of the major gene conferring resistance to carbamates in the Isolan-R strain on chromosome II.[94] Most of the multifactorial resistance in the organophosphorus-R strain was also due to contributions from chromosome II, but chromosome III played a significant supporting role. These results were confirmed by research with other Isolan-R strains of flies, where resistance to Isolan, carbaryl, or carbaryl + piperonyl butoxide was due to factors governed mainly by semidominant gene(s) on chromosome II, with partial assistance from chromosome III and/or chromosome V.[95-97] A Japanese strain of housefly resistant to carbaryl also owed its survival mostly to factor(s) on chromosome II, with some help from chromosomes III and IV.[98] Finally, Baygon or carbaryl insensitivity in flies with multiple resistance to carbamates, organophosphates, and DDT was coded by major genes on chromosomes II[97], or on chromosomes II and III with additional contributions from chromosome V.[98]

The importance of chromosome II in conferring carbamate resistance in the housefly was thus firmly established. Next, genetic confirmation of the mechanism responsible for resistance was needed. Although evidence was mounting for a close relationship between Isolan resistance and lowered aliesterase activity,[71,95,96] most studies pointed toward enhanced metabolism as the critical factor, and therefore this property was anticipated to be linked to autosome II. With this in mind, a study was initiated comparing a Baygon-pressured strain of flies with two susceptible strains carrying various chromosome markers.[74] As expected, an enhanced mixed-function oxidase system yielding more rapid oxidative detoxication of Baygon and Matacil in resistant flies was proven to be regulated by partially dominant genes on autosome II. Additional support from chromosome V was also attributed to effects on metabolic rates, but contributions from chromosome III were shown to be unrelated to enzyme activity (or cuticle penetration).

Thus, factors governing increased mfo activity and carbamate resistance in houseflies are primarily located on chromosome II, but further study is needed to ensure that both phenomena are truly coded by the same gene. Interestingly, certain types of organophosphate, DDT, and naphthalene resistance have also been linked to chromosome II in the fly, although several different gene loci (and several different mechanisms) are probably involved.[92,96-105] However, enhanced detoxication via a more active mfo system is certainly one way by which insects could evolve cross resistance to a wide variety of toxic molecules.

Support for this hypothesis exists from results with two different resistant strains of houseflies.[76] A Baygon-R strain, with most of the resistance and mfo activity associated with chromosome II as described above, was found to be cross resistant to aldrin, allethrin, diazinon, and malathion. In each case lowered toxicity was due to greater mfo-mediated deactivation of the respective insecticide from factors controlled by chromosome II. Another strain, resistant to DDT and to organophosphates, followed a similar pattern, but this time resistance to aldrin, DDT, diazinon, and Baygon and increased oxidase activity were all linked to gene(s) on chromosome V. Substrains from each resistant strain containing only one major R autosome indicated that oxidase activity was inherited through only one chromosome, but that overall resistance to insecticides was passed on by more than one chromosome. Again, the proof that a single gene, be it on chromosome II or on chromosome V, codes for both enhanced oxidative detoxication and resistance to insecticides is lacking. Nevertheless, it seems very probable that, when the loci are pinned down, they will be identical.

The only other species which have been used in studying the genetics of carbamate resistance are the fruit fly, southern house mosquito, and green rice leafhopper. In case of the fly, pressure by carbaryl yielded 8- to 80-fold resistance, due mainly to a partially dominant gene on chromosome II,[106] the same location responsible for DDT, HCH, and parathion resistance.[107,108] In mosquito larvae, resistance to Baygon was shown to be autosomal, semidominant, and polygenic in nature.[3,58] Subsequent genetic analysis placed the greatest contribution to resistance on chromosome II, with help from chromosome III.[5] For the leafhopper, where resistance to carbamates is due to insensitive AChE, as described earlier in this chapter, inherited insensitivity to Baygon is transferred from generation to generation by a single partially dominant gene allele.[18]

Obviously, our understanding of the biochemical genetics associated with carbamate resistance in houseflies far outweighs what is known for other insect species, particularly agricultural and health pests. Although it is tempting to further extend this tremendous bank of knowledge with additional genetic analyses of resistance in the fly, the time has come for those working in this area of research to turn their attention to other economically important insects in an attempt to assure ourselves that what is happening in houseflies is relevant to what occurs in other resistant species.

SYNERGISM

It is almost impossible to thoroughly discuss methylcarbamate insecticide metabolism, mode of action, or resistance without mention of the effects of synergists on these processes. Thus, it is fitting that this book conclude with a brief overview of the role of synergists in carbamate development and future use. For the purposes of this discussion, a synergist will be defined as any chemical which in itself is not toxic to insects as dosages used, but, when combined with an insecticide, greatly enhances the toxicity of the insecticide. This, of course, differs from potentiation, which occurs when two toxic materials applied together elicit a response greater than that expected from the sum of the individual toxicants (also referred to as synergism by some).

In theory, the use of synergists in combination with insecticides could solve many problems. For example, since a synergist reduces the amount of toxicant required for insect kill, much smaller quantities of pesticide would be required for field applications. This would result in considerable reduction in the amount of environmental insult caused by a particular insecticide (assuming that synergists are not as hazardous as insecticides) and in lower cost for pest control (assuming that the synergist at the recommended dose is cheaper than the insecticide). In addition, a synergist often broadens the spectrum of an insecticide to include many naturally tolerant species and, more important, resistant strains of insects, thus extending the usefulness of the insecticide (assuming that the broadened spectrum is limited mostly to destructive insects).

Partly because of a failure in the above parenthetical assumptions, the theoretical potential of synergists has been a commercial disappointment until now, with the exception of pyrethroid-synergist aerosols. Nevertheless, it appears that synergists may have a larger part to play in pest control, especially in combination with methylcarbamate insecticides. This is based on the findings reviewed earlier in this chapter showing that insect resistance and natural tolerance to carbamates are due primarily to enhanced rates of detoxication and that synergists act mostly by inhibiting the enzymes effecting these detoxications.[109-118] The feasibility for commercial development of carbamate-synergist pesticides will be examined after a brief look at the chemistry and biological activity of the synergists themselves.

Methylenedioxyphenyl Compounds

Of the several different groups of chemicals able to synergize carbamate insecticides, the most widely studied are the methylenedioxyphenyl (also called methylenedioxybenzene or 1,3-benzodioxole) derivatives. Their history dates back to around 1942, when sesamin (Figure 10.8) was isolated and identified as one of the active ingredients of sesame oil, a patented synergist for enhancing the toxicity of pyrethrin insecticides.[119,120] Many other natural and synthetic compounds belonging to the methylenedioxyphenyl (MDP) family followed, with the ultimate commercial development in the United States of piperonyl butoxide (pb), sulfoxide, propyl isome, and tropital (Figure 10.8).[116] The first of these is the major seller at about 800,000 lb/year;[121] a few other MDP compounds have enjoyed minor commercial success.[122]

The first realization that MDP materials would increase the observed toxicity of carbamate insecticides occurred in 1958 with the discovery that Isolan, Pyrolan, carbaryl, and 3-*t*-butylphenyl N-methylcarbamate were more lethal to houseflies, bean aphids, and German cockroaches when applied in combination with traditional pyrethrin synergists such as sesamex, piperonyl butoxide, sulfoxide, or propyl isome (Figure 10.8).[123,124] This phenomenon was demonstrated as a marked change in the dosage-mortality curves. For example, in the presence of synergist, the ld-m lines for carbaryl and for 3-*t*-butylphenyl N-methylcarbamate against houseflies shown in Figure 10.4 were straightened, steepened, and displaced considerably to the left, resulting in a 50-fold reduction in LD_{50}.

Several other studies followed through the next few years to firmly establish the effective synergism of dimethyl- and methylcarbamates by a variety of MDP compounds against naturally tolerant insect species and species resistant to the action of organophosphorus and organochlorine insecticides.[41,46,47,49,124-131] But perhaps most important were reports that MDP synergists restored the toxicity of carbamates toward carbamate-resistant strains of insects.[29-31,35,43,70,75,85,132] Thus, resistant houseflies which were able to tolerate individual treatments of 100 μg of carbamate were killed by approximately 1 μg of insecticide when pb (5 μg) was added (Table 10.2).

This flurry of activity led to a reevaluation of the toxicity of many aryl methylcarbamate structures by coadministration with pb or other MDP derivatives. Results such as those illustrated in Table 10.3 are typical.[133] Note that all of the insecticides are approximately equal in their ability to inhibit acetylcholinesterase, and

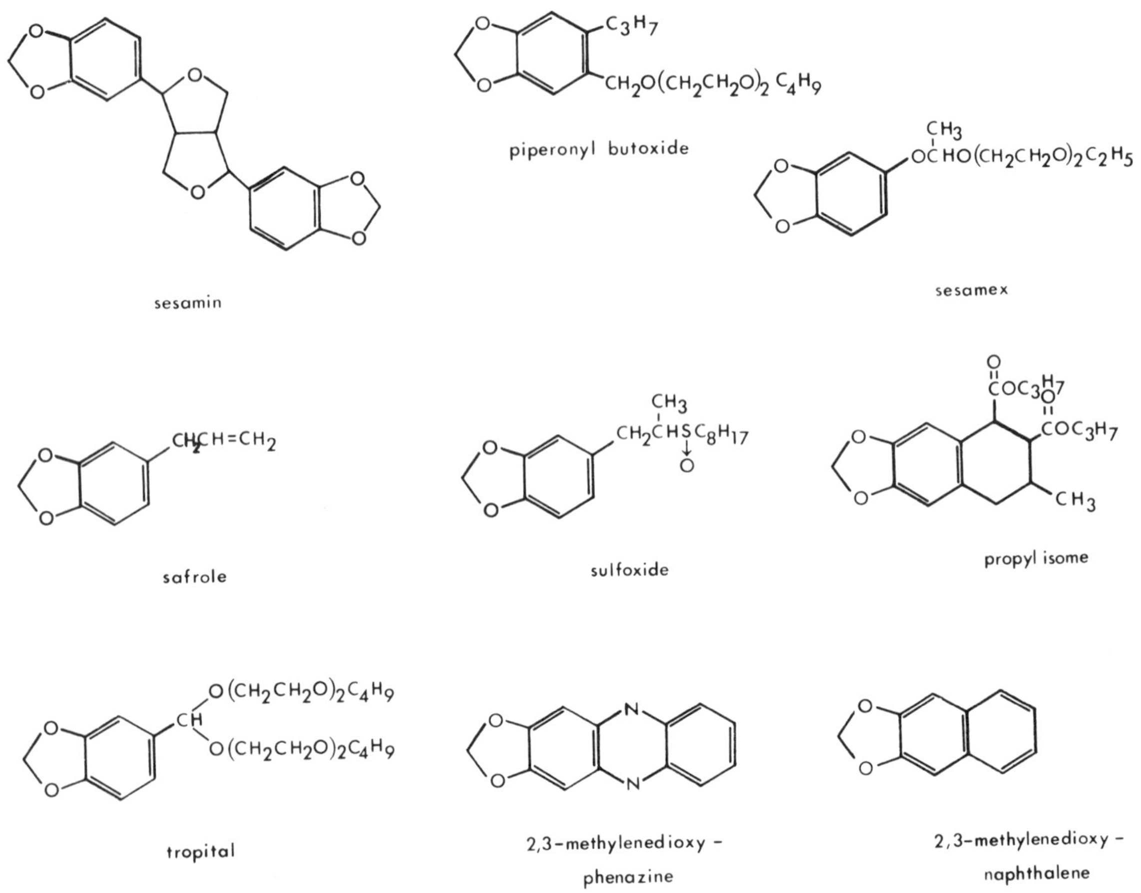

FIGURE 10.8. Structures of methylenedioxyphenyl synergists.

TABLE 10.2

Toxicity of Aryl Methylcarbamate Insecticides Alone and in Combination with Five Parts of Piperonyl Butoxide against Susceptible and Carbamate-resistant Strains of Houseflies[75,85,118,132]

	Average topical LD_{50} (µg/fly)					
Fly strain	UC 10854 alone	UC 10854 + pb	Baygon alone	Baygon + pb	Carbaryl alone	Carbaryl + pb
Susceptible	1.9	0.25	0.37	0.13	100	0.24
UC 10854-resistant	100	1.1	100	0.60	100	1.9
Baygon-resistant	–	–	100	1.0	–	–

consequently all should be equally toxic to the flies. On the contrary, LD_{50} values for unsynergized methylcarbamate esters ranged from 6.5 to more than 500 µg per fly. In the presence of synergist, however, differences in toxicities were virtually erased, allowing the intrinsic activity of the insecticide to manifest itself. Later, data from several studies with a wide range of substituted phenyl N-methylcarbamates indicated a definite correlation between I_{50} and synergized LD_{50},[110,132,134] a factor which also lends credence to AChE inhibition as the carbamate mode of action (see Chapter 3). More recently, the relationship between inhibition and synergized toxicity was narrowed to a correlation (r = 0.875) between affinity (K_a) of the methylcarbamate for

TABLE 10.3

Inhibition of Housefly Head Acetylcholinesterase and Toxicity of Various Methylcarbamate Insecticides to Houseflies in the Presence or Absence of Piperonyl Butoxide

		Topical LD_{50} (μg/fly)	
Methylcarbamate	AChE I_{50} (M)	Carbamate alone	Carbamate + pb
Carbaryl	9.0×10^{-7}	>500	12
Baygon	6.9×10^{-7}	26	7
UC 10854	3.4×10^{-7}	90	9
2-*s*-butoxyphenyl	3.1×10^{-7}	50	7.5
2-*i*-propoxy-5-methoxy	5.6×10^{-7}	6.5	3.0

Adapted from Reference 133, with permission.

AChE and synergized toxicity for a series of carbofuran analogs.[135,136]

Though piperonyl butoxide was the synergist of choice for many of the above investigations, there were indications that some MDP compounds performed better than others with certain carbamates. This prompted an examination of the structural requirements for biological activity, and the importance of an intact methylenedioxy moiety became readily apparent, though it was shown that substituents attached to the phenyl ring also contributed to synergistic activity.[134,137] A more extensive study involved the synthesis and analysis of 62 MDP compounds in combination with three different aromatic methylcarbamate insecticides.[138] Results such as those for the first four compounds in Table 10.4 confirmed that the integrity of the planar methylenedioxy ring could not be altered. Even the very minor modification produced by exchanging the methylene hydrogens with deuterium has been shown to result in a significant loss of synergistic activity.[109,110] However, substitution of an oxygen by sulfur in the MDP nucleus was permissible (Table 10.4).

The nature of the side chain attached to the phenyl ring affected the degree of synergism in roughly the same order with all three carbamates (Table 10.4). Presumably, the substituents influence lipid solubility, and hence cuticle penetration, transportation to the site of action, and binding at the site of action. It is also possible that the side chains control in-vivo stability of the synergists and thus modify activity by preventing metabolic destruction. However, a later study failed to reveal any significant correlation between rate of cuticular penetration or metabolism and degree of synergism for a large number of MDP compounds, though the least effective synergists were most readily metabolized.[139]

Multiple additions of aliphatic groups, halogens, or nitro groups to the phenyl nucleus in the 3,4- or 4,5-positions also yielded effective synergists.[139] By incorporating these substitutions into another ring system, a variety of tricyclic analogs, such as 2,3-methylenedioxyphenazine[140] and 2,3-methylenedioxynaphthalene (Figure 10.8),[109] were prepared and found to greatly enhance the toxicity of methylcarbamate insecticides. For example, the latter synergist lowered carbaryl's toxicity to 0.114 μg per housefly, which represents a reduction of almost 440-fold.[139] Structure–activity studies with methylenedioxynaphthalenes indicated that aryl substituents lowered synergistic potential, though movement of the methylenedioxy moiety from the 2,3-position to the 1,2-position enhanced activity.[111,139,141] As expected, sulfur could replace oxygen in the MD moiety, but any alteration at the methylene position, including replacement of hydrogens with deuterium, destroyed all synergistic activity.

By what mechanism does a synergist cause a carbamate to become hundreds of times more active as an insecticide? As indicated earlier, the additives usually act by preventing the breakdown of the insecticide, thus allowing greater accumulation of poison within the insect's body. This has been demonstrated repeatedly with houseflies and several carbamates; that is, in-vivo metabolism of the ester has been severely retarded following pre- or coadministration of MDP compounds.[60,70,75,76,85,132] A typical example is shown in Figure 10.5 for Baygon in R and S flies.

TABLE 10.4

Effects of Various Modifications of MDP Structures on Their Synergistic Activity when Combined with Methylcarbamate Insecticides against Houseflies[138]

Synergist	Degree of synergism (5:1 syn:ins.) Carbaryl	3,4-Dimethoxyphenyl N-methylcarbamate	Zectran
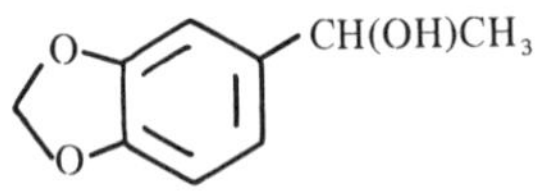	90	53	3.7
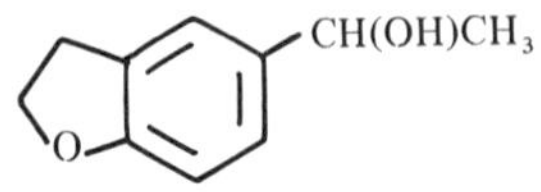	19	7	1.2
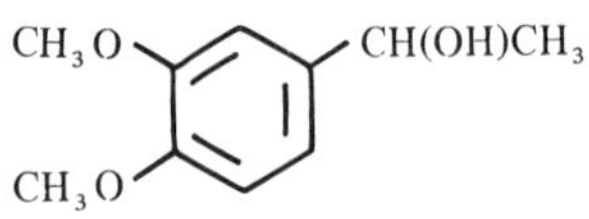	3	4	1.1
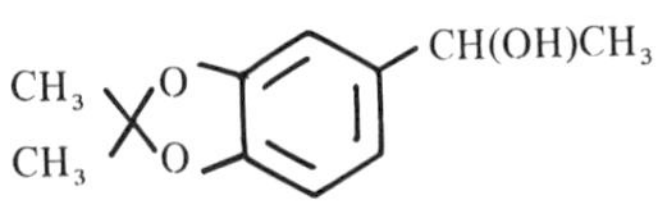	1	1	1.0
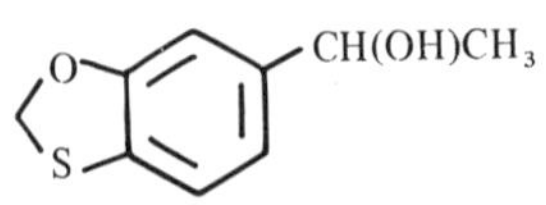	67	30	3.0
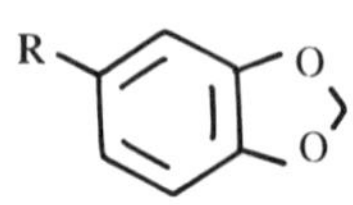			
H	1	–	1.1
OH	5	4	1.4
CH_2OH	12	9	1.3
CH_3	32	43	2.8
$CH(CH_3)OH$	90	53	3.7
OCH_3	82	77	4.2
$CH(CH_3)OC(O)C_6H_5$	129	56	3.9
$CH(CCl_3)OC(O)C_6H_5$	180	118	11.7

More specifically, it has been established that MDP synergists inhibit the mixed-function oxidase(s) responsible for catalysis of oxidative detoxication of carbamate insecticides in such insects as housefly,[74,75] cockroach,[142] mosquito,[143] blowfly,[144] cabbage looper,[78,79,145] and European corn borer.[38] The structure of the synergist, particularly the nature of phenyl ring substituents, can influence the degree to which mfo activity is blocked, as illustrated in Table 10.5.[145] There was an 80- and 160-fold range, respectively, in inhibition of the oxidative breakdown of carbaryl in looper fat body and gut homogenates when only three different synergists were compared. On the other hand, a study with a series of MDP structures indicated that inhibition of housefly microsome oxidations was independent of ring substituent, although the substrates used were not methylcarbamates.[146]

The exact mechanism of insect mfo inhibition by MDP synergists remains a mystery, but several theories have been advanced. These have been thoroughly debated in several recent articles;[113,116-118,121,147,148,148a] only a few will be covered here. The most likely explanation is that MDP compounds compete with carbamates (and other xenobiotics) for the active site(s) on the microsomal mfo system; that is, they serve as alternate substrates, thereby preventing binding and subsequent metabolism (detoxication) of the insecticide. Obviously, the synergists must have a higher affinity for the active site and/or a lower turnover rate than the carbamates. Most of the evidence for this proposal is based on the fact that MDP compounds are oxidatively metabolized (at the MD moiety and on side chains) by the same insect enzyme preparations which metabolize insecticides.[116,149-152] In addition, the rigid structural requirements of an intact MD group and a "handle" on the phenyl moiety (discussed earlier) suggest that "fit" on an active site is involved in mode of action.

A second theory proposes the formation of an electrophilic benzodioxolium ion, resulting from a hydride transfer from the methylene carbon (Figure 10.9, ion form), which acts as the inhibitor by acylating a neucleophilic grouping near the active site of the mfo system.[153] Since such an ion would be quite reactive, the hydride transfer must take place *in situ* at or very near the site of action, that is, within the microsomes. Support for this proposal comes from the inactivity of methylene-substituted MDP compounds (Table 10.4), particularly the inactivity of deuterium analogs,[109,110,141,154] and from molecular orbital studies, which account for the possibility of such a reaction.[155]

A third proposed mechanism involves the formation of a fairly stable methylenedioxy free radical (Figure 10.9) following enzymic homolytic cleavage of one of the hydrogen atoms from the methylene group.[156] The radical then acts as the inhibitor of oxidative detoxication. Proof lies in a mathematical study of structure–activity relationships for a series of MDP compounds. In addition, some support comes from the fact that MDP molecules have been shown to inhibit the classical formation of free radicals by Fenton's reagent.[157]

The ultimate target for either the intact methylenedioxyphenyl molecule or for an MDP metabolite, ion, or radical appears to be cytochrome P-450, the terminal oxidase in the microsomal mfo system of insects (see Chapter 6). This, of course, is the enzyme responsible for activating molecular oxygen and inserting it into carbamate (and other) substrates. Two recent reports describe a definite correlation between cytochrome P-450

TABLE 10.5

The Effect of Synergists on the Oxidative Detoxication of Carbaryl by Cabbage Looper Gut and Fat Body Homogenates

	Inhibition (molar I_{50})	
Synergist	Fat body	Gut
Piperonyl butoxide	1.0×10^{-6}	5.0×10^{-7}
Safrole	3.6×10^{-5}	7.5×10^{-6}
Isosafrole	8.0×10^{-5}	8.0×10^{-5}

Adapted from Reference 145, with permission.

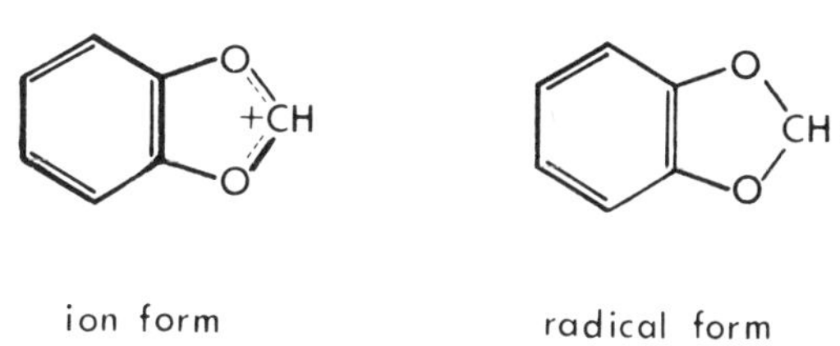

FIGURE 10.9. Forms of methylenedioxyphenyl compounds which may attack a mixed-function oxidase system.

inhibition by pb, sesamex, or sulfoxide in housefly microsomes and decreased oxidative metabolism of carbamate insecticides.[80,158] There is also considerable spectral evidence to show that MDP compounds combine with P-450 in a manner which prevents binding of carbamate substrates.[148] Interestingly, the spectral evidence favors reaction of a free radical, carboxolium ion, or carbanion to form a stable complex with the cytochrome.

Organothiocyanates

Two studies are principally responsible for creating a potential role for aryl thiocyanates as synergists of carbamate insecticides. The first demonstrated the activity of structures such as Thanite (Figure 10.10), a compound which synergizes phenyl N-methylcarbamate, Isolan, UC 10854, Zectran, and carbaryl against both S and carbamate-R houseflies with about the same effectiveness as piperonyl butoxide.[159] Metabolic inhibition was determined as the mode of action, since pretreatment of flies with Thanite reduced in-vivo breakdown of Zectran, Baygon, and phenyl N-methylcarbamate.[85,159] However, a combination of pb and Thanite with 3,4-dimethoxyphenyl N-methylcarbamate gave greater than additive synergism, suggesting that the two synergists were inhibiting metabolism of the insecticide by reactions at different sites along the mfo system.

A second study evaluated the effects of incorporating the thiocyanate grouping into a series of benzyl, phenyl, and alkyl molecules.[160] Several of the benzyl thiocyanates synergized carbaryl's toxicity to susceptible houseflies, though many possessed inherent toxicity of their own. Shifting atoms around to form the corresponding isothiocyanates (–NCS) or preparation of phenyl thiocyanates yielded toxic compounds with no synergistic activity. The most promising material was 4-nitrobenzyl thiocyanate (Figure 10.10), whose mode of action probably involves

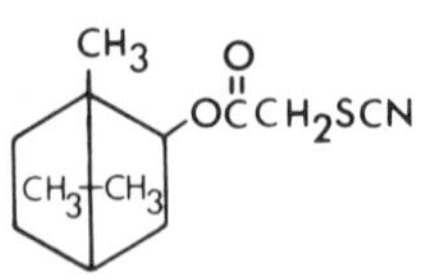

Thanite

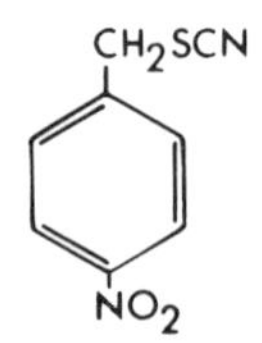

4-nitro benzyl thiocyanate

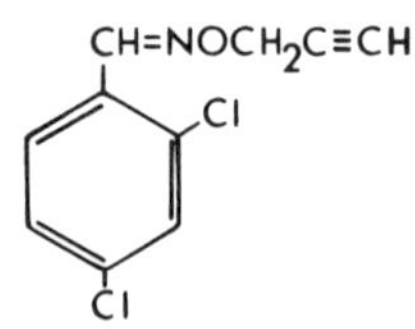

2,4-dichlorobenzaldoximino propynyl ether

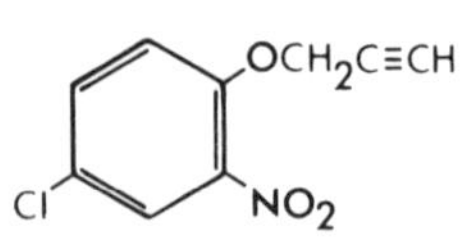

RO 5-1557

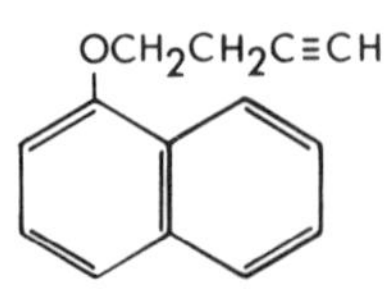

1-naphthyl 3-butynyl ether

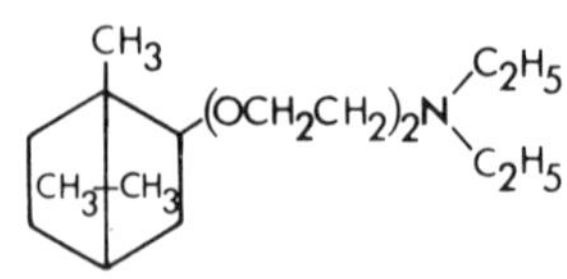

GD-11

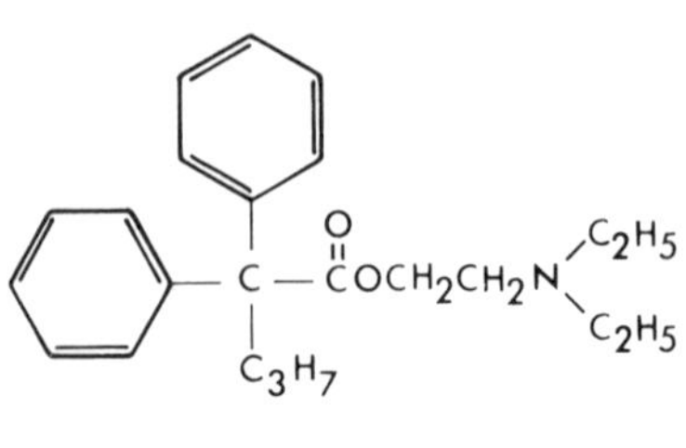

SKF 525-A

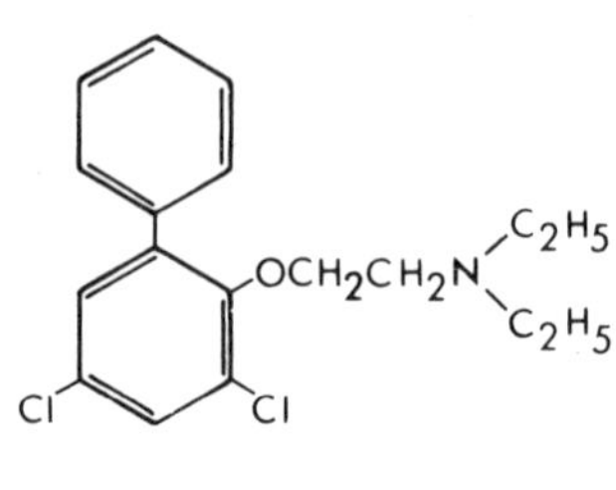

Lilly 18947

FIGURE 10.10. Structures of organothiocyanate, propynyl ether, and aryloxyalkylamine synergists.

inhibition of carbamate metabolism by filling in for the insecticide as an alternate substrate (which is slowly metabolized).[160] It is also possible that the thiocyanates form coordination complexes or chelate with the iron component of the mfo system.

Propynyl Ethers

A relatively recent class of synergists was patented in 1966, namely, the phenyl 2-propynyl ethers (also called propargyl phenyl ethers or propargyloxybenzenes).[161] A representative of this group is RO 5-1557 (Figure 10.10). Many of the ethers, when evaluated in combination with several carbamate insecticides against susceptible houseflies, performed better than pb, sulfoxide, or safrole.[162] They also enhanced the toxicity of dimetilan, Isolan, and Mobam to carbamate-resistant houseflies and synergized carbaryl against several other insect species.

Structure–activity analysis of a series of phenyl 2-propynyl ethers with carbaryl indicated that substitution of the aryl nucleus with a nitro group or two to three halogens gave the most active synergists.[163] Formation of thioethers or insertion of aryl, alkyl, or alkoxy groups into the ring lowered activity. However, compounds such as 3-methyl-4-methylthiophenyl 2-propynyl ether have been shown to successfully synergize carbaryl when applied in combination to houseflies.[164]

Naphthyl propynyl ethers administered with carbaryl to susceptible strains of flies also exhibit synergistic activity.[165] Among 30 derivatives, the most potent material synthesized was 1-naphthyl 3-butynyl ether (Figure 10.10), which enhanced carbaryl's toxicity 176 times. Structural modifications displayed the need for an intact C≡C moiety, though a terminal location was not vital. Only modest losses in synergistic activity followed replacement of the ethereal oxygen atom with a sulfur atom, and, similar to methylenedioxy-naphthyl compounds, nuclear substitution lowered synergistic activity.

A departure from the basic aromatic structure, while maintaining the propynyl ether group, provided synergists such as 2,4-dichlorobenzal-doximino propynyl ether (Figure 10.10).[121] It had been shown previously, however, that this type of structure, without the acetylenic linkage, was active as a synergist for carbaryl against several insect species.[166]

The mode of action of the ethers is probably inhibition of oxidative detoxication, which is indicated by a recent study showing that the rate of metabolism of Mobam in both R and S houseflies was reduced in the presence of 2,4,5-trichlorophenyl 2-propynyl ether.[167] In addition, topical application of propynyl ethers to houseflies prolonged the lifetime of Baygon in vivo.[160] The fact that propynyl ethers have been shown to be metabolized by insect mfo systems[116,165] suggests an alternate-substrate mechanism, but the possibility of the acetylenic moiety chelating or coordinating with the iron of cytochrome P-450 may more readily explain synergism by these compounds.

Aryloxyalkylamines

For many years, SKF 525-A and Lilly 18947 (Figure 10.10) have been known to extend the activity of certain drugs by preventing their metabolism in vivo. Thus, it was logical to test these materials as potential synergistis of carbamate insecticides, and it was not surprising to find that carbaryl, Pyrolan, Isolan, and 3-*t*-butylphenyl N-methylcarbamate showed increased activity in the presence of the Lilly compound.[168] Later, SKF 525-A was also demonstrated to synergize carbaryl or UC 10854 when coadministered to houseflies.[132]

Analog synthesis produced many more active aryloxyalkylamines, but the degree of synergism varied with carbamate structure, implying that different detoxication mechanisms were blocked by different synergists.[169,170] An interesting part of these studies was the discovery that the increase in carbamate toxicity to flies at higher temperatures was erased in the presence of synergists.[170] This indirectly suggests that the synergists inhibited insecticide metabolism, a suggestion which was fortified by the discovery that Baygon metabolism in houseflies was reduced by Lilly 18947.[85] An interesting structure, combining some of the features of Thanite with those of aryloxyalkylamines, is GD-11 (Figure 10.10), which synergizes a variety of carbamate insecticides applied to houseflies.[171]

Benzothiadiazoles and Imidazoles

If the methylenedioxy ring of MDP synergists is replaced by other heterocylic rings, one of the groups of biologically active chemicals formed are the benzothiadiazoles.[172,173] The unsubstituted parent (Figure 10.11), unlike unsubstituted MDP,

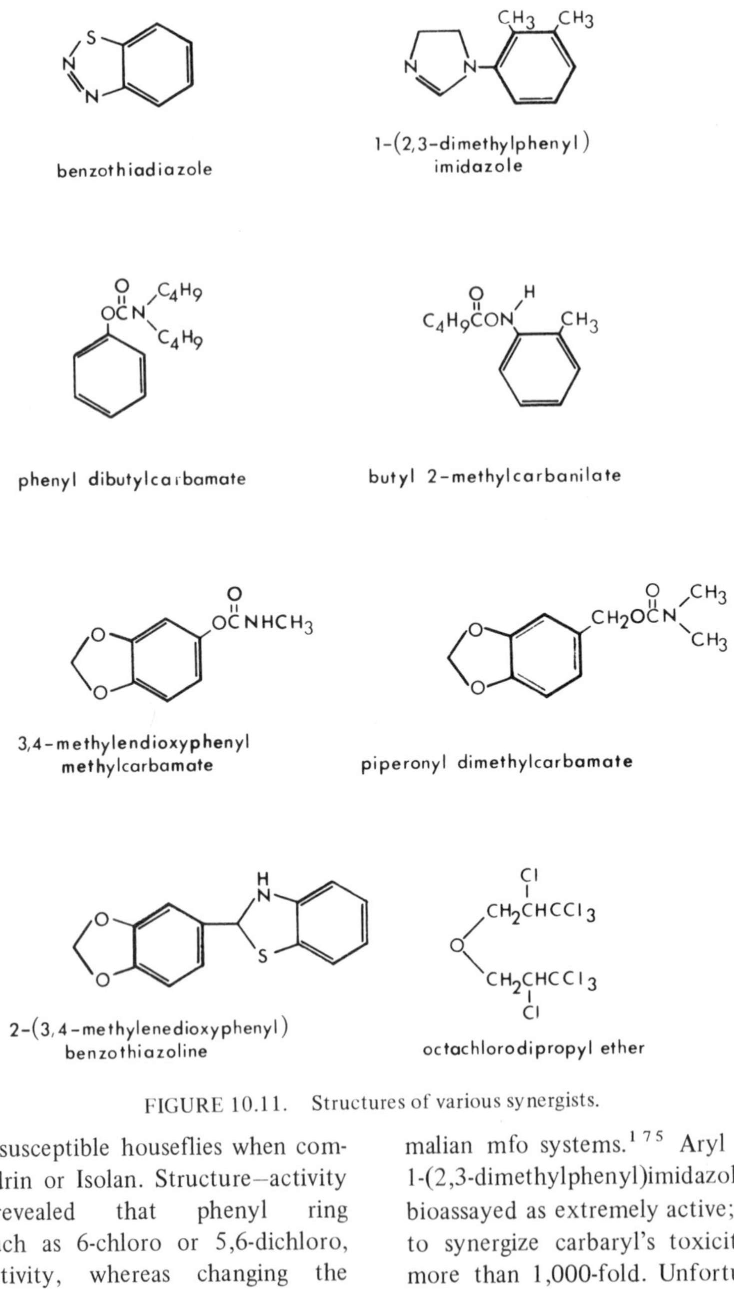

FIGURE 10.11. Structures of various synergists.

is synergistic in susceptible houseflies when combined with Landrin or Isolan. Structure–activity relationships revealed that phenyl ring substitutions, such as 6-chloro or 5,6-dichloro, accentuated activity, whereas changing the 1,2,3-benzothiadiazole ring to the 2,1,3-analog or other heterocyclic combination canceled activity. No evidence is available to support a mechanism for carbamate toxicity enhancement, but one investigation has shown that a series of benzothiadiazoles inhibits armyworm gut mfo activity.[174] The formation of a stable cation, similar to the MDP cation, does not appear to explain this inhibition.[116]

The most recent group of synergists are the imidazoles, discovered because they possess extreme potency as inhibitors of insect and mammalian mfo systems.[175] Aryl structures, such as 1-(2,3-dimethylphenyl)imidazole (Figure 10.11), bioassayed as extremely active; i.e., they were able to synergize carbaryl's toxicity to the housefly more than 1,000-fold. Unfortunately, their near-equal inhibition of mammalian detoxications will probably prevent commercial exploitation. Aliphatic imidazoles have also displayed synergistic activity with certain methylcarbamate insecticides.[176]

Analog Synergism

Analog synergism is the ultimate in alternate-substrate synergism; that is, a nontoxic carbamate competes with a toxic carbamate for the same sites on the mfo system. The potential value of such a combination was discovered early in carbamate history when it was shown that dimetilan enhanced the toxicity of carbaryl or pyrolan to German cockroaches and houseflies.[177] However, this was, in reality, potentiation, since dimetilan when applied alone, was toxic to the insects. Later on, true synergism was demonstrated with a number of toxic and nontoxic carbamate mixtures applied to many different insects.[134,178,179]

One extensive study using houseflies involved evaluation of almost 100 noninsecticidal carbamates as synergists for Isolan and carbaryl.[179] Definite structural requirements for Isolan synergism became evident, and most activity was associated with phenyl esters of alkyl or dialkyl carbamates. Aliphatic esters or unsubstituted carbamates were inactive as synergists. Compounds such as phenyl dibutylcarbamate (Figure 10.11) were able to enhance the toxicity of Isolan to Isolan-R strains of flies fivefold, and butyl 2-methylcarbanilate (Figure 10.11) was more effective than pb in combination with Isolan or carbaryl against carbamate-resistant flies.

Autosynergism

The perfect synergist would appear to be an "autosynergist", that is, a compound which synergizes its own toxicity. This would involve forging a "synergophore" and "toxophore" into the same molecule in such a way that the former does not interfere with the action of the latter. Such a combination was uncovered during a structure–activity study of alkoxyphenyl methylcarbamates, namely, 3,4-methylenedioxyphenyl N-methylcarbamate (Figure 10.11).[127,133] The compound "barely" inhibited AChE, yet proved to be toxic to houseflies. This, together with the fact that the carbamate was not synergized by pb, suggested that it was acting as its own synergist.[110]

In a similar manner, 2- and 3-(2-propynyloxy) phenyl N-methylcarbamates have also demonstrated autosynergistic properties.[111,132] However, incorporation of a thiocyanate[134] or thiadiazole group[118] into phenyl N-methylcarbamate failed to result in self-synergism. Various other MDP carbamates and piperonyl carbamates (Figure 10.11), though not active as autosynergists, were as effective as pb when used in combination with carbaryl to kill houseflies[109] or in combination with pyrethrins against flies and cockroaches.[180]

Miscellaneous Compounds

The list of miscellaneous nontoxic materials which synergize the toxicity of carbamate insecticides to insects grows longer each year. Examples include 2-(3,4-methylenedioxphenyl) benzothiazoline (Figure 10.11),[181] which is a combination of benzothiadiazole and MDP, various alkyl and aryl boronic acids,[134] trio-*o*-cresyl phosphate,[132] S,S,S-tributyl phosphorotrithioate,[28,179] triphenyl phosphate,[28] Arochlor 1254 (a polychlorinated biphenyl),[182] and the herbicide atrazine.[183] In addition, potentiation of insect toxicity occurs with carbamate insecticides mixed with octachlorodipropyl ether (Figure 10.11),[184] DDT, HCH,[185] endosulfan,[186] malathion, parathion, and several other organophosphorus insecticides.[187,188] An understanding of the significance of synergism and potentiation between pesticides or between environmental pollutants is becoming more critical as an increasing number of these interactions are uncovered.

Synergist Effectiveness and Selectivity

Once the structural requirements for a synergist have been unraveled, several other factors must be weighed in order to obtain maximum methylcarbamate insecticide synergism. One of these is the relative quantity of each ingredient needed for treatment. Usually, the higher the synergist: carbamate ratio, the greater the effectiveness of the combination.[75,139,159,160] Of course, eventually a maximum ratio is reached beyond which additional synergist no longer enhances toxicity; that is, insecticide detoxication is completely suppressed.[189]

This general trend has led to routine screening of synergists at a 5:1 or 10:1 synergist:carbamate ratio. However, there are structures which can markedly enhance carbamate toxicity at very low ratios.[109,162] Consider, for example, 2,3-methylenedioxynaphthalene (Figure 10.8), which has been shown to synergize carbaryl's toxicity to the housefly at a concentration as low as 0.001 that of carbaryl.[109] Obviously, these ratios must be considered from an economic standpoint if synergists are to become more important commercially.

The structure of the insecticide, particularly its susceptibility to metabolic destruction, is another important factor. Presumably, this is why synergists such as Lilly 18947 are more effective when paired with carbamate insecticides than when paired with other classes of insecticide,[168] or why myristicin enhances carbaryl toxicity, but not pyrethrum toxicity.[190] Even among carbamates, the degree of synergism of different structures in response to the same synergist varies considerably, probably because of the metabolic stability of substituents on the aromatic nucleus of the carbamate.[109] However, since cuticle penetration, toxicity, and susceptibility to the mfo enzymes are dependent on the overall lipophilic nature of the carbamate molecule, this could also explain some differences in the degree of synergism.[191] Thus, there are many carbamates which are nontoxic even in the presence of synergists.

Holding the carbamate and synergist constant, but varying the insect species, also yields a wide range of response. A recent survey of 74 species encompassing 8 orders and 37 families uncovered an amazing 36,000-fold difference in the toxicity of carbaryl between the most susceptible insect (0.11 μg/g) and the most tolerant (4,000 μg/g).[192,193] Topical treatment with a 5:1 pb: carbaryl ratio revealed a range in the magnitude of synergism from 1.5- to 400-fold. However, even in the presence of synergist, there was a tremendous variation in toxicity, suggesting that factors other than oxidative detoxication, such as cuticle penetration, excretion, hydrolysis, etc., played supporting roles in selectivity.

Perhaps carbaryl would have been more equally toxic to all species if a different application method had been used in the above study. For example, it has been shown that MDP and thiocyanate synergists can greatly enhance toxicity of a carbamate ester by aiding its penetration through insect cuticle following topical application.[194] Thus, the 400-fold degree of synergism afforded by topical application of Thanite in combination with carbaryl was nullified when the mixture was sprayed on as an oil emulsion, where both ingredients penetrated cuticle with ease.

The age, sex, and life stage of an insect can also markedly influence the degree of synergism. Presumably, this is due to the fact that mfo activity varies considerably as an insect matures,[195,196] and hence methylcarbamate toxicity varies accordingly. However, these differences in carbamate susceptibility can be minimized by blocking mfo enzymes with such synergists as pb, sesamex, and 2,3-methylenedioxynaphthalene.[197,198]

Even more important than the factors governing synergism selectivity among insects is the selectivity that must be present between insects and other forms of life. Obviously, a synergist:carbamate combination should not be more toxic to these nontargets than the carbamate alone. Fortunately, there are several factors which minimize this event, particularly with mammals. First of all, many methylcarbamate insecticides are readily detoxified in higher animals, primarily via hydrolysis (see Chapter 7), and this metabolic mechanism is not affected by most synergists. Second, insect mfo systems appear to be much more sensitive to inhibition by synergists than do mammalian mfo systems.[112,116,117,146,147,151,199] Third, mammals are better able to metabolize synergists, so that they persist for shorter periods in vivo.[116,117,149-152,200]

An example of the dramatic difference in toxicity which the above factors impart was found recently for the carbaryl:2,3-methylenedioxyphenyl (1:5) combination.[200] Topically applied to houseflies, the mixture gave an LD_{50} of 5 mg (carbaryl)/kg, but when fed to mice, no symptoms of toxicity were evident after ingestion of 750 mg (carbaryl)/kg. In 12 hrs, most of a dose of the synergist has been metabolized within the mouse, whereas after 24 hr in the fly, most of the synergist remained intact. Differences of similar magnitude, though somewhat less pronounced, were obtained with a carbaryl:2-propynyl-1-naphthyl combination.[165]

Future Use

Although appropriate combinations of pyrethroid insecticides and MDP synergists have provided safe and efficient control of certain insects for the past 30 years, successful practical application of synergists on a large scale for management of insect pests has not justified economically the extensive research and development described on the preceding pages. It was proposed at a recent international conference that scientists have reached a plateau with respect to defining the biochemical interactions and potential usefulness of synergists, and now efforts should be actively directed toward establishing whether or not these compounds have a place in the real world of pest control.[201]

Which insect pests should constitute the target for synergist:carbamate combinations? Today's environmental concern makes it unwise to conduct research into broadening the spectrum of a field-applied carbamate insecticide so that more naturally tolerant insect species are killed. The goal should be the management of carbamate-resistant strains of insects. Two approaches could be taken: control of already resistant strains, or prevention of resistance build-up. The former involves finding a suitable synergist-insecticide duo which controls the immune insect under actual field conditions, and the latter involves introduction of a new carbamate insecticide structure formulated only in combination with a synergist to prevent development of resistance to the new insecticide.

The effectiveness of simultaneous insecticide-synergist pressure in preventing resistance development would require fairly sophisticated long-term field evaluation. However, its success is theoretically predictable, based on laboratory studies where insects exposed to carbamates in the presence of piperonyl butoxide responded with only minimal changes in tolerance after many generations of opportunity.[29,30,47] An example of this phenomenon is illustrated by the data in Table 10.6. Virtually total resistance to carbaryl was obtained by houseflies after eight generations of insecticide pressure, but copressure by pb yielded only a fourfold increase in carbaryl tolerance after 50 generations. Whether this type of protection would occur under practical use conditions is certainly worthy of more experimentation.

Having established the target insects and method of approach, the choice of synergist looms

TABLE 10.6

Development of Resistance in Houseflies Selected with Carbaryl or with a 1:5 Ratio of Carbaryl plus Piperonyl Butoxide[29]

	Carbaryl LD_{70} (μg/fly)	
Number of generations	Carbaryl alone	Carbaryl + pb
0	3	0.12
2	5	0.13
4	15	0.16
6	40	0.19
8	70[a]	0.22
20	70	0.38
30	–	0.48
50	–	0.50

[a]Limit of carbaryl's solubility in topical test solution.

as a critical decision. As discussed above, the efficacy of a particular compound depends on many factors, such as structure of carbamate, species of insect, compatibility, etc., but indications are that effective synergists could be found. Since it is likely that the quantity of synergist relative to carbamate will be large, the cost of the additive must be low to be feasible for large-acreage applications. At the present time, this would appear to rule out the MDP class of chemicals, since most of these are based on multistep syntheses involving the intermediate safrole, an ingredient which is available from botanical sources in amounts of less than 1,400,000 lb annually.[121] Although economical alternate sources for the MDP group are being sought,[202] it is difficult to foresee ready availability of the millions of pounds which would be needed for public-health and agricultural uses.

Organothiocyanates could be cheap to produce in quantity, but their effectiveness with methylcarbamates is questionable. Perhaps the most likely candidates are the propynyl ethers, since many of the ingredients are potentially available in unlimited quantities.[121] Added to this is the fact that some of these materials are effective synergists at low synergist:insecticide ratios. In addition, there are surely other low-cost synergist structures awaiting commercial exploitation, so that economics should not be a limiting factor.

Type of formulation, compatibility, shelf life, phytotoxicity, and environmental stability are some of the myriad of other delicately balanced parameters which would have to be considered, regardless of the synergist chosen for development. In addition, and of greatest significance, the question of human safety will have to be answered. When applied alone, the currently registered MDP synergists are of low mammalian toxicity, with rat oral LD_{50} values from 2,000 mg/kg to 15,000 mg/kg.[116] The no-effect levels also indicate minimum toxicity, with few abnormalities produced in rats fed very high levels of synergist.[116,203] Though certain MDP structures are known tumorigenic agents, the synergists registered for commercial use are apparently non-carcinogenic, though there is some evidence that pb, sulfoxide, and propyl isome produce malignant tumors of the lymphatic system in mice.[116,203,204]

Assuming that the synergists *per se* are nontoxic, what are their effects on other toxic agents consumed by mammals? Reports have labeled various synergists as drug extenders, carcinogen potentiators, inhibitors of steroid metabolism, and synergists of poisons other than insecticides.[116] The relevance of these findings, however, must be considered, keeping in mind the high levels of synergist used to obtain the results. To truly evaluate potential hazards, more attention should be given to interactions caused by levels and types of synergist residues likely to occur under actual field usage. Based on the available data and on the "exempt from tolerance" classification for currently registered MDP synergists when applied to growing crops, the wide-scale use of synergists without undue harmful side effects is certainly feasible.[116,201,205]

Thus, the potential exists, as it has since the introduction of carbamate insecticides. Although there is no doubt that synergists can and will continue to assist in the elucidation of carbamate metabolism, mode of action, and resistance mechanisms, it is time to take them out of the laboratory and put them to the test in the field. Granted that the obstacles in the way of successful registration of a carbamate-synergist insecticide for large-scale application are indeed formidable, applied entomologists, toxicologists, and formulation chemists should work together toward a critical scientific evaluation of this promising method of insect control.

REFERENCES

1. **Brown, A. W. A.,** Insect resistance in arthropods, *W.H.O. Monogr. Ser.*, No. 38, 1958.
2. **Hoskins, W. M. and Gordon, H. T.,** Arthropod resistance to chemicals, *Annu. Rev. Entomol.*, 1, 89, 1956.
3. **Georghiou, G. P.,** Genetic studies on insecticide resistance, *Adv. Pest Control Res.*, 6, 171, 1965.
4. **Crow, J. F.,** Genetics of insect resistance to chemicals, *Annu. Rev. Entomol.*, 2, 227, 1957.
5. **Georghiou, G. P.,** Genetics of resistance to insecticides in houseflies and mosquitoes, *Exp. Parasitol.*, 26, 224, 1969.
6. **Tsukamoto, M.,** Biochemical genetics of insecticide resistance in the housefly, *Residue Rev.*, 25, 289, 1969.
7. **Hoskins, W. M.,** Use of the dosage–mortality curve in quantitative estimation of insecticide resistance, *Misc. Publ. Entomol. Soc. Am.*, 2, 85, 1960.
8. **Tsukamoto, M.,** The log dosage–probit mortality curve in genetic researches of insect resistance to insecticides, *Botyu-Kagaku,* 28, 91, 1963.
9. **Georghiou, G. P.,** The evolution of resistance to pesticides, *Annu. Rev. Ecol. Syst.*, 3, 133, 1972.
10. **Brooks, G. T.,** *Chlorinated Insecticides,* Vol. 2, CRC Press, Cleveland, 1974.
11. **Brown, A. W. A.,** Pest Resistance to pesticides, in *Pesticides in the Environment,* Vol. 1, Part 2, White-Stevens, R., Ed., Marcel Dekker, New York, 1971.
12. **Brown, A. W. A. and Pal, R.,** *Insect Resistance in Arthropods,* 2nd ed., World Health Organization, Geneva, 1971.
13. **Tahori, A. S., Ed.,** *Insecticide Resistance, Synergism and Enzyme Induction,* Gordon and Breach, London, 1971.
14. **Smith, A. G.,** Control of apple leaf rollers in Nelson, *Orchardist N. Z.*, 34, 315, 1963.
15. **Adkisson, P. L. and Nemec, S. J.,** Comparative effectiveness of certain insecticides for killing bollworms and tobacco budworms, *Tex. Agric. Exp. Stn. Bull.*, B-1048, 1966.
16. **Graves, J. B., Clower, D. F., and Bradley, J. R., Jr.,** Resistance of the tobacco budworm to several insecticides in Louisiana, *J. Econ. Entomol.*, 60, 887, 1967.
17. **Evans, D. E.,** Resistance to carbamate insecticides in *Aeneolamia varia saccharina* (Distant), *Trop. Agric.* Trinidad, 50, 153, 1973.
18. **Iwata, T. and Hama, H.,** Insensitivity of cholinesterase in *Nephotettix cincticeps* resistant to carbamate and organophosphate insecticides, *J. Econ. Entomol.*, 65, 643, 1972.
19. **Fahmy, M. A. H. and Fukuto, T. R.,** Insecticidal activity of N-acylcarbamates on susceptible and carbaryl-resistant strains of the Egyptian cotton leafworm, *J. Econ. Entomol.*, 63, 1783, 1970.
20. **Wharton, R. H. and Roulston, W. J.,** Resistance of ticks to chemicals, *Annu. Rev. Entomol.*, 15, 381, 1970.
21. **Bennett, G. W. and Spink, W. T.,** Insecticide resistance of German cockroaches from various areas of Louisiana, *J. Econ. Entomol.*, 61, 426, 1968.
22. **Georghiou, G. P.,** Resistance to carbamate and organophosphorus insecticides in *Anopheles albimanus, Am. J. Trop. Med. Hyg.*, 21, 797, 1972.

23. **McEwen, F. L. and Hervey, G. E. R.**, An evaluation of newer insecticides for control of DDT-resistant cabbage loopers, *J. Econ. Entomol.*, 49, 385, 1956.
24. **McEwen, F. L. and Splittstoesser, C. M.**, Resistance to organophosphate insecticides in the cabbage looper in New York, *J. Econ. Entomol.*, 63, 646, 1970.
25. **Davis, A. C. and Kuhr, R. J.**, Laboratory and field evaluation of methomyl's toxicity to the cabbage looper, *J. Econ. Entomol.*, 67, 681, 1974.
26. **Lowry, W. L., McGarr, R. L., Robertson, O. T., Berger, R. S., and Graham, H. M.**, Bollworm and tobacco budworm resistance to several insecticides in the lower Rio Grande valley of Texas, *J. Econ. Entomol.*, 58, 732, 1965.
27. **Rhabb, R. L. and Guthrie, F. E.**, Resistance of tobacco hornworms to certain insecticides in North Carolina, *J. Econ. Entomol.*, 57, 995, 1964.
28. **Ariaratnam, V. and Georghiou, G. P.**, Selection for resistance to carbamate and organophosphorus insecticides in *Anopheles albimanus*, *Nature* (London), 232, 642, 1971.
29. **Moorefield, H. H.**, Insect resistance to the carbamate insecticides, *Misc. Publ. Entomol. Soc. Am.*, 2, 145, 1960.
30. **Georghiou, G. P., Metcalf, R. L., and March, R. B.**, The development and characterization of resistance to carbamate insecticides in the housefly, *Musca domestica*, *J. Econ. Entomol.*, 54, 132, 1961.
31. **Georghiou, G. P. and Garber, M. J.**, Studies on the inheritance of carbamate resistance in the housefly (*Musca domestica* L.), *Bull. W.H.O.*, 32, 181, 1965.
32. **Georghiou, G. P. and Metcalf, R. L.**, Dieldrin susceptibility: partial restoration in *Anopheles* selected with a carbamate, *Science*, 140, 301, 1963.
33. **Tadano, T. and Brown, A. W. A.**, Development of resistance to various insecticides in *Culex pipiens fatigans* Wiedemann, *Bull. W.H.O.*, 35, 189, 1966.
34. **Green, L. R. and Dorough, H. W.**, Reactivation of carbamate-inhibited cholinesterases from susceptible and resistant houseflies, *Tex. Agric. Exp. Stn. Progr. Rep.*, No. 2835, 1970.
35. **Georghiou, G. P., Metcalf, R. L., and Gidden, F. E.**, Carbamate-resistance in mosquitoes. Selection of *Culex pipiens fatigans* Wiedemann (= *C. quinquefasciatus* say) for resistance to Baygon, *Bull. W H.O.*, 35, 691, 1966.
36. **Kumar, V. and Morrison, F. O.**, Carbamate and phosphate resistance in adult granary weevils, *J. Econ. Entomol.*, 60, 1430, 1967.
37. **Harding, J. A. and Dyar, R. C.**, Resistance induced in European corn borers in the laboratory by exposing successive generations to DDT, diazinon, or carbaryl, *J. Econ. Entomol.*, 63, 250, 1970.
38. **Kuhr, R. J. and Davis, A. C.**, Toxicity and metabolism of carbaryl in the European corn borer, *Pestic. Biochem. Physiol.*, 5, 330, 1975.
39. **Hoskins, W. M. and Nagasawa, S.**, Cross-resistance in Sevin-selected houseflies and a summary of resistance among the carbamate insecticides, *Botyu-Kagaku*, 26, 115, 1961.
40. **Ikemoto, H.**, Some notes on Sevin resistance in the housefly, *Musca domestica domestica* and *M. domestica vicinia*, *Botyu-Kagaku*, 29, 68, 1964.
41. **Cole, M. M. and Clark, P. H.**, Toxicity of various carbamates and synergists to several strains of body lice, *J. Econ. Entomol.*, 55, 98, 1962.
42. **Clark, P. H. and Cole, M. M.**, Laboratory studies on resistance of the body louse to insecticides, *J. Econ. Entomol.*, 57, 205, 1964.
43. **Clark, P. H. and Cole, M. M.**, Resistance of body lice to carbaryl, *J. Econ. Entomol.*, 60, 398, 1967.
44. **Adams, C. H. and Cross, W. H.**, Insecticide resistance in *Bracon mellitor*, a parasite of the boll weevil, *J. Econ. Entomol.*, 60, 1016, 1967.
45. **Croft, B. A. and Meyer, R. H.**, Carbamate and organophosphorus resistance patterns in populations of *Amblyseius fallacis*, *Env. Entomol.*, 2, 691, 1973.
46. **Georghiou, G. P.**, Carbamate insecticides: the cross-resistance spectra of four carbamate-resistant strains of the housefly after protracted selection pressure, *J. Econ. Entomol.*, 55, 494, 1962.
47. **Georghiou, G. P.**, Carbamate insecticides: toxic action of synergized carbamates against twelve resistant strains of the house fly, *J. Econ. Entomol.*, 55, 768, 1962.
48. **McDonald, I. C. and Cochran, D. G.**, Carbamate cross resistance in a carbaryl-resistant strain of the German cockroach, *J. Econ. Entomol.*, 61, 670, 1968.
49. **Eldefrawi, M. E., Miskus, R., and Hoskins, W. M.**, Resistance to Sevin by DDT- and parathion-resistant houseflies and sesoxane as Sevin synergist, *Science*, 129, 898, 1959.
50. **Atkins, E. L., Jr. and Anderson, L. D.**, DDT resistance in honey bees, *J. Econ. Entomol.*, 55, 791, 1962.
51. **LaBrecque, G. C., Wilson, H. G., and Smith, C. N.**, Effectiveness of two carbamates against DDT- and malathion-resistant house flies, *J. Econ. Entomol.*, 52, 178, 1959.
52. **Cochran, D. G.**, Further studies on cross-resistance in the German cockroach, *Bull. W.H.O.*, 24, 557, 1961.
53. **Weismann, R. and Kocher, C.**, Untersuchungen über ein neues, gegen resistente *Musca domestica* L. wirksames Insektizid, *Z. Angew. Entomol.*, 33, 297, 1951.
54. **Meltzer, J.**, Unspecific resistance mechanisms in the housefly, *Musca domestica* L., *Indian J. Malariol.*, 12, 579, 1958.
55. **Hoskins, W. M. and Nagasawa, S.**, Cross resistance in Sevin-selected house flies and a summary of resistance among carbamate insecticides, *Botyu-Kagaku*, 26, 115, 1961.

56. **Ishii, T. and Sherman, M.**, Resistance of a Hawaiian strain of the German cockroach to several insecticides, *J. Econ. Entomol.*, 58, 46, 1965.
57. **Meltzer, J.**, Multiresistance in the house fly, *Musca domestica* L., induced by selection with insecticides, *Meded. Landbouwhogesch. Opzoekingsstn. Staat Gent,* 21, 459, 1956.
58. **Georghiou, G. P.**, Development and characterization of resistance to *o*-isopropoxyphenyl methylcarbamate in the mosquito *Culex pipiens quinquefasciatus* Say, *Nature,* 207, 883, 1965.
59. **Iwata, T. and Hama, H.**, Green rice leafhoppers resistant to carbamate insecticides, *Botyu-Kagaku,* 36, 174, 1971.
60. **Eldefrawi, M. E. and Hoskins, W. M.**, Relation of the rate of penetration and metabolism to the toxicity of Sevin to three insect species, *J. Econ. Entomol.*, 54, 401, 1961.
61. **Forgash, A. J. and Hansens, E. J.**, Cross resistance in diazinon-resistant strain of *Musca domestica* (L.), *J. Econ. Entomol.*, 52, 733, 1959.
62. **Forgash, A. J. and Hansens, E. J.**, Further studies on the toxicity of insecticides to diazinon-resistant *Musca domestica, J. Econ. Entomol.*, 53, 741, 1960.
63. **Forgash, A. J. and Hansens, E. J.**, Resistance levels in diazinon-pressured and nonpressured polyresistant houseflies, *Econ. Entomol.*, 55, 679, 1962.
64. **Forgash, A. J. and Hansens, E. J.**, Effect of selection on cross resistance in diazinon-resistant *Musca domestica, J. J. Econ. Entomol.*, 60, 1241, 1967.
65. **Kawahara, S., Kiritani, K., and Sasaba, T.**, Selective activity of rice-pest insecticides against the green rice leafhopper and spider, *Botyu-Kagaku,* 36, 121, 1971.
66. **Schaefer, C. H. and Wilder, W. H.**, Insecticide resistance and cross-resistance in *Aedes nigromaculis, J. Econ. Entomol.*, 63, 1224, 1970.
67. **Schaefer, C. H. and Wilder, W. H.**, A new carbamate insecticide, RE-11775 [*m-sec*-butylphenyl N-methyl-N-(phenylthio)carbamate], with promising activity against organophosphorus-resistant mosquitoes, *J. Econ. Entomol.*, 63, 480, 1970.
68. **Bajpai, V. N. and Perti, S. L.**, Resistance to malathion, *Pesticides,* 3, 43, 1969.
69. **Busvine, J. R. and Feroz, M.**, Biochemistry of resistance: introduction with remarks on resistance of *Cimex lectularius,* in *Insecticide Resistance, Synergism and Enzyme Induction,* Tahori, A. S., Ed., Gordon and Breach, London, 1971.
70. **Georghiou, G. P. and Metcalf, R. L.**, The absorption and metabolism of 3-isopropylphenyl N-methylcarbamate by susceptible and carbamate-selected strains of house flies, *J. Econ. Entomol.*, 54, 231, 1961.
71. **Plapp, F. W., Jr., Chapman, G. A., and Bigley, W. S.**, A mechanism of resistance to Isolan in the housefly, *J. Econ. Entomol.*, 57, 692, 1964.
72. **Metcalf, R. L., Fukuto, T. R., Collins, C., Borck, K., Burk, J., Reynolds, H. T., and Osman, M. F.**, Metabolism of 2-methyl-2-(methylthio)propionaldehyde O-(methylcarbamoyl)oxime in plant and insect, *J. Agric. Food Chem.*, 14, 579, 1966.
73. **Tsukamoto, M. and Casida, J. E.**, Metabolism of methylcarbamate insecticides by the $NADPH_2$-requiring enzyme system from houseflies, *Nature,* 213, 49, 1967.
74. **Tsukamoto, M., Shrivastava, S. P., and Casida, J. E.**, Biochemical genetics of house fly resistance to carbamate insecticide chemicals, *J. Econ. Entomol.*, 61, 50, 1968.
75. **Shrivastava, S. P., Tsukamoto, M., and Casida, J. E.**, Oxidative metabolism of C^{14}-labeled Baygon by living house flies and by house fly enzyme preparations, *J. Econ. Entomol.*, 62, 483, 1969.
76. **Plapp, F. W., Jr. and Casida, J. E.**, Genetic control of housefly NADPH-dependent oxidases: relation to insecticide chemical metabolism and resistance, *J. Econ. Entomol.*, 62, 1174, 1969.
77. **Shrivastava, S. P., Georghiou, G. P., Metcalf, R. L., and Fukuto, T. R.**, Carbamate resistance in mosquitoes, The metabolism of propoxur by susceptible and resistant larvae of *Culex pipiens fatigans, Bull. W.H.O.*, 42, 931, 1970.
78. **Kuhr, R. J.**, The metabolic fate of methomyl in the cabbage looper, *Pestic. Biochem. Physiol.*, 3, 113, 1973.
79. **Kuhr, R. J.**, Comparative metabolism of carbaryl by resistant and susceptible strains of the cabbage looper, *J. Econ. Entomol.*, 64, 1373, 1971.
80. **Perry, A. S. and Buckner, A. J.**, Microsomal cytochrome P-450 in resistant and susceptible houseflies, *Life Sci. Part 2,* 9, 335, 1970.
81. **Matthews, H. B. and Casida, J. E.**, Properties of housefly microsomal cytochromes in relation to sex, strain, substrate specificity, and apparent inhibition and induction by synergist and insecticide chemicals, *Life Sci. Part 1,* 9, 989, 1970.
82. **Philpot, R. M. and Hodgson, E.**, Differences in the cytochrome P-450 from resistant and susceptible houseflies, *Chem. Biol. Interact.*, 4, 399, 1971.
83. **Ku, T.-Y. and Bishop, J. L.**, Penetration, excretion, and metabolism of carbaryl in susceptible and resistant German cockroaches, *J. Econ. Entomol.*, 60, 1328, 1967.
84. **Dorough, H. W.**, Fate of Furadan insecticide in susceptible and carbamate-resistant houseflies, *Tex. Agric. Exp. Stn. Progr. Rep.*, No. 2832, 1970.
85. **Metcalf, R. L., Osman, M. F., and Fukuto, T. R.**, Metabolism of C^{14}-labeled carbamate insecticides to $C^{14}O_2$ in the housefly, *J. Econ. Entomol.*, 60, 445, 1967.

86. **Mehendale, H. M. and Dorough, H. W.,** Conjugative metabolism and action of carbamate insecticides, in *Insecticides,* Tahori, A. S., Ed., Gordon and Breach, London, 1972.
87. **Mehendale, H. M. and Dorough, H. W.,** In vitro glucosylation of 1-naphthol by insects, *J. Insect Physiol.,* 18, 981, 1972.
88. **O'Brien, R. D.,** *Insecticides: Action and Metabolism,* Academic Press, New York, London, 1967.
89. **Plapp, F. W., Jr. and Bigley, W. S.,** Carbamate insecticides and aliesterase activity in insects, *J. Econ. Entomol.,* 54, 793, 1961.
90. **Hama, H. and Iwata, T.,** Sensitive aliesterase to a carbamate insecticide, propoxur, in the resistant strains of the green rice leafhopper, *Appl. Entomol. Zool.,* 7, 177, 1972.
91. **Hama, H. and Iwata, T.,** Insensitive cholinesterase in the Nakagawara strain of the green rice leafhopper, *Nephotettix cincticeps,* as a cause of resistance to carbamate insecticides, *Appl. Entomol. Zool.,* 6, 183, 1971.
91a. **Ayad, H. and Georghiou, G. P.,** Resistance to organophosphates and carbamates in *Anopheles albimanus* based on reduced sensitivity of acetylcholinesterase, *J. Econ. Entomol.,* 68, 295, 1975.
92. **Plapp, F. W., Jr.,** On the molecular biology of insecticide resistance, in *Biochemical Toxicology of Insecticides,* O'Brien, R. D. and Yamamoto, I., Eds., Academic Press, New York, London, 1970.
93. **Wagoner, D. E.,** Linkage group–karyotype correlation in the house fly determined by cytological analysis of X-ray induced translocations, *Genetics,* 57, 729, 1967.
94. **Georghiou, G. P.,** Genetic Analysis of Carbamate Resistance in *Musca domestica* L., paper presented at the Pacific Branch of the Entomological Society of America, San Diego, June 1966.
95. **Hoyer, R. F., Plapp, F. W., Jr., and Orchard, R. D.,** Linkage relationships of several insecticide resistant factors in the house fly (*Musca domestica* L.), *Entomol. Exp. Appl.,* 8, 65, 1965.
96. **Plapp, F. W., Jr. and Hoyer, R. F.,** Insecticide resistance in the house fly: resistance spectra and preliminary genetics of resistance in eight strains, *J. Econ. Entomol.,* 60, 768, 1967.
97. **Hoopingarner, R. A. and Krause, D. H.,** The Mackinac Island fly problem. II. Induced insecticide resistance, *Mich. State Univ. Agric. Exp. Stn. Q. Bull.,* 50, 281, 1968.
98. **Plapp, F. W., Jr.,** Inheritance of dominant factors for resistance to carbamate insecticides in the housefly, *J. Econ. Entomol.,* 63, 138, 1970.
99. **Kasai, T. and Ogita, Z.,** A genetic study on Sevin-resistance and joint toxic action of Sevin with γ-BHC against houseflies, *Botyu-Kagaku,* 30, 12, 1965.
100. **Tsukamoto, M. and Suzuki, R.,** Genetic analyses of DDT-resistance in two strains of the housefly, *Musca domestica* L., *Botyu-Kagaku,* 29, 76, 1964.
101. **Tsukamoto, M. and Suzuki, R.,** Genetic analyses of Diazinon resistance in the housefly, *Botyu-Kagaku,* 31, 1, 1966.
102. **Oppenoorth, F. J.,** Biochemical genetics of insecticide resistance, *Annu. Rev. Entomol.,* 10, 185, 1965.
103. **Oppenoorth, F. J.,** Resistance in insects: the role of metabolism and the possible use of synergists, *Bull. W.H.O.,* 44, 195, 1971.
104. **Khan, M. A. Q.,** Some biochemical characteristics of the microsomal cyclodiene epoxidase system and its inheritance in the housefly, *J. Econ. Entomol.,* 62, 388, 1969.
105. **Schafer, J. A. and Terriere, L. C.,** Enzymatic and physical factors in housefly resistance to naphthalene, *J. Econ. Entomol.,* 63, 787, 1970.
106. **Kikkawa, H.,** The genetic study on the resistance to Sevin in *Drosophila melanogaster, Botyu-Kagaku,* 29, 42, 1964.
107. **Tsukamoto, M. and Ogaki, M.,** Gene analysis of resistance to DDT and BHC in *Drosophila melanogaster, Botyu-Kagaku,* 19, 25, 1954.
108. **Kikkawa, H.,** Genetic studies on the resistance to parathion in *Drosophila melanogaster.* I. Gene analysis, *Annu. Rep. Sci. Works Fac. Sci. Osaka Univ.,* 9, 1, 1961.
109. **Metcalf, R. L., Fukuto, T. R., Wilkinson, C., Fahmy, M. H., El-Aziz, S. A., and Metcalf, E. R.,** Mode of action of carbamate synergists, *J. Agric. Food Chem.,* 14, 555, 1966.
110. **Metcalf, R. L.,** Mode of action of insecticide synergists, *Annu. Rev. Entomol.,* 12, 229, 1967.
111. **Metcalf, R. L.,** The role of oxidative reactions in the mode of action of insecticides, in *Enzymatic Oxidations of Toxicants,* Hodgson, E., Ed., North Carolina State University Press, Raleigh, 1968.
112. **Brooks, G. T.,** Aspects of microsomal metabolism of insecticides and the action of synergists, *Meded. Rijksfac. Landbouwetensch. Gent,* 33, 629, 1968.
113. **Wilkinson, C. F.,** Detoxication of pesticides and the mechanisms of synergism, in *Enzymatic Oxidations of Toxicants,* Hodgson, E., Ed., North Carolina State University Press, Raleigh, 1968.
114. **Wilkinson, C. F.,** The role of insecticide synergists in resistance problems, *World Rev. Pest Control,* 7, 155, 1968.
115. **Casida, J. E.,** Insect microsomes and insecticide chemical oxidations in *Microsomes and Drug Oxidations,* Gillette, J. R. et al., Eds., Academic Press, New York, London, 1969.
116. **Casida, J. E.,** Mixed-function oxidase involvement in the biochemistry of insecticide synergists, *J. Agric. Food Chem.,* 18, 753, 1970.
117. **Wilkinson, C. F.,** Insecticide synergists and their mode of action, in *Insecticide Resistance, Synergism, Enzyme Induction,* Tahori, A. S., Ed., Gordon and Breach, New York, 1971.
118. **Wilkinson, C. F.,** Effects of synergists on the metabolism and toxicity of anticholinesterases, *Bull. W.H.O.,* 44, 171, 1971.

119. **Haller, H. L., McGovran, E. R., Goodhue, L. D., and Sullivan, W. N.**, The synergistic action of sesamin with pyrethrum insecticides, *J. Org. Chem.*, 7, 183, 1942.
120. **Haller, H. L., LaForge, F. B., and Sullivan, W. N.**, Some compounds related to sesamin: their structures and their synergistic effect with pyrethrum insecticides, *J. Org. Chem.*, 7, 185, 1942.
121. **Hennessey, D. J.**, The potential of carbamate synergists as pest control agents, in *Biochemical Toxicology of Insecticides,* O'Brien, R. D. and Yamamoto, I., Eds., Academic Press, New York, London, 1970.
122. **Hewlett, P. S.**, Synergism and potentiation in insecticides, *Chem. Ind.*, 22, 701, 1968.
123. **Moorefield, H. H.**, Syngergism of the carbamate insecticides, *Contrib. Boyce Thompson Inst.*, 19, 501, 1958.
124. **Moorefield, H. H.** (Union Carbide Corp.), U.S. Patents 2,904,463–5, September 15, 1959.
125. **Fuchs, W. H. and Zschintzch, J.**, Synergistic action of piperonyl butoxide on the insecticide action of carbamate, *Naturwissenschaften,* 46, 273, 1959.
126. **Eldefrawi, M. E., Miskus, R., and Sutcher, V.**, Methylenedioxyphenyl derivatives as synergists for carbamate insecticides on susceptible, DDT- and parathion-resistant houseflies, *J. Econ. Entomol.*, 53, 231, 1960.
127. **Metcalf, R. L., Fukuto, T. R., and Winton, M. Y.**, Alkoxyphenyl N-methylcarbamates as insecticides, *J. Econ. Entomol.*, 53, 828, 1960.
128. **Shorey, H. H.**, Synergism of carbamate insecticides against cabbage looper and beet armyworm larvae, *J. Econ. Entomol.*, 54, 1243, 1961.
129. **Hadaway, A. B., Barlow, F., and Duncan, J.**, Effects of piperonyl butoxide on insecticide potency, *Bull. Entomol. Res.*, 53, 769, 1963.
130. **Matsubara, H.**, Synergist for insecticides. XXIII. The synergistic effect of synthetic synergists on 1-naphthyl-N-methylcarbamate, *Botyu-Kagaku,* 28, 35, 1963.
131. **Matsubara, H., Ito, H., Kawasaki, M., Kai, K., and Kanamori, S.**, Synergism of barthrin, dimethrin, and 1-naphthyl-N-methylcarbamate with pyrethrum synergists against diazinon-resistant and susceptible houseflies, *Botyu-Kagaku,* 29, 1, 1964.
132. **Metcalf, R. L. and Fukuto, T. R.**, Effects of chemical structure on intoxication and detoxication of phenyl N-methylcarbamates in insects, *J. Agric. Food Chem.*, 13, 220, 1965.
133. **Fukuto, T. R., Metcalf, R. L., Winton, M. Y., and Roberts, P. A.**, The synergism of substituted phenyl N-methylcarbamates by piperonyl butoxide, *J. Econ. Entomol.*, 55, 341, 1962.
134. **Weiden, M. H. J. and Moorefield, H. H.**, Synergism and species-specificity of carbamate insecticides, *J. Agric. Food Chem.*, 13, 200, 1965.
135. **Yu, C.-C., Metcalf, R. L., and Booth, G. M.**, Inhibition of acetylcholinesterase from mammals and insects by carbofuran and its related compounds and their toxicities toward these animals, *J. Agric. Food Chem.*, 20, 923, 1972.
136. **Yu, C.-C., Park, K. S., and Metcalf, R. L.**, Correlation of toxicity and acetylcholinesterase (AChE) inhibition in 2-alkyl-substituted 1,3-benzadioxolyl-4 N-methylcarbamates and related compounds, *Pestic. Biochem. Physiol.*, 4, 178, 1974.
137. **Moorefield, H. H. and Weiden, M. H. J.**, Influence of the methylenedioxyphenyl structure in synergism of a carbamate insecticide for house flies, *Contrib. Boyce Thompson Inst.*, 22, 425, 1964.
138. **Wilkinson, C., Metcalf, R. L., and Fukuto, T. R.**, Some structural requirements of methylenedioxyphenyl derivatives as synergists of carbamate insecticides, *J. Agric. Food Chem.*, 14, 73, 1966.
139. **Wilkinson, C. F.**, Penetration, metabolism, and synergistic activity with carbaryl of some simple derivatives of 1,3-benzodioxole in the housefly, *J. Agric. Food Chem.*, 15, 139, 1967.
140. **Desmarchelier, J., Krieger, R. J., Lee, P. W., and Fukuto, T. R.**, Carbaryl synergism by substituted 1,3-benzodioxoles and related compounds and inhibition of mixed-function oxidase, *J. Econ. Entomol.*, 66, 631, 1973.
141. **Sacher, R. M., Metcalf, R. L., Metcalf, E. R., and Fukuto, T. R.**, Structure activity relations of methylenedioxynaphthyl derivatives as synergists of carbaryl in house flies, *J. Econ. Entomol.*, 64, 1011, 1971.
142. **Gemrich, E. G.**, Enzymatic degradation of 2-chloro-4,5-dimethylphenyl N-methylcarbamate by the fat bodies of *Blaberus giganteus, J. Agric. Food Chem.*, 15, 617, 1967.
143. **Shrivastava, S. P., Georghiou, G. P., and Fukuto, T. R.**, Metabolism of N-methylcarbamate insecticides by mosquito larval enzyme system requiring $NADPH_2$, *Entomol. Exp. Appl.*, 14, 333, 1971.
144. **Price, G. M. and Kuhr, R. J.**, The metabolism of the insecticide carbaryl (1-naphthyl N-methylcarbamate) by fat body of the blowfly larva, *Calliphora erythrocephala, Biochem. J.*, 112, 133, 1969.
145. **Kuhr, R. J.**, Metabolism of carbamate insecticide chemicals in plants and insects, *J. Agric. Food Chem.*, 18, 1023, 1970.
146. **Lewis, S. E., Wilkinson, C. F., and Ray, J. W.**, The relationship between microsomal epoxidation and lipid peroxidation in houseflies and pig liver and the inhibitory effect of derivatives of 1,3-benzodioxol (methylendioxybenzene), *Biochem. Pharmacol.*, 16, 1195, 1967.
147. **Philleo, W. W., Schonbrod, R. D., and Terriere, L. C.**, Mechanism of insecticide synergism: methylenedioxyphenyl compounds as inhibitors of hydroxylation of naphthalene in houseflies, *J. Agric. Food Chem.*, 13, 113, 1965.
148. **Hodgson, E. and Kulkarni, A. P.**, Interactions of pesticides with cytochrome P-450, in *Mechanisms of Pesticide Action,* Kohn, G. K., Ed., American Chemical Society, Washington, D.C., 1974.

148a. **Hodgson, E. and Philpot, R. M.,** Interaction of methylenedioxyphenyl (1,3-benzodioxole) compounds with enzymes and their effects on mammals, *Drug Metab. Rev.*, 3, 231, 1974.

149. **Casida, J. E., Engel, J. L., Esaac, E. G., Kamienski, F. X., and Kuwatsuka, S.,** Methylene ^{14}C-dioxyphenyl compounds: metabolism in relation to their synergistic action, *Science,* 153, 1130, 1966.

150. **Esaac, E. G. and Casida, J. E.,** Metabolism in relation to mode of action of methylenedioxyphenyl synergists in houseflies, *J. Agric. Food Chem.*, 17, 539, 1969.

151. **Wilkinson, C. F. and Hicks, L. J.,** Microsomal metabolism of the 1,3-benzodioxole ring and its possible significance in synergistic action, *J. Agric. Food Chem.*, 17, 829, 1969.

152. **Kamienski, F. X. and Casida, J. E.,** Importance of demethylation in the metabolism in vivo and in vitro of methylenedioxyphenyl synergists and related compounds in mammals, *Biochem. Pharmacol.*, 19, 91, 1970.

153. **Hennessey, D. J.,** Hydride-transferring ability of methylenedioxybenzenes as a basis of synergistic activity, *J. Agric. Food Chem.*, 13, 218, 1965.

154. **Hennessey, D. J., Gibbons, E. V., Durante, A. J., Verma, S. K., Patel, H. C., and LaVoie, E. J.,** Structure, Synthesis, Activity, and Selectivity in Insecticide Synergists, paper presented at the 164th American Chemical Society Meeting, New York, August 1972.

155. **Cloney, R. D. and Scherr, V. M.,** Molecular orbital study of hydride-transfer ability of benzodiheterolines as a basis of synergistic activity, *J. Agric. Food Chem.*, 16, 791, 1968.

156. **Hansch, C.,** Use of homolytic, steric, and hydrophobic constants in a structure–activity study of 1,3-benzodioxole synergists, *J. Med. Chem.*, 11, 920, 1968.

157. **Marshall, R. S. and Wilkinson, C. F.,** The epoxidation of aldrin by a modified Fenton's reagent and its inhibition by substituted 1,3-benzodioxoles, *Biochem. Pharmacol.*, 19, 2665, 1970.

158. **Matthews, H. B. and Casida, J. E.,** Properties of housefly microsomal cytochromes in relation to sex, strain, substrate specificity, and apparent inhibition and induction by synergist and insecticide chemicals, *Life Sci. Part I*, 9, 989, 1970.

159. **El-Sebae, A. H., Metcalf, R. L., and Fukuto, T. R.,** Carbamate insecticides: synergism by organothiocyanates, *J. Econ. Entomol.*, 57, 478, 1964.

160. **Bakry, N., Metcalf, R. L., and Fukuto, T. R.,** Organothiocyanates as insecticides and carbamate synergists, *J. Econ. Entomol.*, 61, 1303, 1968.

161. **Kooy, H., Jr.** (Hoffman-LaRoche), Dutch Patent 6,601,926, August 17, 1966.

162. **Barnes, J. R. and Fellig, J.,** Synergism of carbamate insecticides by phenyl 2-propynyl ethers, *J. Econ. Entomol.*, 62, 86, 1969.

163. **Fellig, J., Barnes, J. R., Rachlin, A. I., O'Brien, J. P., and Focella, A.,** Substituted phenyl 2-propynyl ethers as carbamate synergists, *J. Agric. Food Chem.*, 18, 78, 1970.

164. **Martin, H. and Pissiotas, G.** (CIBA Ltd.), South African Patent 70 06, 998, September 1971.

165. **Sacher, R. M., Metcalf, R. L., and Fukuto, T. R.,** Propynyl naphthyl ethers as selective carbamate synergists, *J. Agric. Food Chem.*, 16, 779, 1968.

166. **Brown, N. C. and Wickham, J. C.** (Cooper, McDougall & Robertson, Ltd.), British Patent 960,241, June 10, 1964.

167. **Morallo, B. D.,** The metabolism of Mobam (4-benzo[b]thienyl N-methylcarbamate) in resistant and susceptible houseflies, *Diss. Abstr. Int. B*, 30, 3226, 1970.

168. **Moorefield, H. H. and Tefft, E. R.,** Evaluation of 2-(3,5-dichloro-2-biphenyloxy)triethylamine as an insecticide adjuvant, *Contrib. Boyce Thompson Inst.*, 20, 293, 1959.

169. **Fahmy, M. A. and Gordon, H. T.,** Selective synergism of carbamate insecticides on houseflies by aryloxyalkylamines, *J. Econ. Entomol.*, 58, 451, 1965.

170. **Enan, O. and Gordon, H. T.,** Temperature effects on toxicity of synergized carbamate insecticides on houseflies, *J. Econ. Entomol.*, 58, 513, 1965.

171. **Hayashi, A. and Fujimura, M.,** Synergistic effects of carbamate insecticides and GD-11 on *Musca domestica* (housefly), *Nippon Oyo Dobutsu Konchu Gakkai-Shi,* 16, 158, 1972.

172. **Kirby, P., Soloway, S. B., and Haddock, E.,** (Shell Internationale Research Mootschoppij, New York), British Patent 1,176,799, January 7, 1970.

173. **Felton, J. C., Jenner, D. W., and Kirby, P.,** Benzothiadiazoles, a novel group of insecticide synergists, *J. Agric. Food Chem.*, 18, 671, 1970.

174. **Gil, D. L. and Wilkinson, C. F.,** Inhibition of Microsomal Enzyme Activity by 1,2,3-Benzothiodiazoles, paper presented at the 164th American Chemical Society Meeting, New York, August 1972.

175. **Wilkinson, C. F., Hetnarski, K., and Hicks, L. J.,** Substituted imidazoles as inhibitors of microsomal oxidation and insecticide synergists, *Pestic. Biochem. Physiol.*, 4, 299, 1974.

176. **Adolphi, H., Steimmig, A., and Spaenig, H.** (Badische Anilin- & Sodafabrik A.-G.), French Patent 1,486,817, June 30, 1967.

177. **Gordon, H. T. and Eldefrawi, M. E.,** Analog-synergism of several carbamate insecticides, *J. Econ. Entomol.*, 53, 1004, 1960.

178. **Jaeger, A. and Piessker, H.** (Schering A.-G.), German Patents 1,171,200, May 27, 1964; 1,300,725, August 7, 1969; 1,792,329, September 21, 1972.

179. **Plapp, F. W. and Valega, T. M.,** Synergism of carbamate and organophosphate insecticides by noninsecticidal carbamates, *J. Econ. Entomol.,* 60, 1094, 1967.
180. **Thayer, H. I., Hartle, R. J., and Mallis, A.,** N-Alkylcarbamates as insecticides and pyrethrins synergists, *J. Agric. Food Chem.,* 13, 43, 1965.
181. **Epstein, P. F.** (Stauffer Chemical Co.), U.S. Patent 3,641, 245, February 8, 1972.
182. **Plapp, F. W., Jr.,** Polychlorinated biphenyl: an environmental contaminant acts as an insecticide synergist, *Environ. Entomol.,* 1, 580, 1972.
183. **Lichtenstein, E. P., Liang, T. T., and Anderegg, B. N.,** Synergism of insecticides by herbicides, *Science,* 181, 847, 1973.
184. **Georghiou, G. P. and Metcalf, R. L.,** Synergism of carbamate insecticides with octachlorodipropyl ether, *J. Econ. Entomol.,* 54, 150, 1961.
185. **Jacobi, E., Lust, S., van Schoor, A., and Zima, O.,** (Emil Merck Chemische Fabrik), German Patent 956, 638, January 24, 1957.
186. **Demozay, D.** (PEPRO – Société pour le Developpement et la Vente de Specialities Chimques), French Patent 2,109,020, June 30, 1972.
187. **Jones, H. A., Garmon, J. A., and Dickinson, B. C.** (Food Machinery and Chemical Corp.), U.S. Patents 2,961, 371, November 22, 1960; 2,990,315-23, June 27, 1961; 2,995,487–8, August 8, 1961.
188. **Kaemmerer, K. and Stendel, W.** (Farbenfabriken Bayer A.-G.) British Patent 929,716, June 26, 1963.
189. **Hewlett, P. S. and Wilkinson, C. F.,** Synergism between carbaryl and some 1,3-benzodioxole compounds in houseflies, *J. Sci. Food Agric.,* 18, 279, 1967.
190. **Lichtenstein, E. P. and Casida, J. E.,** Myristicin, an insecticide and synergist occurring naturally in the edible parts of parsnips, *J. Agric. Food Chem.,* 11, 410, 1963.
191. **Weiden, M. H. J.,** Insecticidal carbamoyloximes, *J. Sci. Food Agric. Suppl.,* p. 19, 1968.
192. **Brattsten, L. B. and Metcalf, R. L.,** The synergistic ratio of cabaryl with piperonyl butoxide as an indicator of the distribution of multifunction oxidases in the Insecta, *J. Econ. Entomol.,* 63, 101, 1970.
193. **Brattsten, L. B. and Metcalf, R. L.,** Synergism of carbaryl toxicity in natural insect populations, *J. Econ. Entomol.,* 66, 1347, 1973.
194. **Sun, Y.-P. and Johnson, E. R.,** Quasi-synergism and penetration of insecticides, *J. Econ. Entomol.,* 65, 349, 1972.
195. **Wilkinson, C. F. and Brattsten, L. B.,** Microsomal drug metabolizing enzymes in insects, *Drug Metab. Rev.,* 1, 153, 1972.
196. **Kuhr, R. J.,** Mechanisms of insecticide metabolism in insects: tissue, cellular, and subcellular studies, in *Mechanisms of Pesticide Action,* Kohn, G. K., Ed., American Chemical Society, Washington, D.C., 1974.
197. **El-Aziz, S. A., Metcalf, R. L., and Fukuto, T. R.,** Physiological factors influencing the toxicity of carbamate insecticides to insects, *J. Econ. Entomol.,* 62, 318, 1969.
198. **Benke, G. M. and Wilkinson, C. F.,** In vitro microsomal epoxidase activity and susceptibility to carbaryl and carbaryl-piperonyl butoxide combinations in house crickets of different age and sex, *J. Econ. Entomol.,* 64, 1032, 1971.
199. **Chakraborty, J. and Smith, J. N.,** Enzymic oxidation of some alkylbenzenes in insects and vertebrates, *Biochem. J.,* 102, 498, 1967.
200. **Sacher, R. M., Metcalf, R. L., and Fukuto, T. R.,** Selectivity of carbaryl–2,3-methylenedioxynaphthalene combination. Metabolism of the synergist in two strains of houseflies and in mice, *J. Agric. Food Chem.,* 17, 551, 1969.
201. **Hennessey, D. J.,** The design of synergists for large-scale application, in *Insecticide Resistance, Synergism, Enzyme Induction,* Tahori, A. S., Ed., Gordon and Breach, New York, 1971.
202. **Bonthrone, W. and Cornforth, J. W.,** Methylenation of pyrocatechols, *J. Chem. Soc. C,* 9, 1202, 1969.
203. **Mrak, E. M., Chairman,** *Report of the Secretary's Commission on Pesticides and their Relationship to Environmental Health,* U.S. Department of Health, Education and Welfare, Washington, D.C., 1969.
204. **Innes, J. R. M., Ulland, B. M., Valerio, M. G., Petrucelli, L., Fishbein, L., Hart, E. R., Pallotta, A. J., Bates, R. R., Falk, H. L., Gart, J. J., Klein, M., Mitchell, I., and Peters, J.,** Bioassay of pesticides and industrial chemicals for tumorigenicity in mice: a preliminary note, *J. Natl. Cancer Inst.,* 42, 1101, 1969.
205. **Brown, N. C.,** A review of the toxicology of piperonyl butoxide, *Pyrethrum Post,* 11, 66, 1971.

INDEX

A

A 363, metabolism, 175–177
Acetylcholine
  hydrolysis of, 71–74
  in physostigmine, 3–4
  structure, figure, 90
Acetylcholinesterase
  in cholinesterase inhibition by carbamates, 48–58, 108–109, 115
  definition, 44
  inhibition by carbamate insecticides, 71–97, 117–119
  modifications in species of insects, 56–58, 271–272
  reaction of carbamate insecticides with, figure, 45
  reaction of phenyl N-methylcarbamate with, figure, 72
N-Acetyl zectran
  metabolism, 175
AChE, see Acetylcholinesterase
Achromobacter, metabolizing carbaryl, 158
Acylamidophenyl, anticholinesterase agent, 86
Acylcarbamates
  anticholinesterase activity, 77–78
  mode of action, 77–79
N-Acyl-N-methylcarbamates, synthesis of, 25–26
Affinity constant, see K_a
Aldicarb
  characteristics of, 5–6, 18
  in cow milk, 188
  degradation
    in farm animals, 203–204
    in fish, 205
    in fresh water, 224
    in fruits, 212
    in plants, 211–213
    in poultry, 204–205
    in soil, 219–220
    in vegetables, 212–213
  metabolism
    in animals, 187–188
    in insects, 188–190
    in plants, 189–190
  radiosynthesis of, figure, 36
  residue analysis, 244, 248
  resistance, 267
  stereoisomers, figure, 91
  structure, figure, 3, 90
  toxicity, 104, 105
Aldicarb oxime, see Aldicarb
Aliphatic oximes, insect poisons, 16
Alkylcarbamates
  anticholinesterase activity, 77
  structure, 76–77
Aminocarb, metabolism, 175–177
Ammonium carbamate, as commercial insecticide, 15
Anticholinesterase, see also Cholinesterase inhibition by carbamates
  relationship of insecticidal activity to, 93–94
  substrate enzyme interaction, 71–74
Antidotes for carbamate poisoning, 59–60, 112–115
Arochlor 1254, synergist, 283
Aromatic carbamates
  aryl methylcarbamate esters, 15
  insecticidal activity of, 74–89
  structures, figure, 16
  synthesis of, figure, 26
Aromatic esterase
  responsible for carbamate hydrolysis in insects, 126
Aromatic methylcarbamate insecticides, residue analysis, 251
Arprocarb, see Baygon
Aryloxyalkylamines, synergism, 281
Atrazine, synergist, 283
Atropine
  action on physostigmine, 71
  antidote for carbamate poisoning, 59–60, 109, 113–115, table, 114
Azobenzene carbamates, development, 85–86

B

B 37344, metabolism, 177
B 39007, metabolism, 172–173
B 44646, metabolism, 175–177
Banol
  behavioral toxicology, table, 116
  metabolism, 164–166
  metabolism, figure, 165
  residue analysis, 250
  resistance, 266
Bassa, characteristics of, 6
Baygon
  adsorptive isotherms, 216
  behavioral toxicology, 115–116
  characteristics of, 6, 18
  control agent, 105
  differential in toxicity, 106–107
  degradation
    in air, 227–228
    in fresh water, 223
    in plants, 214
    in soil, 221–222
    photodecomposition, 229–230
  hydrolysis, 126–127
  in public-health programs, 265–266
  metabolism, 172–173
  metabolism, figure, 269
  nonhydrolytic metabolism, 129–130
  poisoning by, 109
  residue analysis, 246, 247, 250, 251, 252
  resistance, 262, 265, 266, 267, 268, 273, 274
  species-specificity, 104
  synergism, 280, 281
  tested on humans, 62–63, 109
  tested on rats, 61
  toxicity, 78, 104, 108

toxicity to non-target organisms, 117, 118
tables, 117–119
treatment for insect-borne diseases, 63–64
Behavioral toxicology of carbamates, 115–116
Benzothiadiazoles, synergism, 281–282
Biological degradation, see Degradation
Biotransformations, 125–126
Blattanex, see Baygon
Bound residues, 132
Bufencarb, see Bux
Butacarb
characteristics of, 6
degradation in farm animals, 204
metabolism, 171–172
3-(2-Butyl)phenylthio-N-methylcarbamate, resistance, 268
Bux
characteristics of, 6–7, 18
degradation in soil, 220–221
metabolism, 169–170
resistance, 268
toxicity to non-target organisms, 118
table, 117–119

C

C-3, metabolism, 160–162
C 8853, photodecomposition, 231
CAM-1, see Cholinesterase antagonist monitor, in residue analysis
Calabar bean, toxicology, 2–4
Carbamate compounds, substituents
acyl group, 77–79
acylamido group, 86
alkenyls, 83–84
alkoxy analog, 82–83
alkyl group, 76–77, 81–83
alkylthio substituents, 83
amino group, 85–86
asymmetry in, 80
charge in, 79–80
dimethylamino group, 85
O,O-dimethylphosphorothioyl group, 79
halogens, 84–85
phenyl substitution in, 83–89
thiophenyl group, 79
Carbamate esters, see also Carbamates
activity, 71
discovery and development, 4
Carbamate fungicides and herbicides, structures, figure, 17
Carbamate insect repellents, structure, figure, 75
Carbamate insecticides
detoxication of, see Detoxication
environmental stability studies, 201–236
Carbamate metabolism, see Metabolism
Carbamate poisoning, see also Detoxication, and Toxicology
antidotes, 59–60, 112–115
dermal contact, 59
oral ingestion, 59
respiratory poisoning, 59
signs and symptoms, 45, 58–66, 108–109
calliphorid flies, 65
dogs, 60
houseflies, 65–66
humans, 58–59, 60, 62–64
insects, 64–66
lepidopterous larvae, 65
looper larvae, 65
mammals, 58–64
rats, 60–61
selective toxicity, 41, 71, 104, 117–119
symptom-toxicity investigations, 59–66, 109–110
Carbamates, see also specific headings, e.g. Detoxication, Metabolism, etc.
chemical and physical properties, table, 18
cholinesterase inhibition by, see Cholinesterase inhibition by carbamates
compounds, see Carbamate compounds
definition, 15
degradation of, see Degradation
detoxication of, see Detoxication
discovery and development, 4–12
enzyme inhibition, 44–47, 94–95
general chemistry, 15–16
hydrolysis, 19–21, 125–127
alternate mechanism, 20–21
mechanism I, 20
mechanism II, 20
hydrolytic stability of, 94–95
inhibition of housefly head, AChE, tables, 76, 77, 81, 82, 83, 84, 86, 87, 88, 92, 93
interaction with AChE, 43–58
kinetic interaction with esterases, 48–55, 57
metabolism of, see Metabolism
microsomal metabolism of, 129–132
mode of action, 41–66
radiosynthesis of, 33–36
residue analysis, see Residue analysis
resistance to, see Resistance
selective toxicity in, 41–43, 117–119
tables, 117–119
structure, 3
figure, 72
synergism, see Synergism
toxicity, see Carbamate poisoning, and Toxicology
toxicology, see Toxicology
Carbamic acid
alkyl esters of, 15
aryl esters of, 15
salts of, 15
Carbaminates, see Carbamates
Carbamult, see Promecarb
Carbamylation constant, definition, 47
Carbamylcholine as antidote for carbamate poisoning, 115
Carbanolate, metabolism, 164–166
Carbaryl
adsorptive isotherms, 216
behavioral toxicology, 116
binding in soil aqueous system, 216
biochemical and pharmacological effects, 110–112
characteristics of, 7, 18
degradation

in air, 227–228
in dairy animals, 202–203
in fish, 205
in fresh water, 223–224
in fruits, 206–207
in plants, 206–210
in poultry, 204–205
in sea water, 225
in soil, 215–217
in vegetables, 207–210
levels in food, 233–234
photodecomposition, 227–230
removal of residues by processing foods, 234–236
differential in toxicity, 106
discovery and development, 4, 146
effects on soil microbes, 216–217
glucuronide conjugation of, 134, 148
glutathione conjugates, 152–153
hydrolysis, 19–20, 125–126, 147
mercapturic acids, 152–153
metabolism, 146–160
bacteria in, 158–159
complexity of, 146–148
in cows, 154–155
in insects, 155–156
in plants, 157–158
in rats, 155
in soil, 158–160
in urine, 150–154, figure, 152
in vertebrates, 148–155
metabolites, 27–29
figure, 28
1-napthol in, 156–157
nonhydrolytic metabolism, 129–130, 147
nonhydrolytic pathway, 147
figure, 161
oxidative mechanisms, 129–132, 147–148
poisoning by, 109–110
table, 109, 110
radioactive residues in cow milk, 154
radiosynthesis of, 34–35
figure, 34
residue analysis, 244, 246, 247–249, 250, 251, 252, 253
residues in laying hens, 155
residues in plants, figure, 157
resistance, 262–274, 275, 280, 283
soil insecticide, 209
species-specificity, 104
structure, 3
sulfate conjugation, 135, 148
synthesis of, 24
toxicity, 104, 108, 284
toxicity to non-target organisms, 117, 118
tables, 117–119
treatment for insect-borne diseases, 63–64
water-soluble metabolites, 147
Carbofuran
characteristics of, 7–8, 18
degradation
in farm animals, 203–204
in fish, 205
in soils, 217–218
photodecomposition, 232
in plants, 210–211
in poultry, 204–205
excretion and metabolism by animals, 178–180
in ecosystems, 225–226
metabolism
in animals, 178–180
in insects, 180–181
in plants, 181–182
metabolites of, figure, 179
residue analysis, 246, 247, 250, 251, 252
species-specificity, 104
toxicity, 104, 107
toxicity to non-target organisms, 118, 119
tables, 117–119
Carpolin, see Carbaryl
Carzol, see Formetanate
Catechol, carbaryl metabolite, 158
Chlorxylam, metabolism, 164–166
Cholinesterase antagonist monitor, in residue analysis, 245–246
Cholinesterase inhibition by carbamates, see also Anticholinesterase
carbamate toxicity, 41
depression, 59, 60–64, 110–112
table, 110
enzyme-substrate complex, 71–74
function of peripheral, 54
general reaction mechanism, 44–46
in nerve synapses, 43–44, 54
ions, effect of, 52
kinetics of animal cholinesterase inhibition, 48–53
kinetics of insect cholinesterase inhibition, 53–58
learning process, effects on, 115–116
table, 116
pH, effect of, 51–52
reaction constants, tables, 46, 49–50, 51, 52, 55, 57
temperature, effect of, 51
Collembola, reaction to carbofuran, 217
Colorimetric methods of residue analysis, 243–244
Conjugates
glucoside formation, 134
glucuronide formation, 133–134
mercapturic acid formation, 135–136
phosphate formation, 135
structure, 132–136
sulfate formation, 135
toxicological significance, 132–136
Cosban, metabolism, 163–164
Curaterr, metabolism, 178–182
Cytochrome P-450, oxidase, 128–132, 270, 279

D

Decarbamylation of enzyme, definition, 47
Degradation
air, 222–223, 227–228
animals, 201–205
human exposure, 232–236

levels of carbaryl in foods, 233–234
removal of residues by processing foods, 234–236
photodecomposition, 228–232
plants, 205–213
soils, 214–221
water
ecosystems, 225–226
fresh, 223–225, 227
sea, 225
Denapon, metabolism, 146–160
Detoxication, see also Carbamate poisoning, and Toxicology
conjugation
in insects, 134–135
in mammals, 132–136
in plants, 134–135
enhanced, 269–270
hydrolysis
in insects, 126–127
in mammals, 125–126
oxidation
in insects, 129, 131–132
in mammals, 127–128, 130–132
in plants, 129
Dicarzol, see Formetanate hydrochloride
Dimetan
discovery and development, 4
structure, 3
Dimethylcarbamates
acetylcholinesterase, activity in, 44–58, 75–76
carbamate insecticide chemicals, 16
discovery, 4, 74–75
metabolism, 144–145
structure, figure, 75
synthesis of, 24–25
toxicity, 75–76
N,N-Dimethylcarbamates, see Dimethylcarbamates
3,5-Dimethylphenyl N-methylcarbamate, metabolism, 163–164, 221
Dimetilan
adsorptive isotherms, 216
degradation
in fresh water, 223
photodecomposition, 230–231
discovery and development, 4
metabolism, 145
structure, 17
synergism, 281, 283
synthesis of, figure, 25
toxicity, 107, 108
Dioxacarb, characteristics, 8
Dithiocarbamates, herbicides, 16
Double-cycle carbamates, see Double-ring carbamates
Double-ring carbamates, anticholinesterase activity, 88–89
Dowco 139, metabolism, 173–175
Draza, see Mesurol
DS-15647, metabolism, 193
DuPont 1179, metabolism, 190–191

E

Ecosystems to emulate environmental conditions, 225-226
Elocron, see Dioxacarb
Enzymes in hydrolysis of carbamate structures, 125–127
Enzymic methods of residue analysis, 244–246
Eserine, see Physostigmine
Esterases
kinetic interaction with carbamates, 48–55, 57
responsible for carbamate hydrolysis, 125–127
Ethyl carbamate, see Urethane
Etrofol, see Hopcide
Etrofolan, see Mipcin

F

Famid, see Dioxacarb
Florisil-column chromatography in residue analysis, 248
Fluorescence methods of residue analysis, 246-247
Food, carbamates in, 232–236
Formetanate
anticholinesterase activity, 86
characteristics of, 8, 18
degradation
in soil, 221
photodecomposition, 231
metabolism, 186
Fundal Forte 750, metabolism, 186
Furadan, see Carbofuran
Fusarium solani, metabolizing carbaryl, 159

G

Gas chromatography methods of residue analysis, 247–251
Geigy insecticides, discovery and development, 4
GLC analysis of residues, 247–251
Glucoside formation, 134
Glucuronide conjugation of phenols and alcohols, figure, 133
Glucuronide formation, 133–134

H

H-5727, metabolism, 167
HRS-1422
characteristics of, 9
metabolism, 168–169
Hopcide, characteristics of, 9
Human exposure to carbamates, 109–110, 232–236
Hydrolysis mechanisms, carbamate, 19–21, 125–127
Hydroquinone, carbaryl metabolite, 158
Hydroxyalkylphenyl N-methylcarbamates, metabolites, 30, 131

I

I_{50}, hydrolytic stability and, 93–97
Imidazoles, synergism, 281–282
Insect-borne diseases, treated with carbamates, 63–64
Insecticides
carbamate, see also Carbamates, and Methylcarbamates
degradation of, see Degradation
environmental stability studies, 201–236
human exposure to, 232–236
in public health programs, 265–266
residue analysis, see Residue analysis
commercial, see also Trade names, 5
organochlorine, environmental stability, 201
organophosphorus
activity, 71
toxicology, 103–119
synthetic, history and development, 1–5
toxicology, see Toxicology, and Carbamate poisoning
IPFC, metabolism, 167
Isolan
degradation in soil, 221
discovery and development, 4
metabolism, 145–146
resistance, 266, 267, 268, 273, 274
structure, 3, 17
synergism, 275, 280, 281, 282, 283
Isoprocarb, see Mipcin

K

K_a, equilibrium constant, 46–47
Karbosep, metabolism, 146–160

L

Landrin
carbamate metabolites, figure, 167
characteristics of, 9, 18
degradation
in soil, 220–221
photodecomposition, 232
metabolism, 166–167
residue analysis, 246–247, 251, 252
structure, 16
synergism, 282
synthesis of, figure, 31
Lannate, see Methomyl
LC, see Liquid chromatography
LD_{50}
definition, 104, 261
hydrolytic stability and, 93–97
toxicology, 106–107, 108
Learning process, affected by carbamate poisoning, 115–116
Lilly 18947, synergism, 281
Liquid chromatography methods of residue analysis, 252
Log dosage – mortality (ld – m) curves for test populations of insects, figure, 261–264

M

Macbal, characteristics of, 5
Mass spectroscopy methods of residue analysis, 252–253
Matach, toxicity, 104
Matacil
characteristics of, 9
degradation
in fresh water, 224
in plants, 213
photodecomposition, 227–230
metabolism, 175–177
residue analysis, 246, 247, 252
resistance, 267, 274
structure, 177
toxicity to non-target organisms, 117
MCA 600, metabolism, 184
Meobal
characteristics of, 9
metabolism, 162–163
residue analysis, 250
resistance, 266
structure, 16
Mercaptodimethur, see Mesurol
Mercapturic acid formation, 135–136
Mesurol
characteristics of, 9, 18
degradation
in fresh water, 224
in soil
photodecomposition, 227–230
metabolism, 177
residue analysis, 246, 247, 248, 251, 252
resistance, 266, 267
structure, 16
Metabolism
dimethylcarbamate, figure, 145
enhanced detoxication, 269
monomethylcarbamates, 146–193
synergism in, 274–286
target selectivity and stability, determined by, 125
Metacrate, see Tsumacide
Metalkamate, see Bux
Methiocarb, see Mesurol
Methomyl
characteristics of, 10, 18
degradation
in plants, 213
in soil, 221–222
metabolism, 190–191
residue analysis, 248, 252
stereoisomers, figure, 91
toxicity to non-target organisms, 117
tables, 117–119
N-Methylcarbamate, see Methylcarbamates
Methylcarbamates
acetylcholinesterase and other cholinesterases, reaction with, 44–46
air oxidation, 17
autosynergism, 283
carbamate insecticide chemicals, 15–16, 71

chemical and physical properties of, 18
degradative reactions, see also Degradation, 27–33
detoxication, see Detoxication
discovery and development, 4–12
hydrolysis rate, 76
metabolism, see also Metabolism, 146–193
metabolite conjugation, see also Metabolism, 132–136
metabolite hydrolysis, see also Metabolism, 125–127
metabolite oxidation, see also Metabolism, 127–132
mode of action, 41–66
oxidation, figure, 130
photodegradation, 17
radiosynthesis of, 33–36
residue analysis, see Residue analysis
resistance to, see Resistance
ring substituents and hydrolytic stability, 19–20
stability, 17
structures, 16, 17
synergism, see Synergism
synthesis of, 21–24
method A, 22–23
method B, 23–24
method C, 24
synthesis of metabolites, 27–83
toxicity, see Carbamate poisoning, and Toxicology
toxicology, see Toxicology
O-(Methylcarbamoyl)oxime
degradation in soil, 221
insecticidal activity, 89–93
structure, 90
figure, 17
Methylenedioxyphenyl synergists, structure, 275–279
figure , 276
Metmercapturon, see Mesurol
Mexacarbate, see Zectran
Microsomal metabolism of carbamates, 127–132
Mipcin
characteristics of, 10
Mobam
characteristics of, 10, 18
degradation in farm animals, 203
metabolism, 184
radioactive, 185
residue analysis, 250
resistance, 266, 267
structure, 16
synergism, 281
synthesis of, 24
Mode of action of carbamates, 41–66
Monomethylcarbamates, see Methylcarbamates
MPD, see Methylenedioxyphenyl synergists, structure
MPMC, metabolism, 162–163
MS, see Mass spectroscopy
MTMC, metabolism, 160–162

N

Nabam
discovery and development, 4
structure, 3
NAC, metabolism, 146–160
1-Naphthol
degradation
in fresh water, 224
in sea water, 225
metabolism, 156–157
residue analysis, 249, 251, 252
Nerve transmission
cellular, 42–43
chemical, 43–44
figure, 43
Nervous system
cholinergic transmission, 43–44
insects, 42–44, 56
autonomic systems, 42
central nervous system, 42
peripheral nervous system, 42
somatic system, 42
vertebrates, 41–44
autonomic system, 42
central nervous system, 41–42
schematic representation, 42
somatic system, 41–42
NIA 10242, metabolism, 178–182
NIH shift in microsomes, 129
Nudrin, see Methomyl

O

Octachlorodipropyl ether, synergist, 283
Organophosphorus insecticides, see Insecticides, organophosphorus
Organothiocyanate synergists, structure, 280–281
figure, 280
Ortho 5353, metabolism, 130–131
Osbac, see Bassa
Oxamyl, characteristics of, 10–11
Oxidative pathways in N-methylcarbamates, 130–131
Oxime carbamates, see also O-(Methylcarbamoyl)oximes
geometrical isomerism, 90–91
structure, figure, 93
structure – activity relationships, 90

P

Parathion, poisoning by, 109
table, 110
Phenols, residue analysis, 243
Phenyl carbamates, see also Phenyl methylcarbamates
Phenyl methylcarbamates
Alkyl, Alkoxy, Allcylthiol, 82–84
amino substitution, 84
hologen substitution, 84–85
insecticidal activity, 74–89
multiple substitution, 86–88
nitro-substitution, 85
structures, figure, 16
synergism, 274–293
synthesis, 29–31
Phenyl N-methylcarbamates, see Phenyl methylcarbamates

Phosphate formation, 135
Photodecomposition of carbamates, 228–232
Photolysis of pesticide chemicals, 228–232
Physostigmine
  activity, 71
  anticholinesterase agent, 48
  structure, 3
  toxicology, 2–5
  vagusstoff in, 3–4
Pinolene, added to insecticides, 208
Piperonyl butoxide, synergist, 277
Pirimicarb, characteristics of, 11
Pirimor, see Pirimicarb
Promecarb
  characteristics of, 11
  residue analysis, 248
Propoxur, see Baygon
Propynyl ethers, synergism, 281
Prostigmine
  AChE inhibition, 46, 71
  structure, 72
  toxic aspects, 71
Pseudomonas, metabolizing carbaryl, 158
Pyramat
  metabolism, 146
  synthesis of, figure, 25
Pyridinium oximes as antidote for carbamate poisoning, 113–115
Pyrolan
  adsorptive isotherms, 216
  degradation
    in fresh water, 223
    photodecomposition, 230–231
  resistance, 264, 267, 268
  synergism, 275
Pyruvate, carbaryl metabolite, 158

R

Radiocarbon in carbaryl-treated soils, 160
Radiosynthesis of carbamates, 33–36
RE 5030, metabolism, 171–172
RE 5365
  metabolism, 170–171
  photodecomposition, 231–232
RE 11775
  degradation
    in fresh water, 224
    photodecomposition, 231–232
  metabolism, 170–171
Residue analysis
  bound and unextractable residues, 132
  colorimetric methods, 243–244
  enzymic methods, 244–246
  fluorescence, 246–247
  gas chromatography
    derivatization methods, 248–251
    GLC analysis, 247–251
    intact carbamates, 247–248
  liquid chromatography, 252
  mass spectroscopy, 252
  methodology, problems of, 243–253
  thin-layer chromatography, 251, 252
Resistance, 259–274
  cross resistance, 261, 266–268
  definition of, 259
  development in the field, 259–262
  genetics, 272
  log dosage–mortality (ld–m) curves, 261
  mechanisms
    behavioral, 260
    biochemical, 260
    enhanced detoxication, 269
    enhanced excretion, 270
    enhanced penetration, 270
    modified acetylcholinesterase, 271
    physiological, 260
  multiple resistance, 261, 266–268
  negatively correlated resistance, 268–269
  occurrence of
    field conditions, 262
    laboratory conditions, 264–266
  revision back to susceptibility, 262

S

S-1046, metabolism, 162–163
Schering 36056, metabolism, 186
SD 8530, metabolism, 166–167
SD 17250-W, metabolism, 191–193
Selective toxicity
  carbamates, 41–43, 117–119
  dimethylcarbamates, 4, 75
  establishment of, 125
  metabolic processes in, 143–144
  studies of, 143–144
  studies of, see also Carbaryl, metabolism of, 143
  synergism in, 284–286
Sendran, see Baygon
Sevin, see also Carbaryl
  discovery and development, 4
  metabolism, 146–160
  poisoning by, 109
SKF 525-A, synergism, 281
Soil fungi, metabolizing carbaryl, 159
Soil microbes, effects of carbaryl on, 216–217
Specific toxicity, see Selective toxicity
Spectrophotofluorometry in residue analysis, 246
N-Substituted carbamates, structure, figure, 79
Substituted heterocyclic N,N-dimethylcarbamate insecticide chemicals, figure, 17
Substituted phenyl N-hydroxymethylcarbamates
  figure, 28
  synthesis of, 27–28
Substituted phenyl N-methylcarbamates, see also Phenyl methylcarbamates
  double-ring, 88–89
  ring-hydroxylated metabolites, figure, 30
Sulfate formation, 135
N-Sulfenyl-N-methylcarbamates, synthesis, 25–26
  figure, 27
Suncide, see Baygon
Synergism, 274–286
  analog synergism, 282–283

aryloxyalkylamines, 281
autosynergism, 283
benzothiadiazoles, 281–282
imidazoles, 281–282
methylenedioxyphenyl compounds, 275–279
organothiocyanates, 280
propynyl ethers, 281
role of in carbamate development, 274–275
synergist effectiveness and selectivity, 283–284
synergist structures, figure, 276, 280, 282
synergist use, 284–286
Synthesis of methylcarbamates, see Methylcarbamates, synthesis of
Synthetic carbamate insecticides, see Carbamates

T

Talcord, metabolism, 191–193
Temik, see also Aldicarb
characteristics of, 5–6
metabolism, 187–188
Tetraethylammonium chloride as antidote for carbamate poisoning, 113
Tetraethylthiuram monosulfide
discovery and development, 4
structure, 3
Thanite, synergist of carbamate insecticides, 280
Thin-layer chromatography methods of residue analysis, 252
Thiocarbamates
anticholinesterase activity, 74
herbicides and fungicides, 16
Thiocarboxime, metabolism, 191–193
Thioureidophenyl N-methylcarbamates, as anticholinesterase agent, 86
Thiram
discovery and development, 4
structure, 3
Tirpate, synthesis of sulfoxide, 31–32
figure, 32
TLC, see Thin-layer chromatography
Tolerance, see Resistance, definition of
N-(2-Toluenesulfenyl)carbofuran, metabolism, 182–184
Toxicology, see also Carbamate poisoning, and Detoxication
biochemical and pharmacological effects, 110–112
contact action, 105
defined, 103
insects, 103–106
behavior patterns disrupted by, 105–106
carbamate exposure, 105–106
contact toxicity characteristics, 104–105
degree of toxicity, 103–104
latent toxicity, table, 105–106
oral and injected, 104–105
species-specificity, table, 104
variation in susceptibility, 105
variation in toxicity, 104
mammals, 106–116
acute toxicity, table, 106
antidotes, 59–60, 112–115, tables, 114
behavioral toxicology, 115–116
dermal toxicity, 106, 109–110
diet, effects of, 108, 116
differences in susceptibility, 107
injected, 107
LD_{50} values, 106–107, 108, tables, 107, 108
mode of action, see also Mode of action, 108–109
oral toxicity, 106
protein and, 108
resistance of the organism, 108
signs and symptoms, see also Carbamate poisoning, signs and symptoms, 108
solvent or carrier effect on, 107–108
man, 62–64, 109–110
oral ingestion, 109
penetration through skin, 109–110
non-target organisms, 117–119
tables, 117–119
plant-systemic activity, 105
signs and symptoms, see also Carbamate poisoning, signs and symptoms, 58–64, 109–110
stomach poisoning, 105
S,S,S-Tributyl phosphorotrithioate, synergist, 283
Trio-*o*-cresyl phosphate, synergist, 283
Triphenyl phosphate, synergist, 283
Tsumacide
characteristics of, 11
metabolism, 160–162
resistance, 266
structure, 3
Tyrosinase, theory of hydrolysis, disproved, 131

U

UC 7744, see Carbaryl
UC 10854
characteristics of, 11
metabolism, 167–168
photodecomposition, 227–230
resistance, 266, 267, 268, 273
structure, 16
synergism, 280
tested on insects
tested on rats, table, 61
treatment for insect-borne diseases, 63–64
UC 21149, metabolism, 187–190
UC 34096, metabolism, 186–187
Unden, see Baygon
Unextractable residues, 132
Urethane
formation of, 15
uses, 15

V

Vagusstoff, see Physostigmine
Velocity constant, see Carbamylation constant
Vigor tolerance, see Resistance, definition of
Vydate, see Oxamyl

W

WL

WF 21959, metabolism, 191–193

X

XMC, metabolism, 163–164

Z

Zectran
- characteristics of, 12
- degradation
  - in air, 227–228
  - photodecomposition, 229–230, 232
- differential in toxicity, 106–107
- metabolism, 173–175
- poisoning by, 109
- radiosynthesis of, figure, 35
- residue analysis, 246, 247, 248, 252
- resistance, 266, 267, 268
- structure, 16
- synergism, 280
- toxicity to non-target organisms, 118, 119
  - tables, 117–119